OXFORD

GCSE Maths for OCR

SPECIFICATION A

Jayne Kranat
Series editor
Mike Heylings

OXFORD
UNIVERSITY PRESS

Great Clarendon Street, Oxford OX2 6DP

Oxford University Press is a department of the University of Oxford. It furthers the University's objective of excellence in research, scholarship, and education by publishing worldwide in

Oxford New York

Auckland Cape Town Dar es Salaam Hong Kong Karachi Kuala Lumpur Madrid Melbourne Mexico City Nairobi New Delhi Shanghai Taipei Toronto

With offices in

Argentina Austria Brazil Chile Czech Republic France Greece Guatemala Hungary Italy Japan Poland Portugal Singapore South Korea Switzerland Thailand Turkey Ukraine Vietnam

First published 2010

British Library Cataloguing in Publication Data

Data available

ISBN: 978-0-19-913929-3
10 9 8 7 6 5 4 3 2 1

Printed in Spain by Cayfosa (Impresia Iberica)

We are grateful to the following for permission to reproduce copyright material:

Front cover: Tatiana53/Shutterstock; Oscar E. Gutierrez/iStock; Krivoy Rog/iStock; nel4/iStock;

p2-3: J A Giordano/CORBIS SABA; p8: Denise Kappa/Dreamstime.com; p9: Christopher Rawlins/Dreamstime.com; p14-15: Harish Tyagi/epa/Corbis; p19: Peter Chafer/Shutterstock; p20: Picsfive / Dreamstime.com; p21: Julia Allen/Dreamstime.com; p23: Corepix/Dreamstime.com; p24: OUP/Photodisc; p25: Maigi/Dreamstime.com; p26: Shutter1970/Dreamstime.com; p29: Graeme Dawes/Shutterstock; p31: Dima Kalinin/Shutterstock; p34-35: Czintos Ödön/Dreamstime.com; p37: Andresr/Shutterstock; p38: OUP/Digital Vision; p43: Matt Trommer/Shutterstock; p52-53: Jakub Krechowicz/Dreamstime; p60: Joe Gough/Dreamstime.com; p61: Christian Jung/Dreamstime.com; p64-65: MARKABOND/Shutterstock; p78-79: 4774344sean/Dreamstime; p81: xc/Shutterstock; p83: Volkantaner/Dreamstime.com; p85: Arkadiy Yarmolenko/Shutterstock; p87: OUP/Graphi-Ogre; p90b: Stuart Key/Dreamstime.com; p90t: Mikael Damkier/Dreamstime.com; p96-97: Dgareri/Dreamstime; p118-119: Regien Paassen/Shutterstock; p122: OUP/Photodisc; p132-133: The London Art Archive/Alamy; p152-153: OUP; p157: James Steidl/Shutterstock; p166-167: Ivan Cholakov/Dreamstime; p180-181: Doug Steley A/Alamy; p194-195: Juan Fuertes/Shutterstock; p198: Demid/Dreamstime.com; p203: CoolR/Shutterstock; p206-207: Andres Rodriguez/Dreamstime; p218-219: Ivan Kmit/Dreamstime; p232-233: Barbara Helgason/Dreamstime.com; p221: OUP/Corbis; p224: Jian Zhang/Dreamstime.com; p226: Fernando Delvalle/Dreamstime.com; p248-249: Bhe017/Dreamstime; p250: Jerome Whittingham/Dreamstime.com; p252: Ronfromyork/Dreamstime.com; p254: Fernando Delvalle/Dreamstime.com; p255: David Guyler/Shutterstock; p256: OUP/Gareth Byrne; p257: Magdalena Bujak/Shutterstock; p258m: Baevskiy Dmitry/Shutterstock; p258b: Ziga Camernik/Shutterstock; p262-263: OUP; p268: OUP/Photodisc; p277: Alexander Hare/Dreamstime.com; p280-281: Jeremy Broad/Dreamstime; p290-291: Doug Steley A/Alamy; p304-305: Joggie Botma/Dreamstime; p324-325: Kuzma/Shutterstock; p326: Stephen Aaron Rees/Shutterstock; p329: Studioceja/Dreamstime.com; p331: Pierre Landry/Dreamstime.com; p333: OUP/Corbis/Digital Stock; p335: Alena Root/Dreamstime.com; p357: GeorgeS/Shutterstock; p327: Alexander Hare/Dreamstime.com; p338-339: Soleg1974/Dreamstime; p351: Peter Chafer/Shutterstock; p354-355: Slobodan Djajic/Dreamstime.

CASE STUDY 1 RECYCLING
Feng Yu/Dreamstime.com; iStockphoto.com; Christophe Testi/Dreamstime.com; Lobke Peers/Shutterstock; Sailorr/Dreamstime.com; Zentilia/Dreamstime.com.

CASE STUDY 2 HOLIDAY
iStockphoto.com; Ben Renard-wiart /Dreamstime.com; Bruce Hempell /Dreamstime.com; Rider of the storm/Dreamstime.com; Elifranssens/Dreamstime.com; Ben Renard-wiart/Dreamstime.com; Grondin Julien/Dreamstime.com; Ben Mcleish/Dreamstime.com; c./Shutterstock.

CASE STUDY 3 BUSINESS
iStockphoto.com; Christophe Testi/Dreamstime.com; Alex Varlakov/Dreamstime.com; Yang Jay/Dreamstime.com; Design56/Dreamstime.com; Tatiana Popova/Shutterstock; Condor 36/Shutterstock.

CASE STUDY 4 SANDWICH SHOP
Christophe Testi/Dreamstime.com; Andrzej Tokarski/Dreamstime.com; Christopher Elwell/Dreamstime.com; 350jb/Dreamstime.com; Branislav Senic/Dreamstime.com; Yeko Photo Studio/Shutterstock; Studiotouch/Shutterstock.

CASE STUDY 5 WEATHER
Science Photo Library

CASE STUDY 6 RADIO MATHS
Feng Yu/Dreamstime.com; Digital Vision/Getty Images; Giraffarte/Dreamstime.com.

CASE STUDY 7 SPORT
Jim Cooper/Getty Images; Danny De Bruyne/Dreamstime.com; Feng Yu/Shutterstock; Mike Flippo/Shutterstock.

CASE STUDY 8 ART
Andry A/Alamsyah/Alamy; Eduardo Miller/Shutterstock; Michael Mcdonald/Dreamstime.com; www.lucypringle.co.uk; Zaznoba/Dreamstime.com.

The publisher would also like to thank Anna Cox for her work in creating the case studies. The charts on pages 368-9 are reproduced courtesy of the Meteorological Office; the bar chart on page 228 is reproduced courtesy of Defra; the data on page 50 is reproduced courtesy of the IAAF.

About this book

In official partnership with OCR, this book has been specifically written to help you achieve your highest potential in your GCSE Mathematics course for OCR Specification A. It is designed for students who are expecting to enter for the Foundation tier.

The authors are experienced teachers and examiners who have an excellent understanding of the OCR specifications and so are well qualified to help you successfully meet your objectives.

The book is made up of chapters that are based on OCR Specification A, and is organised clearly into the three units that will make up your assessment:

Unit A	(25% of the qualification)	**pages 2 – 129**
Unit B	(25% of the qualification)	**pages 130 – 243**
Unit C	(50% of the qualification)	**pages 244 – 369**

The functional elements of mathematics and **problem-solving** are integrated throughout.

- In particular there are **rich tasks**, which provide an investigative lead-in to the topics – you may need to study some of the techniques in order to be able to complete them properly
- There are also **case studies**, which allow you to apply your GCSE knowledge in a variety of engaging contexts.

Levels of demand are signposted in the exercises, so you can gauge the difficulty level of each topic and monitor your progress throughout the course.

Also integrated into this book are the new **assessment objectives:**

AO1 recall and use their knowledge of the prescribed content
AO2 select and apply mathematical methods in a range of contexts
AO3 interpret and analyse problems and generate strategies to solve them

The **summary assessments** contain questions in the style of the new examinations, with a particular emphasis on these new assessment objectives, so that you can be thoroughly prepared.

Finally, you will notice an asterix that looks like this: *
This shows opportunities for **Quality of Written Communication**, which you will be assessed on in Units B and C. Please note that the use of this asterix is simply to give an indication of the types of question that may contain a QWC element, and these examples should not be considered to be exclusive or exhaustive.

Best wishes with your GCSE Maths – we hope you enjoy your course and achieve success!

Contents

UNIT A

Finding your way around this book

	NUMBER	ALGEBRA	GEOMETRY	DATA
UNIT A	**A1** Integers and decimals **A4** Factors, mutiples and ratio	**A5** Sequences **A7** Formulae and equations	**A3** Constructions **A8** Bearings and Pythagoras	**A2** Summary statistics **A6** Representing and interpreting data
UNIT B	**B9** Fractions, decimals and percentages **B15** Estimation and indices **B17** Percentages and proportional change	**B11** Straight lines and inequalities **B14** More straight line graphs	**B10** Circles, angles and lines **B12** Transformations **B16** Similarity and enlargement **B18** 2-D and 3-D shapes	**B13** Bivariate data and time series
UNIT C	**C22** Everyday arithmetic and bounds	**C19** Algebraic manipulation **C21** Graphs	**C20** 3-D shapes **C23** Area and volume	**C24** Study of chance

1 Integers and decimals

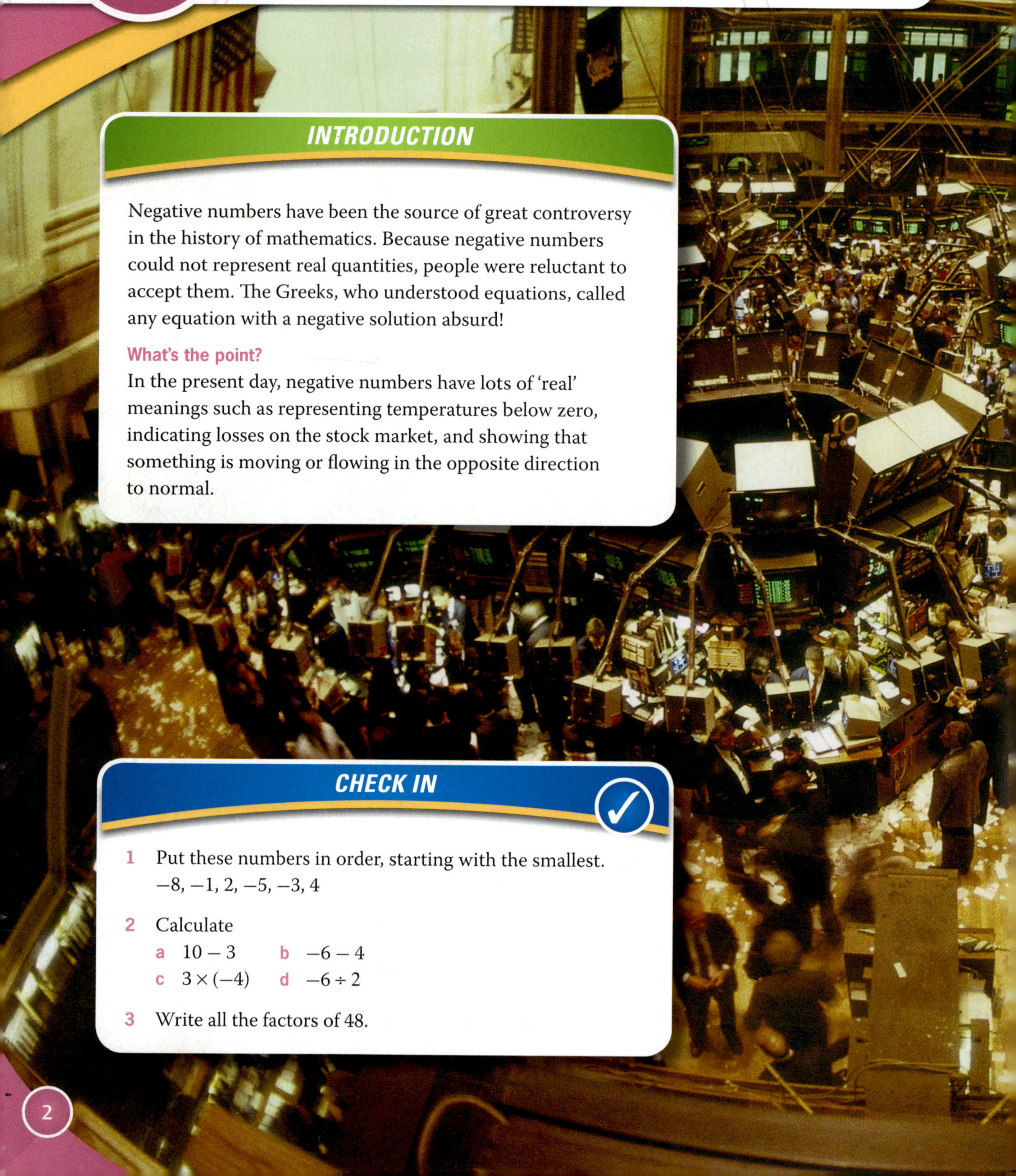

INTRODUCTION

Negative numbers have been the source of great controversy in the history of mathematics. Because negative numbers could not represent real quantities, people were reluctant to accept them. The Greeks, who understood equations, called any equation with a negative solution absurd!

What's the point?

In the present day, negative numbers have lots of 'real' meanings such as representing temperatures below zero, indicating losses on the stock market, and showing that something is moving or flowing in the opposite direction to normal.

CHECK IN

1 Put these numbers in order, starting with the smallest.
$-8, -1, 2, -5, -3, 4$

2 Calculate

a $10 - 3$ b $-6 - 4$

c $3 \times (-4)$ d $-6 \div 2$

3 Write all the factors of 48.

Orientation

What I should know	What I will learn	What this leads to
Key stage 3	■ Calculate with negative numbers ■ Round numbers to significant figures ■ Use a calculator effectively	A9

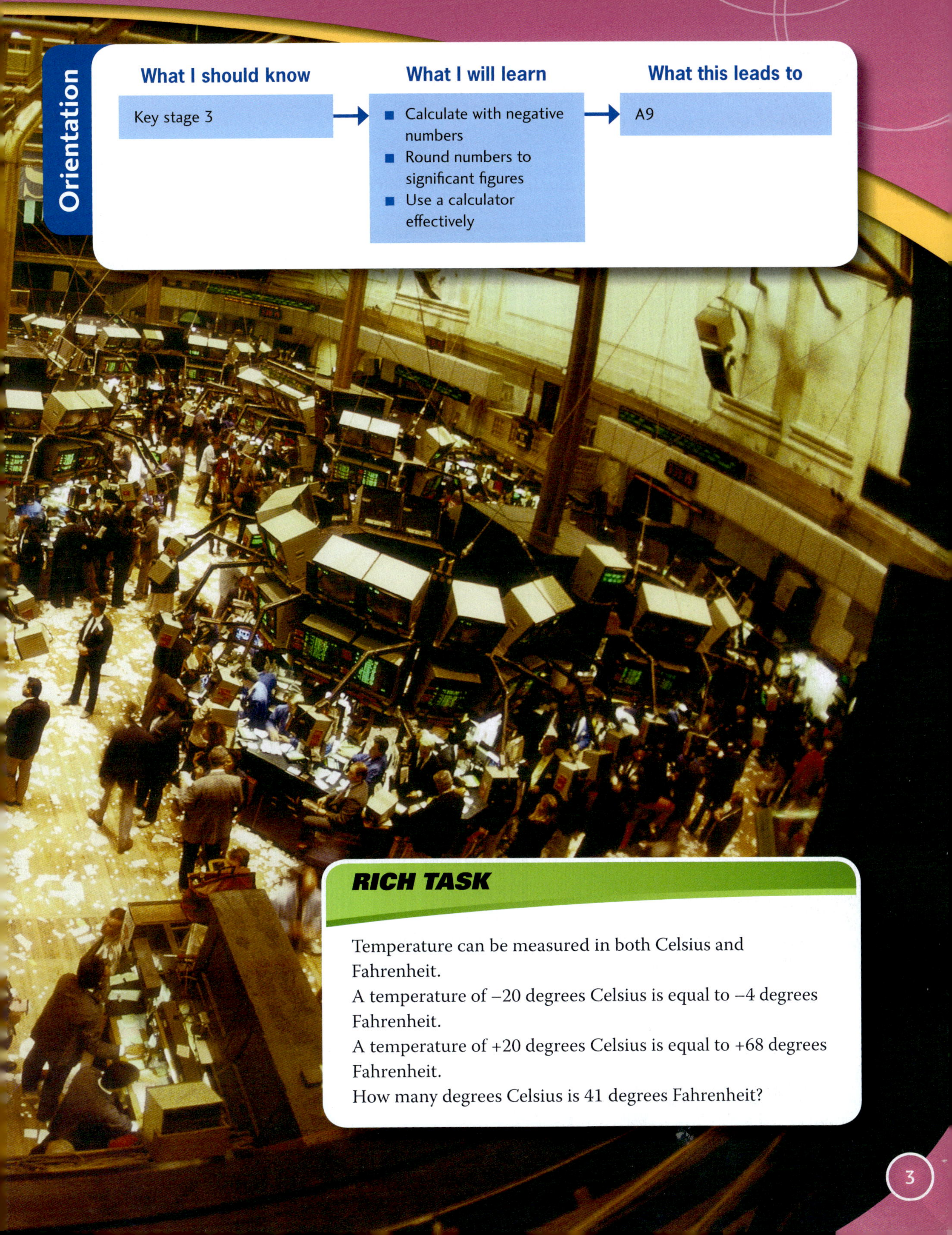

RICH TASK

Temperature can be measured in both Celsius and Fahrenheit.
A temperature of –20 degrees Celsius is equal to –4 degrees Fahrenheit.
A temperature of +20 degrees Celsius is equal to +68 degrees Fahrenheit.
How many degrees Celsius is 41 degrees Fahrenheit?

Using numbers

A1.1

- Understand integers and place value

The standard international bar code number contains 13 digits.

The final digit of a bar code is the check digit.
The check digit lets the scanner determine whether it scanned the number correctly or not.

To find the check digit:

→ **read the other digits from right to left**	**5, 4, 3, 2, 1, …**
→ **find the sum of digits in the odd positions**	**5 + 3 + 1 + 3 + 1 + 9 = 22**
→ **multiply the total by 3**	**22 × 3 = 66**
→ **find the sum of digits in the even positions**	**4 + 2 + 4 + 2 + 0 + 5 = 17**
→ **find overall total**	**66 + 17 = 83**
→ **total + check digit = next multiple of 10.**	**83 + 7 = 90 ✓**

RICH TASK

Find out whether the check digits on these international bar codes are correct.

3 200000 003774

0 123456 789010

4 903726 884816

Numbers are everywhere: newspapers, leaflets, the internet, television, radio.
You need to understand number and place value to make sense of the information.

EXAMPLE

In 2009, the Italian earthquake in L'Aquila left 1179 people injured and over sixty-six thousand people homeless. L'Aquila is 1920 km from London.

a Write in words the number 1179.

b Write in figures the number sixty-six thousand.

c Write down the place value of 9 in the number 1920.

ANSWER

a One thousand one hundred and seventy-nine

b 66 000

c 9 represents nine hundred in 1920

Exercise A1.1

LOW

1 Write these numbers in words.

a 1307 b 29 006 c 300 000 d 605 030

2 Write these numbers in figures.

a Eight thousand and forty-three

b Seventy million

c Two hundred thousand and fifty-one

d Two thousand and ten

3 Write down the value of the bold digit in these numbers.

a 3**8** 000 b 30**4**2 c 10 0**7**0

*4 **RICH TASK**

a Write down the smallest and largest two-digit number that you can make with the digits 9 5

i Subtract the smallest two-digit number from the largest number.

ii Using the two digits in your answer, repeat the process until you cannot go any further.

iii Choose another set of two digits to start the process. What is the final answer each time?

b Write down the largest and smallest three-digit number that you can make with the digits 1 8 2

i Subtract the smallest three-digit number from the largest number.

ii Using the three digits in this answer, repeat the process until you obtain the same three-digit number.

iii Choose another set of three digits to start the process. Why can the final three-digit number always be divided exactly by 99?

Directed numbers

A1.2

- Add, subtract, multiply and divide integers

Judy has 10 cards with numbers on them.
Five cards each have the number 1 and the other five cards each have the number −1.

If you add up the numbers on all 10 cards you have a total of zero.

1	1	1	1	1
−1	−1	−1	−1	−1

1	1	1	1	
−1	−1	−1	−1	−1

Suppose one of the cards marked with 1 is taken away.
What is the total of the remaining nine cards?

The total now is −1. The sum is 0 − 1 = −1.

Suppose three of the remaining cards marked with −1 are taken away.
What is the total of the remaining 6 cards?

The total now is 2. The sum is −1 − (−3) = 2.

1	1	1	1
−1	−1		

→ **Adding a negative number gives a smaller number** **5 + −2 = 5 − 2 = 3**
→ **Subtracting a negative number gives a bigger number** **5 − −3 = 5 + 3 = 8**

Number patterns can show what happens when multiplying and dividing by negative numbers.

6 × 3 =	18
6 × 2 =	12
6 × 1 =	6
6 × 0 =	0
6 × −1 =	−6
6 × −2 =	−12

2 × 3 =	6
2 × 2 =	4
2 × 1 =	2
2 × 0 =	0
2 × −1 =	−2
2 × −2 =	−4

As the number you multiply by decreases by 1, the answer decreases.

Dividing is the **inverse** of multiplying.
Rearranging 6 × −2 = −12 you get
−12 ÷ 6 = −2 and −12 ÷ −2 = 6

→ **Multiplying or dividing a positive number by a negative number gives a negative answer 5 × −2 = −10**
→ **Multiplying or dividing a negative number by a negative number gives a positive answer −12 ÷ −3 = 4**

Exercise A1.2

LOW

Unit A

1 Work out

a $4 + -3$ b $7 + -9$ c $-8 + 4$ d $-9 + 10$

e $-10 + -2$ f $-12 + -3$ g $-4 + -5$ h $-7 + -3$

i $4 - -3$ j $7 - -8$ k $-8 - 11$ l $-9 - 3$

m $-10 - -2$ n $-12 - -7$ o $-4 - -9$ p $-7 - -15$

2 Work out

a 4×-3 b 7×-5 c -8×4 d -9×3

e -10×-2 f -12×-3 g -4×-5 h -7×-3

i $18 \div -2$ j $15 \div -5$ k $-8 \div 4$ l $-9 \div 3$

m $-12 \div -4$ n $-12 \div -3$ o $-40 \div -5$ p $-27 \div -3$

q $5 - 4(6 \times -3)$ r $27 - 3(60 \div 12)$ s $4 - (3^2)$

Take care with the brackets.

3 Erin is travelling from London to Paris, from Paris to Rome, and from Rome to Athens.
The map shows the temperature at each city Erin visits.

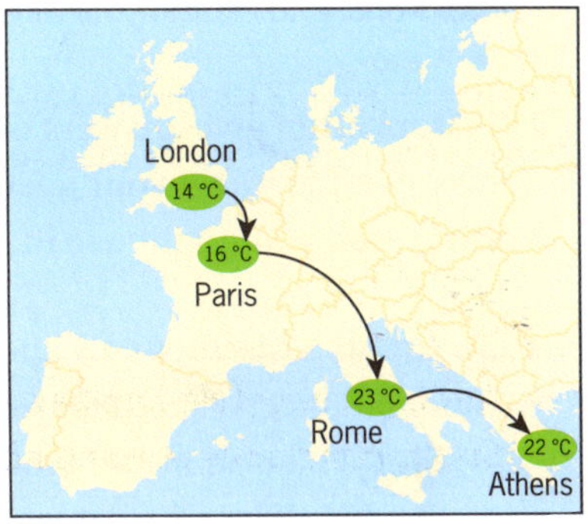

a Copy and complete the table to show the temperature at each city and the difference in temperature between the two cities.

The first one has been completed for you.

From	Temperature °C	To	Temperature °C	Temperature difference °C
London	14	Paris	16	2
Paris		Rome		
Rome		Athens		

b How much colder is it in London than it is in Athens?

4 ☆ **PROBLEM** ☆

There are 12 questions in a quiz.
A correct answer scores 3 points. An incorrect answer scores −2 points.
A question not answered scores nothing. It is possible to have a negative total.

a What are the maximum and minimum points that you could score?

b A pupil answers 10 of the questions. 4 are correct.
Explain why this pupil's total score is zero.

c Write down two different ways in which a pupil could have a total score of 12 points.

Rounding: significant figures and decimal places

A1.3

- Round numbers to a specified degree of accuracy

Cooking temperatures in old cookery books are often given in Fahrenheit, °F.
To change the cooking temperature to Celsius, °C:

first subtract 32 from °F → then multiply by 5 and divide by 9 → then round your answer to the nearest 10.

RICH TASK

Would 350 °F and 360 °F have the same cooking temperature in °C? Show working to explain your answer.

Calculators often give answers of nine digits or more.
You may need to **round** your answer. There are different types of rounding, that you might use in different situations.

→ **To round a number to a specified number of decimal places or significant figures (for example 3.147):**

		2 d.p.	2 s.f.
Step 1	**Look at the first digit to be discarded**	3.147	3.147
Step 2	**If it is 5 or greater, increase the previous digit by 1**	3.15	
Step 3	**If less than 5, the previous digit stays the same**		3.1

p. 222

EXAMPLE

Round

a 187 to the nearest 10
b 3.8946 to 2 decimal places (2 d.p.)
c 5362 to 3 significant figures (3 s.f.)
d 0.046 25 to 3 s.f.

ANSWER

a Nearest 10, so don't require units (7)
This is 5 or more so round the 8 up.
187 is 190 to the nearest 10.

100	10	1
1	8	(7)

b 2 d.p. so leave 2 digits after the decimal point.
3.8946 = 3.89 to 2 d.p.

1	.	$\frac{1}{10}$	$\frac{1}{100}$	$\frac{1}{1000}$	$\frac{1}{10000}$
3	.	8	9	(4)	6

c 3 s.f. so use the first 3 non-zero digits.
5362 = 5360 to 3 s.f.

1000	100	10	1
5	3	6	(2)

d 3 s.f. so use the first 3 non-zero digits.
The next digit is 5 or more so round the 2 up.
0.046 25 = 0.0463 to 3 s.f.

1	.	$\frac{1}{10}$	$\frac{1}{100}$	$\frac{1}{1000}$	$\frac{1}{10000}$	$\frac{1}{100000}$
0	.	0	4	6	2	(5)

Exercise A1.3

MEDIUM LOW

Unit A

1 Write these distances to the nearest 10 m.

a 632 m b 215 m c 198 m d 199 m

2 Write these weights to the nearest 100 g.

a 7724 g b 5493 g c 3056 g d 2609 g

3 Write these numbers **i** to the nearest 10 **ii** to the nearest 100.

a 40 729 b 309 c 586 d 150 902

4 Which of the following distances are the same to one significant figure?

248 km 302 km 254 km 350 km

5 Write down the following numbers correct to the number of decimal places stated.

a 19.372 to 2 d.p. b 19.372 to 1 d.p.

c 0.007 519 to 5 d.p. d 0.007 519 to 3 d.p.

e 153.2617 to 3 d.p. f 153.2617 to 2 d.p.

g 34.1572 to 2 d.p. h 4.9043 to 2 d.p.

For **2** decimal places, count **2** digits after the decimal point.

6 Write down the following numbers correct to the number of significant figures stated.

a 76 286 to 2 s.f. b 76 286 to 3 s.f.

c 23 403 to 3 s.f. d 57 961 to 3 s.f.

e 46 320 to 2 s.f. f 23.403 to 4 s.f.

g 23.405 to 3 s.f. h 7.183 62 to 4 s.f.

For **2** significant figures, count from the first non-zero digit.

7 ☆ **PROBLEM** ☆

A tower block is 200 m high to the nearest 100 m, and 170 m high to the nearest 10 m.
Write down one possible height for this tower block.

A man's weight is 100 kg to the nearest 100 kg, and 50 kg to the nearest 10 kg.
Write down one possible weight for this man.

Effective use of your calculator

A1.4

- Enter complex calculations and use function keys for reciprocals, squares and powers

A scientific calculator will usually work out sums in the standard **order of operations**.

	a	b
For example, to calculate	$a = \frac{(4.2 - 1.7)^2}{10}$	$b = \frac{19.1 \times 3}{\sqrt{12 \times 3}} + 2$
First work out the **brackets**	$a = \frac{2.5^2}{10}$	$b = \frac{19.1 \times 3}{\sqrt{36}} + 2$
Next work out **powers** or **roots**	$a = \frac{6.25}{10}$	$b = \frac{19.1 \times 3}{6} + 2$
Then **multiplication** and **division**	$a = 0.625$	$b = \frac{57.3}{6} + 2$
Finally **addition** and **subtraction**		$b = 9.55 + 2$ $= 11.55$

Multiply the numbers under the $\sqrt{\ }$ first as well.
This is a 'hidden bracket'.

There are some special keys on your calculator that will make calculating simpler.

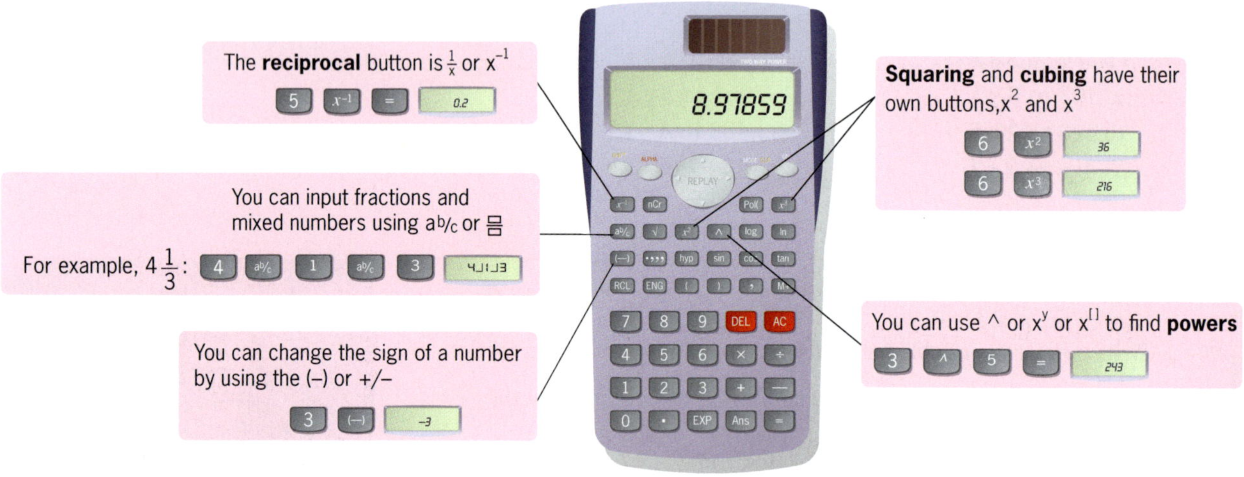

→ **The reciprocal of a number is often useful in calculations.**

The reciprocal of n is $\frac{1}{n}$ | **The reciprocal of 7 is $\frac{1}{7}$**

The reciprocal of $\frac{a}{b}$ is $\frac{b}{a}$ | **The reciprocal of $\frac{4}{5}$ is $\frac{5}{4}$**

Exercise A1.4

MEDIUM

Unit A

1 Calculate (to 2 d.p. where necessary)

a $12+\frac{14.7}{0.3}$ b $5.7^2-\sqrt{75}$ c $\frac{1}{0.8^2}$ d $\frac{2^3}{5^2-2}$

e $\frac{7.6}{12.5-6.8}$ f $\left(\frac{5}{0.4}\right)^2$ g $\frac{45}{(0.3)^2}$ h $\sqrt{36+(4.5)^2}$

2 Calculate the following by

i writing down all the figures shown on your calculator

ii giving your answer to the specified number of decimal places or significant figures.

a $\frac{7.92\times 1.71}{4.2+3.6}$ correct to 1 d.p. b $\frac{4.72-1.4}{-1.4^2}$ correct to 2 d.p.

c $\frac{3.14-8.16}{8.25\times 3.18}$ correct to 2 d.p. d $\frac{16.5\times 10.3}{8.25+5.15}$ correct to 1 d.p.

e $\sqrt{4.5+3.2^2}$ correct to 3 s.f. f $\sqrt{25.3-2.7^3}$ correct to 2 s.f.

3 Calculate the reciprocal of the following numbers.

a 4 b 9 c 2.5

d 3.6 e 0.2 f 0.16

g $\frac{1}{2}$ h $\frac{1}{4}$ i $\frac{2}{5}$

4 Find these quantities by using the power key on your calculator.

a 6^4 b 5^7 c 10^5

d 8^6 e 4.5^4 f 0.4^6

*5 ☆ **ACTIVITY** ☆

Investigate the keys on your calculator.
Choose four keys that you were previously unfamiliar with and write brief instructions for each one, with an example.

Summary and assessment

CHECK OUT

You should now be able to

- use the four operations with directed numbers
- round numbers using
  - significant figures
  - decimal places
- use a calculator effectively.

Exam-style question

In this number pyramid, each number is found by adding the two numbers in the blocks below.

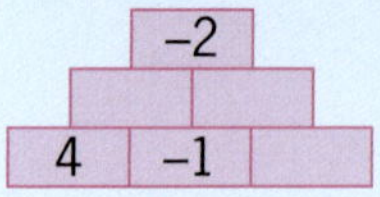

Work out the missing numbers and complete the pyramid.

First look to see which block you can work out. Show your working.

Charlie's answer:

$4 + -1 = 3$
Missing number in 1st block of 2nd row is 3

$3 + ? = -2 \qquad ? = -5$
Missing number in 2nd block of 2nd row is -5

$-1 + ? = -5 \qquad ? = -4$
Missing number in 3rd row is -4

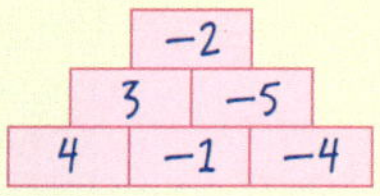

Use your answer to work out another block. Use a ? or a letter to show your working.

Complete the pyramid as you work out the sums.

Now find the last missing number.

Exam practice

1 The number e = 2·718 281 8 correct to 7 decimal places.
Write the number e correct to

a 5 decimal places

b 2 decimal places. *(2 marks)*

2 a Work out $\frac{7 \cdot 2 + 9 \cdot 3}{(3 \cdot 1)^2}$.

Write down all the numbers on your calculator display.

b Give your answer correct to 1 significant figure.

(2 marks)

3 In this number pyramid, each number is found by multiplying the two numbers in the blocks below.

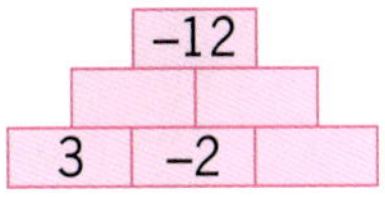

Work out the missing numbers and complete the pyramid.

(3 marks)

2 Summary statistics

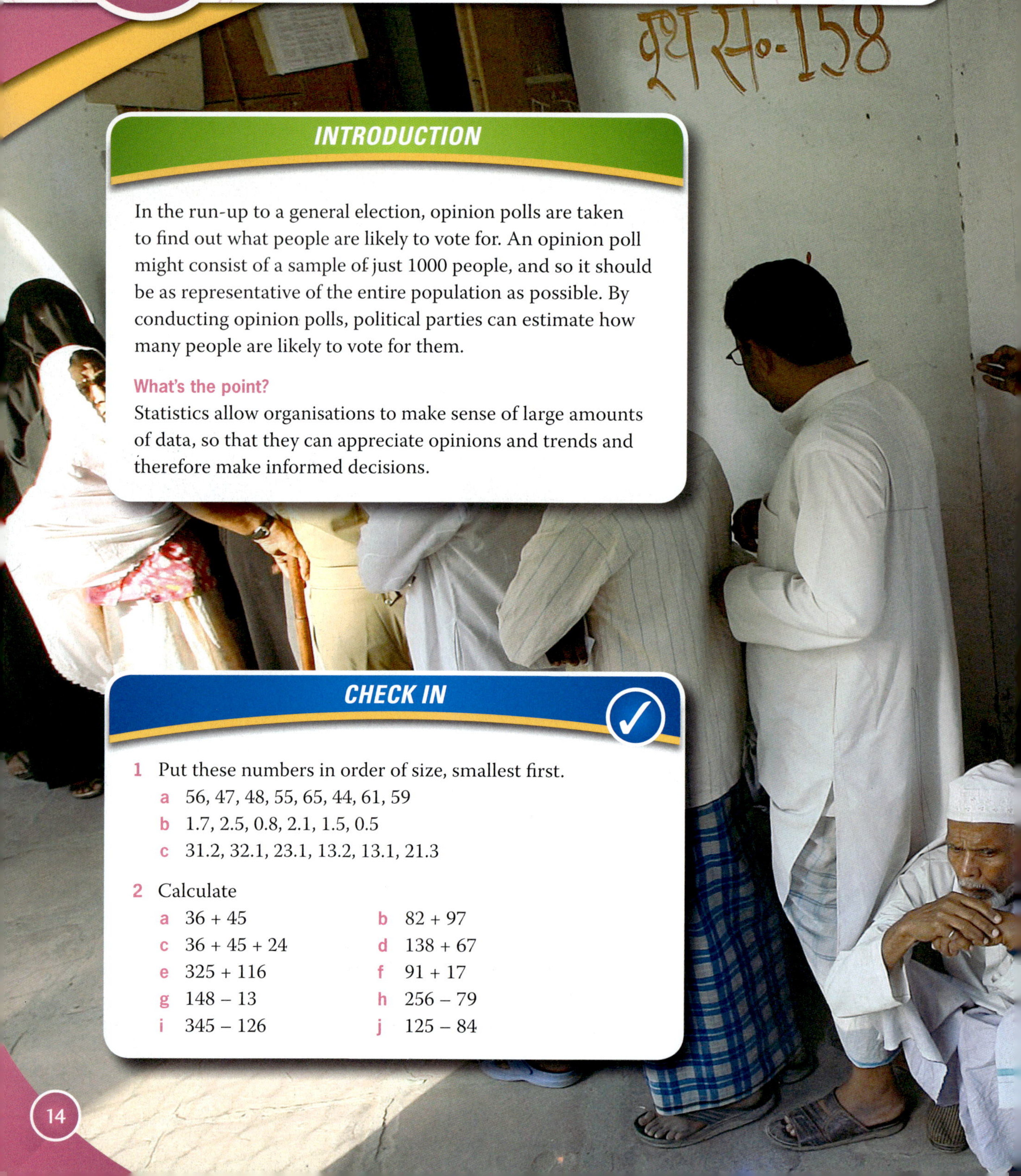

INTRODUCTION

In the run-up to a general election, opinion polls are taken to find out what people are likely to vote for. An opinion poll might consist of a sample of just 1000 people, and so it should be as representative of the entire population as possible. By conducting opinion polls, political parties can estimate how many people are likely to vote for them.

What's the point?

Statistics allow organisations to make sense of large amounts of data, so that they can appreciate opinions and trends and therefore make informed decisions.

CHECK IN

1 Put these numbers in order of size, smallest first.

- a 56, 47, 48, 55, 65, 44, 61, 59
- b 1.7, 2.5, 0.8, 2.1, 1.5, 0.5
- c 31.2, 32.1, 23.1, 13.2, 13.1, 21.3

2 Calculate

- a 36 + 45
- b 82 + 97
- c 36 + 45 + 24
- d 138 + 67
- e 325 + 116
- f 91 + 17
- g 148 − 13
- h 256 − 79
- i 345 − 126
- j 125 − 84

Orientation

What I should know	What I will learn	What this leads to
Key stage 3	■ Use the data handling cycle ■ Use stem and leaf diagrams ■ Design a questionnaire	A6

RICH TASK

What is the average time taken to travel to school for pupils at your school?
Do the boys take less time than the girls?
Do the pupils who live nearer the school take less time than the pupils who live further away?
Investigate this and write a short report on your results.

The handling data cycle

A2.1

- Understand and use the handling data cycle
- Extract and use information from two-way tables

Thao wants to investigate: just how good is your memory?
She collects 25 objects and lays them out in a square grid.

ACTIVITY

Look at the objects for 30 seconds. Now look away. How many objects can you recall?

Planning an investigation is the first part of the **handling data cycle**.
The second part is **collecting data**.

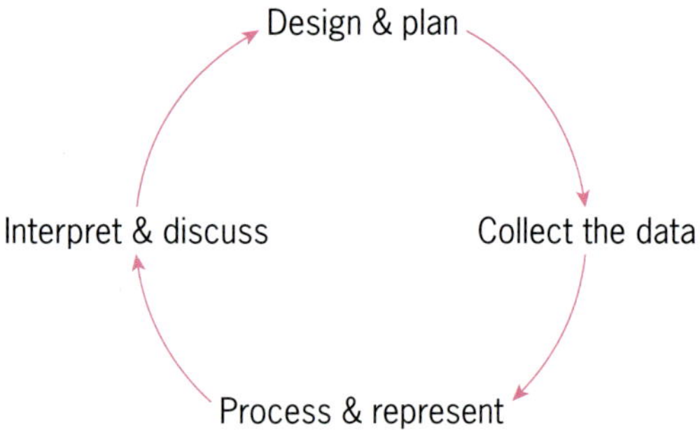

This is part of the **data-collection sheet** Thao used to collect data on memory.

Gender Male(M) Female(F)	Age	Nine objects time (seconds)	25 objects number recalled	1st object recalled
M	15	74	13	iPod

Thao records the time taken for a small grid of nine objects and also the number recalled for all 25 objects.

Thao plans to collate some of her results in a **two-way table**.

Gender \ Number recalled	0–5	6–10	11–15	16–20	21–25
Male					
Female					

Thao will write the tally in the boxes for each category.

Unit A

You can collect data:

- **in surveys where you ask questions of everyone, or from a sample of the population**
- **in experiments such as observations in science or geography.**

Exercise A2.1

LOW

1 This is the data-collection sheet that Clare uses to record information about poems in a book.

Number of verses \ Lines per verse	5	6	7	8	9	10
1						
2						
3						

a What are **i** the least and **ii** the greatest number of verses in any one poem?

b How many lines do **i** the shortest and **ii** the longest poem have?

c Copy the diagram and mark the position of a poem with 2 verses and 8 line per verse.

d There are 16 poems in the book. Could every cell in the table be filled? Explain your answer.

2 Design a suitable data-collection sheet in the form of a two-way table for the following surveys to use.

a Katy is doing a survey to find out how often people go to the cinema and how much they spend.

b James wants to know which flavour crisps he should stock in the school tuck shop. He also wants to know which year groups prefer which flavours.

c Merlin wants to find out how far people would travel to see their favourite band perform and how much they would spend on a ticket.

3 The table shows the distribution of blood groups amongst the population of some countries in the mid 1980s.

People from Wales are thought to be closely related to the Scots and Irish.

Country \ % blood group	A	B	AB	O
England	42	8	3	47
Ireland	32	11	3	54
Scotland	35	11	3	51

a What percentage of Welsh people would you expect to have blood group B?

b What other problems could you investigate from this information?

Types of data and sampling

A2.2

- Understand and use the handling data cycle

Here is some of the **primary data** collected by Thao in her survey on memory.

This is **continuous data** – you can measure it on a scale

This is **discrete data** – you can count it

This is **qualitative data** – you can describe it

Gender Male (M) Female (F)	Age	Nine objects Time (seconds)	25 objects Number recalled	First object recalled
M	15	74	13	iPod
M	15	103	20	Rabbit
F	15	64	15	Rabbit
F	15	58	9	iPod
M	15	58	18	Rabbit
M	15	50	16	Elastic band
M	15	35	13	Ball
F	15	64	14	Button
F	15	60	13	Phone

→ **Primary data is data that you collect yourself.**
→ **Secondary data is data that someone else has collected.**

Thao may have asked only her friends.

When she collected her data, Thao should have considered that:

→ **Data can be biased if not everyone has an equal chance to be part of the survey.**
→ **In a random sample everyone has an equal chance of being chosen.**

Thao might have picked names at random from a hat, or numbered everyone and chosen random numbers.

→ **Increasing the sample size generally leads to better results.**

Thao should have included more than nine people in her survey.

EXAMPLE

James carries out a survey to find out if people in his town enjoy sport. He stands outside a football ground and surveys people's opinions as they go in to watch a match. Write down two reasons why this is not a good sample to use.

ANSWER

1 People who watch football usually enjoy sport.
2 More men than women go to watch football, so the survey could be gender-biased.

Unit A

Exercise A2.2

LOW

1 Say whether the following data are discrete or continuous.
   a How long you take to finish in a cross-country race, and your finishing position in the race.
   b The weight of a parcel and the cost of its postage.
   c The number of eggs and the amount of sugar needed in a cake recipe.
   d The stars in our galaxy and the distance of each star from Earth.

2 Claire looks up information on the internet about the number of accidents in the home.
   a Is this primary or secondary data?
   b Describe another way that Claire can find information about accidents in the home and state whether this would be primary or secondary data.

3 'Do you get an allowance or have a part-time job?'
   a How you could find information from your year group to answer this question?
   b Describe how you would find a representative sample of your year group to ask.
   c Why might it not be enough to ask only 2% of your year group?
   d What problems might you have if you asked 75% of your year group?

Think about how you can avoid **bias**.

4 Thao believes that people can be trained to improve their memory and recall. Design an experiment and/or survey for Thao to use to test this theory. In your design remember to include the following where appropriate:
- a suitable sample size to use
- how you would choose people to take part
- a suitable data-collection sheet.

Summarising small data sets

A2.3

- Calculate mean, range and median of small discrete data sets

Data can be summarised using an average and a measure of spread.

An **average** is a single value used to describe a data set.

> → **The mode is the value that occurs most often.**
> → **The median is the middle value when the data is arranged in order.**
> → **The mean is calculated by adding all the values and dividing by the number of values.**

To find the median with an even number of values:

- Arrange in order, eg
  3 6 7 9 10 13
- Take the two middle values
  7 9
- Add them together and divide by 2
  (7 + 9) ÷ 2 = 8
- The median is 8

Spread is a measure of how widely **dispersed** (or spread out) a data set is.

> → **Range = highest value – lowest value**

EXAMPLE

Louise collected data on the number of times her friends went swimming in one month:

4	7	22	1	6	
2	1	5	6	6	4

Work out **a** the range **b** the mode **c** the mean **d** the median.

ANSWER

a Range = 22 – 1 = **21** 4 7 **22** **1** 6 2 1 5 6 6 4

b Mode = **6** 4 7 22 1 **6** 2 1 5 **6** **6** 4

c Mean = (4 + 7 + 22 + 1 + 6 + 2 + 1 + 5 + 6 + 6 + 4) ÷ 11
= 64 ÷ 11
= **5.8**

d 1 1 2 4 4 5 6 6 6 7 22 Arrange in order
1 1 2 4 4 **5** 6 6 6 7 22 Find the middle number
Median = **5**

The mean does not have to be an integer even if all the data values are integers.

The mode, median and mean are all averages, but they can be different.

Exercise A2.3

LOW

1 15 students took part in a sponsored swim. The number of complete lengths swum were:

45, 63, 72, 63, 63, 24, 54, 73, 99, 65, 63, 72, 39, 44, 63

Work out **a** the range **b** the mode **c** the mean **d** the median.

It is not always possible to find the mode.

2 These marks were obtained by a group of students in a test.

13, 76, 22, 54, 37, 22, 21, 19, 59, 37, 84

Work out **a** the range **b** the mode **c** the mean **d** the median.

3 Gareth searched the internet for prices on a new clutch for his car. He found the following prices, in £, for the same clutch.

89, 87, 64, 88, 82, 88, 85, 83, 81, 89, 90

Work out **a** the range **b** the mode **c** the mean **d** the median.

4 These are the house prices, in £1000s, for houses in the North-east of England sold by an estate agent one month.

153, 174, 129, 132, 167, 153, 199, 162, 134, 128, 127, 127, 127, 164, 127

a Work out
  i the range **ii** the mode
  iii the mean **iv** the median.

b Explain why the mode is not the best measure of average to use.

5 ***RICH TASK***

a Find a set of five numbers with mode 7 and median 7.
Is your answer unique?
What other answers can you find?

b Find two sets of five numbers with range 32 and median 18.
Describe any features that the two sets of numbers have in common.

Stem-and-leaf diagrams

A2.4

- Produce stem-and-leaf diagrams

A **stem-and-leaf diagram** lists small data sets in order.
Thao draws this **stem-and-leaf diagram** for the time taken to recall nine objects by the 11-year-olds in her survey.

12	1
11	3 6
10	6 9
9	1 2 5
8	1 3 4 4
7	2 3 6 8 9
6	0 0 2 7 7
5	6 6 6
4	4 6

Key: 7 | 2 means 72 seconds

When you draw a stem-and-leaf diagram you need to:

Choose a **stem** appropriate for your data. The stem numbers can increase either upwards or downwards, as long as you are consistent.

Order the **leaves.**

Write a **key** to explain the data.

Thao summarises the data:

Range 121 – 44 = 77
Mode 56
Median 78
Mean 2127 ÷ 27 = 78.8

EXAMPLE

These data are the times taken, in seconds, to swim 25 m by a class of students:

118	114	138	126	127	100	105	128	118	133	111	109
107	113	101	96	119	124	112	103	119	140	107	

a Draw a stem-and-leaf diagram for these data.
b Use the diagram to work out **i** the range **ii** the median.

ANSWER

a

9	6
10	0 1 3 5 7 7 9
11	1 2 3 4 8 8 9 9
12	4 6 7 8
13	3 8
14	0

Use the tens digit as the stem

Use the units digit as the leaves

Median

Always give a key

Key: 12 | 4 means 124 seconds

b **i** Range = 44 (140 – 96 = 44). **ii** Median = 114.

Exercise A2.4

MEDIUM LOW

For the following data sets

a draw a stem-and-leaf diagram to show the information

b use your diagram to find the **i** range **ii** median.

REMEMBER ...

a include a key

b write your units.

1 Percentage achieved in a statistics test:

78 82 74 45 69 75 93 54 61 70 48 66
62 51 77 59 51 89 81 52 63 71 59

2 Minutes per day spent playing computer games:

40 26 75 84 33 39 28 66 67 71 80
37 52 47 63 49 41 44 58 69 43
73 55 59 43 61 38 29 30 77 60

3 Time taken in minutes to solve a crossword puzzle:

12 24 21 16 8 9 3 31 18 27 35
41 26 12 17 6 5 19 29 32 37 40
15 22 10 33 11 7 20 27 29 34

To find the median when there are an even number of items, you find the two middle values and find the halfway point between them.

4 Weight in kg of Year 10 boys:

47 51 63 39 42 57 36 37 49 32 60
54 56 45 52 48 61 58 56 59 70 66 56

5 Height in cm of a group of Year 10 girls:

153 147 160 146 162 158 159 171 149 152 150
163 172 167 165 155 155 157 154 168 150 172 152

6 IQ scores of a Year 10 tutor group in a comprehensive school:

101 112 125 109 98 107 108 117 121 116 94
91 105 106 114 118 126 131 92 88 129
89 116 103 108 127 110 117 104 119 133

7 Time taken to the nearest minute to change a car wheel:

10 17 3 19 22 27 16 9 6 30 23
21 12 21 9 25 23 18 33 8 32 15 11

Unit A

Back-to-back stem-and-leaf diagrams

A2.5

- Compare distributions and make inferences

38 students answer a survey question about time spent on the internet one evening. The data is presented in a **back-to-back stem-and-leaf diagram**.

Girls		Boys
8 7	3	
9 7 5 4 4 3 2 1	4	1 4 8
8 6 6 5 3 2 1	5	3 3 4 6 7 7
2 1	6	1 3 5 6 9
	7	0 2 2 5 9

Key: 1 | 5 | 3 means Girls 51 minutes, Boys 53 minutes

The two sets of data are shown with a common stem.

Both data sets are in order, with the lowest values closest to the stem.

RICH TASK

Is Mike correct?
How can you tell?

Mike says:

The data shows that on average boys spend longer on the internet, but the time spent by the boys is more varied.

→ **You can use stem-and-leaf diagrams and summary statistics to make comparisons.**

EXAMPLE

Compare these history test results for a class of boys and girls.

Girls		Boys
5 4	3	9
9 7 7 6 1	4	3 6 7
8 7 4 2	5	2 4 6 8 9
	6	1 9

Key: 1 | 4 | 3 means 41% for girls, 43% for boys

ANSWER

The boys on average scored a higher percentage in history than the girls. There is a greater range of results for the boys than the girls. The **shape** of the diagram shows a larger 'hump' for boys, implying that boys generally scored better in history than girls.

	Girls	Boys
Median	47	54
Range	58 − 34 =24	69 − 39 =30

Exercise A2.5

MEDIUM

Unit A

1 Find summary statistics and compare reaction times of a sample of boys and girls represented in the stem-and-leaf diagram.

Boys		Girls
8 8 8 6 2 2	3	
7 5 4 2	4	0 3 4 6 8
9 8 7 6 3 2	5	2 2 5 6 9
8 6 4 2 1	6	2 3 3 5 6 7
2 1	7	2 3 4 6 7
	8	0 2

Key: 2 | 4 | 0 means
Boys 4.2 second; Girls 4.0

2 These are the survey results for the times (in minutes) taken for a sample of students to complete two jigsaw puzzles.

Puzzle P:	13	17	19	10	8	22	31	11	24	27
	6	37	18	12	29	14	8	17	9	
Puzzle Z:	28	21	29	15	12	9	32	17	18	11
	19	16	8	33	24	14	25	17	23	

a Draw a back-to-back stem-and-leaf diagram to represent these data.

b Work out the **i** range **ii** mode **iii** mean **iv** median.

c Use the diagrams and summary statistics to compare the difficulty of the two jigsaw puzzles.

3 a Draw a back-to-back stem-and-leaf diagram to represent the IQ scores of two classes, X and Y.

X:	105	123	131	117	118	104	98	96	103	112	110
	117	126	129	123	109	108	115	99	89	121	
	134	106	105	122	124	116	110	118	115	130	
Y:	118	119	104	121	126	118	109	97	114	129	130
	107	116	87	93	128	121	118	113	103	102	
	114	107	131	99	106	124	132	119	126	114	

*b Find summary statistics to present a convincing argument to show that students from one class have, in general, the higher IQ.

Summary statistics from frequency tables A2.6

- Calculate mean, range and median with discrete data

Thao draws these tables to show the number of objects recalled by 15-year-olds and 11-year-olds.

Number recalled 15-year-olds	13	14	15	16	17	18	19
Frequency	3	2	7	4	3	6	2

Number recalled 11-year-olds	11	12	13	14	15	16	17
Frequency	1	1	2	3	5	8	7

RICH TASK

Thao says that the recall of both age groups is the same.
On what evidence is she basing her conclusion?
How can you use the tables to present a convincing argument that 15-year-olds can recall more objects than 11-year-olds?

You can find summary statistics from data presented in **frequency tables**.

→ **Mode = value with the highest frequency**
→ **Median (of n values) = $\frac{1}{2}(n+1)$th value**
→ **Mean (of n values) = $\frac{\text{sum of all values}}{n}$**

EXAMPLE

The table shows the length of answers in a crossword puzzle.
For these data, work out the

a mode
b median
c mean
d range.

Word length	Frequency
4	3
5	5
6	7
7	8
8	3
9	1

ANSWER

a Mode = 7.

Words with **7** letters have the highest frequency (8).

EXAMPLE (CONT.)

b Median = 6.

Total number of words = 27 (total frequency)

Middle word length → $\frac{1}{2}$ (27 + 1) = 14th word length in order

4 4 4 5 5 5 5 5 6 6 6 6 6 **6** 6 7 …

$\frac{1}{2}(n + 1)$

$= \frac{1}{2} \times 28 = 14$

c Mean = total number of letters ÷ total number of words = 168 ÷ 27 = 6.2

Mean = 168 ÷ 27

= 6.2.

Word length	Frequency	Word length × frequency
4	3	4 × 3 = 12
5	5	5 × 5 = 25
6	7	6 × 7 = 42
7	8	7 × 8 = 56
8	3	8 × 3 = 24
9	1	9 × 1 = 9
Total	27	168

There are three words with four letters, so 12 letters altogether.

There are 168 letters in total, and 27 words in total.

d Range = 5.

Longest – shortest word length **9 – 4** = 5

Exercise A2.6

MEDIUM LOW

1 The data sets give information about the length of words in two different crosswords. Work out these summary statistics for each crossword.

i mode **ii** median **iii** mean **iv** range.

a

Word length	Frequency
4	8
5	4
6	9
7	5
8	5

b

Word length	Frequency
3	4
4	6
5	9
6	6
7	4

2 The tables show the recall of the 15-year-old boys and girls in Thao's survey.

Number recalled Boys	Frequency
13	2
14	1
15	2
16	2
17	1
18	4
19	1

Number recalled Girls	Frequency
13	1
14	1
15	5
16	2
17	2
18	3
19	1

Thao believes that girls have better recall than boys.

a Explain, using summary statistics, whether she is correct.

b Can you find evidence that the boys have better recall than the girls?

Grouped data

A2.7

- Estimate the mean and identify the modal class for large data sets

p. 82

Large data sets of continuous data can be presented in a **grouped frequency table**.

EXAMPLE

The table shows the times taken, to the nearest minute, by a group of students to solve a crossword puzzle.

a For these data, work out

i the **modal class**

ii an estimate for the mean

iii an estimate for the range.

b John says that the range for this data was actually 17. Explain why John may be correct. What could the possible times be?

Time, t, minutes	Frequency
$5 < t \le 10$	2
$10 < t \le 15$	14
$15 < t \le 20$	13
$20 < t \le 25$	6
$25 < t \le 30$	1

ANSWER

a **i** Modal class = $10 < t \le 15$

Look for the highest frequency (14), so $10 < t \le 15$ is the modal class.

ii
- Use the middle value, or **midpoint,** of each class as a representative value.
- Multiply by frequency to work out total times.

Mean = total time ÷ total number of students = 580 ÷ 36 = 16.1

Time, t, minutes	Frequency	Mid-point	Mid-point × frequency
$5 < t \le 10$	2	7.5	7.5 × 2 = 15
$10 < t \le 15$	14	12.5	12.5 × 14 = 175
$15 < t \le 20$	13	17.5	17.5 × 13 = 227.5
$20 < t \le 25$	6	22.5	22.5 × 6 = 135
$25 < t \le 30$	1	27.5	27.5 × 1 = 27.5
Total	36		580

iii Range = 30 – 5 = 25 Longest possible = 30 minutes Shortest possible = 5 minutes

b Range = longest – shortest time.
In the last class the time could be 26, 27, 28, 29 or 30.
In the first class the time could be 6, 7, 8, 9 or 10.
The possible times to achieve a range of 17 are 26 and 9 (26 – 9 = 17) or 27 and 10.

- An estimated value for the mode cannot easily be found from a grouped frequency table, only the class in which it lies. This class is called the **modal class**.
- The mean from grouped data is only ever an estimate.

Exercise A2.7

MEDIUM

Unit A

1 Alfie kept a record of his monthly mobile phone bills for one year.

a Write down the class interval that contains the mode.

b Calculate an estimate for the mean cost of Alfie's mobile phone bill.

c Give a reason why the mean is a better average to use than the modal class.

d Find possible times if the actual range is 34.

Phone bill, £ (B)	Frequency
$10 < B \le 20$	5
$20 < B \le 30$	4
$30 < B \le 40$	2
$40 < B \le 50$	1

*2 A garden centre gave a discount to its staff.
It kept a record of how much each staff member spent during one month.

a Calculate an estimate for the mean amount of money spent at the garden centre by its staff.

b Write down the class interval that contains the mode.

c Is it possible for the actual range of monies spent to be £159, £160 or £161?

Explain how you worked out your answer.

Monies spent, £ (M)	Frequency
$0 < M \le 40$	11
$40 < M \le 80$	8
$80 < M \le 120$	8
$120 < M \le 160$	9
$160 < M \le 200$	3
$200 < M \le 240$	1

*3 David carried out a survey to find the distance travelled to work by 120 teachers and 120 office workers.

Teachers

Distance, d, km	Frequency
$0 < d \le 5$	12
$5 < d \le 10$	33
$10 < d \le 15$	48
$15 < d \le 20$	20
$20 < d \le 25$	7

Office workers

Distance, d, km	Frequency
$0 < d \le 5$	2
$5 < d \le 10$	21
$10 < d \le 15$	51
$15 < d \le 20$	28
$20 < d \le 25$	18

Work out summary statistics for each data set and comment on David's conclusion that teachers live closer to their work place than office workers.

Questionnaires

A2.8

- Design and use questionnaires

Here is part of a questionnaire:

> *Please complete the following questionnaire*
>
> *1 Do you own or rent your home?*
>
> ☐ *Own* ☐ *Rent*
>
> *2 Are you happy with the refuse collection services supplied by the council?*
>
> ☐ *Yes very* ☐ *Yes they are okay* ☐ *Mostly happy* ☐ *No opinion*
>
> *3 When was the last time you contacted the council offices?*
>
> ☐ *In the last week* ☐ *In the last month* ☐ *In the last year* ☐ *Never*

✯ DISCUSSION ✯

- ➤ What problem might this questionnaire be trying to address?
- ➤ Who do you think is asking the questions?
- ➤ Do you think it is possible to answer all or any of the questions?
- ➤ What other responses could be included in each question? Why?
- ➤ Can you tick more than one answer box in any of the questions?

The handling data cycle can provide answers to problems only if relevant data is collected. A questionnaire can yield primary data from individuals.

You can write open questions or give a set of closed responses to choose from.

> *What's your favourite type of music?*
>
> ☐ *Rock* ☐ *Soul* ☐ *Hip hop* ☐ *Other* ← These are suggested responses.
>
> *If 'Other', please specify* .... ← This is an open response.

- Answer boxes should have a possible choice for everyone and should not overlap or have gaps.

How old are you?
☐ Under 10 ☐ 11–15 ☐ 15–18 ☐ Over 18

This is not a good question as a 10-year-old cannot answer and a 15-year-old can tick two boxes.

- Questions should use clear language, be **unambiguous**, free from **bias**, and also both useful and relevant to your survey.

1. Do you listen to a lot of music?
☐ A lot ☐ Sometimes ☐ Never
2. Do you agree that Kasabian are the best band of 2009?
☐ Yes

This is a ambiguous – 'a lot' means different things for different people.

This is a leading question – questions that begin with 'Do you agree …' tend to be biased.

Exercise A2.8

MEDIUM

*1 A radio station make some changes to the style of their breakfast radio show. They ask this question in a questionnaire:

'What do you think of the new breakfast radio show?'
☐ Fantastic ☐ Good

a Write down what is wrong with this question.
This is another question on the questionnaire:

'How long do you listen to the radio?'
☐ Less than 10 minutes ☐ 1 hour ☐ 1–2 hours

b Write down three things that are wrong with this question.
c Write a question to find out how long people listen to the breakfast radio show on a weekday. Remember to include some response boxes.

*2 **RICH TASK**

You are going to organise a school disco, for school years 7–9, to raise money for charity.
Write a questionnaire to find out:
- what type of music you should play
- what kind of refreshments you should have to sell
- what else you might do to raise money for charity.

Summary and assessment

A2

CHECK OUT

You should now be able to

- classify types of data
- find a sample
- find the mean, the median, and the mode
- find the range
- draw and use stem-and-leaf diagrams
- write questions for questionnaires
- apply the handling data cycle.

Exam-style question

Look at these five numbers

3 6 8 4 9

a Show that the mean of these numbers is six.

b Explain why the median of these numbers is six.

c Four numbers have a mean of six and a median of six, but none of the numbers is six. What could the four numbers be?

Parvati's answer:

a $3 + 6 + 8 + 4 + 9 = 30$
$30 \div 5 = 6$
Mean $= 6$

b To find median numbers need to be in order
Order is: 3 4 6 8 9
6 is in the middle of the ordered numbers so 6 is the median

c 4 numbers, mean $= 6$;
Total of these numbers is 24
Median of 4 numbers is half way between 2nd & 3rd; half way between 5 and 7 is 6
$5 + 7 = 12$, $24 - 12 = 12$;
Other numbers must be above & below 6 & add to give 12
Try 3 5 7 9

'Show that' means you need to show all the working out.

Explain questions can have maths as well as words.

Use the information given to try different ideas.

There may be more than one answer.

Exam practice

1 These are the names of eight people that work in an office.

Barry	Charlie	Adam	Jon
Ellen	Eddie	Sophie	Charlie

a What name is the mode?

b One person leaves and a new person begins work in the office. The name that is the mode now is Ellen. What are the names of the person that leaves and the person who joins the office?

(*2 marks*)

2 Edith drew this table to show the length of each track on her mp3 player.

Time (m minutes)	Number of tracks
$2 \le m < 3$	7
$3 \le m < 4$	40
$4 \le m < 5$	38
$5 \le m < 6$	13
$6 \le m < 7$	2

a Calculate the mean length of the tracks on Edith's mp3 player.

b Which class contains the median?

(*5 marks*)

3 Mr Graham gave his students a test.
He summarised their test percentages in a back-to-back stem-and-leaf diagram.

Boys		Girls
6	8	8 9
7	7	2
6 6 0	6	2 3
1	5	1 7 7
2	4	3

Key: 2 | 4 | 3 means
Boys 42%, Girls 43%

Comment on whether the boys or the girls performed better in the test, justifying your answer.

(*4 marks*)

3 Constructions

INTRODUCTION

When a company grows in size and opens new branches in different towns, it often needs to find a suitable location for its distribution warehouse. In an effort to save transport costs, it will often try to choose a site which has good transport links and which is as close as possible to each branch.

What's the point?

By applying some basic geometry, companies can make their operations more efficient, thereby saving themselves a lot of money.

CHECK IN

1 Using a protractor, measure these angles.

a

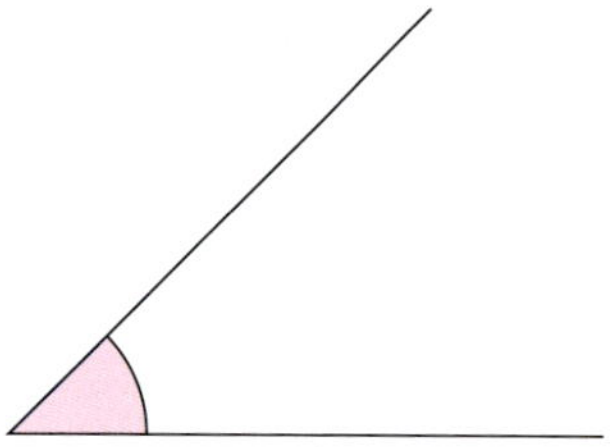

b

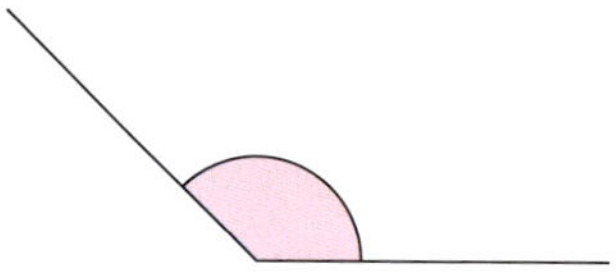

2 Using compasses, draw a circle with a diameter of 4.6 cm.

3 Measure this line

a in millimetres   b in centimetres.

Orientation

What I should know	What I will learn	What this leads to
Key stage 3	■ Read scales accurately ■ Convert between metric and imperial measurements ■ Do accurate scale drawings	A8

RICH TASK

Can you construct this shape accurately? Investigate constructions made from nested shapes.

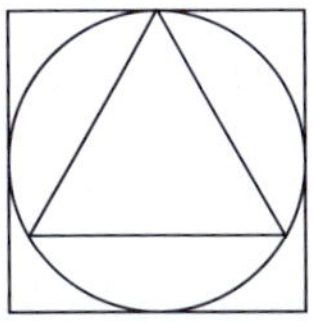

Instruments and scales

A3.1

- Read instruments, understand scales and make sensible estimates

Data is either counted (such as goals in a football match) or measured (such as current in an electrical circuit).

When data is measured, an instrument is used. The data is then read from either a scale or a digital display (such as from a clock face or a digital watch).

Scales are marked in different ways. You may have to use integers, decimals or fractions when interpreting them.

Mass, *Kg*

The scale is marked every 10 kg.
The reading is 80 kg.

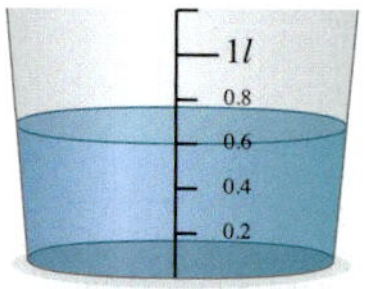

Volume, Litres

The scale is marked every 0.2 litres.
The reading is 0.6 litres.

p. 325

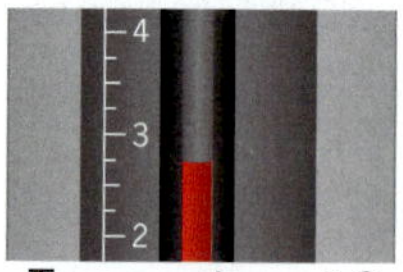

Temperature, °C

The scale is marked every $\frac{1}{4}$ degree.
The reading is $2\frac{3}{4}$ °C.

Sometimes you can read between the marked values.

Current, *amps*

The arrow is pointing midway between 2.6 and 2.7.
The reading is 2.65 amps.

Exercise A3.1

LOW

1 A digital gas meter shows the number of gas units. This meter has a red decimal point. The readings on the first and last day of March are given here. How much gas was used in March?

Meter is a machine for measuring.
Metre is a length of 100 cm.
Watch your spelling!

2 Read these scales. Take a reading for each lettered arrow.

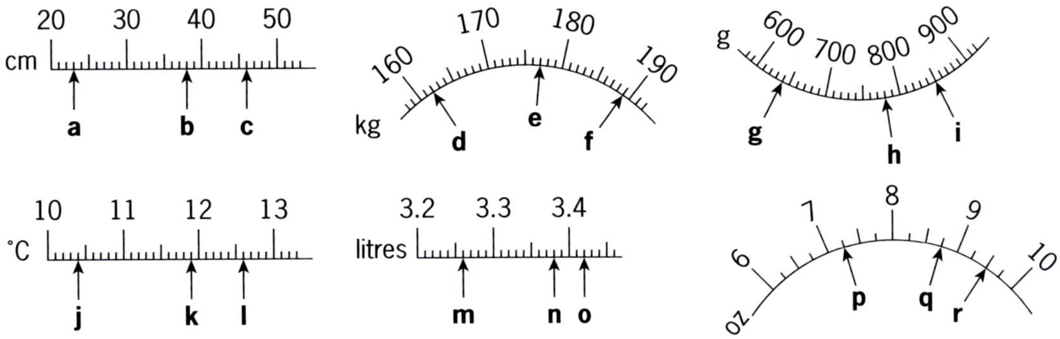

3 The times on these clock faces could be either a.m. or p.m. In each case, write the two times (morning and evening) that they could be on a 24-hour digital clock.

4 Anna's train leaves at 23 minutes to 6 in the evening. What will the 24-hour digital clock on the station platform say at this time?

5 Write the readings for these lettered arrows.

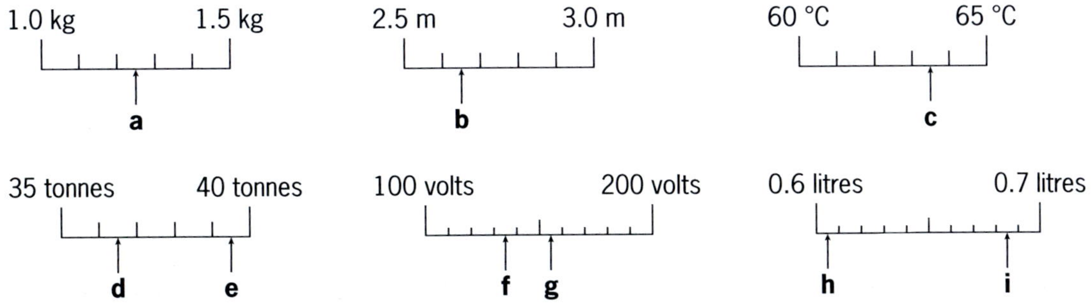

The metric and imperial systems

A3.2

- Convert measurements, know approximate equivalents between metric and imperial systems, and make estimates

☆ DISCUSSION ☆

On a typical school day, what estimates of times, distances, weights, areas, temperatures and so on do you use?

The metric system of measurement is a decimal system because conversions between units are all based on powers of 10. This is the most widely used system of measurement in the world.

Length	Mass	Volume and capacity
1 cm = 10 mm	1 gram = 1000 mg	1 litre = 1000 ml
1 metre = 100 cm	1 kg = 1000 grams	1 litre = 1000 cm^3
1 metre = 1000 mm	1 tonne = 1000 kg	1 ml = 1 cm^3
1 km = 1000 metres		

In full, mm, mg and ml are millimetre, milligram and millilitre. milli- means 'thousandth'.

In full, cm is centimetre. centi- means 'hundredth'. Centilitres (cl) are used on wine bottles.

In full, km and kg are kilometre and kilogram. kilo- means 'thousand'.

The imperial system is a traditional system of measurement and has evolved in Britain over the course of a thousand years.

Length	Mass	Volume and capacity
1 foot = 12 inches	1 pound = 16 ounces (oz)	1 gallon = 8 pints
1 yard = 3 feet	1 stone = 14 pounds (lb)	

There is some controversy in the UK over which system of measurement to use. With both systems frequently being used, it is good to be able to convert between the two!

Length	Mass	Volume and capacity
1 foot ≈ 30 cm	1 pound (lb) ≈ $\frac{1}{2}$ kg	1 gallon ≈ $4\frac{1}{2}$ litres
1 mile ≈ $\frac{8}{5}$ km	1 kg ≈ $2\frac{1}{4}$ pounds	1 pint ≈ $\frac{1}{2}$ litre
5 miles ≈ 8 km		

The symbol ≈ means 'approximately equal to'.

EXAMPLE

A long-distance runner collapsed after 9.65 km. While running, he lost 1800 grams in mass and he drank 3 pints of water. Convert these measurements to metres, kilograms and millilitres.

ANSWER

9.65 km = 9.65 × 1000 = 9650 metres.

1800 grams = 1800 ÷ 1000 = 1.8 kg.

3 pints ≈ 3 × $\frac{1}{2}$ litres = 1.5 litres = 1.5 × 1000 = 1500 millilitres.

Exercise A3.2

LOW

Unit A

1 a The man in this diagram is about 2 metres tall. Estimate, without measuring, the height of

  i the street lamp
  ii the tallest tree
  iii the office block.

  b Make new imperial estimates based on the man's height of 6 feet.

2 a The lorry in this sketch is 14 metres long. Estimate, without measuring,

  i the length of the car
  ii the height of the lorry
  iii the height of the telegraph pole
  iv the height of the pylon.

  b Make imperial estimates based on the lorry's length of 40 feet.

3 A race is run over five laps of a track where each lap is 600 metres. What distance is the race in kilometres?

4 The largest standard size of paper (called A0) is a rectangular sheet 1189 mm by 841 mm. What are these measurements in metres?

5 a Find approximate metric equivalents for

  i 2 feet ii 15 miles iii 5 lbs iv 6 gallons v 4 pints.

  b Find approximate imperial equivalents for

  i 100 cm ii 40 km iii 8 kg iv 18 litres v 2 litres.

6 Amit Patel weighed 87 kg before his diet and 75 kg after his diet. Convert these two masses to pounds (lb) and find how many pounds he lost, giving your answer to the nearest integer.

*7 ☆ RESEARCH ☆

Research the history of the metric system. Where, when and why was it invented?
Which countries have not adopted it as their official system of measurement?

Scale drawing

A3.3

- Use a ruler and protractor accurately to make scale drawings

You can use
- → **a ruler to measure lengths to the nearest millimetre (or 0.1 cm)**
- → **a protractor to measure angles to the nearest degree.**

EXAMPLE

A yacht sails from a harbour H to a port P as shown on this diagram.

a How far did the yacht sail?

b What is the shortest distance between H and P?

Draw an accurate diagram using ruler and protractor to a scale of 1 cm = 1 km.

ANSWER

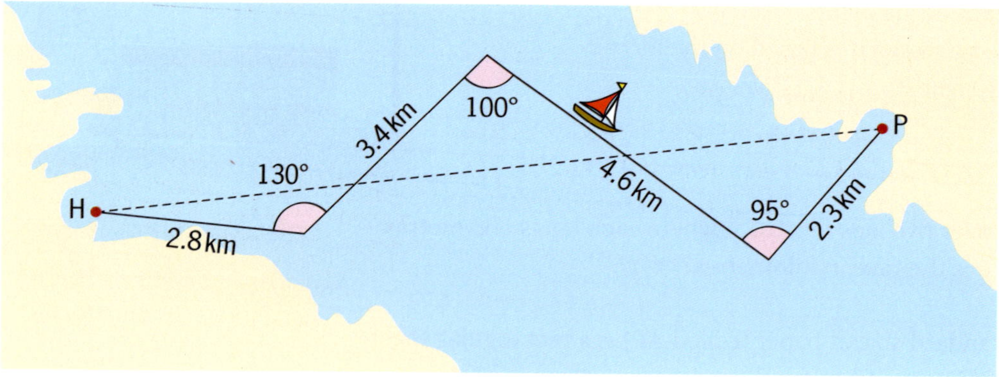

a The yacht sailed 2.8 + 3.4 + 4.6 + 2.3 = 13.1 km.

b Using a ruler, the distance HP is 10.5 km.

You can draw triangles accurately if you know
- → **two sides and the angle between them**
- → **two angles and the side between them.**

RICH TASK

The distance between two trees S and T on the bank of a river is 40 metres. You see a post P on the other bank and measure the angles at S and T as 42° and 64° respectively.
Draw the triangle STP to a scale of 1 mm = 1 metre and find the width of the river, *w*.

Exercise A3.3

MEDIUM

1 Here are sketches of two gas pipelines across open country.
Use a scale of 1 cm = 1 km to draw accurate maps of the pipelines.
Find the shortest distances AB and YZ.

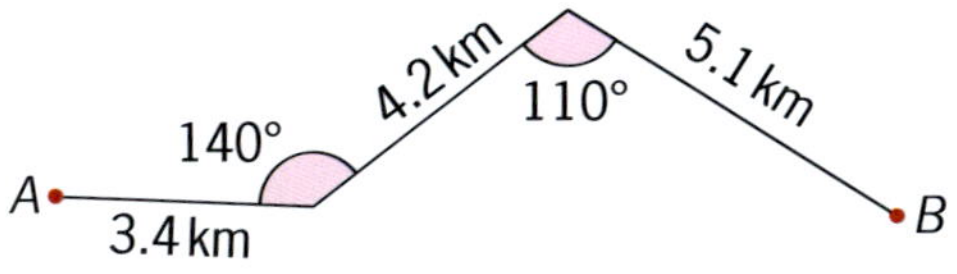

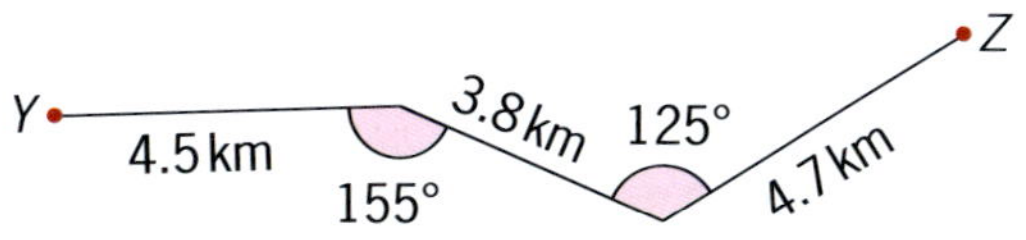

2 Draw these triangles accurately.
Measure the lettered lengths and angles (to the nearest mm and degree).

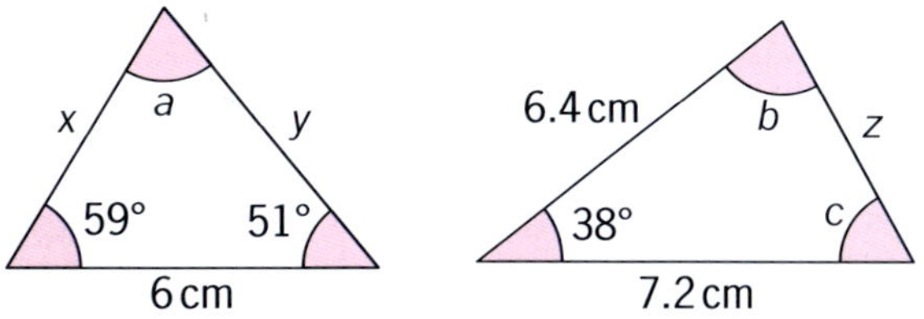

3 The villages Aird, A, and Craig, C, are 7.5 km apart.
The mountain Ben Mor, B, makes angles 62° and 53° with the two villages as shown in this sketch.
Use a scale of 1 cm = 1 km and draw triangle ABC accurately.
Find the distances AB and CB to the nearest 0.1 km.

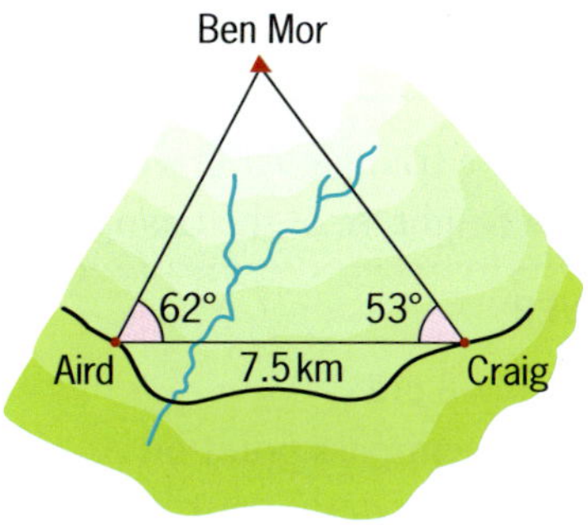

4 A telegraph pole ST is 6.5 metres tall.
It is fixed vertically to point A by a wire AT 3.4 metres from S.
Choose your own scale and find the length of the wire AT and the angle between the wire and the ground.

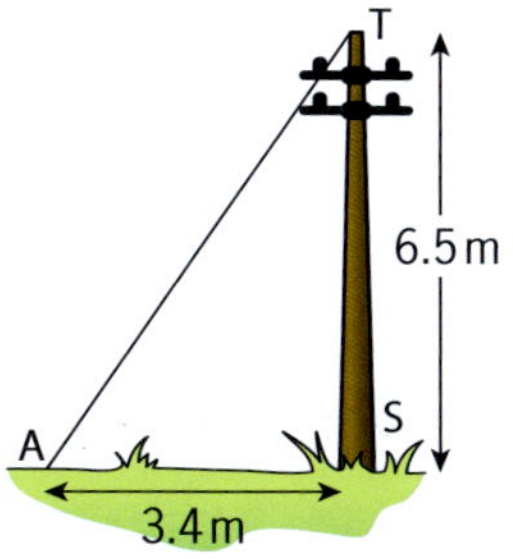

5 **RICH TASK**

The angles of elevation of a mountain top T from points P and Q are 12° and 24° respectively. P and Q are 4.8 km apart. Draw triangle PQT accurately. Find the height h of the mountain.

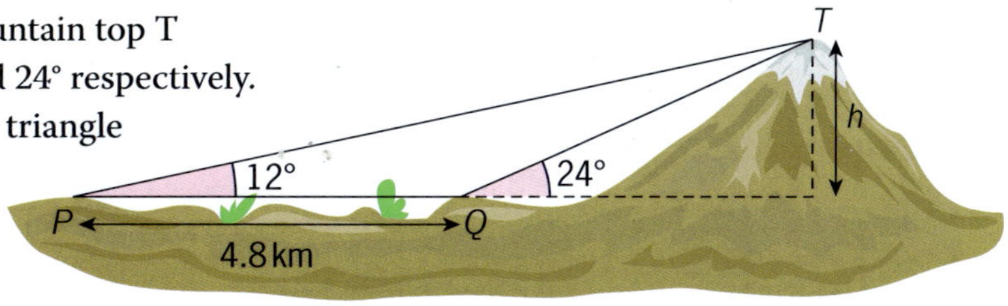

Constructing triangles

A3.4

- Use ruler and compasses to construct triangles

You can draw shapes accurately without needing a protractor if you have the right information.

RICH TASK

A bridge has a cross-section of five equilateral triangles.
Make an accurate drawing of the bridge to a scale of 1 cm = 2 metres.
What is the span of the bridge in metres?

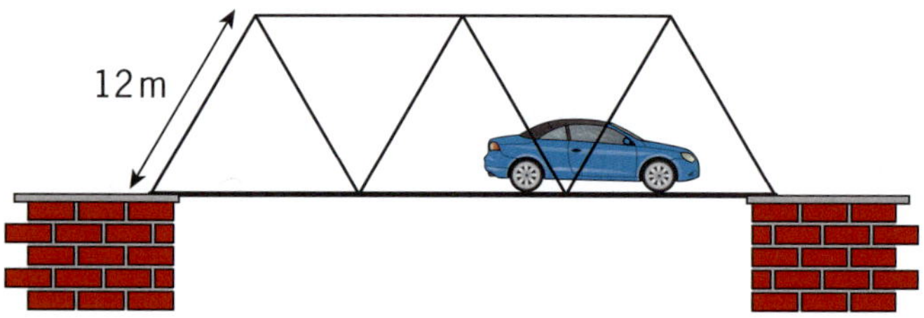

EXAMPLE

Construct a triangle ABC with sides of 8 cm, 7 cm and 6.5 cm.
Find the height, *h*, of the triangle.

ANSWER

Draw AB = 8 cm.
With your compass point on *A* (open at 7 cm), draw arc *Y*.
With your compass point on *B* (open at 6.5 cm), draw arc *Z*.
Label point *C* where *Y* and *Z* cross.
A ruler gives the height *h* as 5.4 cm.

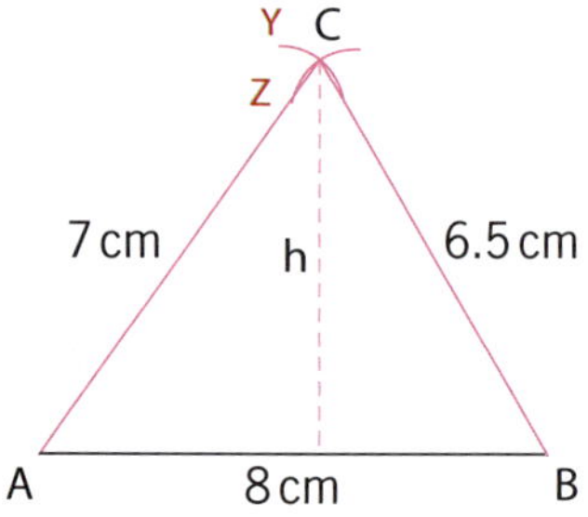

You can construct an angle of 60° without a protractor by constructing an equilateral triangle.

Exercise A3.4

MEDIUM

1 Here are sketches of four triangles.
   a Draw each triangle accurately using ruler and compasses.
   b Measure the three angles of each triangle and find their total.
   c What kind of triangles are Y and Z?

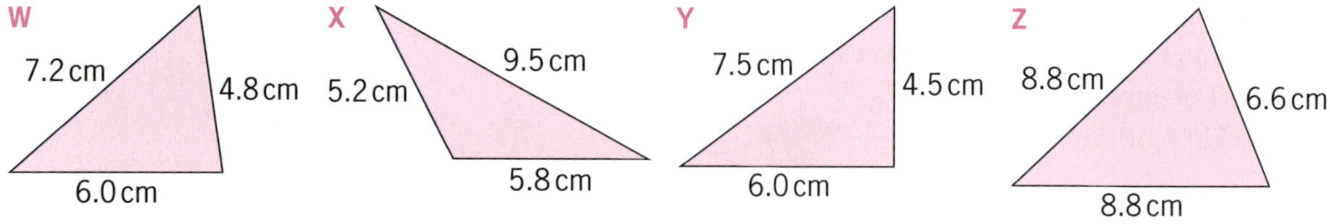

2 A streetlight shines on a road 8 metres wide as shown. Draw the triangle accurately and find angle *a* using a protractor.

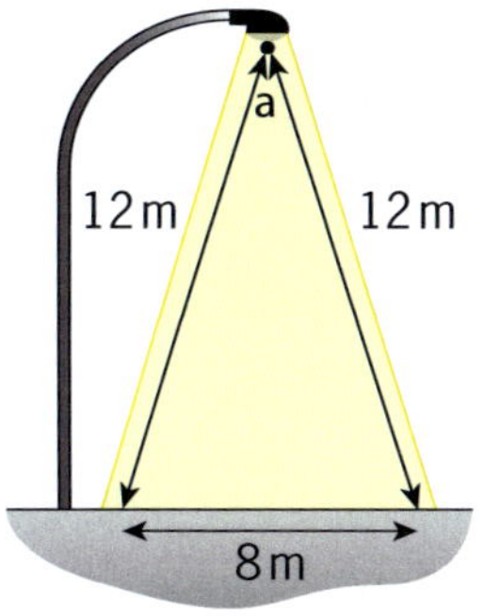

3 A ramp for inspecting a car has the shape shown.
Use a scale of 1 cm = $\frac{1}{2}$ metre to draw the ramp accurately.
How high is the top of the ramp above the ground?

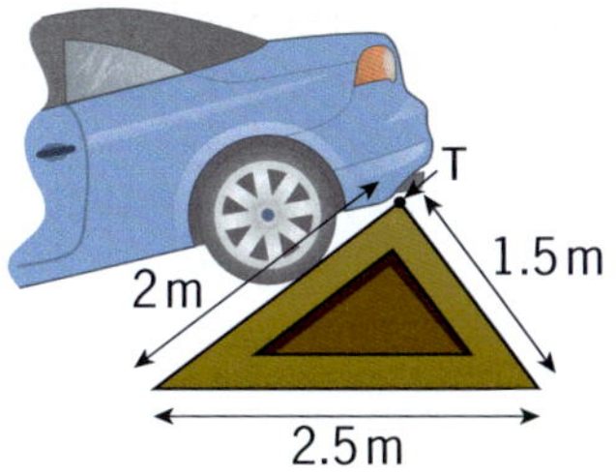

4 a Construct the green equilateral triangle of side 5 cm.
b Construct three other equilateral triangles, one on each of its sides.
Now cut the whole shape out and fold it so that corners P, Q and R meet.
c What 3-D shape will it make?

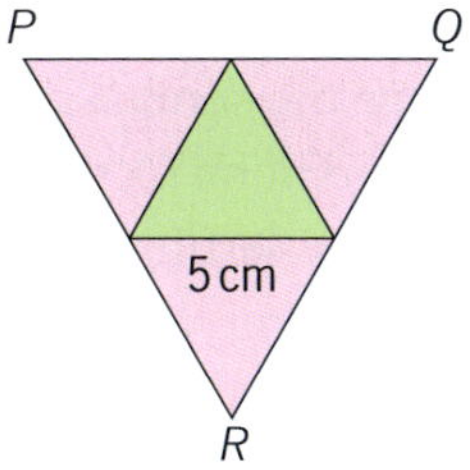

5 a Construct triangle ABC with the sides shown.
b Construct triangle BCD with BD = 5 cm and CD = 10 cm.
c What type of quadrilateral is ABDC ?
d Measure angles *x* and *y*. What do you notice?

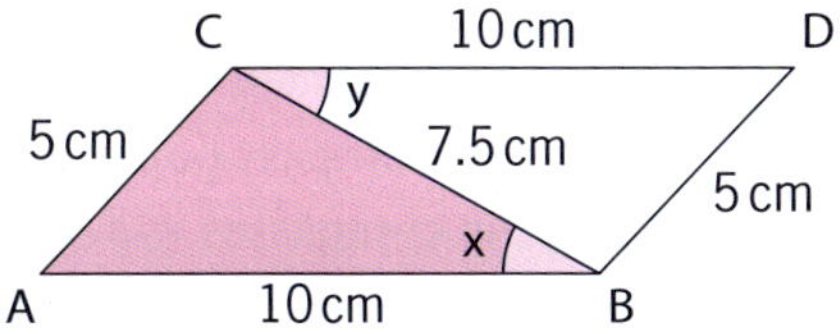

*6 Write your partner a description of a triangle that you want them to draw. Do they draw exactly the triangle that you expected?

7 ☆ CHALLENGE ☆

The Star of David is made from two overlapping equilateral triangles.
Construct the star using your compasses with triangles of side 9 cm.
Invent your own patterns using equilateral triangles and construct them accurately.

DID YOU KNOW...?

Did you know that the Star of David is an ancient symbol important to the Jewish people?

More constructions

A3.5

- Use ruler and compasses to find midpoints and bisectors and to construct perpendiculars

- You can **bisect a line** and find its midpoint.
    - With compass point on end A, draw two arcs P.
    - With compass point on end B, draw two arcs Q (use the same compass setting).
    - Join the intersections X and Y of arcs P and Q.

  The line XY is the perpendicular bisector of AB.
  Point M is the midpoint of AB.

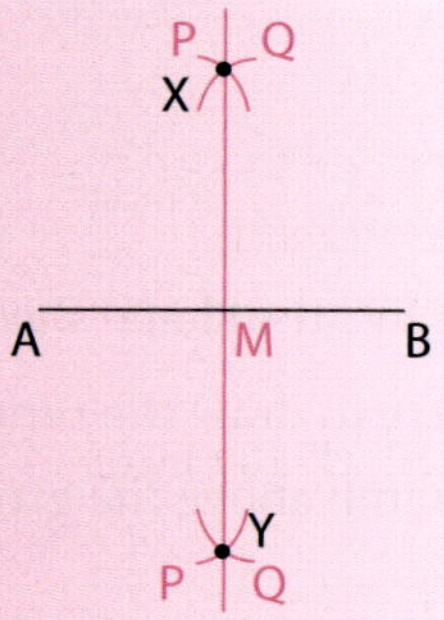

You can check these results with your ruler and protractor.

- You can **bisect an angle**.
    - With compass point on A, draw arcs P and Q.
    - With compass point on P, draw arc R (you can change the compass setting).
    - With compass point on Q, draw arc S (do not change the setting from R).
    - Label the intersection X of R and S.

  The line AX is the bisector of angle A.

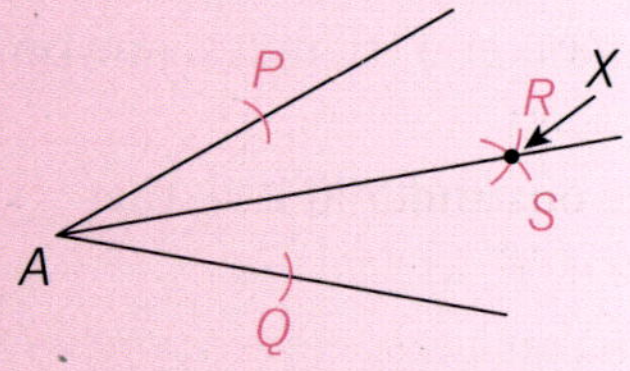

You can check this result with your protractor.

- You can construct a **perpendicular at any point on a line**.

  To draw a perpendicular at point *X*,
    - Start with compass point on X, draw arcs P and Q.
    - Open the compasses wider.
    - With compass point on P, draw arc R.
    - With compass point on Q, draw arc S.
    - Join point X to the intersection Y of arcs R and S.

  The line XY is perpendicular to the line AB.

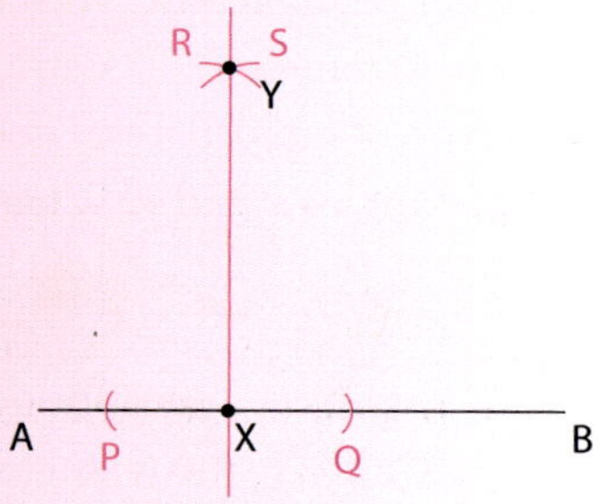

- You can construct a **perpendicular from a point *X* to a line *AB***.
    - X is the point from which the perpendicular is to be drawn.
    - With compass point on X, draw arcs P and Q.
    - With compass point on P, draw arc R.
    - With compass point on Q, draw arc S.
    - Join point X to the intersection Y of arcs R and S.

  The line XY is perpendicular to the line AB.

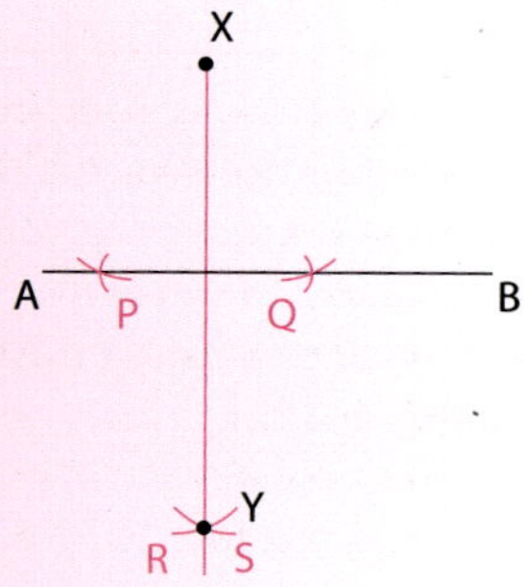

Exercise A3.5

MEDIUM

1 a Draw line AB 8 cm long using a ruler. Construct the perpendicular bisector of the line. Label the midpoint M. Check your results with a ruler and protractor.

b On axes labelled from 0 to 8, construct the perpendicular bisector of the line joining points (1, 3) and (7, 4).

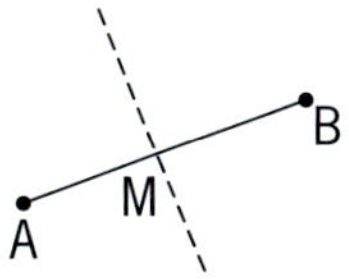

2 a Draw any **acute** angle A. Construct the angle bisector of angle A. Check your result with a protractor.

b Construct the angle bisector of any **obtuse** angle.

3 a Draw line AB 8 cm long. Mark any point X which is above line AB. Construct a line through X which is perpendicular to line AB. Check your result with a protractor.

b On axes labelled from 0 to 8, draw the line through A(2, 2) and B(8, 5). Construct a line through the point X(2, 7) which is perpendicular to line AB.

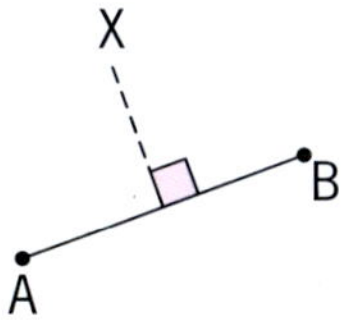

4 a A hiker at X wants to reach the A832. Construct his shortest route.

b A pipe is to be laid equidistant from two roads A and B. Construct the line it takes across the field. (You will need to construct an angle bisector)

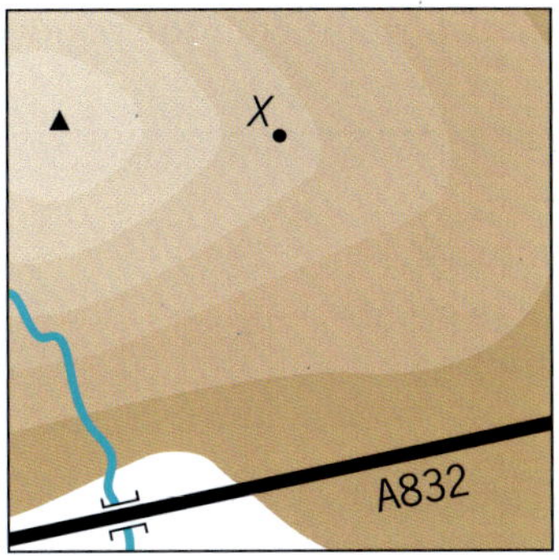

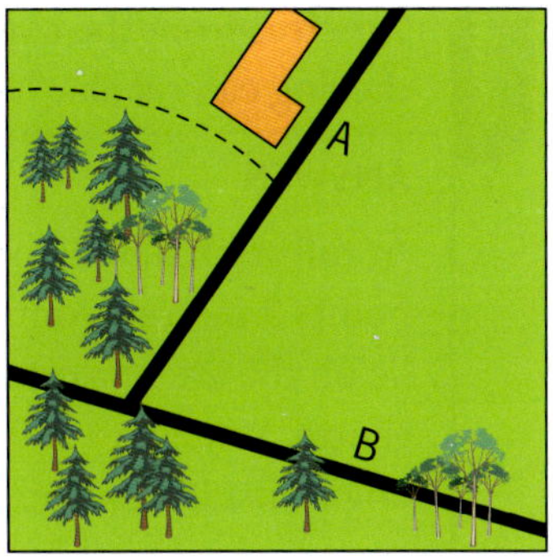

*5 On axes labelled from 0 to 8, draw a line joining the points O(0, 0) and P(7, 4).

a Construct the angle bisector OA of the angle between this line and the x-axis.

b Construct the angle bisector OB of the angle between this line and the y-axis.

c Use a protractor to measure angle BOA between the two angle bisectors. Explain your answer.

*6 ☆ CHALLENGE ☆

Draw any large triangle. Construct the angle bisectors of its three angles. What do you notice? Use your result to label a point as an **incentre** and draw an **incircle**.

Draw a smaller triangle and construct the three perpendicular bisectors of its sides. Label a point as a **circumcentre** and draw a **circumcircle**.

Three trees are growing 11 m, 10 m and 8 m apart. A goat is tethered so it can just reach each tree. How long should the rope be?

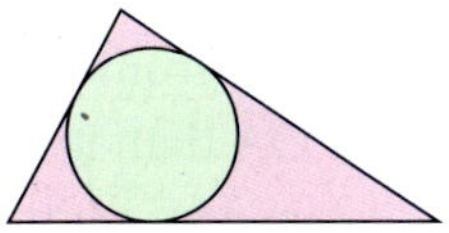
Incircle

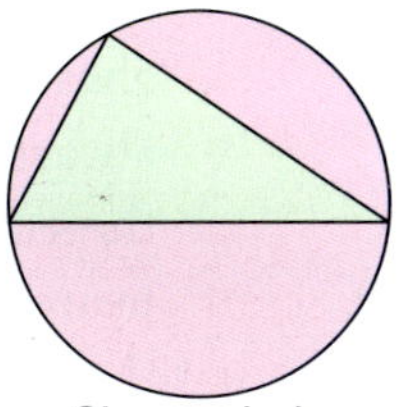
Circumcircle

Loci

- Find loci which give paths or regions in 2 dimensions

→ **A locus is a set of points which obey a rule.**
→ **The points may lie on a line or they may lie inside a region.**

The word 'loci' is the plural of the Latin word 'locus' which means 'a place'.

RICH TASK

Jamie wants to pass through an opening in an electrified fence so that he is always as far from gate post P, as he is from gate post Q. Sketch the path he must take.

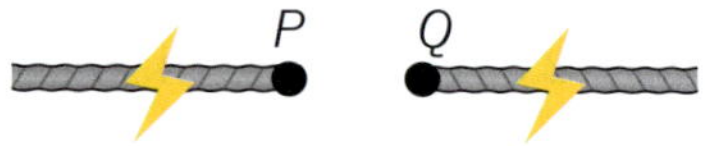

EXAMPLE

A dog is tethered by a lead 3 metres long to a point P which is 2 metres from the corner of two walls. Find the region which the dog can reach.

ANSWER

At full stretch, the dog can reach points 3 metres from P. These points lie on the arc of a circle with P as centre. When **not** at full stretch, the dog is inside this arc. The shaded area shows the region that the dog can reach.

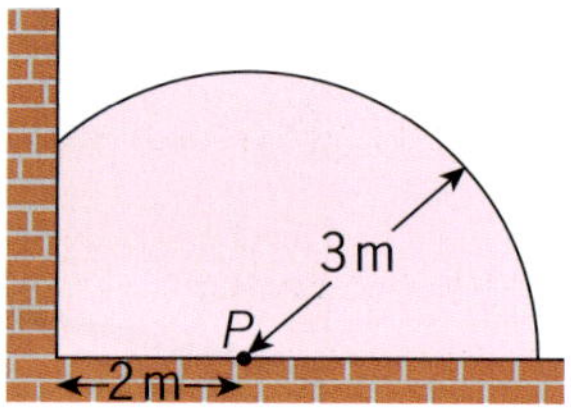

Exercise A3.6

MEDIUM

1 In each case, sketch the locus of point X and describe it in words.
A point X moves on a flat surface so that it is
  - a always 6 cm from a fixed point P
  - b always less than 6 cm from a fixed point P
  - c the same distance from line OA as it is from line OB
  - d nearer to line OA than it is to line OB
  - e the same distance from line KL as it is from line MN
  - f nearer to line KL than to line MN
  - g equidistant from two fixed points R and S 6 cm apart
  - h nearer to point R than it is to point S.

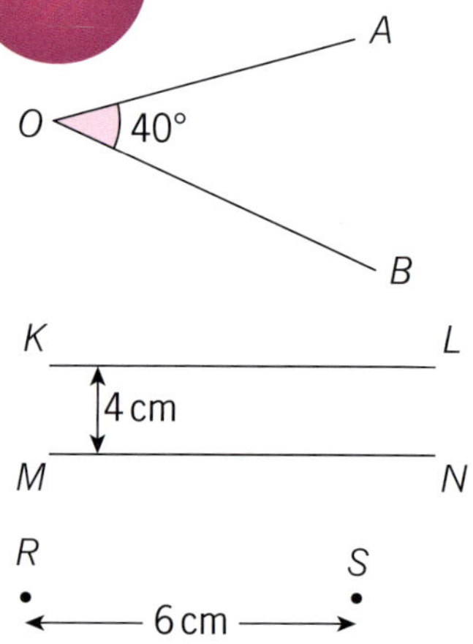

2 a Draw a circle of radius 3 cm. A point X moves so that it is always 4 cm from the circle. Draw the locus of X.

b Repeat for a point Y which is always 3 cm from the circle.

c Repeat for a point Z which is always 2 cm from the circle.

3 A garden has two sprinklers P and Q placed 8 metres apart. P has a range of 6 metres and Q has a range of 5 metres. Use a scale of 1 cm = 1 metre to make an accurate diagram and construct the locus of all points sprinkled by both P and Q.

4 LM and MN are two walls of a house. The owner wants to pave that part of the garden which is either less than 3 metres from wall LM or which is less than 3 metres from point P in wall MN. Make a scale drawing to find the part of the garden which will be paved.

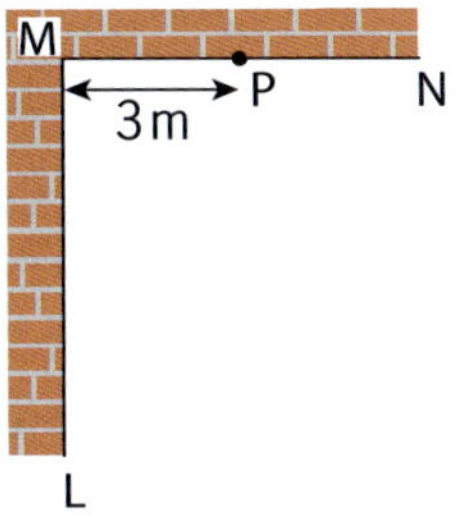

5 A dog's running lead CD is 2 metres long. It can slide along the fixed wire XY. Use a scale of 1 cm = 1 metre and construct the locus of all the points that the dog can reach.

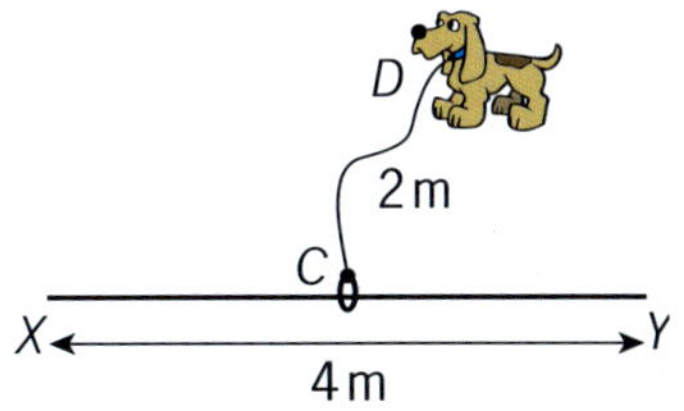

6 ★ CHALLENGE ★

Draw this triangle accurately.
Find the locus of all points inside the triangle which are nearer to RP than they are to QP.
Draw another diagram to show the locus of all points inside the triangle which are nearer to P than they are to R.

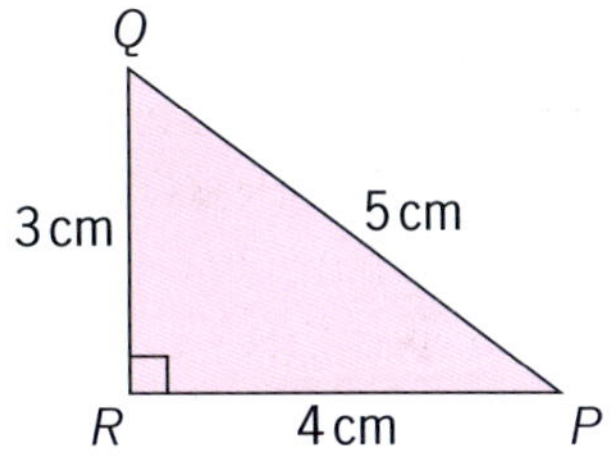

7 Point A is 4 cm from the straight line PQ. A variable point X moves so that it is always as far from the line PQ as it is from point A. What is the locus of X?

8 ★ CHALLENGE ★

Explore the locus of a point on the edge of a rolling object.
Here are two examples.

- Roll a coin along a ruler.
- Imagine a garden roller being rolled up steps.

What are hypocycloids? What is the rotary car engine?
Find animations of them on the internet.

Summary and assessment

A3

CHECK OUT

You should now be able to

- interpret scales and use measurements
- draw triangles and other 2-D shapes using ruler and protractor
- use ruler and compasses for constructions including loci.

Exam-style question

Treasure is buried 3 m from tree T and 7 m from a gatepost G.
The tree and the gatepost are 6 m apart.
Draw a diagram to show where the treasure could be buried.

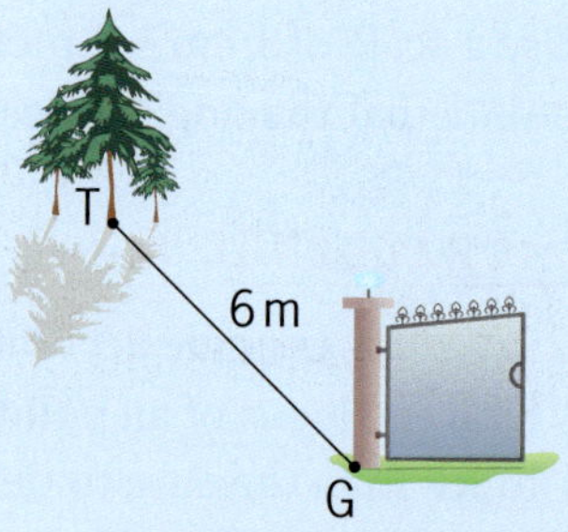

Ben's answer:

Draw circles centred on T and G as described

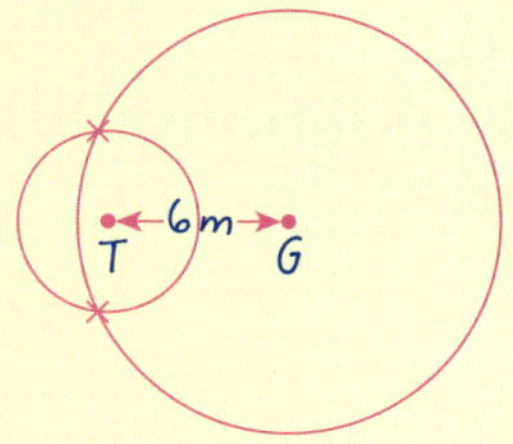

Using the scale 1 cm: 1 m:
3 m from T is 3 cm,
7 m from G is 7 cm.

Mark the points where the circles overlap to show where the treasure is buried.

Construct a circle radius 3 cm centre T, leaving in the construction lines.

Construct a circle radius 7 cm centre G – always leave in the construction lines.

Exam practice

1 Using ruler and protractor, draw the following triangle.
Two sides are 3 cm and 4·5 cm and the angle between these sides is 50°. *(4 marks)*

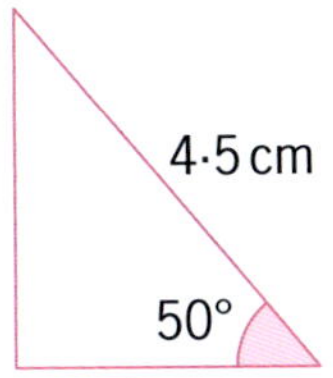

Measure and write down the length of the other side and the size of the other two angles.

2 Use the scales to convert 20 kg to pounds.

(1 mark)

3 Using ruler and compasses only, construct a regular hexagon of side length 4 cm.

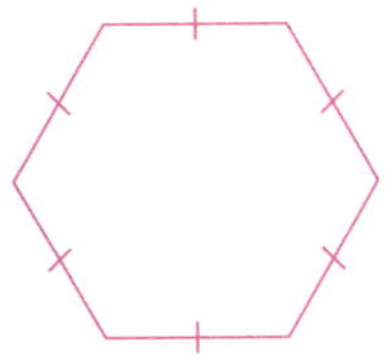

(4 marks)

Case study 1: Sports, measures and results

Mathematics is used widely in sport, particularly when taking measurements and recording results.

Here are the results and reaction times (in alphabetical order) for the 100m Men's Final at the IAAF World Championships in Berlin in August 2009:

Use the photo to order Bailey and Thompson as well as Burns and Chambers. What degree of accuracy is shown here?

Name	Nationality	Time (s)	Reaction (s)
Bailey	ANT	9.93	0.129
Bolt	JAM	9.58	0.146
Burns	TRI	10.00	0.165
Chambers	GBR	10.00	0.123
Gay	USA	9.71	0.144
Patton	USA	10.34	0.149
Powell	JAM	9.84	0.134
Thompson	TRI	9.93	0.119

What degree of accuracy is reported for
a) the result times
b) the reaction times?

Draw a stem-and-leaf diagram to show the result times of this race.

Calculate the
a) range b) median c) mean
of the reported results, giving your answers to an appropriate level of accuracy.

Which average do you think best represents these results? Explain your answer.

Here are the results and reaction times for the 100m Women's Final at the same World Championships:

Name	Nationality	Time (s)	Reaction (s)
Fraser	JAM	10.73	0.146
Stewart	JAM	10.75	0.170
Jeter	USA	10.90	0.160
Campbell-Brown	JAM	10.95	0.135
Williams	USA	11.01	0.158
Ferguson-McKenzie	BAH	11.05	0.130
Sturrup	BAH	11.05	0.137
Bailey	JAM	11.16	0.173

Use diagrams and statistics to compare the Men's and Women's reported results.

File Edit View History Bookmarks Window Help

Fastest ten all-time 100m Women's sprinters as of 20th September 2009

http://berlin.iaaf.org/results/racedate=08-16-2009/sex=M/discCode=100/combCode=hash/roundCode=f/results.html#detM_100_hash_f

Google Google Maps OUP Wikipedia

IAAF WORLD CHAMPIONSHIPS
BERLIN, AUGUST 2009

NEWS PHOTOS VIDEO AUDIO

By Date By Event Entry List Medal Table Placing Table Entry Standards

Any reaction time quicker than 0.1 seconds is considered to be a false start.
How close was the fastest reaction time in the
a) Men's final b) Women's final
to the false start limit?

Do you think that the limit of 0.1 seconds is a suitable value?
Explain your answer.
Find the range of the reaction times for both the Men's and Women's races.
Draw a scatter diagram to show result time against reaction time for
a) the Men's final b) the Women's final.

Do you think there is any link between the reaction times of the athletes and the results of these races?
Explain your answer.
Recalculate the result times for the athletes in
a) the Men's race b) the Women's race
assuming that all of the athletes had a reaction time of 0.1 seconds.
Use diagrams and statistics to show what effect this would have on the results of each race.

4 Factors, multiples and ratio

INTRODUCTION

The concepts of ratio and proportion are vital to art and architecture. The Golden Ratio, *phi*, is a quantity that appears commonly in the natural world (for example in human proportions) and has been used to create aesthetic shapes, such as in buildings and in paintings, that look natural and balanced.

What's the point?

Art and design often draw inspiration from the natural world – an understanding of the mathematical proportions inherent in nature can yield results that are pleasing to the eye.

CHECK IN

1 A car has a fuel consumption of 30 miles per gallon.
How many miles will it travel on 8 gallons of petrol?

2 Gavin is running at 5 mph.
He runs for 3 hours.
How far does he travel?

Orientation

What I should know	What I will learn	What this leads to
A1 + Key stage 3	■ Find prime factors ■ Find and use ratios	B9

RICH TASK

Your height is about the same as three times around your head.
Investigate.

Factors, multiples and primes

A4.1

- Use factors, multiples and prime numbers

ACTIVITY

This is a game for two or more players.
First: take turns to choose one number from this grid.
Then: find all the factors of your chosen number.
Score: the sum of any factors that have not already been used.

1	2	3	4	5	6
7	8	9	10	11	12
13	14	15	16	17	18
19	20	21	22	23	24
25	26	27	28	29	30
31	32	33	34	35	36

The winner: the player with the highest score when all numbers on the grid have been used.

During the game, see if you can answer these questions.

a What number achieved the highest score in your game?
b What type of number has exactly two factors?
c What type of number has an odd number of factors? Why?
d How can you check that all the numbers have been accounted for?

REMEMBER

A **factor** will divide exactly into a given number without a remainder.
Four is a factor of 12.

A **multiple** is in a given number's times-tables. 28 is a multiple of seven.

For example:
Amy chooses 8.
Factors of 8 are 1, 2, 4, 8.
Amy's score is 1 + 2 + 4 + 8 = 15

Ben chooses 12.
Factors of 12 are 1, 2, 3, 4, 6, 12.
Ben's score is 3 + 6 + 12 = 21.

You should have found that

→ **A prime number has only two factors.**
→ **A square number will have an odd number of factors.**

Numbers can be arranged in overlapping circles (called Venn diagrams) to find their **common factors**.

EXAMPLE

Find the common factors of 8 and 12.

ANSWER

Factors of 8 are 1, 2, 4 and 8   Factors of 12 are 1, 2, 3, 4, 6 and 12

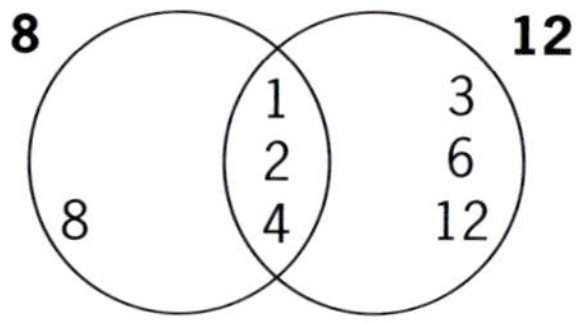

The common factors of 8 and 12 are 1, 2 and 4.

You may have come across **Venn diagrams** before.

Exercise A4.1

MEDIUM LOW

Unit A

1 Select your answer from numbers given in this list:

3 4 5 17 21 23 24 25 28 30

a factors of 36
b factors of 50
c multiples of 3
d multiples of 7
e square numbers
f prime numbers

2 Write down all the factors of

a 15 b 12 c 22 d 25 e 40

3 Write down a factor of 110 that lies between 50 and 60.

4 Write down a multiple of 6 that lies between 40 and 50.

5 Write down a prime number that lies between 30 and 40.

*6 a Write down all the factors of i 22 ii 66
b Explain why all the factors of 22 are also factors of 66.

7 Find the common factors of

a 9 and 15
b 18 and 24
c 21 and 28
d 30 and 36

*8 a Write down the first three multiples of 4.
b Write down the first six multiples of 2.
c Explain why all the multiples of 4 will be multiples of 2.

9 Find two common multiples of

a 6 and 12
b 5 and 6
c 3 and 7
d 6 and 9

10 ★ PROBLEM ★

a Jo's age this year is a multiple of 6.
Next year it will be a multiple of 5. How old might Jo be?

b Amy's age this year is a square number.
Next year it will be a prime number. How old might Amy be?

Prime factors

A4.2

- Use highest common factors, least common multiples and prime factor decomposition

☆ ACTIVITY ☆

This is a similar game to the one on page 54, for two players.
This time: Take turns to choose two numbers from the grid that have not yet been picked.
Score: all factors that are common to both your chosen numbers.
The winner: the player with the highest score when all numbers on the grid have been used.

1	2	3	4	5	6
7	8	9	10	11	12
13	14	15	16	17	18
19	20	21	22	23	24
25	26	27	28	29	30
31	32	33	34	35	36

During the game see if you can answer these questions.

a What number is always a common factor? Why?

b Can you find a way to find the highest common factor between two numbers?

For example:
Cathy chooses 24 and 20.
Factors of 24 are 1, 2, 3, 4, 6, 8, 12, 24.
Factors of 20 are 1, 2, 4, 5, 10, 20.
Common factors are 1, 2, 4.
Cathy scores 1 + 2 + 4 = 7.

Dave chooses 35 and 15.
Factors of 35 are 1, 5, 7, 35.
Factors of 15 are 1, 3, 5, 15.
Common factors are 1, 5.
Dave scores 1 + 5 = 6.

→ **Numbers can be written as a product of their prime factors.**

Prime factors are factors that are also prime numbers.

EXAMPLE

Write 5544 as a product of prime factors.

ANSWER

Start with dividing by 2: $5544 \div 2 = 2772$

Keep dividing until it will not divide exactly: $2772 \div 2 = 1386$

$1386 \div 2 = 693$

Then move to the next prime number (3): $693 \div 3 = 231$

$231 \div 3 = 77$

5 won't work, so try 7: $77 \div 7 = 11$

Stop when you get a prime number answer: 11

Write your number as a product of primes:

$$5544 = 2 \times 2 \times 2 \times 3 \times 3 \times 7 \times 11$$

With the **factor tree method** you can break the number down until you get to prime numbers:

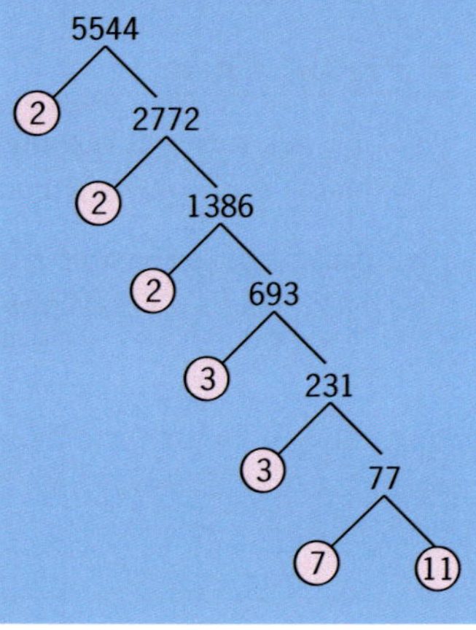

Unit A

You can use the product of prime factors to find the **highest common factor** (HCF) and **least common multiple** (LCM).

> → **The HCF of two (or more) numbers is the largest factor common to both numbers. Four is the HCF of 12 and 16.**
> → **The LCM of two (or more) numbers is the smallest multiple common to both numbers. 42 is the LCM of 14 and 21.**

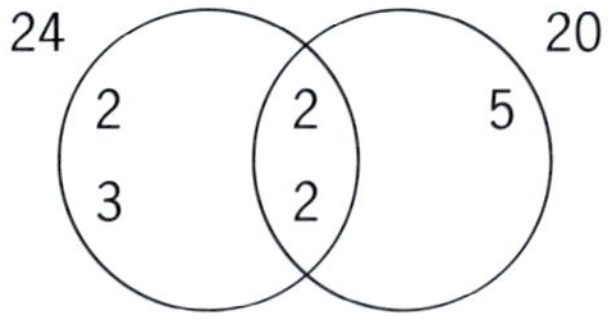

EXAMPLE

Find the HCF and LCM of 24 and 20.

ANSWER

First write 24 and 20 as the product of prime factors:

$24 = 2 \times 2 \times 2 \times 3$ $\qquad$ $20 = 2 \times 2 \times 5$

Arrange the factors in two overlapping circles:

The HCF of 24 and 20 is 4
$(2 \times 2 = 4)$

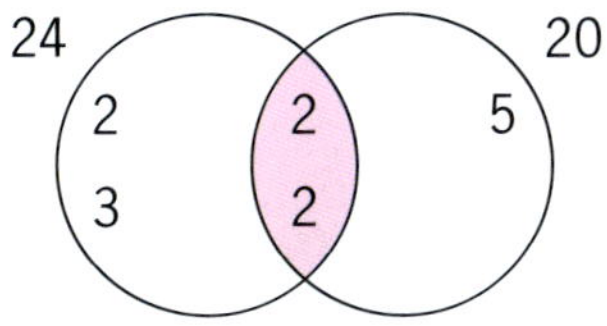

The LCM of 24 and 20 is 120
$(2 \times 3 \times 2 \times 2 \times 5 = 120)$

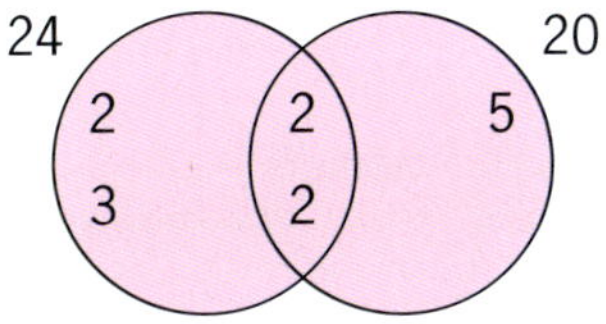

Exercise A4.2

MEDIUM

1 Write the following numbers as a product of prime factors.

a 48 $\quad$ b 27 $\quad$ c 54 $\quad$ d 120
e 72 $\quad$ f 294 $\quad$ g 371

2 Find **i** the least common multiple and **ii** the highest common factor of

a 36 and 56 $\quad$ b 99 and 120 $\quad$ c 94 and 72
d 49 and 28 $\quad$ e 30 and 65

3 ☆ **PROBLEM** ☆

a Two numbers have HCF = 4 and LCM = 60.
One of the numbers is 20. Find the other number.

b Two numbers have HCF = 16 and LCM = 96. One of the numbers is 32. Find the other number.

*4 ***RICH TASK***

A prime number is a number with exactly two factors.
Explain why the square of a prime number has exactly three factors.

Introducing ratio

A4.3

- Use ratio notation, including reduction to its simplest form

This is the **Fibonacci sequence**:

1 1 2 3 5 8 13 21 34 ....

The first two numbers are 1 and 1.
The third number is 1 + 1 = 2.
The fourth number is 1 + 2 = 3, and so on.

RICH TASK

Continue the sequence of Fibonacci numbers and write adjacent pairs as **ratios** in the **unitary form** 1 : *n*.
The first four have been done for you:

1:1 1:2 2:3 = 1:1.5 3:5 = 1:1.666 ...

What happens to the ratio of adjacent Fibonacci numbers?

Use the internet to find out about the **golden ratio**.

→ **A ratio is a way of showing a relationship between two or more numbers.**

→ **A ratio in its simplest form has integer values that cannot be cancelled further.**

→ **You can use the unitary form 1 : *n*, or *n* : 1, to compare ratios.**

EXAMPLE

1 Write these ratios in
i their simplest form ii their unitary form.
a 24:60 b 18:90 c 35:28

2 Seven calculators cost £41.65.
How much do three calculators cost?

ANSWER

1 a i 24:60 = 2:5 Divide both 24 and 60 by 12 (their HCF)
ii 24:60 = 1:2.5 60 ÷ 24 = 2.5
b 18:90 = 1:5 The simplest form is also the unitary form
c i 35:28 = 5:4
ii 35:28 = 1.25:1 Divide smaller number into larger number for unitary form

2 One calculator costs 41.65 ÷ 7 = £5.95.
Three calculators cost 3 × 5.95 = £17.85.

Exercise A4.3

MEDIUM

Unit A

1 Write these ratios in their simplest form

a 2:6 b 6:9 c 14:21 d 25:40 e 18:42

2 Write these ratios in unitary form

a 4:10 b 5:9 c 5:19 d 33:10 e 15:6

3 a Eight gel pens cost £13.20. How much do three gel pens cost?

b Nine pen drives cost £43.74. How much do five pen drives cost?

c Two plant pots cost £16.72. How much do three plant pots cost?

d Five scarves cost £16.75. How much do seven scarves cost?

4 ★ **ACTIVITY** ★

Using the digits

0 1 2 3 4 5 6 7 8 9

Step 1: Choose three of these digits.
Step 2: Add them up and call the total *A*.
Step 3: Make all possible two-digit numbers from the three chosen digits in step 1.
Add them up and call the total *B*.
Step 4: Find the ratio *A*:*B* in the form 1:*n*.

a i Choose your own set of three digits and find the ratio *A*:*B* using steps 1 to 4.

ii Choose a different set of three numbers and find the ratio *A*:*B*.

iii What do you notice about the example and your answers to **i** and **ii**?

b i Choose four of the numbers from the list.
Work through steps 1 to 4 using these numbers and find the ratio *A*:*B*.
(Remember to work systematically to find all possible two-digit numbers in step 3.)

ii Choose a different set of four numbers and find the ratio *A*:*B*.

iii Choose a third set of four numbers and find the ratio *A*:*B*.

iv What do you notice about your answers to **i**, **ii** and **iii**?

c i Choose a set of five numbers from the list and use these to find the ratio *A*:*B*.

ii Compare and comment on your answers to **a i**, **b i** and **c i**.

For example:
Step 1 Choose 2, 8, 5
Step 2 *A* = 2 + 8 + 5 = 15
Step 3 The possible two-digit numbers are 28 25 82 85 52 58
B = 28 + 25 + 82 + 85 + 52 + 58 = 330
Step 4 *A*:*B* = 15:330
=1:22

Use symbols to explain why this might always happen.

Using ratio

A4.4

- Know the links between ratio and fraction notation
- Divide a quantity into a given ratio

RICH TASK

Marvin is exactly four years older than his brother Fred. The ratio of their ages is 3 : 1.

How old are Marvin and Fred? How old will they both be when the ratio of their ages becomes 3 : 2?

You can use ratio to share amounts.

EXAMPLE

Share £75 in the ratio 4 : 1.

ANSWER

$4 + 1 = 5$	First find the number of equal parts
$75 \div 5 = £15$	Find what each equal part is worth
$4 : 1 = £60 : £15$	Multiply each part of the ratio to find each share

EXAMPLE

£3000 is divided between Tom, Dick and Harry in the ratio

T : D : H = 2 : 3 : 5

How much do they each get?

ANSWER

$2 + 3 + 5 = 10 \Rightarrow 3000 \div 10 = £300 \Rightarrow 2:3:5 = 600:900:1500$

Tom gets £600, Dick gets £900 and Harry gets £1500.

EXAMPLE

Harriet and Oliver have a joint birthday party.

The number of guests they are allowed to invite is in the ratio

Harriet : Oliver = 5 : 3.

Harriet invites 20 guests. How many are invited in total?

ANSWER

$20 \div 5 = 4$	First find what each equal share is worth
$3 + 5 = 8$	Work out the total number of equal parts
Total number invited = $8 \times 4 = 32$.	Multiply to find the total

Working with ratios can be linked to work with fractions.

There is more work on fractions in chapter 9.

In the last example, the ratio Harriet : Oliver = 5 : 3.
The fraction of guests invited by Harriet is $\frac{5}{8}$.
The fraction invited by Oliver is $\frac{3}{8}$.

Unit A

Exercise A4.4

MEDIUM

1 Share the following amounts in the given ratios.
- a £48 in the ratio 1 : 5
- b £64 in the ratio 7 : 1
- c £108 in the ratio 2 : 7
- d £96 in the ratio 7 : 5
- e £420 in the ratio 1 : 2 : 4
- f £425 in the ratio 3 : 6 : 8

2 Petra and Sam took part in a sponsored swim.
The amounts they raised were in the ratio 5 : 3. Petra raised £75.50.
How much did Sam raise?

3 Jenny and Mike share prize money in the ratio 3 : 2.
Jenny gets £12.60.
How much does Mike get?

4 In a 400 g tin of tuna there are 250 calories.
How many calories would there be in a 150 g tin of tuna?

5 A 160 g cereal bar has 220 calories.
How many calories will there be in a 100 g cereal bar?

6 A melon medley is made by mixing honeydew and cantaloupe melons in the ratio 3 : 5 by weight. The weight of honeydew melon is 255 g.
What is the total weight of the melon in the melon medley?

7 ☆ **PROBLEM** ☆

Joshua is 12 years old and Reuben is 9 years old.
- a When Joshua is 18 years old, what will the ratio of Joshua's age to Reuben's age be? Write this ratio as simply as possible.
- b When Reuben is 18 years old, what will the ratio of Joshua's age to Reuben's age be? Write this ratio as simply as possible.
- c Find the ages of Joshua and Reuben when the ratio of their ages was 2 : 1.

Summary and assessment

A4

CHECK OUT

You should now be able to

- find factors, multiples and prime numbers
- find least common multiple and highest common factor
- use ratio.

*

Exam-style question

PrintRUs develops 35 mm films in two sizes.
Compact size: 5 inches by 3·5 inches.
Large size: 7 inches by 5 inches.

Are the print sizes in the same ratio to each other?
Justify your answer.

Gary's answer:

Ratio of the compact size
5 : 3.5 = 1 : 0.7

Ratio of large size
7 : 5 = 1 : 0.714

0.7 is nearly the same as 0.714

The print sizes are not exactly in the same ratio.

However, if the sizes had been rounded then the ratios may have been more exact.

Think about what mathematics you need to use.

You justify or explain answers using mathematics as well as words.

Work out the ratios of each size the same way so that you can compare them.

Comment on your sums to answer the question.

Exam practice

1 a Here is a list of number cards.

21	22	23	24	25	26

Choose from the list a number that is

i a multiple of 6

ii a factor of 100

iii a prime number.

b Write 28 as a product of prime factors. *(5 marks)*

2 Two numbers have highest common factor (HCF) of 15 and least common multiple (LCM) of 90.
One of the numbers is 30. What is the other number?

(2 marks)

3 a A bathroom tile has two colours, purple and green, in its design.

The ratio purple : green is 3 : 2.
What fraction of the tile is purple?

b In this tile design, $\frac{5}{8}$ is coloured blue and the remainder is white.
What is the ratio blue : white?

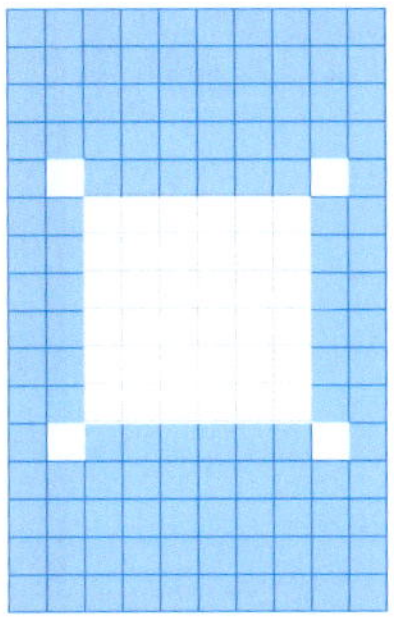

(3 marks)

5 Sequences

INTRODUCTION

There are lots of different types of sequences in mathematics: from the most basic arithmetic sequences such as the set of even numbers, to the curious Fibonacci sequence used to describe the growth of rabbit populations, through to the elegant and complex world of iterative sequences used in chaos theory.

What's the point?

Sequences allow you to find patterns in nature, so that you can understand it better and predict its behaviour.

CHECK IN

1 Work out the difference between
  - a 5 and 8
  - b 3 and 9
  - c 6 and 11
  - d −4 and +2

2 Find the difference between
  - a 10 and 7
  - b 9 and 5
  - c 12 and 7
  - d 3 and −1

3 Write down the first six multiples of
  - a 4
  - b 3
  - c 5
  - d 6

Orientation

What I should know	What I will learn	What this leads to
Key stage 3	■ Generate sequences ■ Find term-to-term and position-to-term rules for sequences	Careers in finance and computing.

RICH TASK

You have 15 circles.
Devise the first three terms of a different sequence using all 15 circles. Describe your sequence in words.

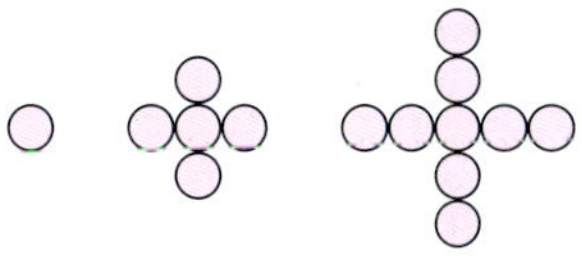

Can you find another sequence that uses all 15 circles? Investigate finding the first three terms of sequences for different numbers of circles.

Generating sequences

A5.1

- Generate the terms of sequences for different situations and describe their behaviour

RICH TASK

Bacteria can grow very quickly when the conditions are right.
If a certain type of bacterium doubles its number of cells every hour, investigate how many cells there will be after several hours if you start with just one cell.
Use the internet to find out about different bacteria and their rates of growth.

→ **A sequence is a set of numbers (called terms) which increase or decrease by following a rule.**

*
EXAMPLE

Ben has £20 in his bank account and his grandma promises to give him £6 every week to save for his holiday.
How much has he altogether in his account during the first few weeks?
Describe how you can quickly work out how much he has when he goes on holiday 12 weeks later.

ANSWER

At the start, Ben has £20.
After 1 week, he has £(20 + 6) = £26. ← he adds on 1 lot of 6. . . .
After 2 weeks, he has £(20 + 6 + 6) = £20 + 2 × 6 = £32. ← 2 lots of 6. . . .
After 3 weeks, he has £(20 + 6 + 6 + 6) = £20 + 3 × 6 = £38. ← 3 lots of 6. . . .

The sequence is 20, 26, 32, 38, ...
The number of 6s that he adds on equals the number of weeks that he has been saving.
So, after 12 weeks, his grandma has given him 12 lots of £6 = 12 × £6 = £72.
His total savings are £20 + £72 = £92.

Exercise A5.1

LOW

Unit A

*1 Nadeem buys a roll of dress material 80 metres long. She uses five metres each evening to make dresses.

a Write a sequence for the number of metres she has on the roll at the start of each day of the first week after she bought it.

b For how many evenings will the roll last? Describe how you work out your answer.

*2 Emma plants two kinds of plants. She starts with five red plants and then builds up the pattern shown here, adding four new plants each time.

Stage 1 Stage 2 Stage 3

a Write down the first four terms of the sequence for the **total** number of plants.

b Write the next three terms without drawing any diagrams. Describe how you found them.

c How many plants will she have used when she gets to stage 20?

*3 A kitchen wall is being covered in tiles. Holly starts with two tiles in the first column and then has three tiles in each column afterwards.

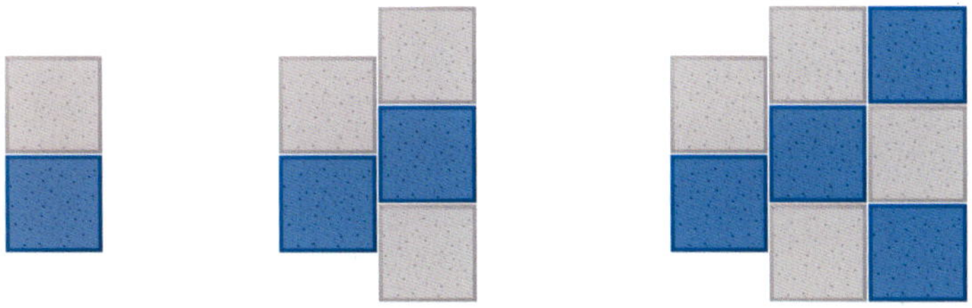

a Copy and complete this table.

b How many tiles will she have used in 20 columns? Find a quick way of working out your answer and explain how you did it.

No. of columns	1	2	3	4	5
No. of tiles used	2				

4 Write the next three terms for each of these sequences. In each case, find the 20th term (without writing out all 20 terms).

a 10, 12, 14, 16, ... b 5, 8, 11, 14, 17, ...

Term-to-term rules

- Find and use a term-to-term rule to generate a sequence

RICH TASK

Aya is making a metal fence using iron rods welded together.

How many rods does she use for the first section?

How many extra rods does she need to make the next section?

What is the total number of rods used so far?

So the sequence for the **total** number of rods begins 7, 11, ...

What are the next two terms of the sequence?

What do you add to one term to make the next term?

The term-to-term rule is

'Start with □ and add □ each time.'

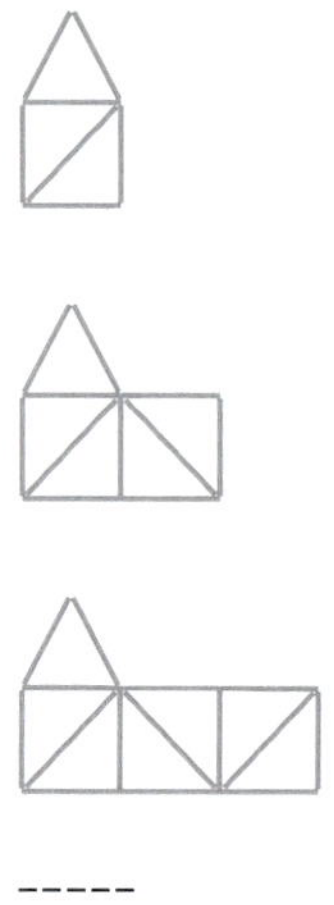

→ **The term-to-term rule of a sequence gives you the first term and then tells you how to find the next term from the previous one.**

EXAMPLE

The first term of a sequence is 4.
The term-to-term rule is given by this number machine.
Generate the first four terms of the sequence.
Describe the term-to-term rule in words.

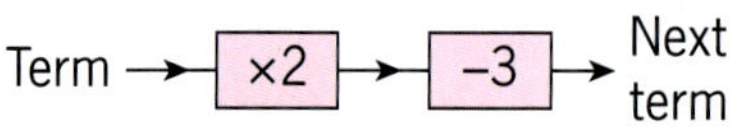

ANSWER

The second term is $2 \times 4 - 3 = 8 - 3 = 5$.
The third term is $2 \times 5 - 3 = 10 - 3 = 7$.
The fourth term is $2 \times 7 - 3 = 14 - 3 = 11$.
The first four terms are 4, 5, 7, 11.
The term-to-term rule is:
'Multiply by 2 and subtract 3.'

Exercise A5.2

MEDIUM LOW

1 You can use straws to make this sequence of patterns.

a Draw the next pattern of the sequence and write the first four terms.

b Find the term-to-term rule in words and use it to write the next three terms.

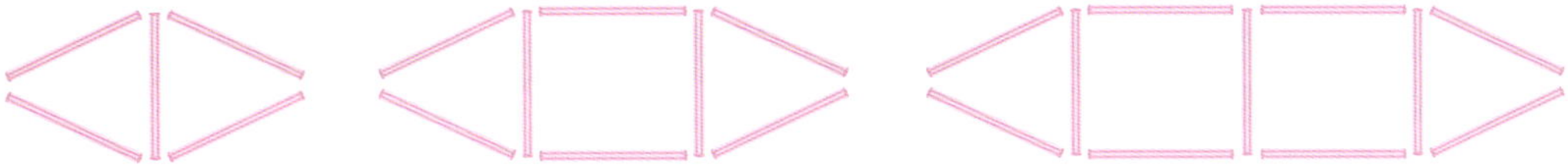

2 Generate the first five terms of the sequences with these rules.

a Start with 3, double and add 2.

b Start with 1, add 1 and treble.

c Start with $\frac{1}{4}$, double and add 1.

d Start with 20, subtract 8 and add 5.

e Start with 600, halve and subtract 20.

f Start with 4, treble and subtract 8.

3 Use this flow diagram to generate the first four terms of the sequence when the first term, x, is given by

a $x = 2$ b $x = 50$ c $x = -30$

Term → +20 → ×2 → Next term

4 **RICH TASK**

Write these operations in any order in this flow diagram.
How many different sequences can you generate?

+4 ×2 −1 Term → ☐ → ☐ → ☐ → Next term

5 For each of these sequences, describe the term-to-term rule in words and find the next three terms.

a 8, 11, 14, 17, ...

b 30, 27, 24, 21, ...

c 1, 3, 7, 15, ...

d 3, 5, 9, 17, ...

e 1, 4, 13, 40, ...

f 10, 7, 4, 1, ...

g 200, 100, 50, 25, ...

h $\frac{1}{2}$, 3, 8, 18, ...

i −7, −4, −1, 2, ...

6 ★ **ICT CHALLENGE** ★

Use a spreadsheet to generate a sequence with the rule 'Start with 5, double and add 1.'

	A	B	C	D	E
1	5	=2 × A1 + 1	=2 × B1 + 1	=2 × C1 + 1	-----
2					

Invent some sequences of your own. Write the rule you use in each case.

Position-to-term rules

A5.3

- Find and use the position-to-term rule for a sequence

RICH TASK

Hayley makes a sequence of patterns using diamond-shaped tiles.
She starts with two tiles and adds three extra tiles each time to make the next pattern.

Hayley uses the three-times table to construct a table of values.
What values are needed in the table for the 4th position?

- A **position-to-term rule** will give her the value of any term, if she knows its position in the sequence.

Hayley's position-to-term rule is 'Multiply the position by 3 and then subtract 1.'

Is she correct when she says that the rule gives 299 tiles for the 100th position?

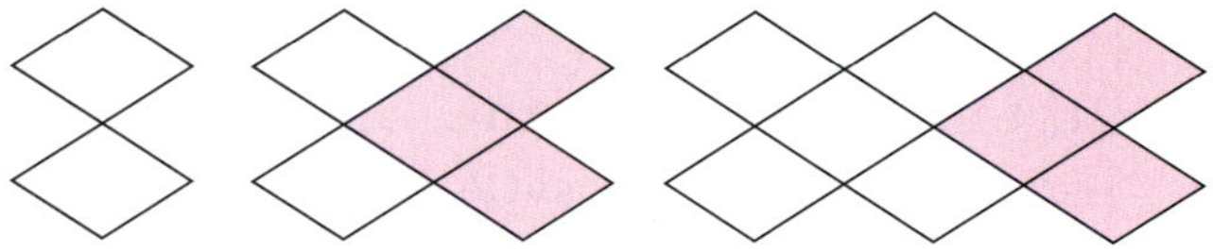

Position	1	2	3	4	×3
3 × table	3	6	9	…	−1
No. of tiles used	2	5	8	…	

The rule can also be shown as a **function machine.**

Position → ×3 → −1 → Term

Exercise A5.3

MEDIUM LOW

1 This sequence of shapes is made using short sticks.

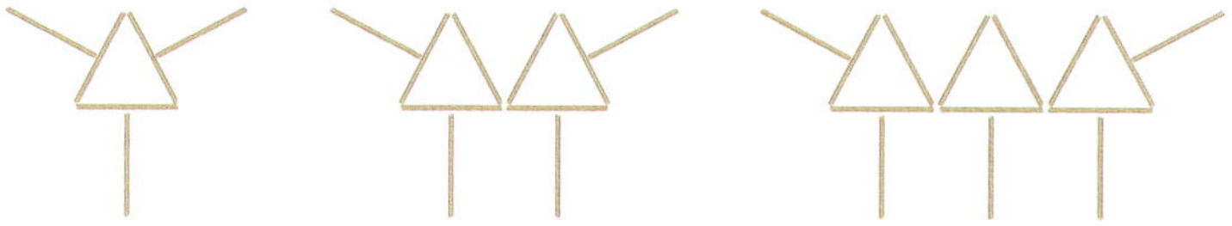

a How many extra sticks are needed to make the next position?

b Draw the diagram for position 4.

c Write down the term-to-term rule in words.

d Copy and complete this table of values and use it to find the position-to-term rule.

e How many sticks are needed to make the 50th position?

Position	1	2	3	4	5
… × table					
Term	6				

2 For each of these sequences

i find the increase from one term to the next, and so make a table of values for the terms in positions 1 to 5

ii use the table to find the position-to-term rule and describe it in words

iii find the value of the terms in the 6th, 7th and 100th positions.

a 3, 7, 11, 15, 19, ...

b 4, 7, 10, 13, 16, ...

c 1, 3, 5, 7, 9, ...

d 5, 9, 13, 17, 21, ...

e 3, 8, 13, 18, 23, ...

f 7, 13, 19, 25, 31, ...

g $3\frac{1}{2}$, 4, $4\frac{1}{2}$, 5, $5\frac{1}{2}$, ...

h 6, 4, 2, 0, −2, ...

3 Find the missing term and the 50th term of these two sequences.

a 7, 13, □, 25, 31, □, ...

b $3\frac{1}{2}$, □, $6\frac{1}{2}$, 8, □, 11, ...

4 Mrs Briggs' electricity meter has readings taken every day.
They are:
00127 00131 00135 □ 00143 00147 .....

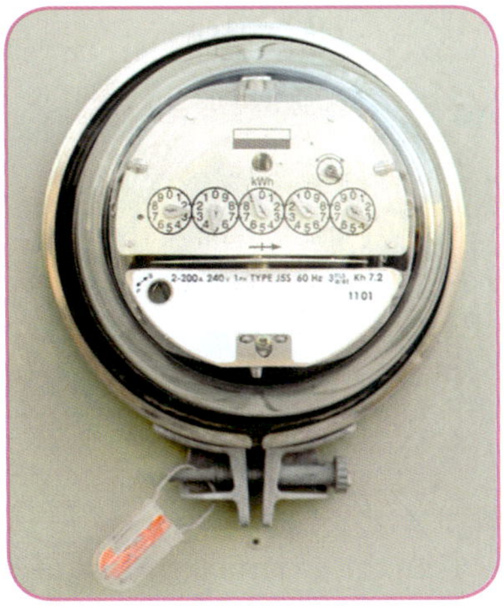

a If the readings follow a pattern, find the missing reading of the sequence.

b If the readings continue to follow the same pattern, find the reading after a month of 30 days.

5 **RICH TASK**

Dileep notices a tap dripping water into a beaker in the science lab.
Every 1 minute, he records the depth of water, in millimetres, as:
15 18 21 24 27 .....
Find the position-to-term rule and so find how deep the water will be after 30 minutes.

6 **RICH TASK**

A rev counter on a motor counts the number of revolutions.
The counter reads 325 when the motor is turned on. It is then read every minute, with these results:
331 337 343 349 .....
What makes you think the motor is turning at a constant speed?
Find the position-to-term rule and the reading on the counter after an hour.

The general term

A5.4

- Find and use the general term or *n*th term

RICH TASK

This sequence of shapes is made from coloured dots. Find the position-to-term rule and explain why the rule works by looking at the pattern of dots. How many dots will there be in the *n*th position?

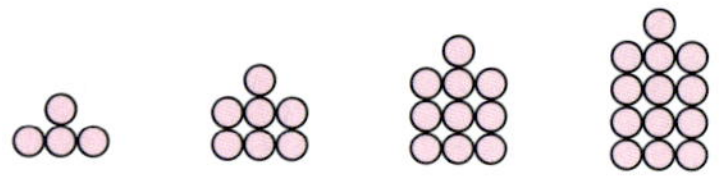

Position 1 2 3 4

→ **The general term is an expression that you can use to calculate the value of any term. It is written in terms of *n* and is often called the *n*th term.**

EXAMPLE

Zak is stacking a supermarket shelf with tins. Each box contains six tins. There are always two tins left over.

Position 1 Position 2 Position 3

Find the general term for the number of tins, and the 20th term of the sequence.

ANSWER

The table of values and function machine for the sequence are:

Position	1	2	3	4	.....	n
6 × table	6	12	18	24	.....	$6n$
Term	8	14	20	26	.....	$6n + 2$

×6, +2

6 × table because there are 6 extra tins each time.

Position number → ×6 → +2 → Term

The general term is $6n + 2$.
The expression is like this, because there are n boxes of 6 tins plus an extra 2 loose tins.
The 20th term is $6 \times 20 + 2 = 120 + 2 = 122$.

The general term can also be written as $T(n)$.
So $T(n) = 6n + 2$ and $T(20) = 120 + 2 = 122$.

T stands for term.
$T(n)$ stands for *n*th term.

Exercise A5.4

MEDIUM

1 Here is a sequence made from square tiles.

 a How many squares are added to get from one term to the next?

 b Copy and complete this table of values.

 c Write an expression for the nth term of the sequence. Explain why the rule works by looking at the pattern of tiles.

 d Find the value of the 6th term and the 50th term.

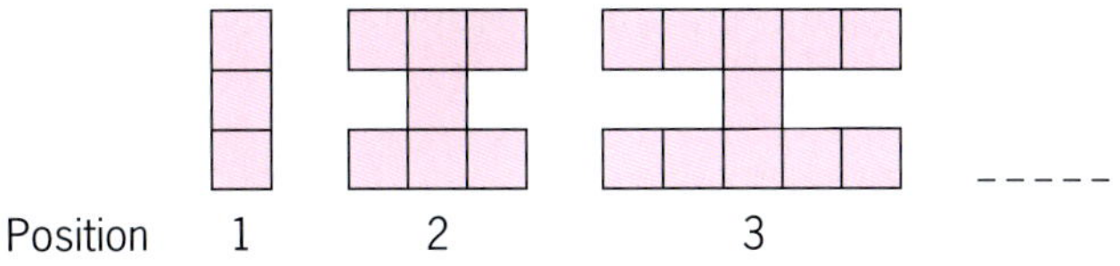

Position	1	2	3	4	5	.....	n
.... × table							
Term							

2 Here are several patterns. For each one, construct a table of values describing how many objects are used in each term.
Write an expression for the nth term and explain how the expression relates to the pattern. Find the value of the 100th term.

a

b

c

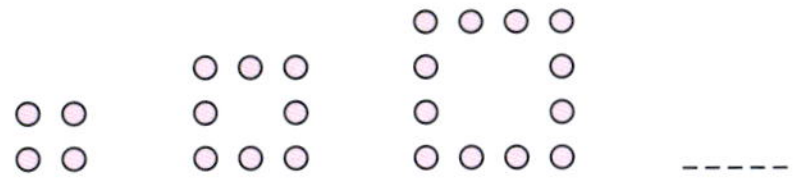

d

3 For each of these sequences

 i construct a table of values

 ii write the position-to-term rule in words

 iii find an expression for the nth term

 iv find the value of the 6th term and the 40th term.

 a 7, 9, 11, 13, 15, ...
 b 2, 5, 8, 11, 14, ...
 c 5, 6, 7, 8, 9, ...
 d 3, 8, 13, 18, 23, ...
 e 7, 13, 19, 25, 31, ...
 f 6, 11, 16, 21, 26, ...
 g −1, 3, 7, 11, 15, ...
 h 10, 17, 24, 31, 38, ...
 i −4, −1, 2, 5, 8, ...
 j 8, 6, 4, 2, 0, ...

4 Find the missing term □, the 6th term, and the 100th term in each of these sequences.

 a 6, 11, 16, □, 26, ...
 b 12, 22, □, 42, 52, ...
 c 3, 7, 11, □, 19, ...
 d 5, 13, 21, □, 37, ...

5 ***RICH TASK***

Mrs Parker lays a line of paving slabs. She puts one potted plant at each end of the line and two potted plants on each slab. Write a sequence for the number of plants she needs for a line of 1, 2, 3, 4, 5, ... slabs. How many slabs are in the line if she uses 40 plants?

Common sequences

A5.5

- Explore some common sequences

Not all sequences have a general term of the type $T(n) = an + b$.

EXAMPLE

Certain bacteria grow by doubling the number of cells every hour.
A scientist starts with 10 cells in a dish.
Write a sequence for the number of cells, $T(n)$, after n hours when $n = 0, 1, 2, 3, 4$.

ANSWER

At the start, the number of cells is 10.
After 1 hour, $T(1) = 10 \times 2 = 20$.
After 2 hours, $T(2) = 10 \times 2 \times 2 = 40$.
After 3 hours, $T(3) = 10 \times 2 \times 2 \times 2 = 80$.
After 4 hours, $T(4) = 10 \times 2 \times 2 \times 2 \times 2 = 160$.
The sequence is 10, 20, 40, 80, 160, ...

Exercise A5.5

MEDIUM

1 a i Show that $T(n) = 2n + 8$ generates a sequence of even numbers.
ii Show that $T(n) = 8 - 2n$ also generates a sequence of even numbers.
iii Write down a major difference between these sequences.
b i Show that $T(n) = 2n + 9$ generates a sequence of odd numbers.
ii Show that $T(n) = 9 - 2n$ also generates a sequence of odd numbers.
iii Write down a major difference between these sequences.
c Make one alteration to the formula $T(n) = 3n + 5$ so that it will generate a sequence of odd numbers.
d A farmer makes a fence using 2 metres of wire between each pair of posts. Write the first five terms of the sequence $T(n) = 2n - 2$. Explain why the terms of this sequence give the number of metres of wire he needs, as he puts in n posts.

2 Here is a pattern made from dots.

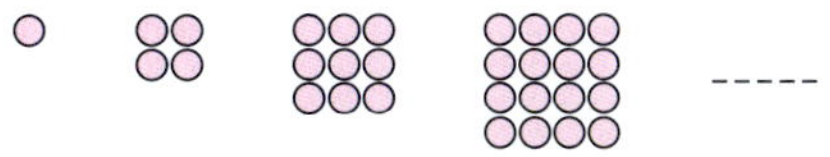

- **a** Write down the sequence of the first six square numbers.
- **b** Write down the general term, T(n), for the nth square number.
- **c** Find the value of **i** T(9) **ii** T(12) **iii** T(20)
- **d** If T(n) = 64, what is the value of n?

3 Here is a pattern made from dots.

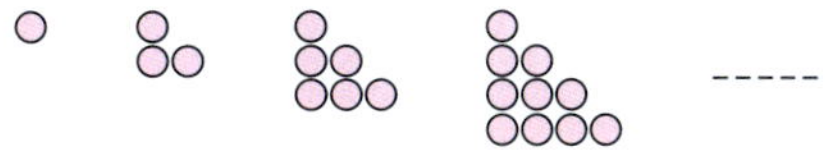

- **a** Use this diagram to write down the sequence of the first four triangle numbers.
- **b** Copy and complete these statements.

  T(1) = 1

  T(2) = 1 + 2 = ...

  T(3) = 1 + 2 + 3 = ...

  T(4) = 1 + 2 + 3 + ... = ...

  T(5) = 1 + 2 + 3 + ...+ ..... = ...
- **c** Show that the formula $T(n) = \frac{1}{2}n(n + 1)$ works as a general term for the first five triangle numbers.
- **d** Find the value of T(6) and show that the formula also works in this case.

4 **RICH TASK**

Everyone in a group shakes hands with each other just once.
How many handshakes will there be when the group contains

a 2 people **b** 3 people **c** 4 people?

Repeat for larger groups of 5, 6 ..... people.
How do your results relate to question **3** above?
How many handshakes will there be for a group of 100 people?

5 **RICH TASK**

Sue can either get £12 a week pocket money as a lump sum or 10p on Monday, 20p on Tuesday, doubling every day through to Sunday. Which method would you suggest she opts for?

Summary and assessment

A5

CHECK OUT

You should now be able to

- generate terms of a sequence using term-to-term and position-to-term rules
- describe the nth term of a sequence
- generate common integer sequences.

*

Exam-style question

Four numbers in a sequence are given by

$2n + 1$ $\quad$ $2n + 5$ $\quad$ $2n + 9$ $\quad$ $2n + 13$

a Explain why the numbers in this sequence are odd numbers.

b Write down the term-to-term rule for numbers in this sequence.

Explain may mean that you do some mathematics, but you may also need to write an explanation.

It is helpful to show your thinking.

Sinead's answer:

a All the numbers in the sequence are 2 times a number plus an odd number.

2 x a whole number = an even number

An even number + an odd number = an odd number

So the numbers in the sequence will always be odd.

b The difference between each number is 4

$(2n+5) - (2n+1) = 4$

$(2n+9) - (2n+5) = 4$ etc

The term-to-term rule is +4

To explain it is not always enough to give an example.

For $n = 1$ the sequence is 3, 7, 11, 15.

This is not enough for an explanation.

Exam practice

1 a Write down the next two terms in this sequence

4 7 10 13 ____ ____

b Describe, in words, the rules for continuing this sequence

3 6 12 24 .....

c The nth term of a sequence is $n^2 + 1$.

Write down the first three terms of this sequence.

(6 marks)

2 Here is a sequence of shapes made with tiles.

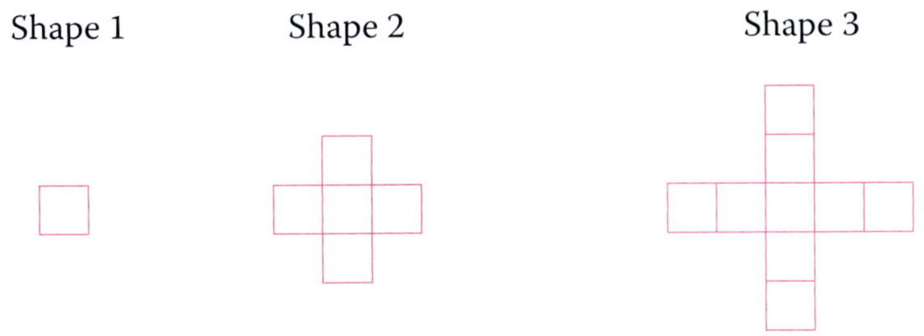

a Altogether how many tiles will there be in

i shape number 6 **ii** shape number 16?

b Write down a rule for finding the total number of tiles, t, for shape number n. *(4 marks*

3 Ellen has left the kitchen sink almost full with water and the kitchen tap is still on.

The distance, in mm, from the water level to the top of the sink, measured each minute is:

37 34 31 28

If the tap is not switched off, how much longer will it take for the sink to overflow? *(3 marks*

6 Representing and interpreting data

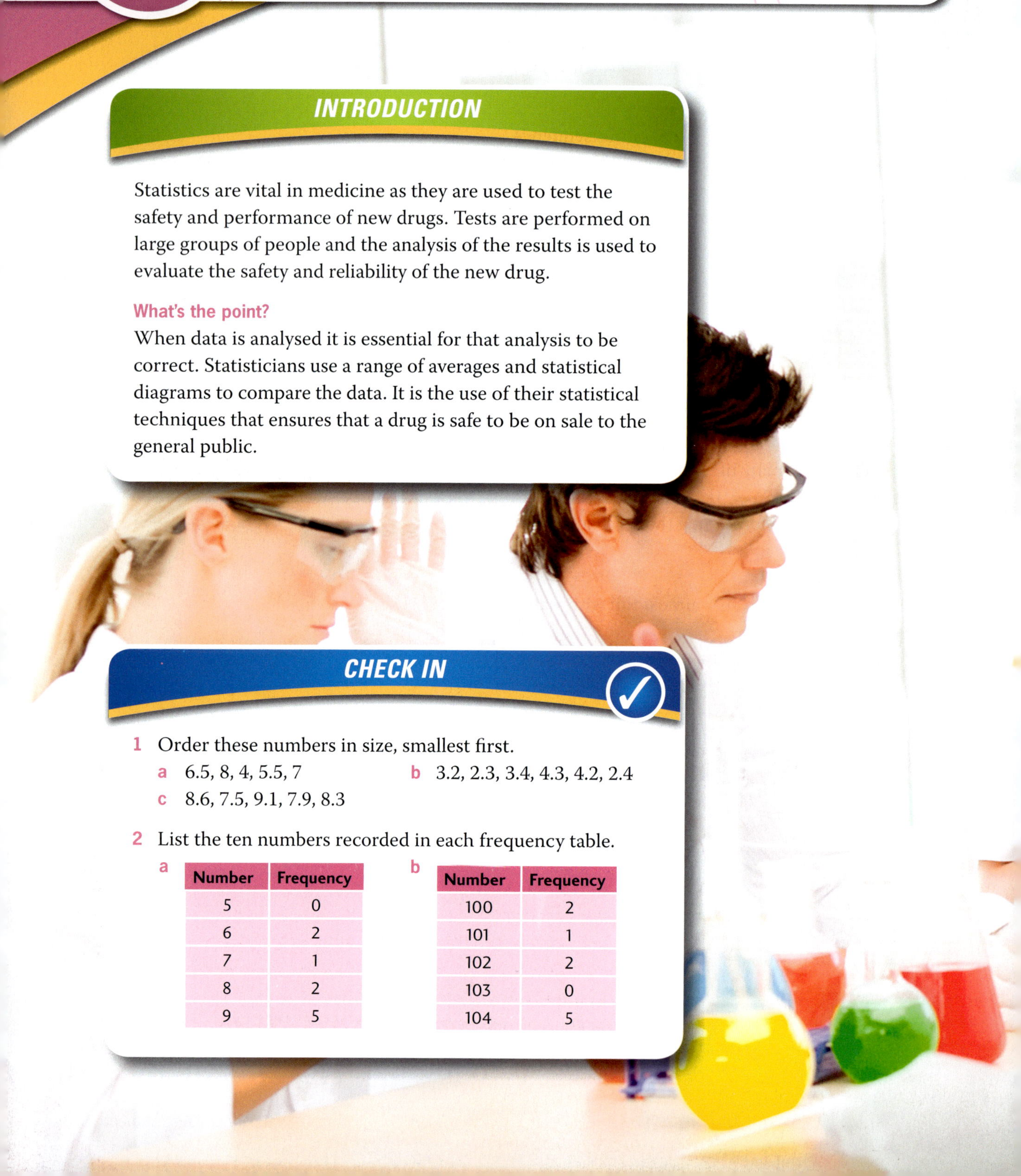

INTRODUCTION

Statistics are vital in medicine as they are used to test the safety and performance of new drugs. Tests are performed on large groups of people and the analysis of the results is used to evaluate the safety and reliability of the new drug.

What's the point?

When data is analysed it is essential for that analysis to be correct. Statisticians use a range of averages and statistical diagrams to compare the data. It is the use of their statistical techniques that ensures that a drug is safe to be on sale to the general public.

CHECK IN

1 Order these numbers in size, smallest first.

a 6.5, 8, 4, 5.5, 7

b 3.2, 2.3, 3.4, 4.3, 4.2, 2.4

c 8.6, 7.5, 9.1, 7.9, 8.3

2 List the ten numbers recorded in each frequency table.

a

Number	Frequency
5	0
6	2
7	1
8	2
9	5

b

Number	Frequency
100	2
101	1
102	2
103	0
104	5

Orientation

What I should know	What I will learn	What this leads to
A2 + Key stage 3	■ Draw diagrams and graphs to represent data ■ Use diagrams and graphs to answer questions	B13

RICH TASK

What is the most likely time during a football game for a team to score a goal?
It is frequently stated by football commentators that teams are most likely to concede a goal within five minutes of scoring a goal themselves. Is this true?
Investigate, and write a report on your results.

Bar charts

A6.1

- Draw and produce bar charts

The **multiple bar chart** shows the number of gold, silver and bronze medals won by the five most successful teams at the 2008 Olympics.

ACTIVITY

Which team won the most gold medals?

Which team won the most silver medals?

Which team won the most bronze medals?

Which team won most medals overall?

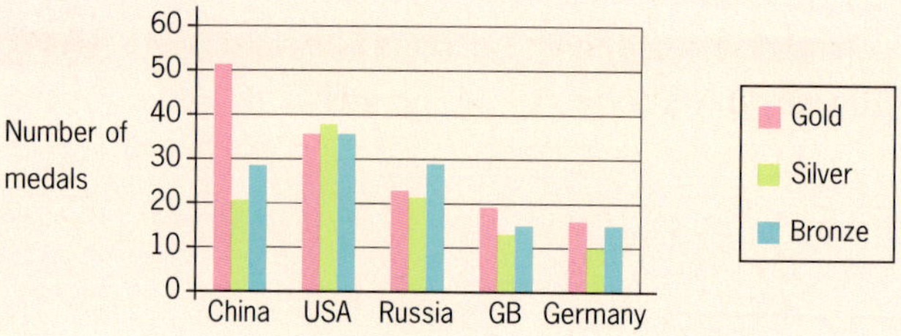

China, with the most gold medals, was placed first.

RICH TASK

Is the number of gold medals the fairest way to decide which country should come first? Use the diagram to explore other ways to decide team position.

EXAMPLE

In Thao's memory survey on page 16 she collected data on the first object recalled. The table summarises the first object recalled by 11-year-olds.

Draw a bar chart to represent these data.

Object	Frequency
Elastic band	13
Rabbit	4
iPod	8
Phone	2

ANSWER

Remember to scale the vertical axis.

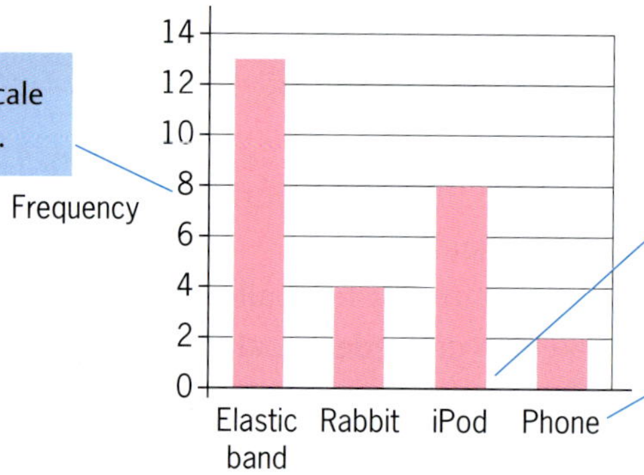

Leave a gap between each bar.

Label each bar on the horizontal axis.

Exercise A6.1

LOW

1 The data shows the transport used by people visiting a sports centre one Saturday. Draw a bar chart to illustrate these data.

Transport	Car	Bus	Walk	Cycle	Other
Frequency	10	25	20	40	5

2 The table summarises Thao's data of first object recalled for 15-year-olds.

a Draw a bar chart to represent these data.

b Write down two comparisons between the recall of the boys and the girls.

	Elastic band	Rabbit	iPod	Phone
Boys	2	1	5	4
Girls	2	4	5	4

*3 Write down three questions that could be answered by studying this bar chart.

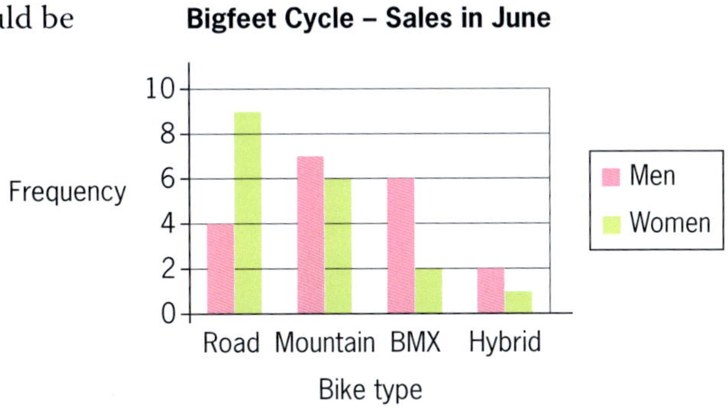

4 The bar charts show the percentage of UK households having use of certain consumer goods in 1990 and 2000.

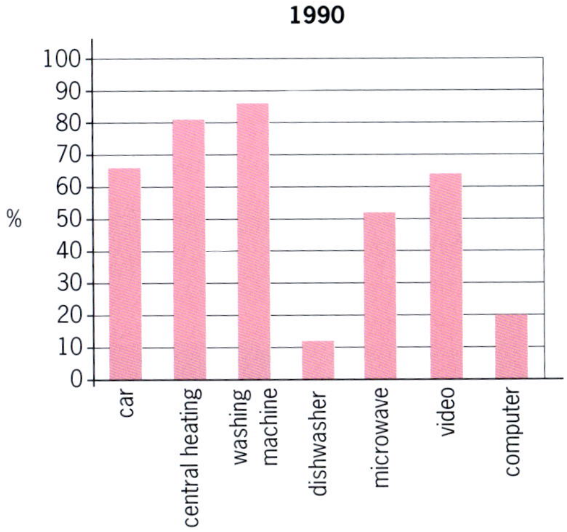

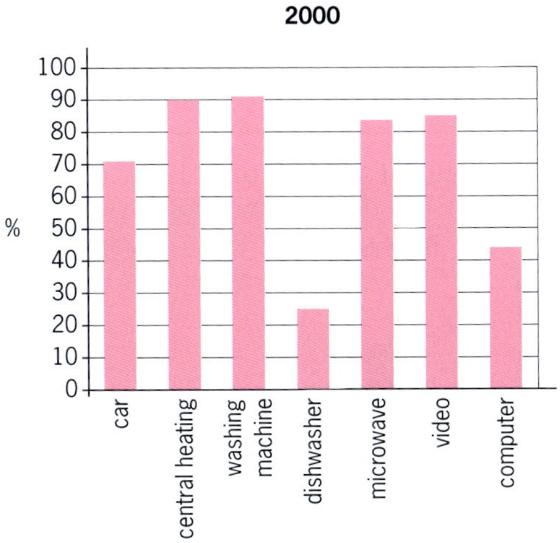

a Over this 10-year period, which of the consumer goods shows

i the least increase **ii** the greatest increase?

b Draw a bar chart to show what you think the graph for 2010 might look like.

Histograms for continuous data

A6.2

- Draw and produce histograms for continuous data

You can draw a **histogram** to show continuous data.

p. 28

EXAMPLE

The table shows the times taken, to the nearest minute, to complete a crossword puzzle.

Time, t, minutes	$0 < t \le 5$	$5 < t \le 10$	$10 < t \le 15$	$15 < t \le 20$	$20 < t \le 25$
Frequency	4	11	6	5	2

Draw a histogram to represent these data.
Use the diagram to write down the **modal class**.

In this class interval, the time is between 20 and 25 minutes.

ANSWER

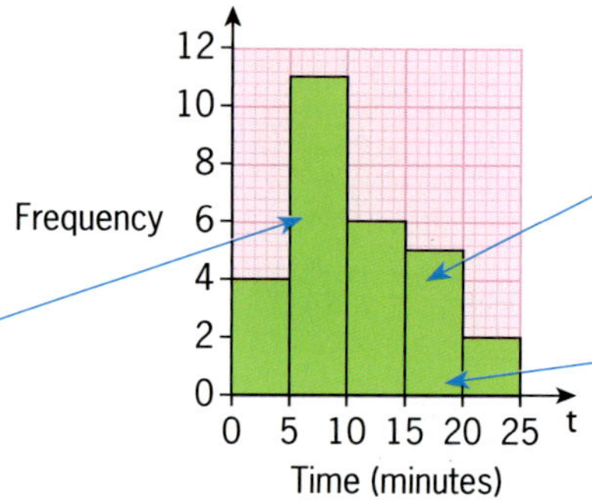

The **modal class** is $5 < t \le 10$, which is the highest bar.

The width of each bar is the same – the histogram has **equal class intervals.**

There are no gaps between the bars – the data is **continuous.**

You can use histograms to compare two sets of data.

EXAMPLE

Compare the weights of samples of Braeburn and Cox apples shown in the histograms.

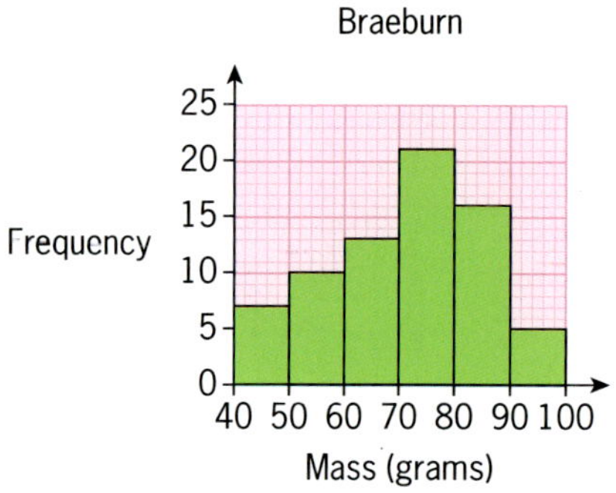

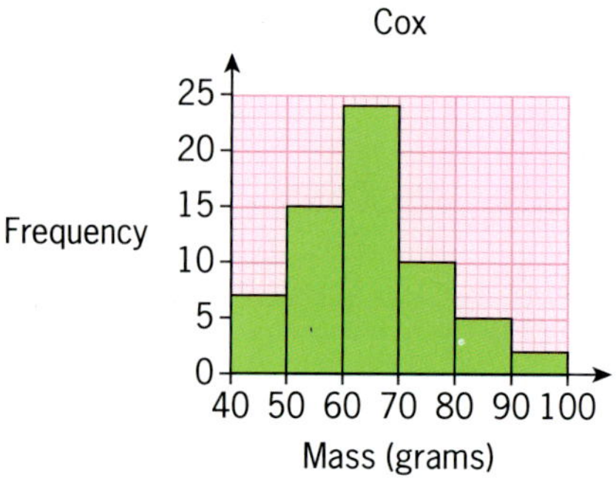

You compare the modal class, not the height of the bar.

ANSWER

The modal class for Braeburn apples, 70–80 g, is more than for Cox apples, 60–70 g.

Exercise A6.2

MEDIUM

1 Draw a histogram to represent the time taken for a sample of students to travel to school as given in the table.

Time, *t*, minutes	$0 \le t < 10$	$10 \le t < 20$	$20 \le t < 30$	$30 \le t < 40$	$40 \le t < 50$	$50 \le t < 60$
Frequency	9	27	42	30	20	12

2 The tables show a summary of the data from a survey of the time taken for a group of boys and girls to complete a sudoku puzzle.

Boys

Time, *t*, minutes	**Frequency**
$0 \le t < 5$	2
$5 \le t < 10$	5
$10 \le t < 15$	12
$15 \le t < 20$	19
$20 \le t < 25$	8
$25 \le t < 30$	4

Girls

Time, *t*, minutes	**Frequency**
$0 \le t < 5$	1
$5 \le t < 10$	5
$10 \le t < 15$	8
$15 \le t < 20$	13
$20 \le t < 25$	17
$25 \le t < 30$	6

a Draw two histograms, one to represent the data for the boys and one for the girls.

b Write down the modal class for the boys and that for the girls.

c In which class does the median lie for the boys, and for the girls?

d Write a sentence to explain whether you think the boys or the girls are quicker at solving the sudoku puzzle. Give reasons for your answer.

3 The histograms show the weights of two samples of strawberries – one grown on a patio and one in a greenhouse.

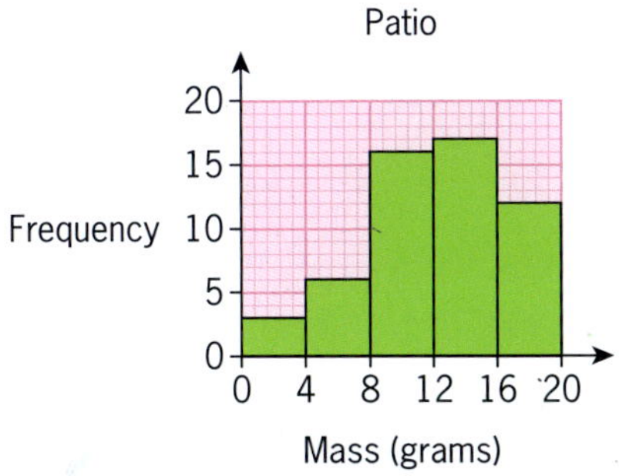

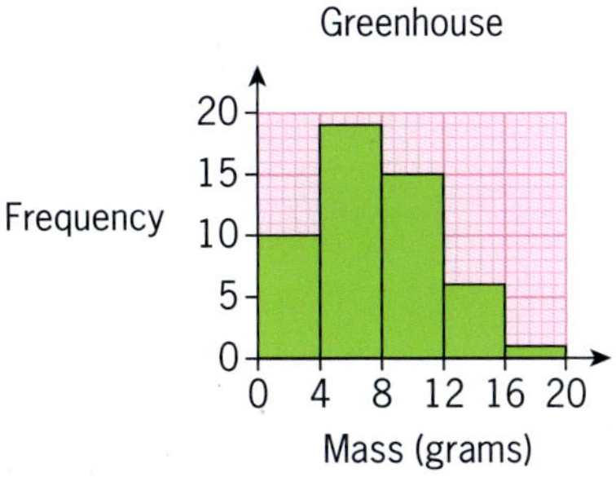

Compare the weights of the two samples, commenting on any similarities and differences.

Unit A

Pie charts

- Draw and produce pie charts for categorical data

Thao draws a **pie chart** to represent the first object recalled by 11-year-olds.

ACTIVITY

Compare this diagram with the bar chart showing the same information on page 80.

What was the **modal** object recalled first by 11-year-olds in Thao's survey?

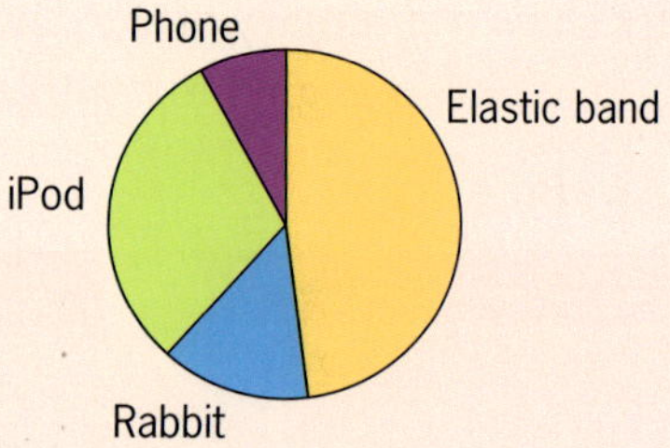

EXAMPLE

A survey collected information on the favourite music type of a sample of office workers. Draw a pie chart to represent the data.

Work out the fraction of the total multiplied by 360.

Music	Frequency	Working	Angle (°)
Classic	1	$\frac{1}{40} \times 360$	9
Indie	12	$\frac{12}{40} \times 360$	108
Jazz	9	$\frac{9}{40} \times 360$	81
Pop	8	$\frac{8}{40} \times 360$	72
Rock	10	$\frac{10}{40} \times 360$	90
Total	40		360

The answer is the angle on the pie chart.

Check that the total degrees add up to 360°.

ANSWER

Favourite music

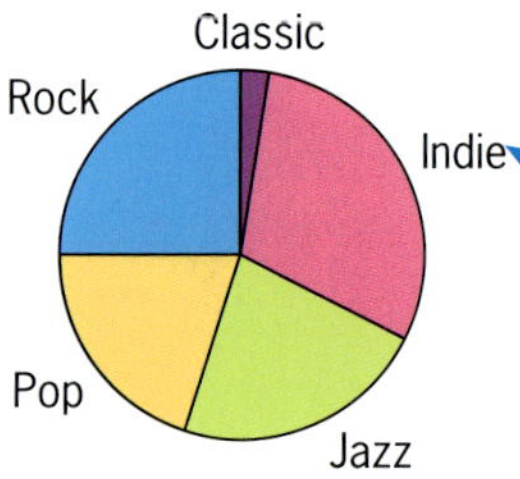

Remember to label each sector or write a key. Angles are not marked on pie charts

EXAMPLE

The pie chart shows the number of livestock in the UK in millions in 2005.

a What angle represents cattle?

b How many cattle were there?

ANSWER

a 5% × 360 = 18°

b 5% × 224.7 = 11.235 million cattle

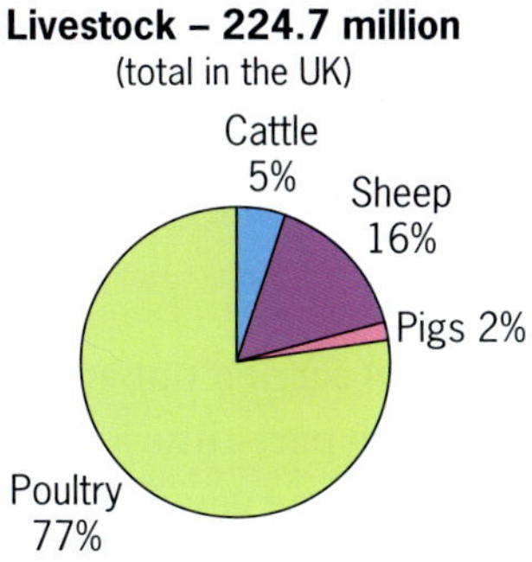

Exercise A6.3

MEDIUM LOW

1 The information shows a typical school day for a student in France.

Activity	Lessons	Break	Lunch
Time, minutes	280	80	90

a Draw a pie chart for these data.

b Draw a pie chart for your own typical school day and compare the diagrams.

2 The table shows the number of farms, to the nearest 1000, in the UK in 2005.

Country	England	Scotland	Wales	Northern Ireland
Number of farms (thousands)	196	51	37	27

a Draw a pie chart to represent these data.

b This pie chart shows the average size of farms, in hectares, in the UK in 2005.

i Which country had the largest area of farmland in the UK in 2005?

ii What percentage of UK farmland was in England?

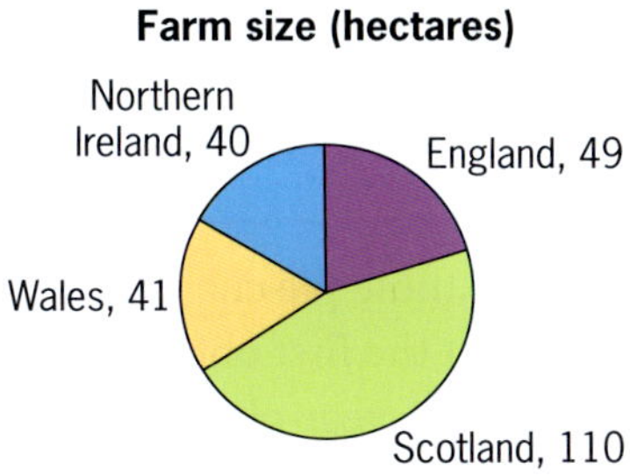

3 The pie chart shows how crop farmers in the UK used their land in 2005.

a How many tonnes of potatoes were grown?

b Work out the **angle** that represents vegetables.

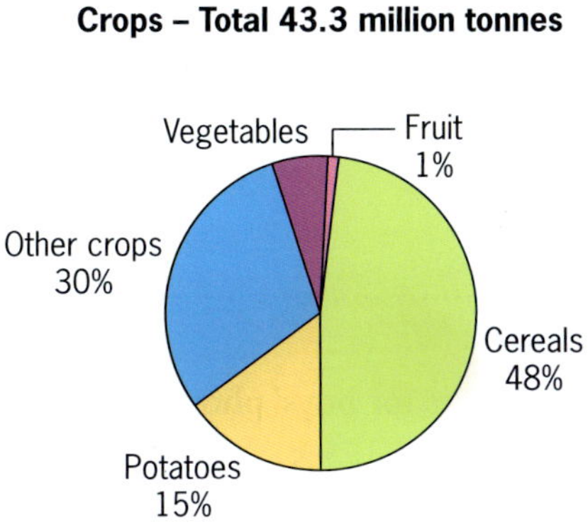

Unit A

Problem solving through diagrams

- Interpret graphs and diagrams and draw conclusions

A survey was carried out among sixth-formers to find out their monthly earnings from part-time jobs. There were equal numbers of boys and girls in the survey.

The pie charts show the results of the survey.

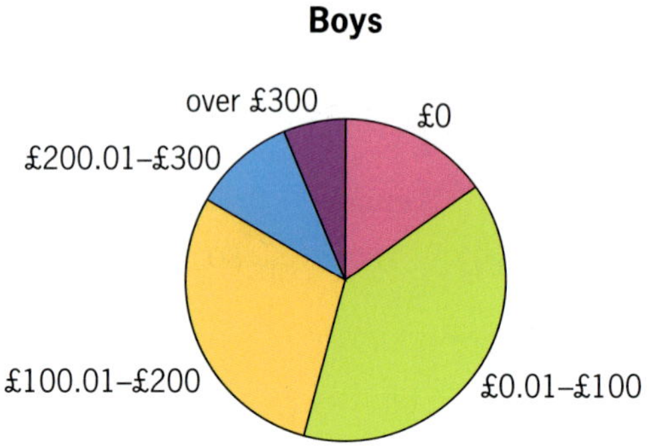

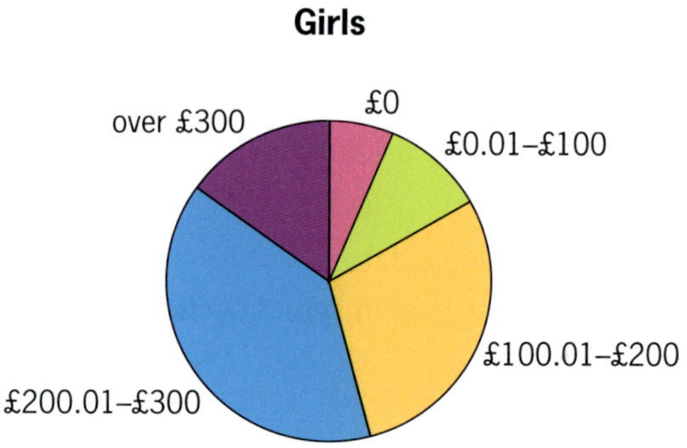

RICH TASK

Use the information to convince a friend that girls earn more than boys.

You can use pie charts to compare information.
However, you also need the totals to make sense of that information.

EXAMPLE

Thao drew these pie charts to represent the first object recalled by 15-year-old boys and girls.
Which objects did the same **proportion** of boys and girls recall first?

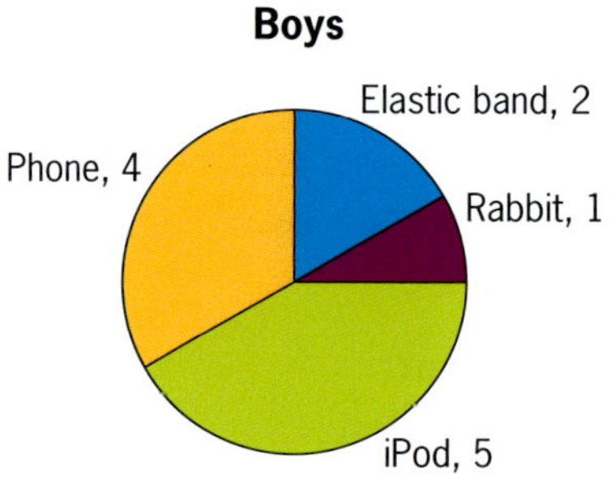

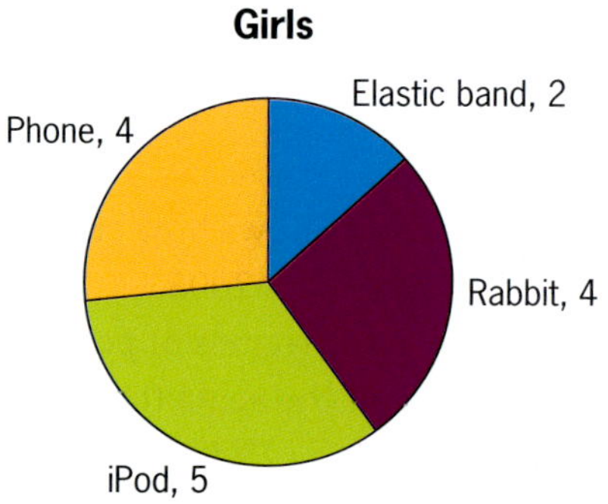

ANSWER

The sectors for boys' phone and girls' iPod have the same size angle.

$4 \div 12 = \frac{1}{3}$ $\quad$ $5 \div 15 = \frac{1}{3}$

The proportions for boys/phone and girls/iPod are the same.

Exercise A6.4

MEDIUM

1 The pie chart shows the domestic water use in England and Wales for 2002.

a What was the most common use of domestic water in 2002?

b How much water was used per day on flushing the toilet?

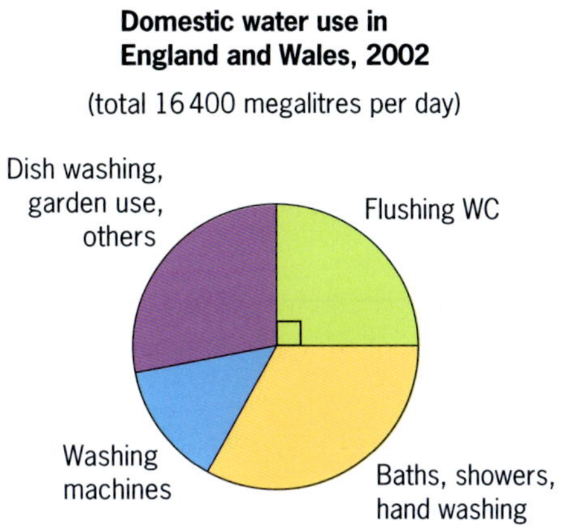

* 2 These pie charts show aspects of international tourism in 2004.

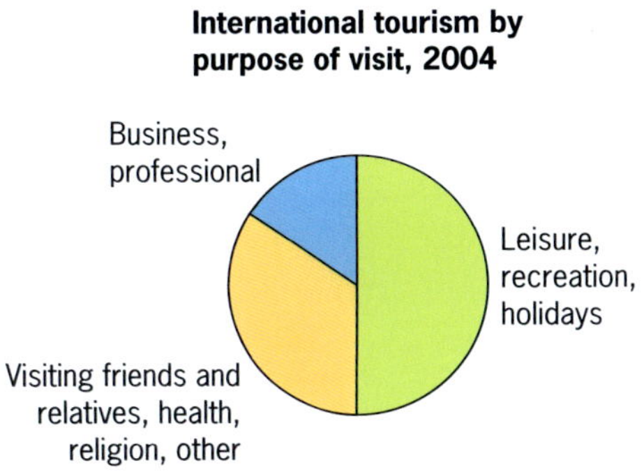

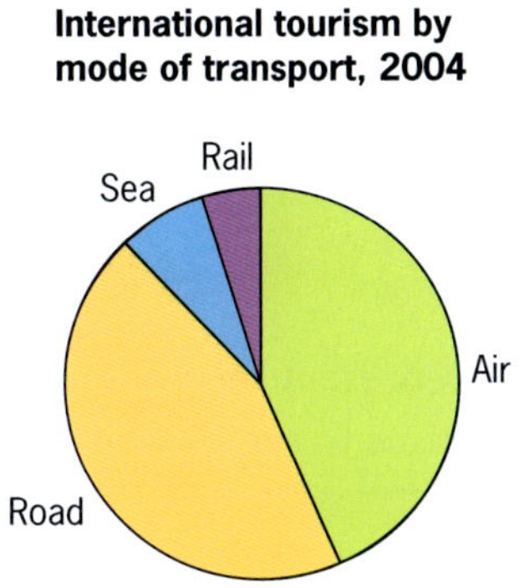

a What was the most common purpose of visit in 2004?

b What was the least popular international mode of transport in 2004?

c Is it possible from these diagrams to work out the mode of transport of the business visits? Explain your answer.

* 3 The tables show the population (in hundred of millions of people, to the nearest hundred million) and land area (in millions of square kilometres, to the nearest million) of the different continents.

Continent	Europe	Asia	Africa	North America	South America	Antarctica
Population	7	37	9	4	3	0
Land area	11	44	30	24	18	13

a Draw pie charts to represent these data (use the same colour for each country).

b Write three questions that you could answer by studying these pie charts.

Frequency polygons

- Draw and produce frequency polygons
- Identify the modal class for grouped data

On page 82 you found out how to draw a **histogram** for continuous data.

You can use a **frequency polygon** to represent continuous data.

EXAMPLE

Thao shows a tray of nine objects for exactly 20 seconds to a large sample of men and women.
She summarises the time they took to recall all nine objects in these tables.

Men

Time, t, seconds	Frequency
$0 < t \leq 20$	5
$20 < t \leq 40$	11
$40 < t \leq 60$	13
$60 < t \leq 80$	20
$80 < t \leq 100$	16
$100 < t \leq 120$	7

Women

Time, t, seconds	Frequency
$0 < t \leq 20$	7
$20 < t \leq 40$	22
$40 < t \leq 60$	18
$60 < t \leq 80$	16
$80 < t \leq 100$	6
$100 < t \leq 120$	3

Draw frequency polygons to represent the times taken by the men and the women. Use the diagrams to compare the times taken.

ANSWER

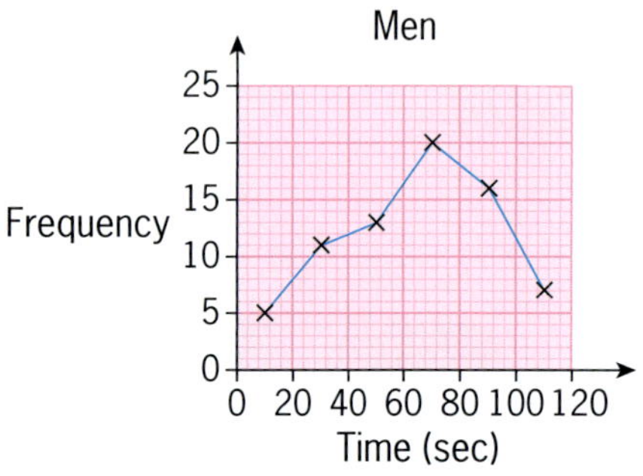

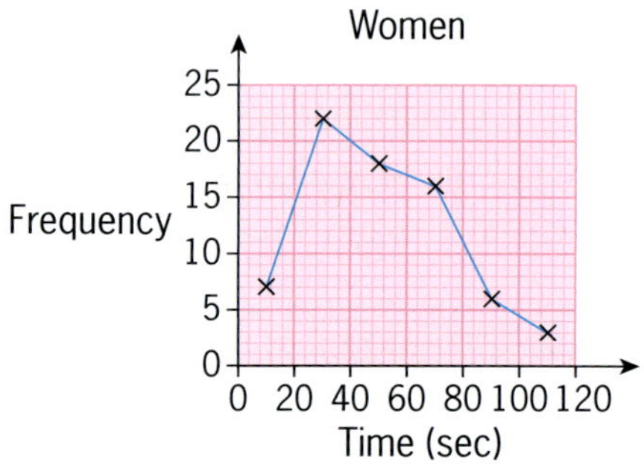

Plot points at the **midpoint** of each class. So for men, points at (10, 5), (30, 11), (50, 13) ... for women, points at (10, 7), (30, 22), (50, 18) ...

The modal class for men, 60–80, is higher than that for women, 20–40. The ranges are broadly similar, as both men and women have times within 0–20 seconds and 100–120 seconds.

You can compare continuous distributions by:

- → modal class – **compare where the graph peaks (not how high the peak is)**
- → range – **compare the ranges of class intervals.**

Unit A

EXAMPLE

Here is part of a report that appeared in a school magazine.

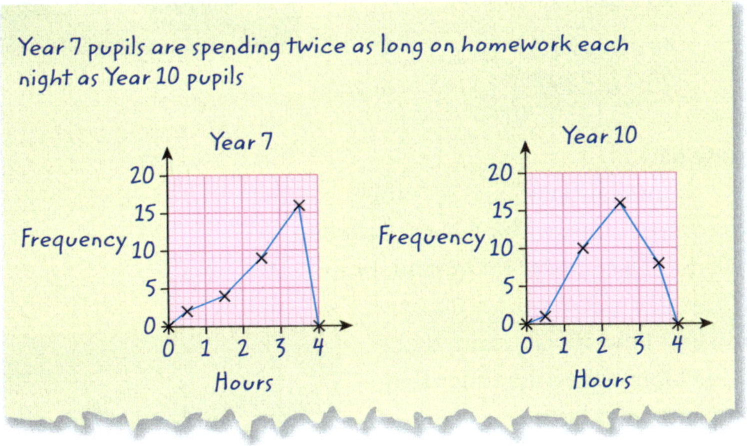

Is this claim fair?

ANSWER

No: the Year 7 modal class is 3 to 4 hours, whereas the Year 10 modal class is 2 to 3 hours. Also, the range of time spent on homework is the same.

Exercise A6.5

MEDIUM

1 David carried out a survey to find the time taken by 120 teachers and 120 office workers to travel home from work.

Teachers

Time, t, minutes	Frequency
$0 < t \le 10$	12
$10 < t \le 20$	33
$20 < t \le 30$	48
$30 < t \le 40$	20
$40 < t \le 50$	7

Office workers

Time, t, minutes	Frequency
$10 < t \le 20$	2
$20 < t \le 30$	21
$30 < t \le 40$	51
$40 < t \le 50$	28
$50 < t \le 60$	18

a Draw frequency polygons for these data.

b Work out for each data set the **i** modal class **ii** range.

c Use your answers to **b** and make comparisons, with reasons, between the time taken by the teachers and office workers to travel home from work.

Continued ➤

2 Jayne kept a daily record of the number of miles she travelled in her car during two months.

December

Miles travelled (m)	Frequency
$0 < m \leq 20$	3
$20 < m \leq 40$	8
$40 < m \leq 60$	10
$60 < m \leq 80$	6
$80 < m \leq 100$	4

January

Miles travelled (m)	Frequency
$0 < m \leq 20$	0
$20 < m \leq 40$	5
$40 < m \leq 60$	12
$60 < m \leq 80$	8
$80 < m \leq 100$	6

a Draw frequency polygons for these data.

b Work out for each data set the **i** modal class **ii** range.

c Use your answers to b and make comparisons, with reasons, between the number of miles Jayne travelled in December and in January.

*3 Two different companies, Duracomp and Powerblast, claim that their batteries are the ones to buy. The tables show the times that samples of batteries from these two companies lasted.

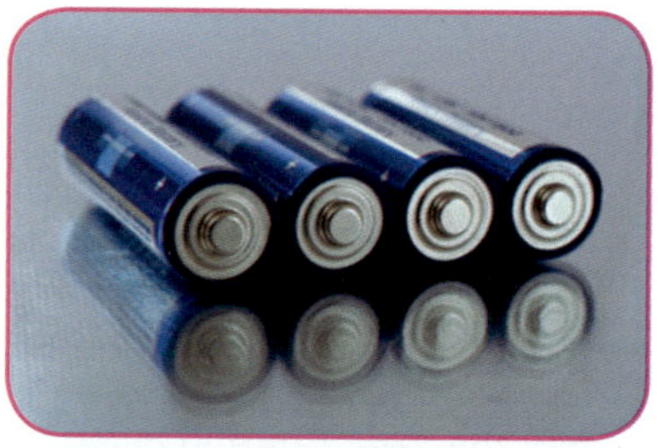

Duracomp

Time, t, hours	Frequency
$0 < t \leq 5$	8
$5 < t \leq 10$	6
$10 < t \leq 15$	7
$15 < t \leq 20$	5
$20 < t \leq 25$	0
$25 < t \leq 30$	4

Powerblast

Time, t, hours	Frequency
$0 < t \leq 5$	3
$5 < t \leq 10$	9
$10 < t \leq 15$	7
$15 < t \leq 20$	8
$20 < t \leq 25$	2
$25 < t \leq 30$	1

a Draw frequency polygons to represent these data.

b By finding different and appropriate measures, comment on the claims made by these two companies, giving reasons for your comments. Decide from which of the two companies you would recommend buying.

ADDITIONAL TASK

Draw a histogram for the Duracomp data, on the same diagram as the frequency polygon. How do the two types of diagram relate to each other?

4 These two graphs were drawn to summarise the football Premiership points score per team for two different years.

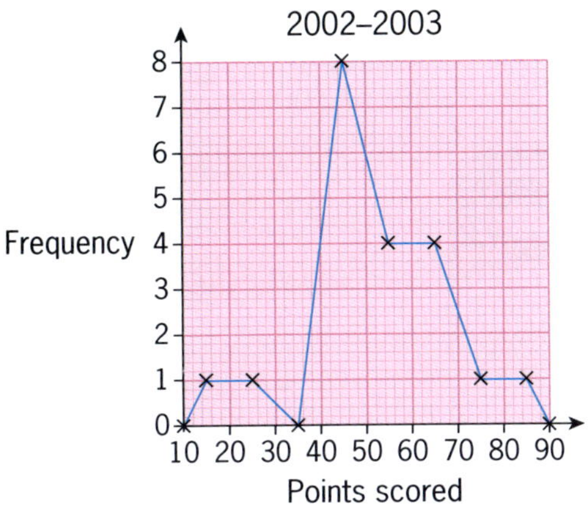

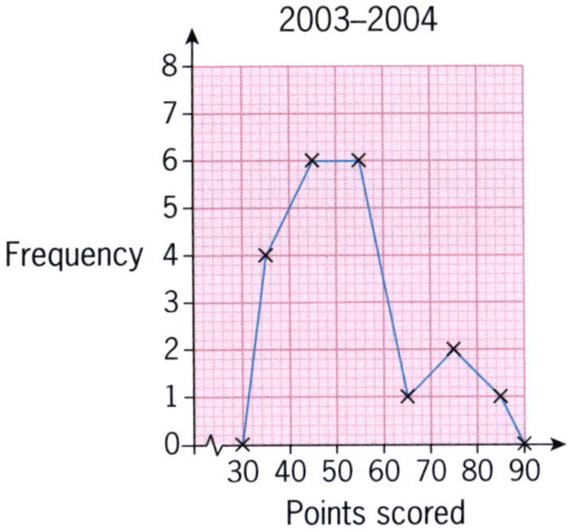

This news report was written about one of these two years.

> $52\frac{1}{2}$ points were achieved on average with a range of 67 by teams in the Premiership this season

Which year was being reported on? Give a reason for your choice.

5 ☆ **CHALLENGE** ☆

The frequency polygons show the ages of a sample of 100 people buying plants at a garden centre on a weekday and a Sunday.

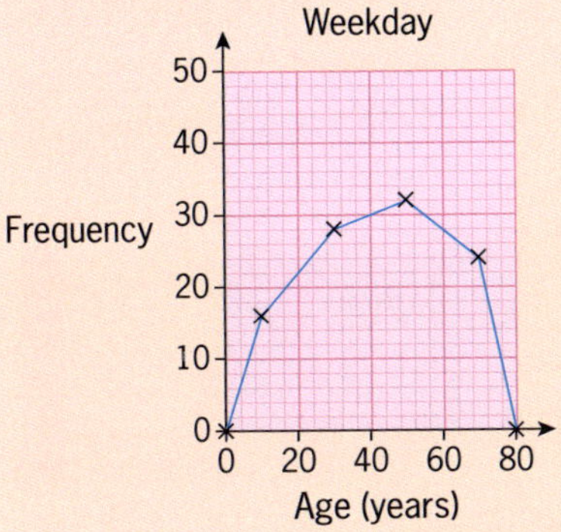

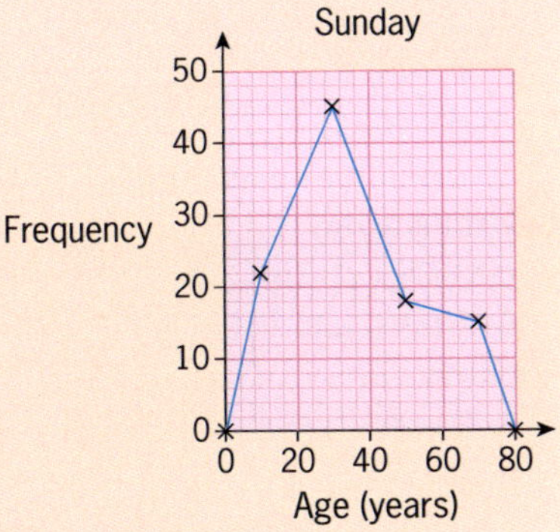

The garden centre believes that people in the 40–60 age range spend twice as much as people in the 20–40 age range, who, in turn, spend twice as much as people in the 60–80 age range.

The amount spent by the 0–20 age range is very small.

Comment, with reasons, whether the garden centre takes more money on a weekday or a Sunday.

Summary and assessment

CHECK OUT

You should now be able to draw and interpret

- bar charts
- frequency polygons
- pie charts
- histograms with equal class widths.

Exam-style question

The bar chart shows the number of pupils at school in a class each day for one week.

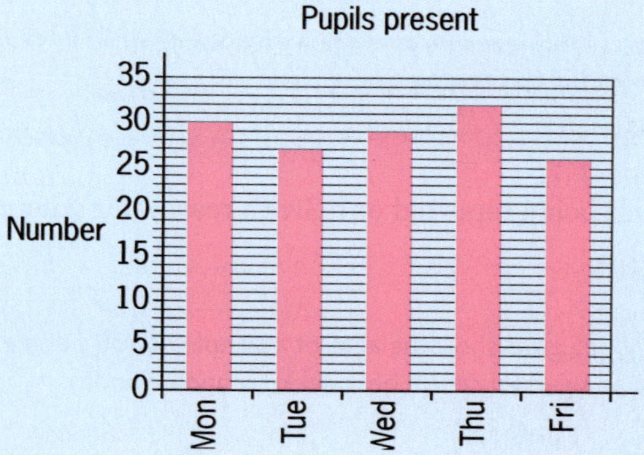

On Thursday of that week there were no pupils absent. Draw a bar chart to show the number of pupils absent in that class each day of that week.

Kyle's answer:

I need to know the number of absentees each day

On Thursday no pupils were absent; 32 is the total number of pupils in that class

Monday 32 − 30 = 2
Tuesday 32 − 27 = 5
Wednesday 32 − 29 = 3
Friday 32 − 26 = 6

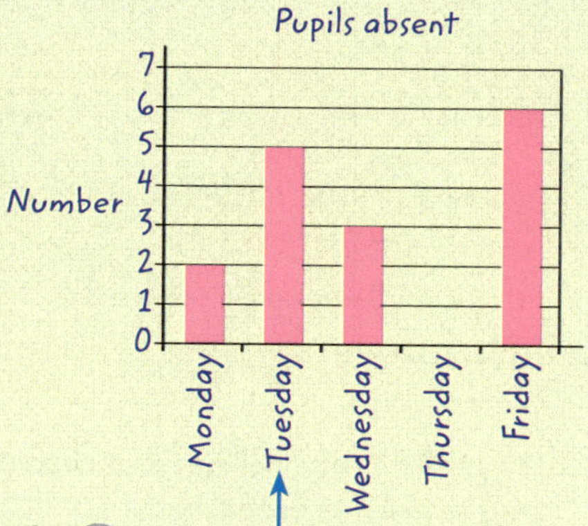

Think about the information you need to find.

Show your thinking and working out.

Use the information given in the question.

Finally, answer the question.

Exam practice

1 The mob run is a cross-country event that all Year 7 and Year 8 pupils take part in at a school. Chris summarised the times taken by the pupils to complete the mob run in this table.

Time (t, minutes)	Frequency
$15 < t \leq 20$	16
$20 < t \leq 25$	60
$25 < t \leq 30$	84
$30 < t \leq 35$	22
$35 < t \leq 40$	8

Draw a histogram to represent these data. *(3 marks)*

2 In a survey, 60 girls were asked how much they spent shopping one weekend.
The pie chart summarises their answers.

a How many of the girls spent between £10.01 and £20?

b What was the modal amount spent?

c Explain why you cannot find the actual range of amounts spent by the girls. *(3 marks)*

Girls' spending

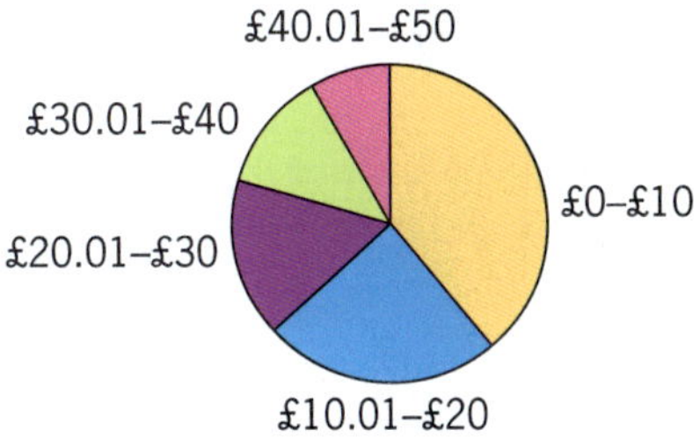

3 Reuben kept a daily record of the time in minutes he cycled during two months.
The table summarises the data for July.

Time (m, minutes)	Frequency
$0 < m \leq 5$	6
$5 < m \leq 10$	11
$10 < m \leq 15$	10
$15 < m \leq 20$	4

a Draw a frequency polygon for these data.

b The frequency polygon represents the data Reuben collected for August.
Write down one similarity and one difference between the daily times for which Reuben cycled during July and August. *(4 marks)*

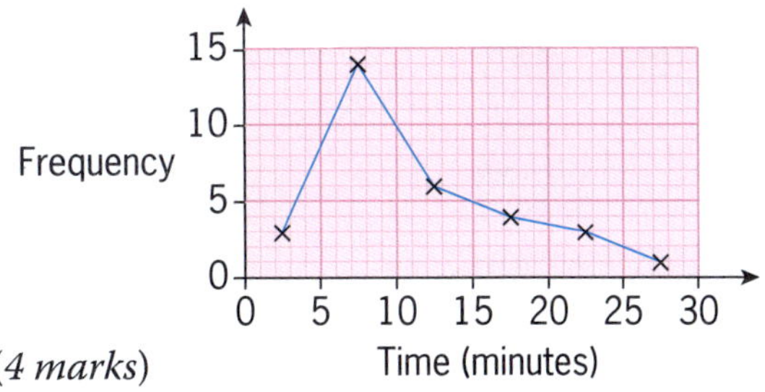

Case study 2: Sandwich shop

The manager of a catering company can use data about customer numbers in order to spot trends in customer behaviour and to plan for the future.

Simply Sandwiches

Customer numbers at 'Simply sandwiches' takeaway over a given two-week period were:

Day	Number of Customers	
	Week 1	Week 2
Monday	50	54
Tuesday	68	60
Wednesday	47	53
Thursday	58	57
Friday	52	56
Saturday	76	70
Total		

Simply Sandwiches

Use the data in the table to construct a bar chart to show how many customers visit the sandwich shop on each day during this two-week period.

Copy and complete the table.

During this two-week period,

a) which is the busiest day at the sandwich shop

b) what is the range of daily customer numbers?

This bar chart shows customer behaviour over a different two-week period. Compare the two sets of data, including reference to the quietest/busiest days and the spread of the data. Give possible explanations and justify your response with reference to the data.

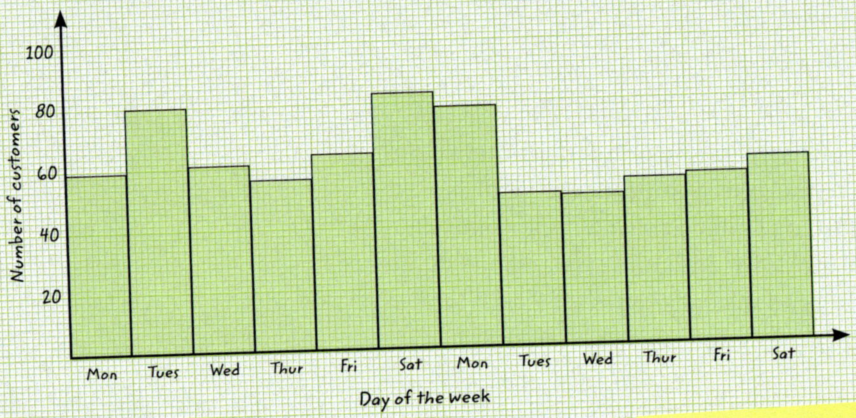

Is data collected over a two-week period enough to be able to estimate customer numbers for any given week? Justify your answer referring to the information.

A manager can use data about customer numbers to help estimate how much stock to order each week. In reality, limitations due to space and the shelf life of products also apply.

In the second week of the two-week period at 'Simply sandwiches', sales of the different varieties of sandwiches were:

Variety	Mon	Tues	Weds	Thurs	Fri	Sat	Total	Average
Ham	14	16	13	14	17	18		
Cheese	9	11	12	10	8	12		
Hummous	6	5	7	4	6	8		
Tuna	7	6	6	8	6	9		
Chicken	18	22	15	21	19	23		
Total								

Copy the table and fill in the missing values.

The manager does a weekly stocktake every Sunday before placing the order for the following week.

The stocktake figures for this week were:

Product	Stock (packs)	Portions per pack	Portions left	Stock needed	Amount to order
Bread	6	20			
Ham	2.5	10			
Cheese	3	10			
Hummous	2	14			
Tuna	1.5	8			
Chicken	1	10			

Copy and complete the table to show an estimate of how much of each product the manager should order to last for the following week.

The stock will be delivered on Wednesday morning.

Estimate how much (if any) of each product the manager will have in stock when the order arrives.

Comment on how well the manager estimated the order for the previous week.

Justify your answers by referring to the data.

7 Formulae and equations

INTRODUCTION

Formula 1 engineers use complex mathematical equations to predict the effect on performance of their cars when they make technical modifications. Mathematicians turn problems in the real world into mathematical equations which they know how to solve.

What's the point?

Learning how to solve equations allows complicated real-life problems to be solved.

CHECK IN

1 Find the value of each expression when **i** $n = 3$ **ii** $n = -2$

a $3n$ b $2n + 1$ c $2 - 4n$

2 Work out

a $4x \div 4$ b $3m \div 3$

c $6n \div 2$ d $8p \div 4$

Orientation

What I should know	What I will learn	What this leads to
Key stage 3	■ Solve equations ■ Derive formulae ■ Rearrange formulae	B11

RICH TASK

This grid of numbers uses each of the numbers from 1 to 9. Every row, column, and diagonal adds up to 15. It is called a magic square.

Using the numbers from 1 to 9, invent your own magic square.

4	9	2
3	5	7
8	1	6

Algebraic language

- Know the difference between an equation, a formula and an expression

→ **A term is a number or a letter, or a combination.**
5, x, and $7y$ are terms.

→ **An expression is a collection of terms separated by + or – signs.**
$5 + x$, $2x - 1$, and $3x + 2y - 5$ are expressions.

→ **An equation contains one or more expressions and an = sign. It is only true for particular values of the unknown letter.**

p. 280

ACTIVITY

Here is a balance. To find the value of x, you write an equation for the boxes and weights on each side.

The equation is $4x + 7 = 15$
What do you subtract from both sides to simplify the balance?
Do you agree that the **solution** of the equation is $x = 2$?

→ **A formula is a statement involving more than one letter usually written as an equation.**
It holds true for lots of different values of the letters.

p. 326

ACTIVITY

The cost £P of a coach journey between Derby and London depends on the number of adults A and the number of children C in the group where $P = \frac{21A + 8C}{2}$.

When there are 2 adults and 1 child, what values do you substitute for A and C?

Show that the total cost for the coach journey is £25.

Exercise A7.1

MEDIUM LOW

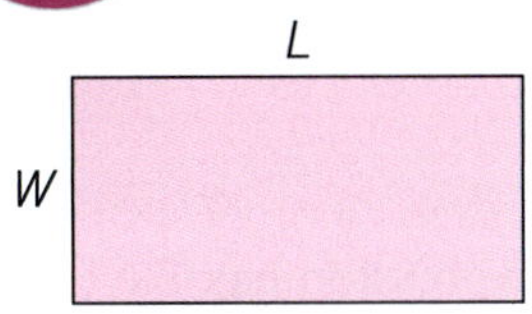

1 In each case, decide if you have an equation, a formula or an expression.

a A rectangle has a length L and a width W. Its perimeter P is given by $P = 2L + 2W$.

b Each box weighs x kg. If 2 boxes and 5 kg balance 9 kg, then $2x + 5 = 9$.

c $3x + 2x + 4$

2 In question 1 above,

a In part a, find P when $L = 4$ and $W = 3$.

b In part b, find the value of x.

c In part c, substitute $x = 2$ and find the value.

3 Match the words in the first box to the descriptions in the second box.

a	The cost £C of a taxi for M miles is $C = 1 + 5M$	Equation
b	The solution of $2x + 4 = 10$ is $x = 3$	Expression
c	Combine terms to make $3x + 4 + 5x - 1$	Formula

4 Would you use a formula, an expression, or an equation to find:

a the time to cook a chicken when you know its weight

b a simpler way of writing $3x + 2x + 5 - 6$

c the value of x when $3x + 1 = 10$?

5 a Substitute $z = 5$ into these formulae to find the values of P and Q.

i $P = 3z + 1$ ii $Q = \dfrac{2z+5}{3}$

b Solve these equations. i $3x + 1 = 7$ ii $3x - 1 = 14$

c Simplify i $3x + 4 + 2x - 1$ ii $6x - 5 - x + 4$

6 Invent two formulae, two equations, and two expressions of your own. Say which are which.

Unit A

Manipulating expressions

A7.2

- Collect 'like terms', expand brackets and find common factors

When an expression has **like terms**, they can be combined to make a simpler expression.

For example, $5x + 2x$ has 'like terms' which combine to give $7x$

$8x^2 - 2x^2$ also has 'like terms' which combine to give $6x^2$.

> Like terms contain the same letters raised to the same power.

EXAMPLE

Simplify $4x + 2y + 6 + x^2 - 2 + 5x - 3y + 7x^2$.

ANSWER

Recognise the 'like terms'. $4x + 2y + 6 + x^2 - 2 + 5x - 3y + 7x^2$

Collect 'like terms'. $= 9x - y + 4 + 8x^2$.

> Simplify $2y - 3y$ using a number line to get $-1y$ which is written as $-y$.

When an expression has brackets, you can **expand** the brackets by multiplication. Use think-lines to help you multiply.

For example, $3(2x - 5) = 3 \times 2x - 3 \times 5 = 6x - 15$

and $x(x + 4) = x \times x + x \times 4 = x^2 + 4x$

> These lines show you which terms to multiply.

The inverse of expanding brackets is **factorising**.

expand

$2(x + 2)$ → $2x + 4$

factorise

For example, $6x - 21 = 3(2x - 7)$ 3 is a factor common to 6 and 21.

$x^2 + 5x = x(x + 5)$ x is common to x^2 and $5x$.

> You can check by expanding:
> $3(2x - 7) = 3 \times 2x - 3 \times 7$
> $= 6x - 21$
> $x(x + 5) = x \times x + x \times 5$
> $= x^2 + 5x$

EXAMPLE

a Expand brackets in the expression $3(2x + 4)$.

b Expand and simplify the expression $x(x + 1) + 2(x - 1)$.

c Factorise the expression $7x - 21$.

ANSWER

a $3(2x + 4) = 3 \times 2x + 3 \times 4$

$= 6x + 12$

b $x(x + 1) + 2(x - 1) = x^2 + x + 2x - 2$

$= x^2 + 3x - 2$

c The common factor is 7.

$7x - 21 = 7(x - 3)$

Exercise A7.2

MEDIUM

1 This necklace is made from beads of length x and y.
How long is the necklace?
Give your answer in terms of x and y.

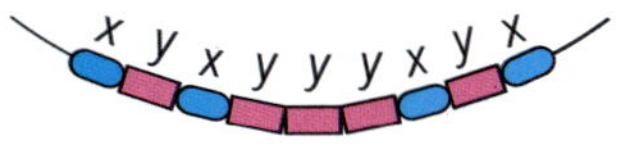

2 Simplify these expressions by collecting 'like terms'.

a $4x + 5y + 2x + 3y$
b $6x + 5x^2 + 2x + 3x^2$
c $9x + 7y - 5x - 4y$
d $8x^2 + 5 - 7x^2 + 3$
e $2m - 3n + 6m + 5n$
f $p + 3q + 4 + 5p - q + 3$
g $a + b + 1 + 5a - 3b + 8$
h $1 + x + x^2 + 6 + 7x + 8x^2$

3 Expand these brackets.

a $2(3x + 5)$
b $5(2x + 4)$
c $6(3y - 2)$
d $4(2z - 3)$
e $5(3 + 4a)$
f $2(5 - 3c)$
g $2(a + 5b)$
h $4(2x - 3y)$
i $3(6a - 2b + c)$
j $3(4 + 2x^2)$
k $5(3y^2 + 2y)$
l $4(1 + 2x + 3x^2)$

4 Expand these brackets and simplify the expressions.

a $2(4x + 3) + 3(5x + 2)$
b $6(2x + 3) + 2(3x - 4)$
c $5(p + 2) + 2(2p - 3)$
d $4(2q - 3) + 5(4 + 2q)$
e $3(m - 4) + 5(2m + 3)$
f $3(2 + 5n) + 2(2n - 3)$

5 Find the common factors and factorise these expressions.

a $6x + 8$
b $9x + 6$
c $8x + 12$
d $5x + 10$
e $3x + 9$
f $6x - 18$
g $12y + 8$
h $15a - 10$
i $4z - 2$
j $8c + 6d$
k $10x + 15y$
l $9s - 6t$

6 Simplify these expressions and factorise your answers.

a $6x + 7 + 4x + 8$
b $3y + 9 + 5y + 3$
c $8z + 5 + 8z + 11$
d $5p + 6q + 3p - 2q$
e $4a + 8b + 5a - 2b$
f $s + 4t + 7s - 2t$
g $9 + 3m - 1 + 5m$
h $6c - 7 + 12c + 1$
i $8 + 9z + 6 - 2z$

7 A rectangle has an area of $12x + 24$. What might its length and width be in terms of x?

8 **RICH TASK**

You have 4 beads of length x cm and 3 beads of length y cm.

a Write the length of the necklace that you can make with all these beads.

b How long can the necklace be if you have some beads left over?

Unit A

Substitution into formulae

A7.3

- Substitute into formulae in different contexts expressed in words and symbols

Formulae occur in many different aspects of life such as science, medicine, engineering and business.

ACTIVITY

In science, the energy E needed to send a current I through a light-bulb with resistance R is given by the formula $E = RI^2$.

To find the value of E when $R = 10$ and $I = 5$, you substitute these values into the formula.

When $I = 5$ what is the value of I^2?
Use your answer to find the value of 10×5^2.
So what is the value of E when $R = 10$ and $I = 5$?

EXAMPLE

Jack has a job which pays him a basic wage and then an hourly rate for every hour of overtime.
So his total earnings = basic pay + (hourly rate × number of hours overtime).

Calculate his total earnings if his basic pay is £210 per week and he then works 5 hours overtime at £6.20 per hour.

ANSWER

His total earnings for this week = 210 + (6.20 × 5) = 210 + 31
= £241

Exercise A7.3

MEDIUM LOW

1 Mrs Stubbs has some potted plants delivered to her home. She pays according to this formula:

Total cost, £ = Delivery charge + 2 × number of potted plants

How much does she pay if

a the delivery charge is £10 and she buys 6 plants

b the delivery charge is £15 and she buys 8 plants?

2 An electricity bill is calculated from this formula.

Total bill = Standing charge + (Cost per unit × Number of units used)

Work out the bill when

a the standing charge is £6.00 and you use 3500 units at 18p per unit

b the standing charge is £8.50 and you use 7450 units at 16.5p per unit.

3 Arnie Fisher works out the fuel consumption of his car in miles per gallon by filling his tank and reading the speedometer (which tells him how many miles he has driven) before and after a journey.

$$\text{Fuel consumption} = \frac{\text{speedometer at end} - \text{speedometer at start}}{\text{Number of litres used}} \times 4.5$$

a Find his fuel consumption on these three journeys X, Y and Z.

b Give a reason why journey Z is likely to give the most accurate result. Can you use the speedometer readings to calculate an even more accurate overall result?

		X	Y	Z
Speedometer readings	at start	32 450	32 510	32 635
	at end	32 510	32 635	32 839
Litres used		6.5	14.5	23.0

4 The voltage drop in an electric circuit is given by the formula

$V = e + Ri$

Find the value of V when

a $e = 12, R = 10, i = 3$

b $e = 6, R = 2, i = 12$

c $e = 0.4, R = 8, i = 2.4$

d $e = 5.2, R = 1.2, i = 3.5$

5 The velocity of a car as it accelerates is given by the formula.

$v = u + at$

Find the value of v when

a $u = 10, a = 2, t = 4$

b $u = 8, a = 6, t = \frac{1}{2}$

c $u = 1.2, a = 0.6, t = 12$

d $u = 4, a = -3, t = 2$

6 The energy of a moving object is given by $E = \frac{1}{2}mv^2$.

Find E when

a $m = 6, v = 4$

b $m = 5, v = 2$

c $m = 3, v = 5$

7 **RICH TASK**

The body-mass index (BMI) for an adult of height h metres and mass m kg is given by $\text{BMI} = \frac{m}{h^2}$.

The adult is underweight if BMI < 18.5.

The adult is overweight if BMI > 25.

What is the BMI for an adult of 1.67 metres and 71 kg?

Use a spreadsheet to calculate the BMI for different adults.

Rearranging formulae

A7.4

- Substitute into and then rearrange a formula

RICH TASK

A building plot has three sides of 12 m, 10 m and 16 m.
The fourth side has two gates x metres wide and a fence 4 m long.
What is the formula for the perimeter P of this plot?
If you simplify the formula by collecting terms, do you get the formula $P = 2x + 42$?

If the builder wants a perimeter of 50 metres, how wide should the gates be?
You need to find x knowing that $P = 50$.
If you substitute $P = 50$, what value of x do you get?

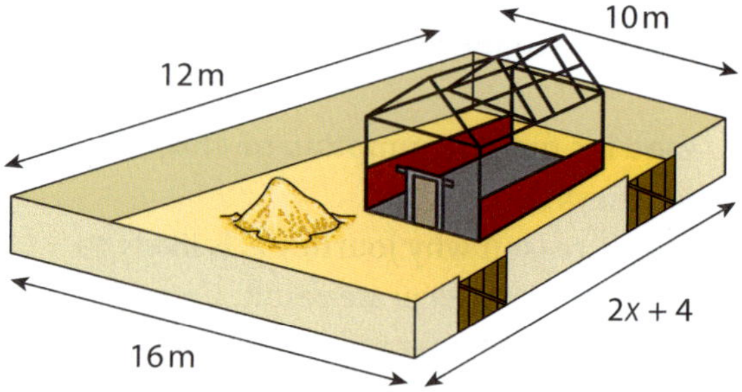

EXAMPLE

The formula to find the area of a triangle is

$$\text{Area} = \frac{1}{2} \times \text{base} \times \text{height}$$

Find the height of a triangle which has an area of 12 cm² and a base of 3 cm.

ANSWER

Substitute the area and base into the formula

$$12 = \frac{1}{2} \times 3 \times \text{height}$$

Now rearrange the formula to leave the height on its own.

Multiply both sides by 2 $\quad 24 = 3 \times \text{height}$

Divide both sides by 3 $\quad \frac{24}{3} = \text{height}$

So the height of the triangle is $\frac{24}{3} = 8$ cm

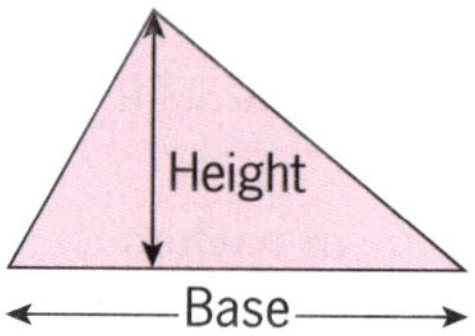

You will learn more about the area of a triangle in unit C.

Exercise A7.4

MEDIUM

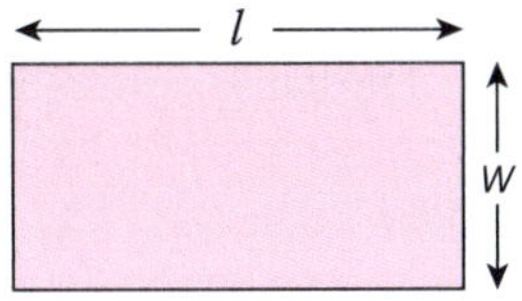

1 The area and perimeter of this rectangle are given by

$A = l \times w$ and $P = 2l + 2w$

Find

a w when $A = 12$ and $l = 6$
b w when $A = 15$ and $l = 10$
c l when $A = 20$ and $w = 5$
d l when $A = 33$ and $w = 6$
e w when $P = 18$ and $l = 5$
f w when $P = 32$ and $l = 6$

2 Voltages, currents and resistances in electrical circuits are related by various formulae. Here are two of them.

$V = RI$ and $E = V + RI$

Find

a I when $V = 10$ and $R = 5$
b I when $V = 18$ and $R = 3$
c R when $V = 20$ and $I = 4$
d R when $V = 48$ and $I = 12$
e V when $E = 18$, $R = 2$ and $I = 4$
f V when $E = 25$, $R = 6$ and $I = 3$
g I when $E = 32$, $V = 12$ and $R = 5$
h I when $E = 50$, $V = 18$ and $R = 8$

3 A spring has a length, L. When you pull on it with a force F, its length extends by a distance, x, where $F = \frac{kx}{L}$.

Find

a x when $k = 6$, $L = 10$ and $F = 3$
b x when $k = 15$, $L = 20$, $F = 3$
c k when $L = 8$, $x = 6$, $F = 24$
d k when $L = 1.2$, $x = 2.5$, $F = 9.5$

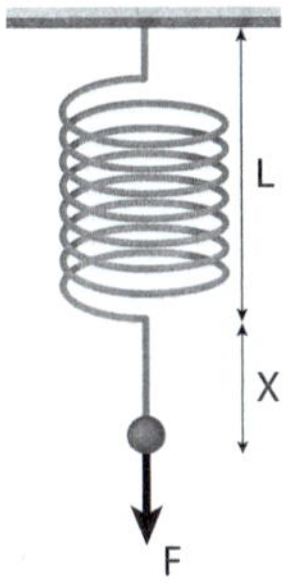

4** ***RICH TASK

Janina gets some of her maths homework wrong. Copy each of her answers, say what she did wrong, and help Janina to find the correct answers.

Find x when $z = 4x + 2y$ given that $z = 90$ and $y = 5$. $90 = 4x + 2 \times 5$ $= 4x + 10$ $90 + 10 = 4x$ $100 = 4x$ $x = 100 - 4 = 96$	Find q when $p = \frac{q+4}{r}$ given that $p = 12$ and $r = 5$. $12 = \frac{q+4}{5}$ $12 - 5 = q + 4$ $7 = q + 4$ $q = 7 + 4 = 11$	Find a when $A = \frac{1}{2}h(a + b)$ given that $A = 20$, $h = 6$ and $b = 1$. $20 = \frac{1}{2} \times 6 + (a + 1)$ $20 = 3 + (a + 1)$ $20 + 3 = a + 1$ $23 = a + 1$ $23 + 1 = a$ $a = 24$

Deriving formulae

A7.5

- Derive formulae in different contexts

RICH TASK

A square of side 10 cm has four small squares of side x cm cut away from its four corners.

Explore the area of the shape which is left for different values of x.

What happens when x is bigger than 5?

What happens if you start with a rectangle 10 cm by 6 cm?

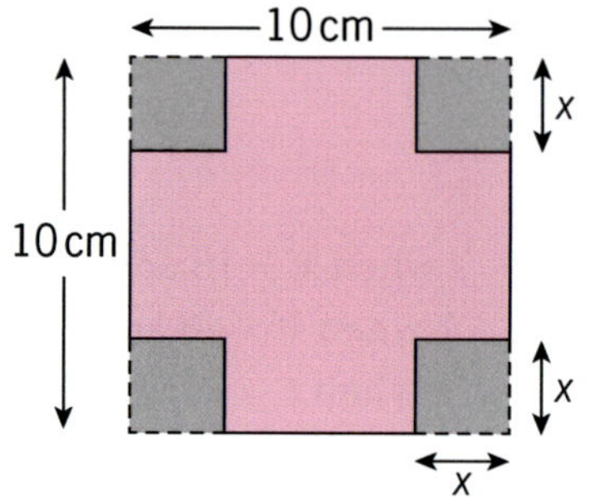

When one variable depends on other variables, you can often **derive** a formula which links them together.

EXAMPLE

A taxi fare is £1 fixed charge plus £2 for each mile of the journey.
Create a flow diagram and also a formula to calculate the fare £F for M miles.
How much is the fare for a journey of 4 miles?

ANSWER

Multiply the mileage by £2 and then add on the basic £1 charge.
So the flow diagram is

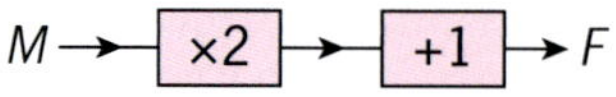

The formula is $F = 2 \times M + 1$

$$F = 2M + 1$$

When $M = 4$, substitute into the formula to get a taxi fare
$F = 2 \times 4 + 1 = £9$

Exercise A7.5

MEDIUM LOW

1 The number of tea-bags used by some people is given by:

'One bag per person and one for the pot'.

a Write a formula for the number, n, of bags used for x people.

b How many bags are needed for a pot of tea for 4 people?
Is this number sensible?

2 Find a formula in terms of x for the area of each of these shapes.
Write your answers as simply as you can. All lengths are in cm.

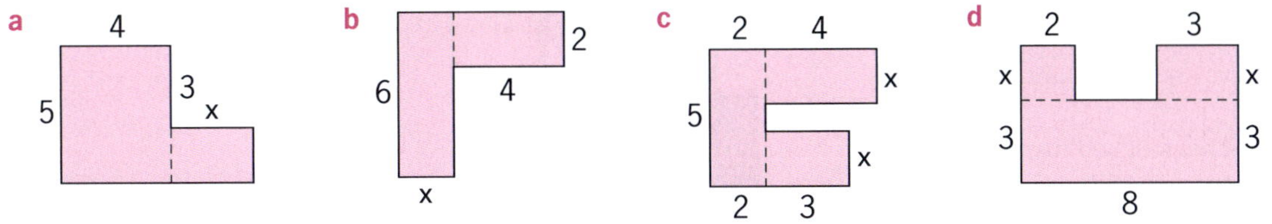

3 The milometer in Garry's car reads 45 670 at the start of a journey.

a What will be the final reading on his milometer if he drives for 4 hours at 60 mph?

b If he drives for x hours at an average speed of 50 miles per hour, write a formula for the final reading on his milometer at the end of the trip.

4 When Ahmed travels by air, he has to arrive at the airport 3 hours before the flight. When he lands, it takes him 2 hours to reach his destination. The aircraft flies at an average speed of 500 miles per hour for x hours.

a What is the total time of his journey if the flight takes 3 hours?

b Write a formula in terms of x for the total time of his journey.

5 **RICH TASK**

Think of any integer. Multiply it by 5 and then add 10.
Multiply your answer by 20 and add on your age.
Subtract 200.
What do you recognise in your answer?
Can you derive a formula to prove how it works?

Equations

A7.6

- Solve equations using inverse operations

RICH TASK

The Smith family goes on holiday by air.
The airline says that no item of luggage should weight more than 15 kg.
They have 3 identical suitcases each weighing x kg.
They have one other suitcase weighing 8 kg.
When all are weighed together, they weigh a total of 50 kg.
Is the Smiths' luggage allowed on the plane?

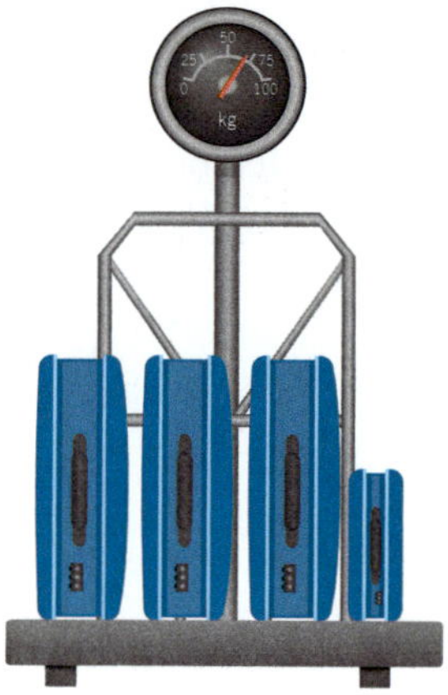

A balance with unknown weights can be written as an **equation**. Whatever you do to one side of the balance, you need to also do to the other side to keep it balanced. This is the same for an equation.

This balance gives the equation $2x + 3 = 15$

Subtract 3 from both sides $2x = 12$

Divide both sides by 2 $x = 6$

The solution is $x = 6$.

You can check your answer by substitution.
$2x + 3 = 2 \times 6 + 3$
$= 12 + 3 = 15$ ✓

In the above example, **inverse operations** were used to get x on its own on one side of the equation.

→ **The inverse of addition is subtraction.**
→ **The inverse of multiplication is division.**

EXAMPLE

Solve the equation $14 = 5x - 6$

ANSWER

Write the balance the other way round.

$5x - 6 = 14$

Add 6 to both sides $5x = 20$

Divide both sides by 5 $x = 4$

The solution is $x = 4$

When solving an equation, it is helpful to get all the unknown's (in this case, x's) on one side and all the numbers on the other.

Think $14 + 6 = 20$

$20 \div 5 = \frac{20}{5} = 4$

Exercise A7.6

MEDIUM LOW

1 Solve these equations using inverse operations.

a $x + 4 = 9$ b $x - 4 = 9$ c $2x = 10$ d $\frac{x}{2} = 10$

e $x + 3 = 4$ f $3x = 15$ g $\frac{x}{3} = 2$ h $x - 7 = 1$

i $x + 4 = 1$ j $\frac{x}{5} = 3$ k $6x = 18$ l $x - 5 = -1$

2 Find the value of x in each of these balances.

a

b

3 Solve these equations.

a $9 = 4x + 1$ b $40 = 7y + 5$ c $12 = 4z + 8$ d $14 = 6p - 4$

e $8 = 3q - 19$ f $2x + 7 = 10$ g $6 + 4n = 11$ h $7a + 8 = 8$

i $4y - 13 = 3$ j $6 = 4 + 4x$ k $9 = 7 + 4a$ l $2z - 32 = 1$

You often use letters other than x in algebra. Treat them in exactly the same way as you would treat an equation with x.

4 A delivery van contains 6 identical boxes, each weighing x kg, and a crate weighing 5 kg. The total mass of the contents is 47 kg. Find the value of x.

5 a I think of a number, n, double it and then subtract 7. If my answer is 11, find n.

b I think of another number, m. I treble it and add 12. Find m if the answer is 33.

6 Adam is x years old. His brother Shem is 5 years older than Adam. Their two ages add up to 29. Write an equation and find how old Adam is.

7 ***RICH TASK***

The number in each shaded box equals the sum of the two boxes beneath it.
Find the values of x and y.

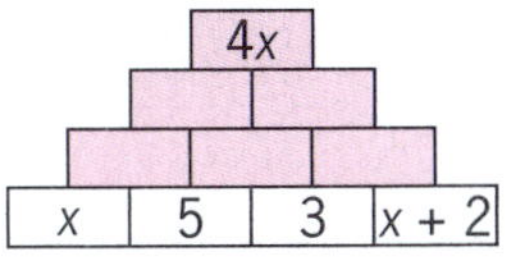

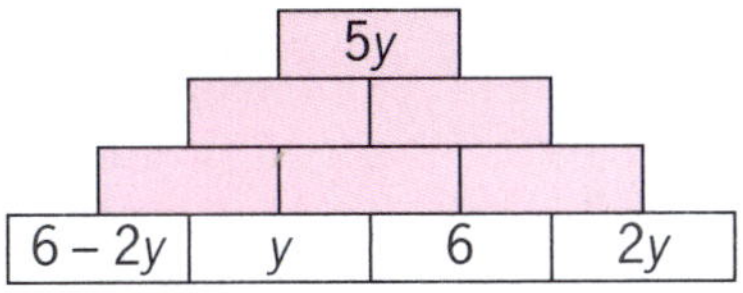

More equations

A7.7

- Solve equations with unknowns on either or both sides of the equation and solve equations with brackets

RICH TASK

Here is a tongue-twister of a puzzle.
A brick and a half weighs a kilo and half a brick.
How much does a brick weigh?

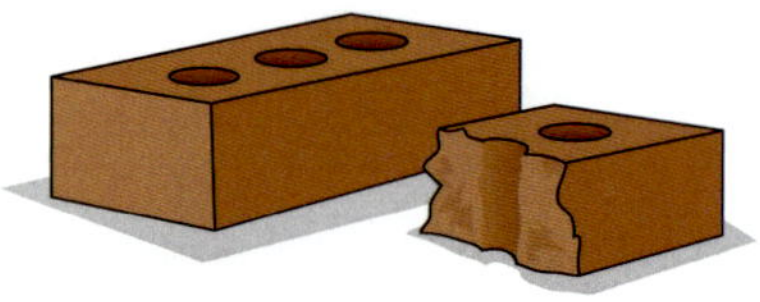

p. 212

Here is a balance with unknowns on both sides of the balance.
Can you find the value of x?

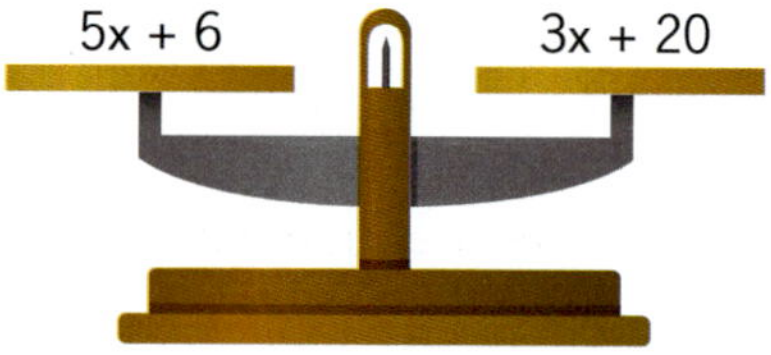

EXAMPLE

Solve these equations **a** $6x - 3 = 2x + 29$ **b** $16 - 4x = 4$

ANSWER

a To have x's only on the left-hand side, subtract $2x$ from both sides.

	$4x - 3 = 29$
Add 3 to both sides	$4x = 32$
Divide both sides by 4	$x = 8$

The solution is x = 8

b Move x to the right-hand side by adding $4x$ to both sides.

both sides.	$16 = 4x + 4$
Take away 4 from both sides	$12 = 4x$
Divide both sides by 4	$3 = x$
The solution is	$x = 3$

When an equation has brackets, expand the brackets and, if possible, simplify the equation before operating on both sides.

$3(x - 2) = 15$
$3x - 6 = 15$
$3x = 21$
$x = 7$

EXAMPLE

Solve the equation $2(3x + 7) + 3(x - 4) = 20$

ANSWER

Use think lines	$2(3x + 7) + 3(x - 4) = 20$
Expand the brackets	$6x + 14 + 3x - 12 = 20$
Collect 'like terms'	$9x + 2 = 20$
Subtract 2 from both sides	$9x = 18$
Divide both sides by 9	$x = 2$

The solution is x = 2.

The solution can be checked by substitution.
$2(3 \times 2 + 7) + 3(2 - 4)$
$= 2(13) + 3(-2)$
$= 26 - 6 = 20$ ✓

Exercise A7.7

MEDIUM

Unit A

1 Solve these equations where there are unknowns on both sides.

a $7x + 3 = 4x + 18$ b $6x + 2 = 4x + 9$ c $8x - 2 = 6x + 1$
d $4x - 6 = x + 1$ e $9x - 8 = 5x + 4$ f $6x - 1 = 4x + 11$
g $3x + 8 = 2x + 17$ h $5x + 6 = 2x + 15$ i $7x - 6 = 5x - 2$
j $9x - 8 = 5x - 3$ k $5x - 1 = 3x + 6$

2 Solve these equations.

a $2(3x + 4) = 20$ b $3(4x + 1) = 15$ c $5(2x - 3) = 25$
d $4(4x - 3) = 36$ e $2(5x + 2) = 14$ f $6(2x - 1) = 30$
g $3(2 + 3x) = 24$ h $5(1 + 2x) = 45$ i $4(6 + x) = 40$
j $7(2x - 5) = 49$ k $4(5 - 2x) = 16$ l $9(9 - 4x) = 36$

Take care with parts **k** and **l**.
The x terms are negative.

3 Solve these equations.

a $3(2x + 5) = 2x + 20$ b $8x - 7 = 2(3x - 1)$ c $9x = 2(4 + 3x)$
d $7(2x - 3) = 5(1 + 2x)$ e $2(7x - 2) = 3(2 + 3x)$ f $8x = 5(x - 1) + 8$
g $4(x - 3) = 3(2x + 1) - 5x$

4 Solve these equations. You will need to collect 'like terms'.

a $4(2x + 1) + 3(3x + 2) = 44$ b $2(5x + 2) + 5(x + 2) = 29$
c $3(4x + 3) + 4(2x - 2) = 1$ d $6(2 + x) + 2(1 - 2x) = 20$
e $5(2x - 3) + 4(5 - 2x) = 9$ f $4(3 - 2x) + 9x - 15 = 0$
g $2(8x - 6) + 9 - 12x = 1$ h $4(x + 2) - 2(x + 3) = 2$
i $7(x - 2) - 3 - 4x = 13$ j $7 - 3x + 2(6x - 4) = 35$

5 Elin has £x and she spends £4. She now trebles how much she has and finds she has £24. Write an equation and find the value of x.

6 Kawai earns £x an hour. She works for 30 hours and gets a bonus of £40. She earns £220 altogether. Write an equation for x and find her hourly rate of pay.

7 Mustafa has £x and he spends £12. He doubles how much he now has with a gift from his father. He then spends £10 and has £16 left. Write an equation and find the value of x. How much does his father give him?

8 ***RICH TASK***

The expression in a rectangle is the sum of the two expressions in the circles on either side of the rectangle. Find expressions for the contents of circles B and C. Write an equation in x and solve it to find the value of x.

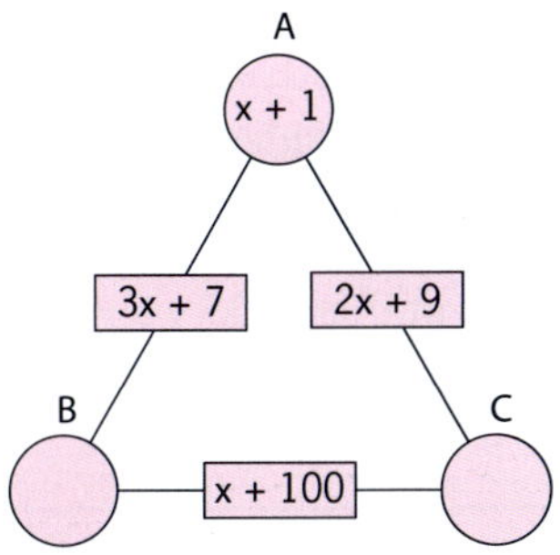

A mixture of equations

A7.8

- Solve a variety of different equations.

ACTIVITY

Simplify the equation $11 - 4x = 2$

Show that the solution is $x = 2\frac{1}{4}$

As a first step, what could you add to both sides to remove the $-4x$?

ACTIVITY

Expand the brackets in the equation

$2(4x + 3) - 3(x - 5) = 36$

Show that this equation simplifies to $5x + 21 = 36$.

Can you now show that the solution is $x = 3$?

ACTIVITY

Some equations have the unknown x as a numerator.

For example, to solve $8 = \frac{x}{3}$,

show that the solution is $x = 24$.

ACTIVITY

- Some equations have the unknown x as a denominator.
  For example, to solve $3 = \frac{12}{x}$,
  show that the solution is $x = 4$.

What is the inverse of *dividing by x*?

Check your answer by working out $\frac{12}{4}$.

EXAMPLE

Solve the equations

a $5(x - 1) - 2(x + 4) = 14$

b $5 = \frac{45}{x}$

ANSWER

a Expand the brackets to get $5x - 5 - 2x - 8 = 14$

Simplify $3x - 13 = 14$

Add 13 to both sides $3x = 27$

Divide both sides by 3 $x = 9$

b $5 = \frac{45}{x}$

Think of this as $45 \div x = 5$

What do you divide 45 by to get 5?

$x = 9$

Remember that $-2 \times +4 = -8$

Remember that $14 + 13 = 27$

Exercise A7.8

MEDIUM

Unit A

1 Solve these equations.

a $5 = \frac{x}{4}$ b $5 = \frac{x}{3}$ c $8 = \frac{x}{2}$

d $5 = \frac{x}{6}$ e $5 = \frac{35}{x}$ f $6 = \frac{18}{x}$

g $20 = \frac{100}{x}$ h $12 = \frac{60}{x}$ i $4 = \frac{12}{x} + 1$

j $5 = \frac{18}{x} + 2$ k $5 = \frac{36}{x} - 4$ l $8 = \frac{40}{x} - 12$

Hint for parts **i–l**:
Combine the number terms first.
So in **i**, $4 = \frac{12}{x} + 1$
$3 = \frac{12}{x}$

2 Solve these equations. Some answers are fractions and some are negative.

a $13 - 4x = 1$ b $12 - 2x = 4$ c $21 - 3x = 3$ d $8 - 2x = 5$
e $16 + 2x = 17$ f $21 + 4x = 26$ g $2x + 7 = 13$ h $3x - 2 = 8$
i $15 - 3x = 2x$ j $20 - 4x = x$ k $7 - x = x$ l $9 - 2x = 2x$
m $2x + 5 = 1$ n $3x + 8 = 2$ o $5x + 6 = 1$ p $15 + 4x = 3$
q $2x + 7 = 4$ r $2x + 15 = 1$ s $4x + 9 = 4$ t $5 - 3x = 9$

3 Solve these equations. Take care with the negative signs.

a $3(3x + 2) + 4(2x + 2) = 48$ b $4(2x + 3) + 3(2x - 3) = 17$
c $5(2x - 1) + 2(3x + 4) = 51$ d $4(3x - 5) + 3(2 - 3x) = 1$
e $3(5x + 2) - 2(3x - 2) = 19$ f $6(3x - 2) - 2(2x - 5) = 26$
g $8(3 - 2x) + 20x - 9 = 31$ h $5x - 2 - 3(6 - 2x) = 2$
i $2(2x + 5) - 3(x - 6) = 26$ j $2(1 + 6x) -2(5x - 4) = 4$

4 Solve these equations.

a $2(x + 3) = 4(2x - 3)$ b $5(x - 2) = 3(2x - 3)$ c $3(1 - x) = 4(7 - 2x)$
d $3(2 + x) = 5(2 + x)$ e $3(1 - 2x) = 4(x - 3)$ f $4(2 - x) = 5(x + 7)$
g $7(1 - x) = 5 - 3x$ h $13 - 5x = 9(2 - x)$ i $3(6x - 1) = 2(1 - x)$
j $8(3 + 2x) = 6(4 - x)$ k $6(2x + 3) = 9(7 - 2x)$ l $2(1 - x) = 6(2 + x)$

5 Mrs Desai had £80 and she spent £x on a dress. Mrs Patel had £120 and she bought two pairs of shoes at £x per pair. They both had the same amount left over. Find the cost of the dress.

6 Tom is x years old. Eve is 8 years younger than Tom. Tom is twice as old as Eve. How old is Tom?

Coordinates and midpoints

A7.9

- Find and use coordinates on a line and in a plane.

A number line starting at zero can extend in both directions.
You can think of the line in 1-D as an **x-axis**. Points *P* and *Q* have *x*-values of 3 and 7.
Is point *M* the midpoint of the line *PQ*?
How is the *x*-value for *M* linked to 3 and 7?

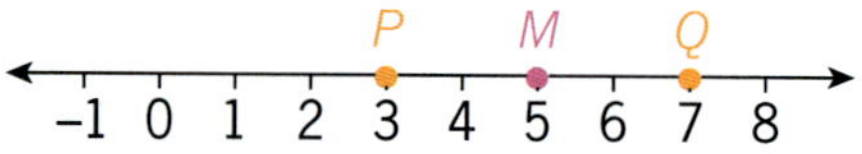

When writing coordinates you always write the *x* value first and then the *y*: (*x*, *y*).

p. 169

Two number lines (or **axes**) are used to fix a point in 2-D.
You can introduce a vertical number line called the **y-axis**.
What are the coordinates (*x*, *y*) for points *R* and *S*?
What are the coordinates for the midpoint *M* of the line *RS*?

Can you work out the coordinates of *M* from the coordinates for *R* and *S*?

Think of finding averages.

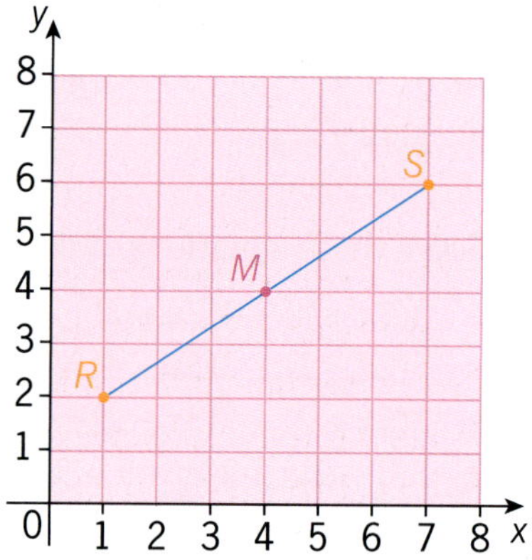

EXAMPLE

What are the coordinates of port *P* and harbour *H*?
Is town *T*, village *V* or airport *A* at the point (–4, –2)?
A lighthouse is exactly halfway between *P* and *H*.
On which island is it?
Can you calculate the coordinates of the lighthouse from the coordinates of *P* and *H*?

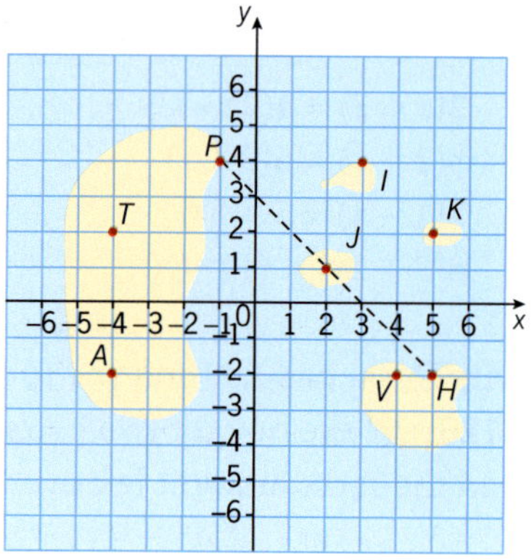

ANSWER

Port *P* is at the point (–1, 4).
Harbour *H* is at point (5, –2).
Point (–4, –2) gives the airport *A*.
By counting squares, the lighthouse is at point (2, 1), which is on island J.
For the lighthouse, $x = \frac{-1+5}{2} = \frac{4}{2} = 2$ and $y = \frac{4+-2}{2} = \frac{2}{2} = 1$
and these values check with the answer (2, 1).

Exercise A7.9

MEDIUM LOW

1 Write down the coordinates of

a the church
b the post office
c the bridge over the river
d the youth hostel
e the school

What features of this place are to be found at the point with these coordinates?

f (3, –2)
g (3, –4)
h (–2, –2)
i (–3, –5)

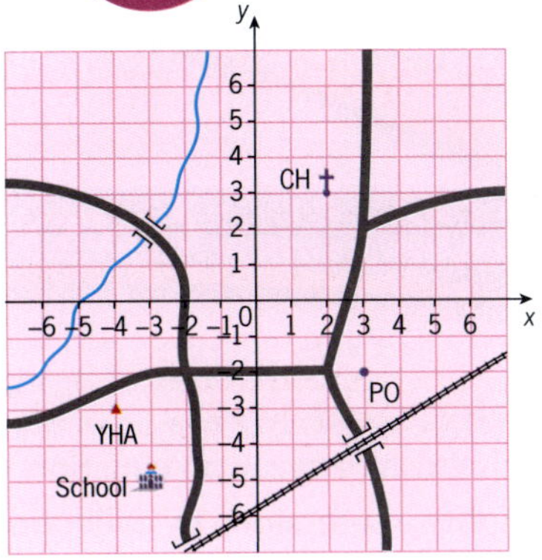

2 This map shows two power lines crossing open country. The pylons holding the power lines are shown by the symbol ^.

a Is pylon B at the midpoint of pylons A and C?
b Is pylon D at the midpoint of pylons C and E?
c Calculate the coordinates of
  i pylon B from those of pylons A and C
  ii pylon D from those of pylons C and E.

The power line PQR needs two extra pylons S and T. Pylon S is to be halfway between P and Q. Pylon T is to be halfway between Q and R.

d Calculate the coordinates of S and T.

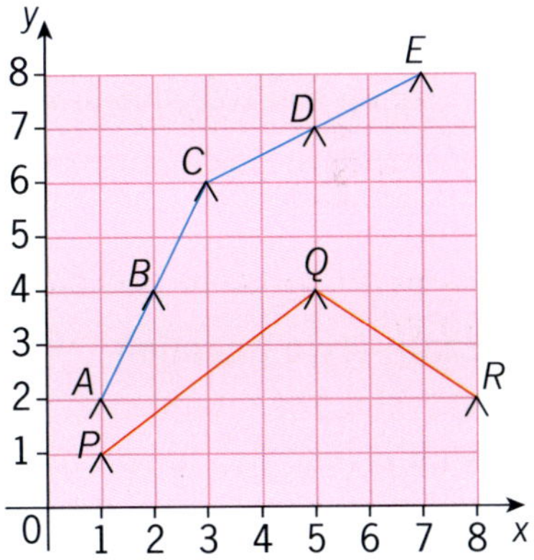

3 a M is the midpoint of the line segment AB. If M is the point (8, 7) and A is the point (6, 4), find the coordinates of point B.

b N is the midpoint of line segment KL. Given K(0, 1) and N(2, 5), find the coordinates of point L.

4 Calculate the midpoint of the points A(50, 80) and C(10, –30). Calculate the midpoint of the points B(110, 20) and D(–50, 30). What do your answers tell you about the quadrilateral $ABCD$?

5 P is the point (18, 4) and Q is the point (10, 24). Find the coordinates of

a the midpoint M of line segment PQ
b the midpoint N of the line segment PM.

How far along line segment PQ is point N?

Summary and assessment

A7

CHECK OUT

You should now be able to

- distinguish between an equation, a formula and an expression
- simplify expressions by collecting like terms
- factorise and expand expressions using brackets
- derive a formula
- change the subject of a formula
- set up and solve linear equations

Exam-style question

Mo thinks of a number x.

Multiplying the number by 5 and subtracting 4 gives the same answer as multiplying the number by 3 and adding 2.

Write down an equation in x and solve it to find the number.

Read the information in the question and write it out using maths symbols.

Alex's answer:

$x \rightarrow \times 5 \rightarrow -4$

$x \rightarrow \times 3 \rightarrow +2$

$5x - 4 = 3x + 2$

$5x - 3x = 2 + 4$

$2x = 6$

$x = 3$

You may need to write the information as a diagram first to help.

Collect like terms, showing your working to solve the equation.

Exam practice

1 a Simplify $a + 2a + b + 3b$

b Factorise **i** $5x - 15$ **ii** $x^2 + 4x$

c Solve $5x - 3 = 20$ *(5 marks)*

2 This is a formula to change British shoe size, *b*, into European shoe size, *e*.

$$e = \frac{5}{4}b + 32$$

Rearrange the formula to make *b* the subject. *(3 marks)*

3 Shoshana is thinking of an expression.
The diagram shows the input and output numbers for this expression

Input		Output
5	→	7
3	→	3
8	→	13

a Could the expression that Shoshana is thinking of be $n + 2$? Explain your answer.

b Write down Shoshana's expression. *(3 marks)*

8 Bearings and Pythagoras

INTRODUCTION

When a ship sails across an ocean it sets course using a bearing. During a typical voyage, constant adjustments are made to the bearing to allow for the effects of wind, currents, and other ships.

What's the point?

If you visit a new town and get lost, you can use your mobile phone and, with directions and landmarks, find your way. But there are some situations, such as crossing an ocean, where there are no landmarks to guide you!

CHECK IN

1 State the sum of the three angles x, y, z.

a

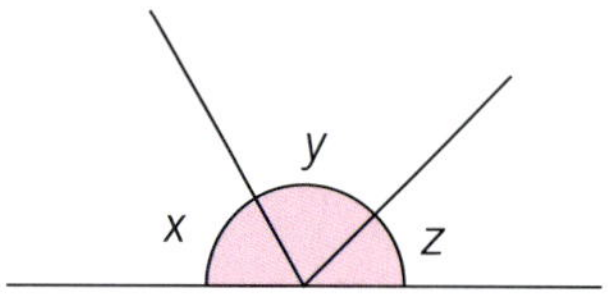

b

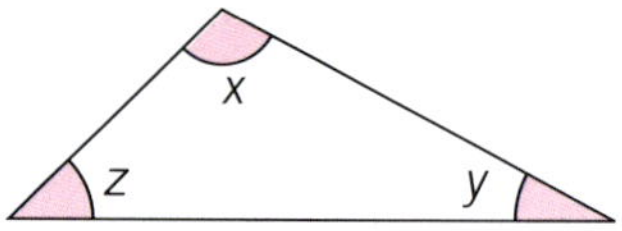

c

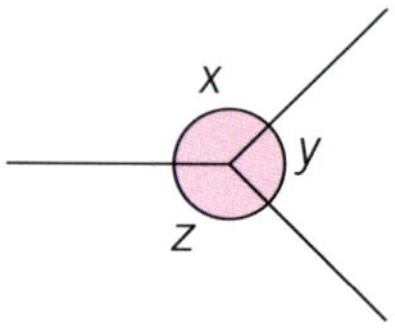

2 Find the value of each unknown angle.

a

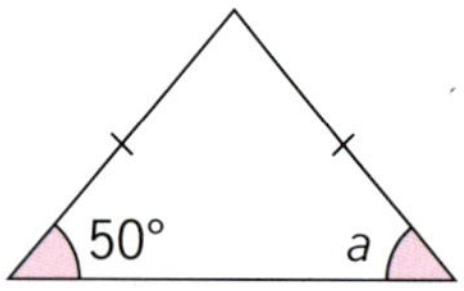

b

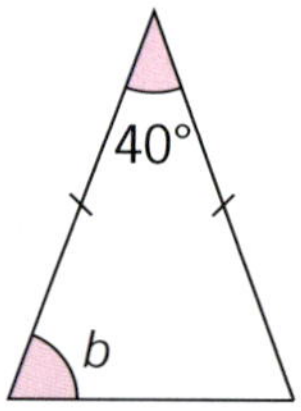

Orientation

What I should know	What I will learn	What this leads to
A3 Key stage 3	■ Use Pythagoras' theorem ■ Specify bearings	Careers in architecture and graphic design

RICH TASK

Two ships, on the same line of latitude, are both travelling at a speed of 10 km/h. The first ship is on a bearing of 030° (angle a) and the second ship is on a bearing of 330° (angle b).

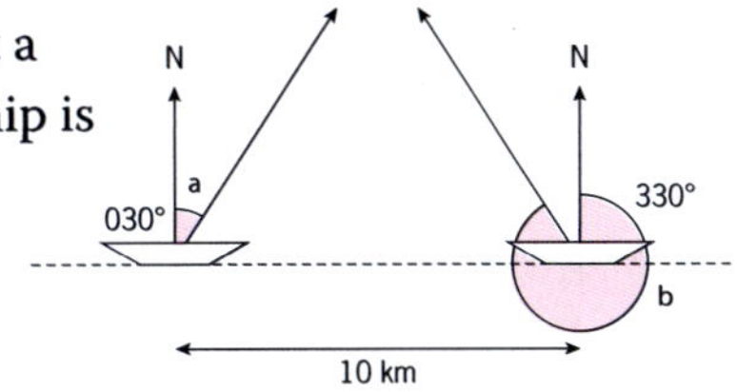

Construct a scale diagram to show where the two ships meet. On the same diagram make a construction to show where the ships would have met if the first ship increased its bearing by 10° and the second ship decreased its bearing by 10°. Investigate.

Maps and bearings

A8.1

- Interpret maps and use bearings

A **bearing** is always measured **clockwise** from the **north**.

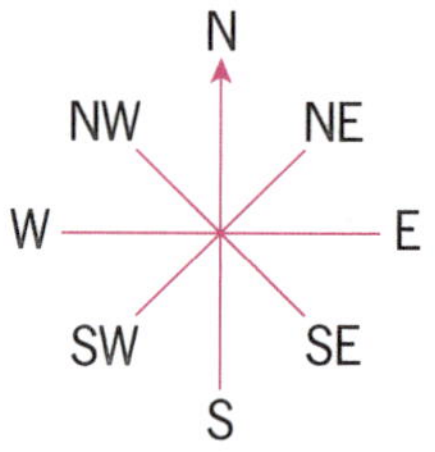

A bearing always needs three figures. For example, a bearing of 45° is written 045°.

EXAMPLE

Give the positions of the church and the pylon from you.

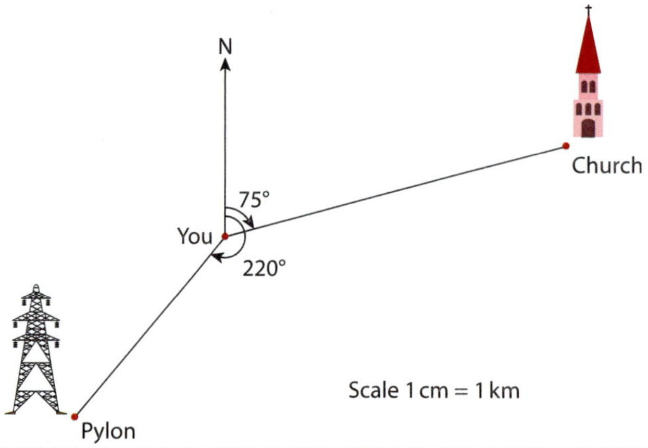

ANSWER

Using the scale, the church is 4.5 km from you on a bearing of 075° and the pylon is 3.0 km from you on a bearing of 220°.

You can also use bearings when you are on a journey.

EXAMPLE

Give directions for an aircraft flying from Cardiff to Oxford via Gloucester.

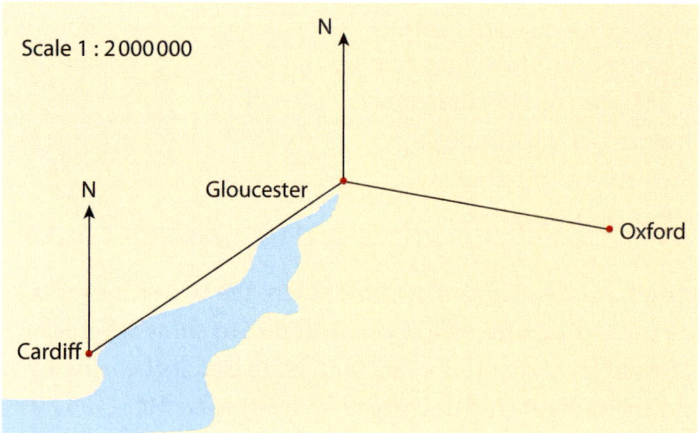

ANSWER

1 cm on the map stands for 2 000 000 cm = 20 000 m = 20 km in reality.
The plane flies 4 × 20 = 80 km on a bearing of 056° from Cardiff to Gloucester, and then flies 3.5 × 20 = 70 km on a bearing of 100° from Gloucester to Oxford.

To convert cm to km,
÷ 100 cm → m
÷ 1000 m → km

Exercise A8.1

MEDIUM LOW

Unit A

1 Place tracing paper over the map and mark the positions of these towns.

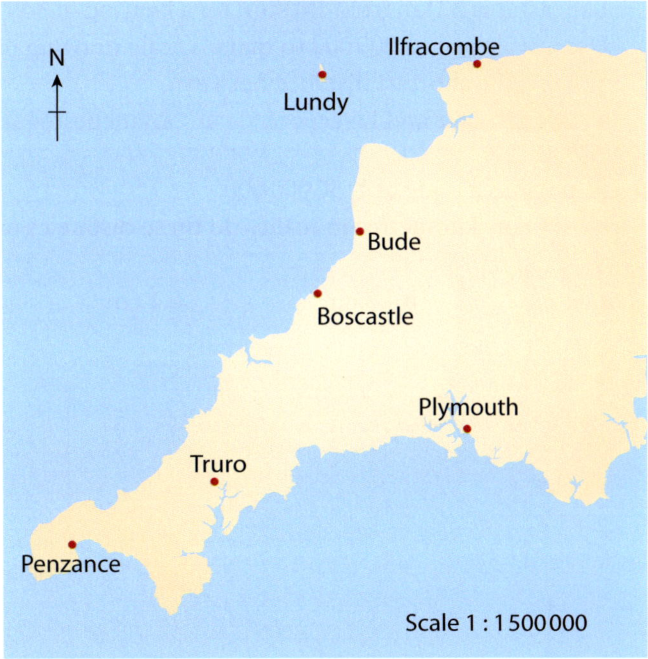

   a Use a protractor to find the bearing from Boscastle of
      i Ilfracombe   ii Plymouth
      iii Truro   iv Penzance
      v Bude   vi Lundy.
   b Find the bearing of
      i Truro from Penzance
      ii Penzance from Truro.
   c How many kilometres does 1 cm on the map stand for?
   d Find the shortest distance in kilometres between
      i Truro and Bude
      ii Penzance and Boscastle
      iii Lundy and Bude
      iv Plymouth and Ilfracombe.

2 Give a 3-figure bearing for these compass directions:
   a due south   b north-east   c south-west.

3 A bus stop is 30 metres from you on a bearing of 080°.
   A post box is 40 metres from you on a bearing of 120°.
   A gate is 35 metres from you on a bearing of 140°.
   Use a scale of 1 cm = 10 metres to find the distance between
   a the bus stop and the gate
   b the gate and the post box.

4 From a hill top, you see a cathedral 20 km from you on a bearing of 046°. You also see a power station 47 km from you on a bearing of 137° and a television mast 32 km from you on a bearing of 165°.
   Make a scale drawing showing these places to a scale of 1 cm = 10 km.
   Find the shortest distance from the power station to
   a the television mast
   b the cathedral.

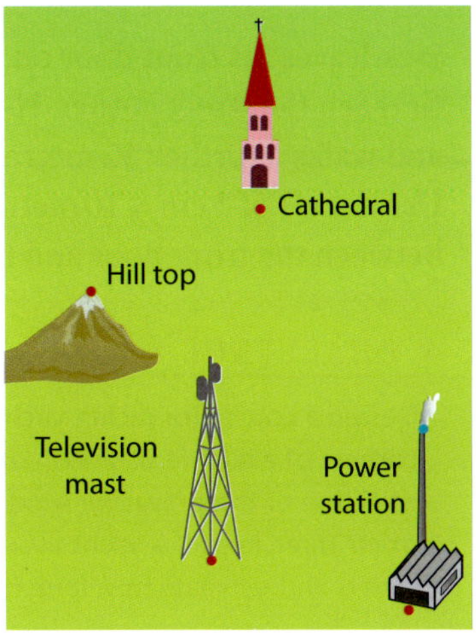

Continued ➤

Unit A

5 You live in Preston, Lancashire.
Manchester is 40 km from Preston on a bearing of 140°.
Liverpool is 45 km from Preston on a bearing of 210°.
Lancaster is 32 km from Preston on a bearing of 345°.
Use a scale of 1 : 1 000 000 to make a scale drawing of these cities.
What is the shortest distance between

a Manchester and Liverpool b Manchester and Lancaster?

6 A map has a scale of 1 : 50 000 000.
How many kilometres in reality do these distances on the map represent?

a 1 cm b 5 cm c 6.4 cm

7

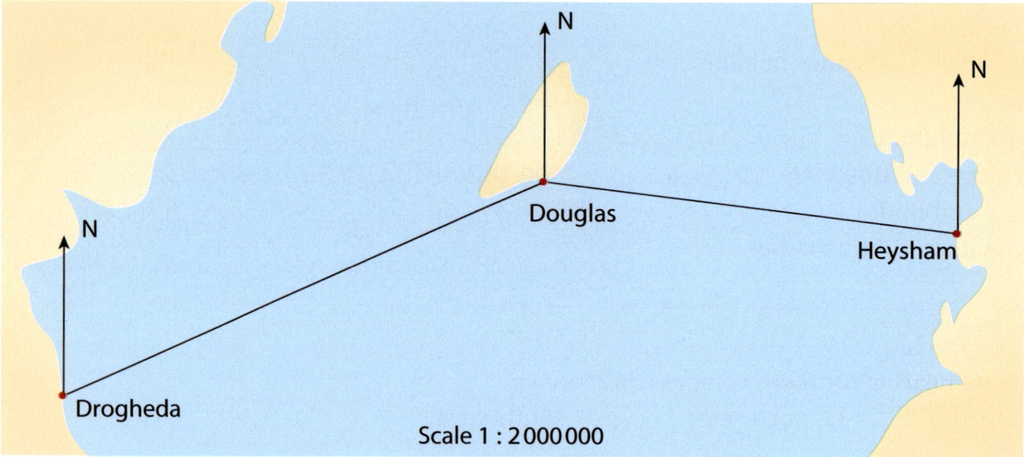

a A fishing boat sails from Drogheda to Heysham via Douglas as on this map. Describe its route, giving distances and bearings for each leg.

b If an aeroplane flew direct from Heysham to Drogheda, what bearing would it take?

8 Jack leaves his front door on a bearing of 100° and walks 65 metres to reach the street corner. He then turns onto a bearing of 040° and walks a further 45 metres to arrive at the bus stop.
Use a scale of 1 cm = 10 metres to find the shortest distance between the front door and the bus stop.

9 ☆ CHALLENGE ☆

Design a course for racing yachts. The yachts must sail round the corners of a square of side 5 km. The first leg is in the general direction of the prevailing wind (say 060°). Give the bearings of the other three legs. If a yacht averages a speed of 10 knots, find what a *knot* is and estimate how long it takes to get round the course.

A8.2 Pythagoras' Theorem

- Understand and use Pythagoras' Theorem in 2-D

RICH TASK

A floor is covered with small triangular tiles.
Patterns are drawn on the floor.
Each pattern has a coloured triangle with squares on its sides.
Count the tiles in each square for each pattern.
Enter your

	Number of tiles inside		
Pattern	a small square	the other small square	the large square
A			
B	8	8	16
C			
D			

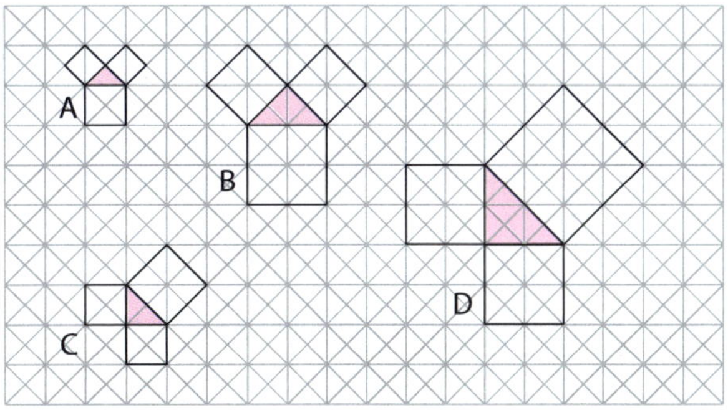

Draw some more patterns of your own like these.
What do you notice about the total number of tiles in the two small squares?

You use Pythagoras' Theorem to calculate the length of one side of a right-angled triangle when you know the length of the other two sides.

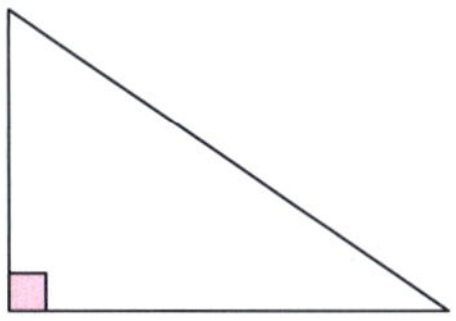

ACTIVITY

Draw this diagram of three squares X, Y and Z accurately.

Split Y into four equal parts as shown.

Cut out X and the four parts of Y.

Can you arrange these five pieces to fit inside square Z?

What can you now say about the areas of X, Y and Z?

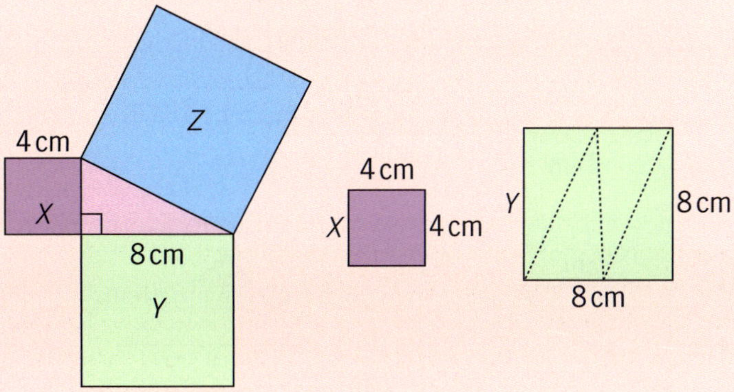

- Pythagoras' Theorem says that: the area of the square on the **hypotenuse** is equal to the sum of the areas of the squares on the other two shorter sides.

  Area of Z = Area of X + Area of Y

  $c^2 = \quad a^2 \quad + \quad b^2$

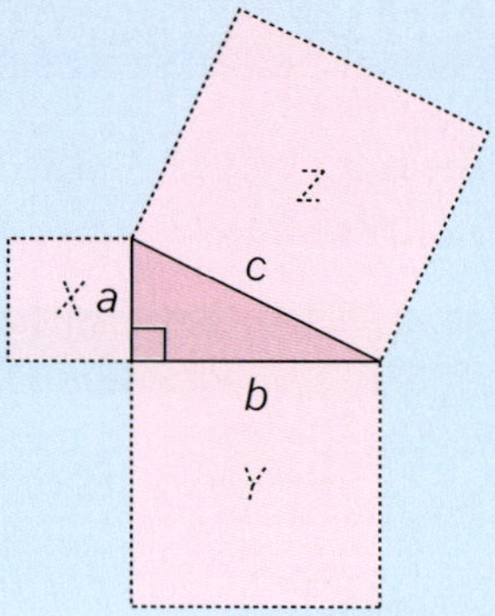

The **hypotenuse** is always the longest side of a right-angled triangle.

Area of Z is $c \times c = c^2$, and so on.

EXAMPLE

Find the length c.

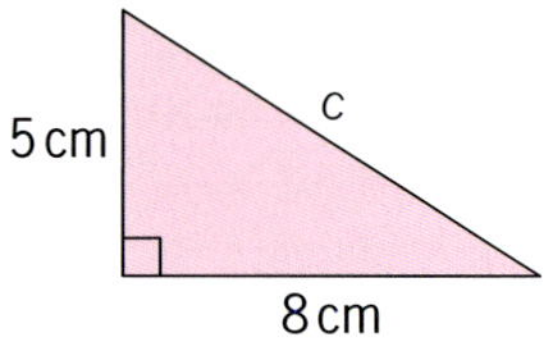

ANSWER

Pythagoras' Theorem gives $c^2 = 5^2 + 8^2$

$= 25 + 64 = 89$

So $c = \sqrt{89} = 9.4$ cm to 3 d.p.

Use the square root key $\sqrt{\ }$ on your calculator.

Exercise A8.2

MEDIUM

1 Calculate the length of each hypotenuse in these triangles.
Give your answers (in cm) as integers or to 1 decimal place when necessary.

a 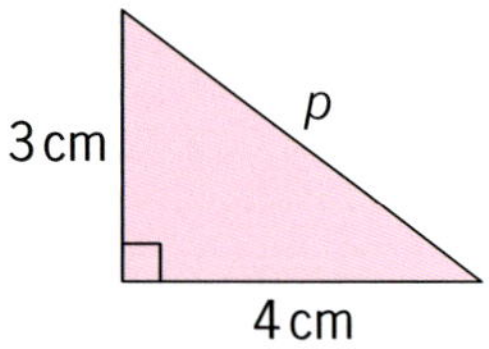

b 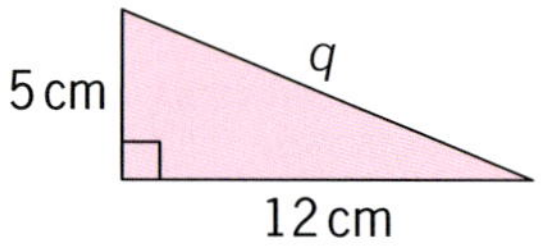

c 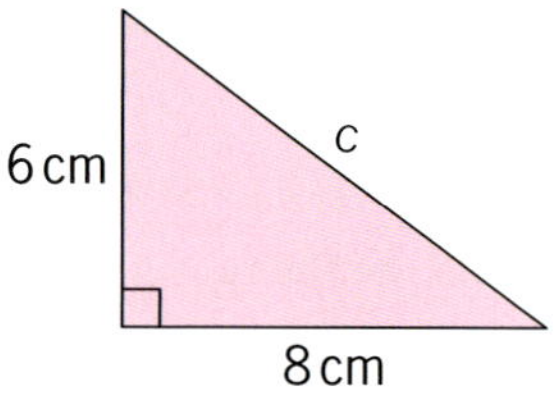

d 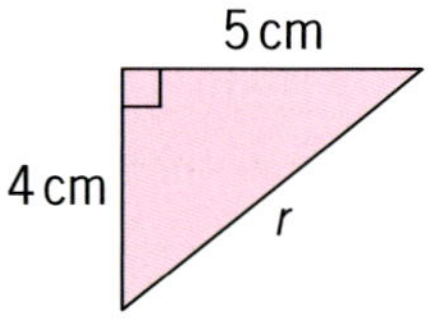

e 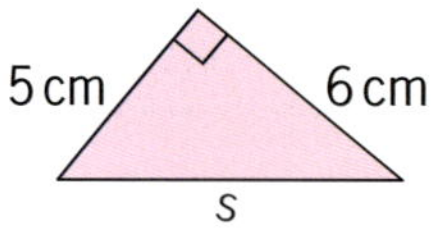

f 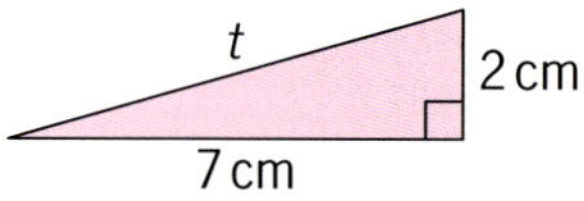

2 A telegraph pole 6 metres tall is held in place by a sloping wire as shown.

Calculate the length, x, of the wire.

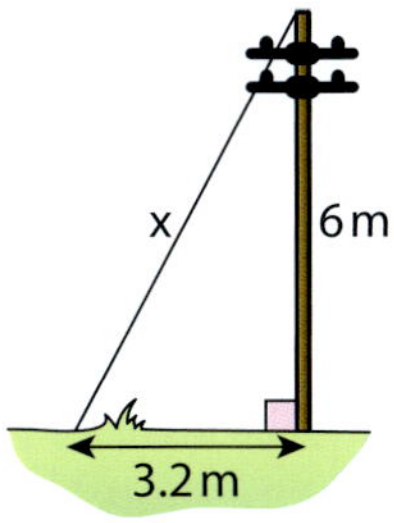

3 A ladder leans against a wall to reach a window.

Calculate the length, y, of the ladder.

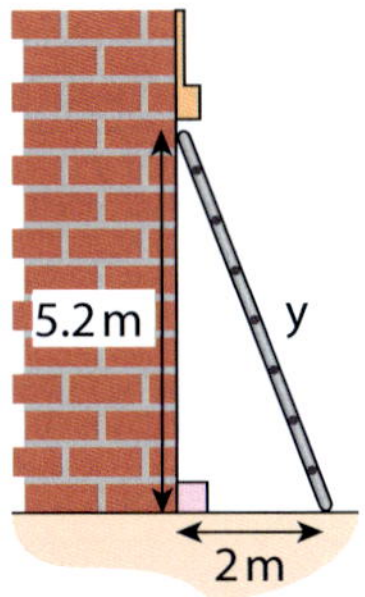

4 A hiker walks 6 miles due north and then 2.5 miles due east.

Calculate the distance, d, that she now is from her start.

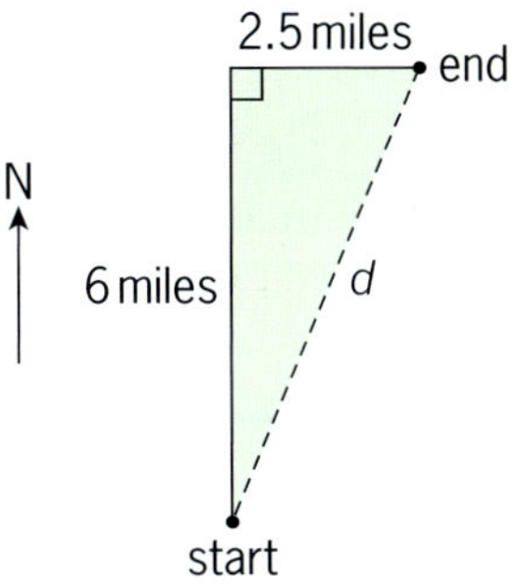

5 A skier drops a vertical distance of 0.9 km as he skis down a steep hill. Calculate the actual distance, d, that he skis down the slope.

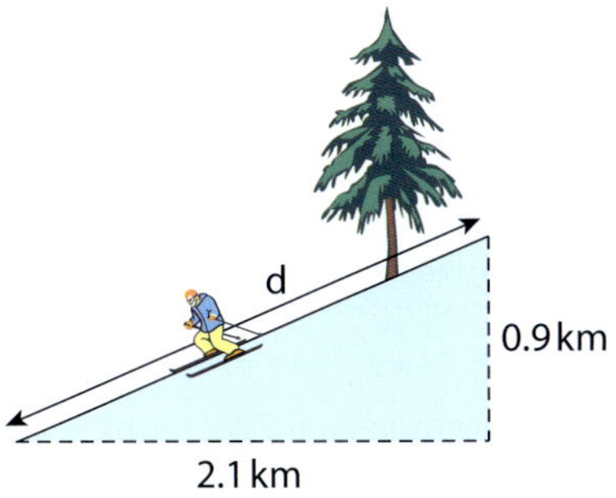

6 Jim and Jan both calculate the length x. Find the errors they made in their working and correct them.

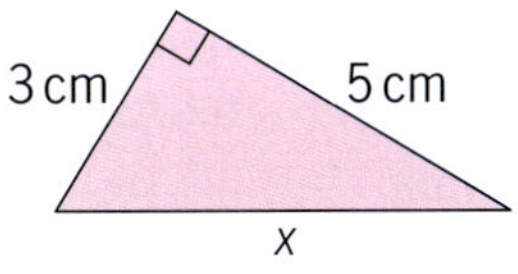

Jim	Jan
$x^2 + 3^2 = 5^2$	$x^2 = 3^2 + 5^2$
$x^2 + 9 = 25$	$= 9 + 25$
$x^2 = 16$	$= 34$
$x = 4$ cm	$x = 34$ cm

Unit A

More about Pythagoras' Theorem

A8.3

- Use Pythagoras' Theorem to calculate any side of a right-angled triangle

You can use Pythagoras' Theorem to find one of the shorter sides of a right-angled triangle when you know the other two sides.

EXAMPLE

Calculate the length a in this right-angled triangle.

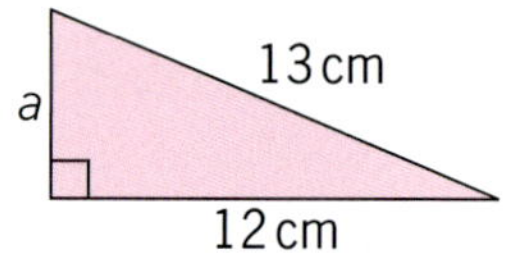

ANSWER

The hypotenuse in this triangle is 13 cm.

Pythagoras' Theorem gives $13^2 = a^2 + 12^2$

$$169 = a^2 + 144$$

Solve this equation by subtracting 144 from both sides.

$$a^2 = 25$$

So $a = \sqrt{25} = 5$ cm.

$13^2 = 13 \times 13 = 169$
$12^2 = 12 \times 12 = 144$

$169 - 144 = 25$

You can use Pythagoras' theorem to solve real-life problems.

EXAMPLE

A hiker at the village of Anlaby, A, is 8.7 km as the crow flies from Byfleet, B.
If he walks 3.6 km due south from A, calculate the distance, x, that he must then walk to reach B, which is due east.

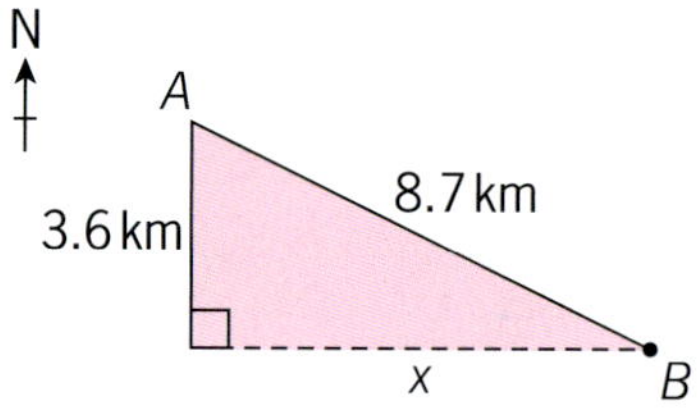

ANSWER

The hypotenuse is 8.7 km long.

Pythagoras' Theorem gives $8.7^2 = x^2 + 3.6^2$

$$75.69 = x^2 + 12.96$$

Subtract 12.96 from both sides $x^2 = 62.73$

So $x = \sqrt{62.73} = 7.9$ km

The hiker has 7.9 km more to walk.

Exercise A8.3

MEDIUM

1 Calculate the length of the lettered sides in these triangles.
Give your answers (in cm) as integers or to 1 decimal place.

a
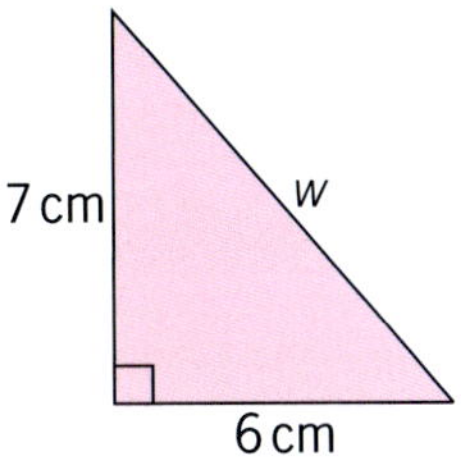

b
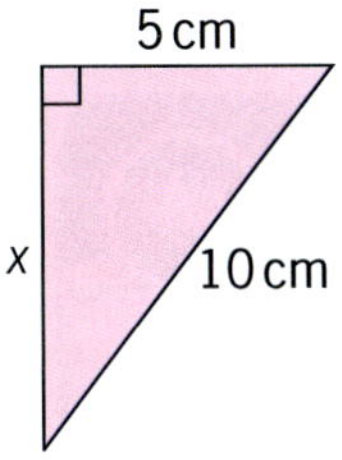

c
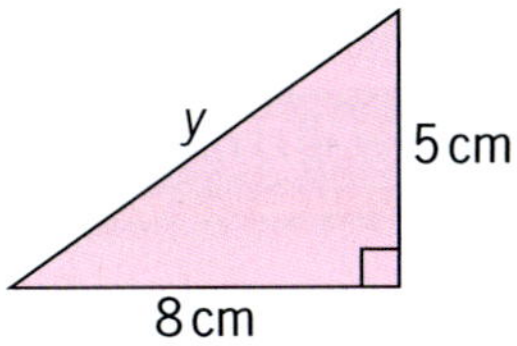

d
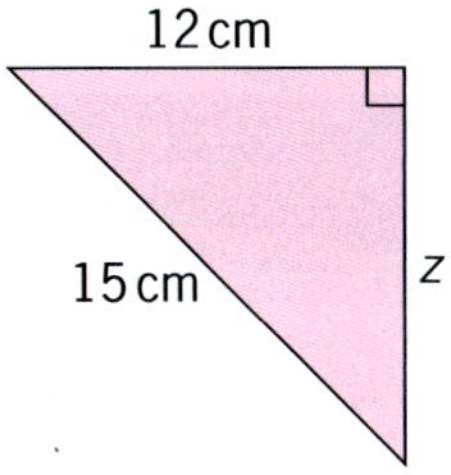

2 Calculate the length of the lettered sides in these triangles.
Give your answers (in cm) as integers or to 1 decimal place.

a
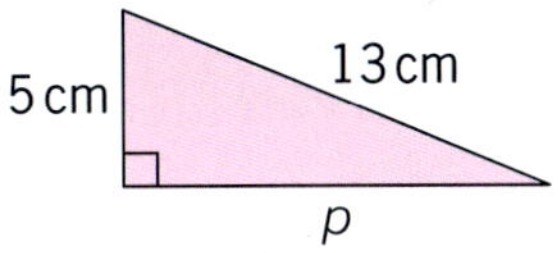

b
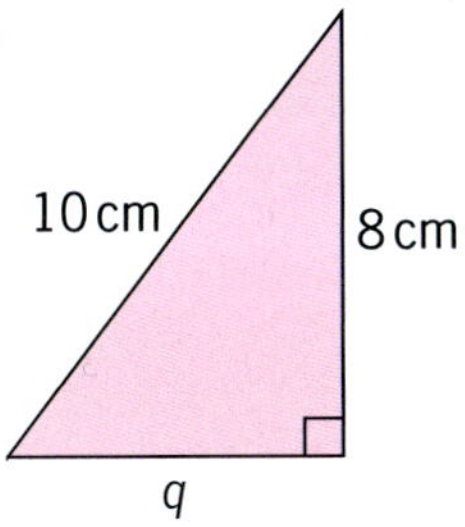

c
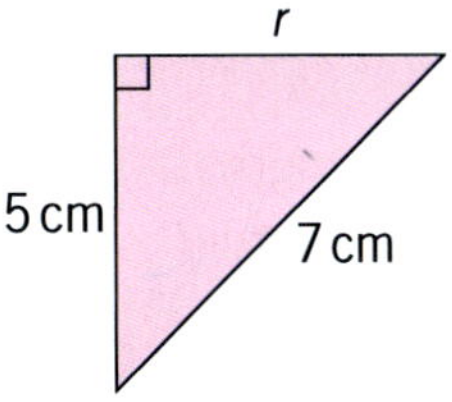

d
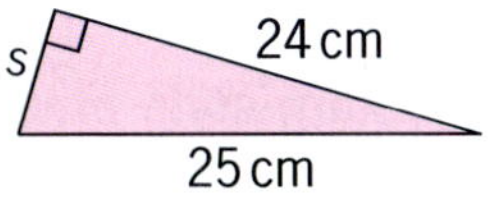

3 A 4-metre long wooden plank leans 2.6 metres up against a wall. Calculate the distance, x, between the bottom of the ladder and the bottom of the wall.

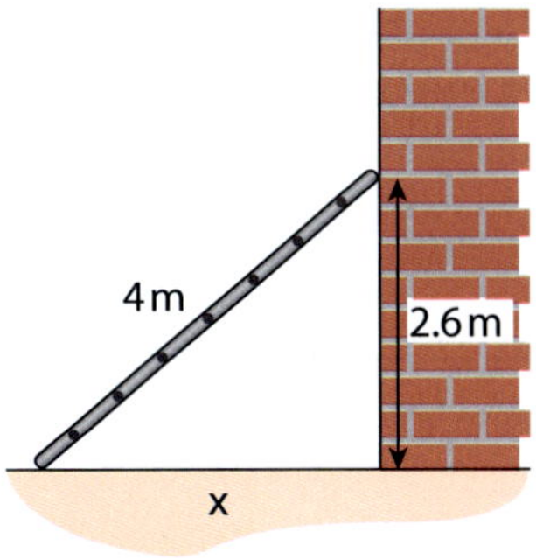

4 An Alpine railway 5.1 km long climbs a mountain as it covers a horizontal distance of 4.9 km.
Calculate the height, h, of the mountain (correct to 2 significant figures).

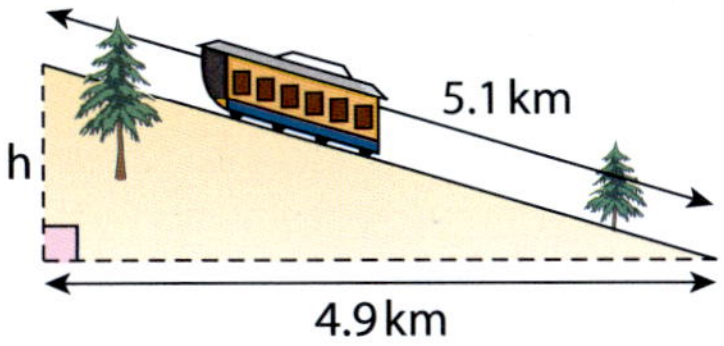

5 A factory chimney AB is 28 metres tall and casts a shadow AC. If BC = 35 metres, find the length of the shadow AC.

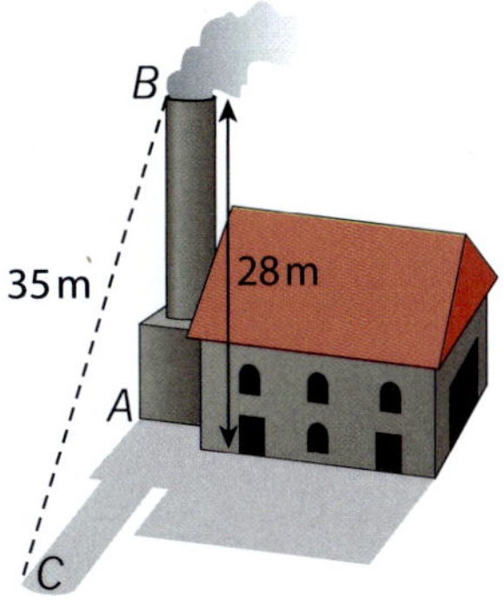

Summary and assessment

CHECK OUT

You should now be able to

- use bearings and scale drawing
- use Pythagoras' Theorem in 2-D.

Exam-style question

A triangle is drawn with side lengths 3·5 cm, 8·4 cm and 9·1 cm.
Is the triangle a right-angled triangle?

Clara's answer:

$3.5^2 + 8.4^2 = 12.25 + 70.56$
$= 82.81$

$9.1^2 = 82.81$

The answers are the same

$3.5^2 + 8.4^2 = 9.1^2$

Yes the triangle is right-angled

A right-angled triangle will fit Pythagoras' Theorem.

Use your results to write out an answer to the question.

Square the two shorter sides given, then add the squared numbers.

Square the longest side.

Exam practice

1 ABC is a right-angled triangle.

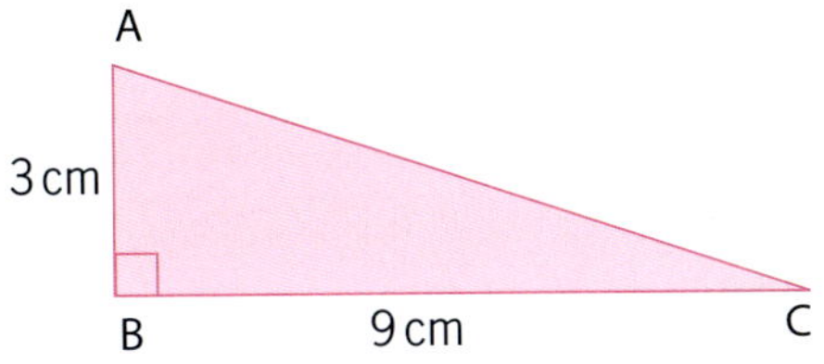

Work out AC. *(3 marks)*

2 The map shows part of the United Kingdom.
The map scale is 1 cm = 50 km

Work out the actual distance and bearing of

a Cambridge from Birmingham

b Chelmsford from Bath

c Truro from Exeter. *(6 marks)*

3 A builder carries out maintenance work at the top of a tower 24 m high. He makes a chute to carry the rubbish from the building. The chute is made from linking tubes 0·8 m in length. The bottom of the chute is 5 m from the base of the tower. How many tubes does the builder need? *(4 marks)*

9 Fractions, decimals and percentages

INTRODUCTION

The very first fractions can be traced back to the ancient Egyptians. They wrote all their fractions as unit fractions (such as $\frac{1}{2} + \frac{1}{3}$ to mean the fraction $\frac{5}{6}$). It was not until the 17th century that fractions as we know them today existed in Europe.

What's the point?

Understanding how fractions work enables us to make a lot of connections in mathematics, for instance in solving real-life problems involving ratio and proportion.

CHECK IN

1 Copy and complete, so that for each statement there is a percentage, a fraction and a decimal.

$10\% = \frac{1}{10} = ?$ $\quad 25\% = ? = 0.25$

$50\% = \ ? \ = ?$ $\quad ? = ? = 0.23$

2 Calculate, using a mental, written or calculator method:

a $3 \div 8$ b $150 \div 180$

Orientation

What I should know	What I will learn	What this leads to
A4	■ Calculate using decimals and percentages ■ Multiply, divide, add and subtract fractions	C17

RICH TASK

In this diagram a square has been divided into different sized pieces.

Calculate the fraction or percentage of the whole square that each piece represents.

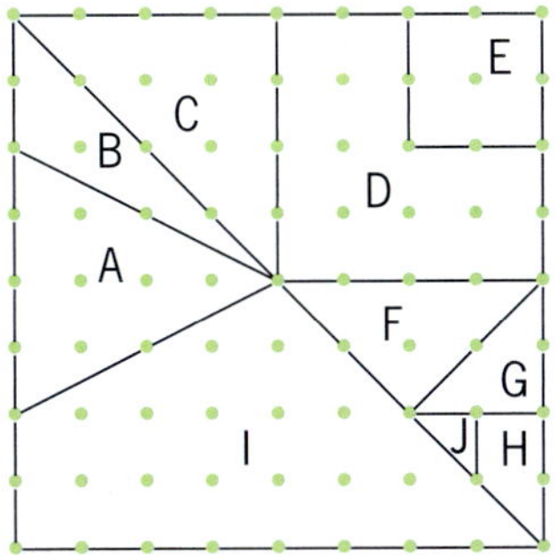

Equivalent fractions

B9.1

- Understand equivalent fractions and simplify a fraction
- Order fractions by rewriting them with a common denominator

Here are two clocks.

The time is **half** past ten.
It is 30 minutes past the hour, there are 60 minutes in an hour, so $\frac{1}{2} = \frac{30}{60}$
The minute hand is at 6, there are 12 numbers on the clock, so $\frac{1}{2} = \frac{30}{60} = \frac{6}{12}$

The time is twenty past eight.
It is **one third** past the hour.
It is 20 minutes past the hour, there are 60 minutes in an hour, so $\frac{1}{3} = \frac{20}{60}$
The minute hand is at 4, there are 12 numbers on the clock, so $\frac{1}{3} = \frac{20}{60} = \frac{4}{12}$

→ **You can find equivalent fractions by multiplying or dividing the numerator and denominator by the same amount.**

→ **A fraction is simplified when it cannot be cancelled further.**

The top number is the **numerator**. The bottom number is the **denominator**.

$\frac{8}{20}$ can be simplified to $\frac{2}{5}$

$\frac{8}{20} \overset{\div 4}{=} \frac{2}{5}$ (÷4 top and bottom)

$\frac{7}{10} \overset{\times 3}{=} \frac{21}{30}$ (×3 top and bottom)

$\frac{7}{10}$ and $\frac{21}{30}$ are equivalent fractions.

EXAMPLE

Simplify **a** $\frac{3}{15}$ **b** $\frac{4}{10}$ **c** $\frac{18}{24}$

ANSWER

a $\frac{3}{15} = \frac{1}{5}$ **b** $\frac{4}{10} = \frac{2}{5}$ **c** $\frac{18}{24} = \frac{3}{4}$

EXAMPLE

Write these fractions in ascending order (smallest first)

a $\frac{1}{6}$ $\frac{1}{2}$ $\frac{1}{5}$ **b** $\frac{1}{2}$ $\frac{3}{8}$ $\frac{5}{6}$ $\frac{3}{4}$

ANSWER

a Think of the fractions on a number line.

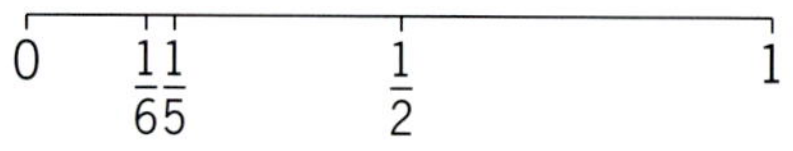

So the order is $\frac{1}{6}$ $\frac{1}{5}$ $\frac{1}{2}$

b Find equivalent fractions that give a common denominator.

- the LCM of 2, 4, 6, and 8 is 24

$\frac{1}{2} = \frac{12}{24}$ $\frac{3}{8} = \frac{9}{24}$ $\frac{5}{6} = \frac{20}{24}$ $\frac{3}{4} = \frac{18}{24}$

Re-order the numerators: $\frac{9}{24}, \frac{12}{24}, \frac{18}{24}, \frac{20}{24}$

So the order is $\frac{3}{8}, \frac{1}{2}, \frac{3}{4}, \frac{5}{6}$

Exercise B9.1

MEDIUM LOW

This exercise is to be completed without a calculator.

1 Simplify these fractions as far as possible.

a $\frac{4}{6}$ b $\frac{10}{30}$ c $\frac{6}{16}$

d $\frac{12}{20}$ e $\frac{14}{21}$ f $\frac{25}{35}$

For questions 2 and 3 part c, split the shapes into equal amounts.

2 Write down the fraction of these shapes that is shaded.

a
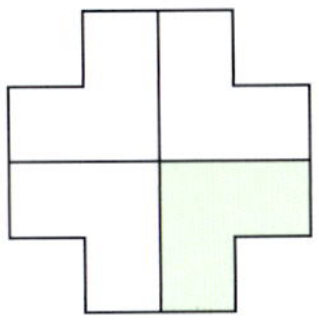

b
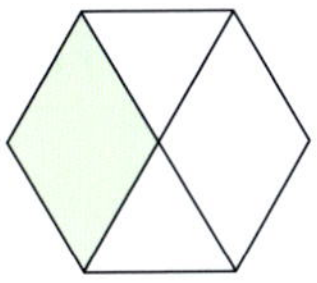

c

3 Write down the fraction of these shapes that is unshaded.

a

b
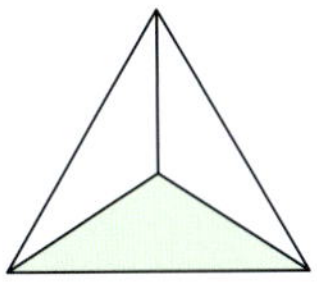

c
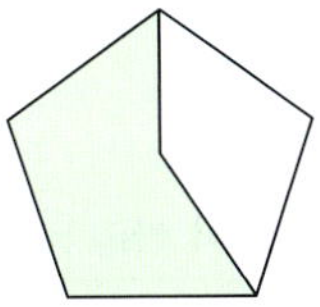

4 Which of these fractions is not equivalent to $\frac{1}{3}$?

$\frac{3}{9}$ $\frac{4}{12}$ $\frac{6}{18}$ $\frac{9}{30}$

5 Which of these fractions are equivalent to $\frac{3}{4}$?

$\frac{30}{40}$ $\frac{6}{8}$ $\frac{15}{18}$ $\frac{9}{12}$

6 Write these fractions in ascending order.

a $\frac{1}{4}$ $\frac{1}{2}$ $\frac{1}{3}$ $\frac{1}{9}$ b $\frac{2}{7}$ $\frac{5}{7}$ $\frac{4}{7}$

c $\frac{3}{8}$ $\frac{5}{6}$ $\frac{1}{2}$ d $\frac{3}{4}$ $\frac{2}{3}$ $\frac{5}{6}$

7 ★ **PROBLEM** ★

Which of these fractions is closest to $\frac{1}{2}$?

$\frac{7}{10}$ $\frac{11}{20}$ $\frac{13}{30}$ $\frac{21}{40}$

Explain how you decided.

Unit B

Fractions, decimals and percentages

B9.2

- Convert simple fractions to percentages, or vice versa
- Recognise that each terminating decimal is a fraction

Decimals, percentages and fractions are just different ways of describing amounts. **Converting** between fractions, decimals and percentages can really help in calculations.

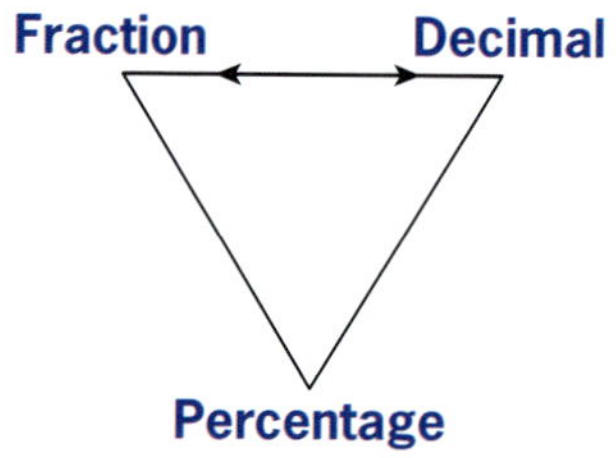

→ **To write a fraction as a decimal, divide the numerator by the denominator.**

$\frac{1}{4} = 1 \div 4 = 0.25$

$\frac{2}{5} = 2 \div 5 = 0.4$

→ **To write a decimal as a fraction, count the number of places after the decimal point.**

0.75 **Two** places after the decimal point So $0.75 = \frac{75}{100} = \frac{3}{4}$

0.6 **One** place after the decimal point So $0.6 = \frac{6}{10} = \frac{3}{5}$

Look at a place value table,

Units	.	$\frac{1}{10}$	$\frac{1}{100}$

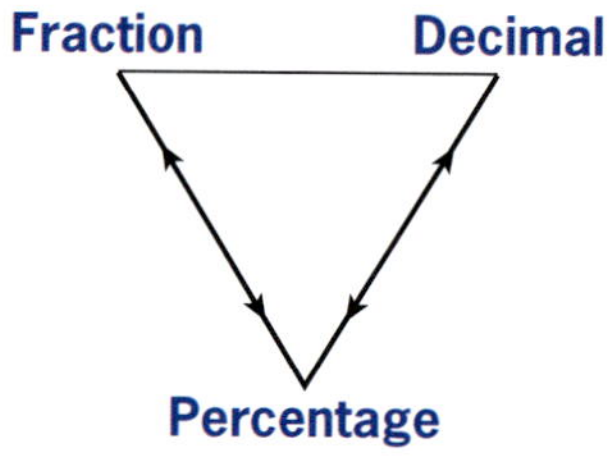

→ **To write a decimal as a percentage, multiply by 100.**

$0.23 = 0.23 \times 100\% = 23\%$

$0.07 = 0.07 \times 100\% = 7\%$

Percentage means part of 100 ⇨ $25\% = \frac{25}{100} \left(= \frac{1}{4}\right)$

→ **To write a percentage as a decimal or a fraction, divide by 100.**

$20\% = 20 \div 100 = 0.2$

$0.2 = \frac{2}{10} = \frac{1}{5}$

$\Rightarrow 20\% = 0.2 = \frac{1}{5}$

$14\% = 14 \div 100 = 0.14$

$0.14 = \frac{14}{100} = \frac{7}{50}$

$\Rightarrow 14\% = 0.14 = \frac{7}{50}$

→ **To compare fractions, decimals and percentages, convert them all to the same form.**

EXAMPLE

Write these in ascending order (smallest first) 0.83 $\frac{3}{8}$ 38%

ANSWER

Convert to percentages:

0.83 = 83% $\frac{3}{8} = 3 \div 8 = 0.375 = 37.5\%$

Ascending order: $\frac{3}{8}$ 38% 0.83

You could convert them all to fractions or decimals instead.

Exercise B9.2

MEDIUM LOW

This exercise is to be completed without a calculator.

1 Write down the percentage of the shape that is shaded

a

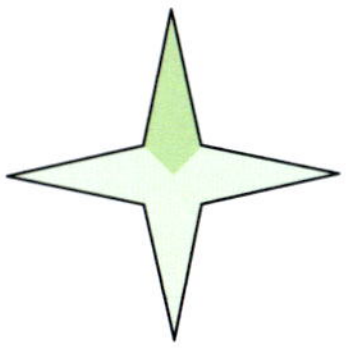

b

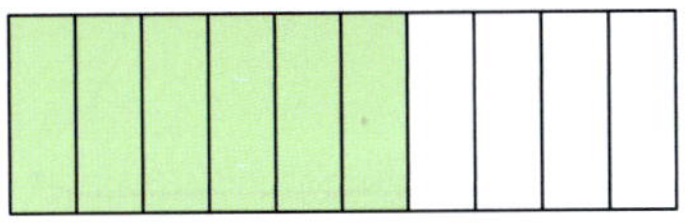

2 Copy and complete the table showing equivalent fractions, decimals and percentages.

Fraction	Decimal	Percentage
$\frac{1}{2}$		
	0.25	
		75%
		10%
$\frac{1}{5}$		
	0.7	
		23%
$\frac{3}{5}$		

▲ Question 2

3 Write in ascending order

a $\frac{1}{4}$ 0.7 40% b 0.45 $\frac{4}{5}$ 40%

c $\frac{2}{5}$ 15% 0.09 d 0.6 $\frac{11}{20}$ 7%

Convert the amounts to the same form.

4 Write $\frac{5}{8}$ as a decimal and as a percentage.

5 Which is bigger, 25% or $\frac{2}{5}$? Show how you decide.

6 Which is smaller, 15% or $\frac{1}{5}$? Show how you decide.

7 Write these as percentages.

a 90 out of 200 b 40 out of 50

8 What percentage is £48 of £150?

9 James scores 18 marks in a test.
The test was marked out of 30.
Write James's score as a percentage.

Calculations with decimals

B9.3

- Add and subtract numbers with up to two decimal places
- Multiply and divide decimal numbers

Place value matters when calculating with decimals.

Adding and subtracting decimals

When you **add** or **subtract** decimals using a standard written method, make sure the corresponding digits line up.

EXAMPLE

Work out **a** 12.2 + 8.14 **b** 15.6 – 6.93

ANSWER

a

$$\begin{array}{r} 12.20 \\ \underline{+8.14} \\ 20.34 \end{array}$$

b

$$\begin{array}{r} 15.60 \\ \underline{-6.93} \\ 8.67 \end{array}$$

Line up the decimal points in columns. You can write a zero in the empty spaces as shown.

Multiplying and dividing decimals

For **multiplying** decimals, you can use the number of decimal places in the question to work out the number of decimal places in the answer.

EXAMPLE

Work out **a** 0.04 × 1.2 **b** 0.5 × 0.6

ANSWER

a Calculate with the whole numbers:
4 × 12 = 48
Total of 3 places after the decimal point:
0.04 × 1.2 = 0.048
So three digits after the d.p. in the answer.

b Calculate with the whole numbers:
5 × 6 = 30
Total of 2 places after the decimal point:
0.5 × 0.6 = 0.30
So two digits after the d.p. in the answer.

For **division**, multiply both decimal numbers so that the denominator becomes a whole number.

EXAMPLE

Work out **a** 5.34 ÷ 0.3 **b** 2.6 ÷ 0.25

ANSWER

a Multiply both numbers by 10: 5.34 ÷ 0.3 = 53.4 ÷ 3
= 17.8

b Multiply both numbers by 100: 2.6 ÷ 0.25 = 260 ÷ 25
= 10.4

Exercise B9.3

LOW

This exercise is to be completed without a calculator.

1 Write these decimals in ascending order (smallest first).

a 0.3 0.205 0.61 0.43
b 0.84 0.701 0.9 0.205
c 0.41 0.405 0.45 0.415
d 0.809 0.98 0.89 0.908
e 0.2 0.25 0.205 0.52 0.5

2 Work out

a 5.6 + 3.7
b 3.5 + 6.94
c 7.4 + 2.09
d 4.2 + 5.05
e 3.97 + 0.23
f 17.25 + 5.64
g 139 + 2.65
h 0.24 + 14.9

3 Work out

a 12.5 – 8.2
b 13.3 – 2.1
c 34.2 – 6.5
d 5.9 – 2.7
e 4.3 – 2.8
f 7.2 – 3.8
g 8.2 – 5.75
h 7.7 – 2.96

4 Work out

a 0.3 × 0.2
b 0.4 × 1.2
c 2.3 × 0.5
d 0.6 × 0.8
e 0.04 × 15
f 12 × 0.06
g 1.8 × 0.12
h 2.3 × 0.15

5 Work out

a 1.2 ÷ 0.4
b 15.6 ÷ 0.3
c 120 ÷ 0.6
d 3.64 ÷ 0.05
e 25.6 ÷ 0.8
f 10.9 ÷ 0.04
g 15 ÷ 0.15
h 0.98 ÷ 0.02

6 ★ **PROBLEM** ★

Choose from this list

0.2 0.25 0.4 0.8 2 5

a two numbers that give the lowest possible product
b two numbers that give an answer 10 in a division sum.

______ ÷ ______ = 10

*7 ★ **PROBLEM** ★

Tommy and Gavin added up 15.6 + 3.78 in different ways.

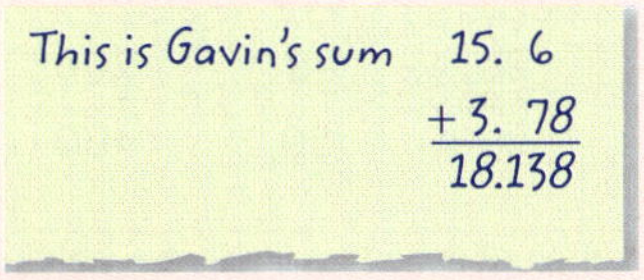

a Explain how Tommy has gone wrong.
b Explain what Gavin has done incorrectly.
c Show Tommy and Gavin how to calculate 15.6 + 3.78 correctly.

Calculations with percentages

B9.4

- Understand the multiplicative nature of percentages as operators

You can use percentages that you know to work out percentages that you don't. **Doubling** and **halving** are useful strategies.

You should already know some basic percentage equivalents:

$50\% = \frac{1}{2}$

$25\% = \frac{1}{4}$

$10\% = 0.1$

$1\% = 0.01$

EXAMPLE

Find **a** 5% of 130 **b** 20% of 86

ANSWER

a 10% of 130 = 13 5% is half of 10% 5% of 130 = $\frac{1}{2}$ of 13 = 6.5

b 10% of 86 = 8.6 Double 10% is 20% 20% of 86 = 2 × 8.6 = 17.2

Partitioning is also a useful strategy. You split the percentage into easier amounts.

p. 246

EXAMPLE

Work out **a** 12% of 60 **b** 49% of 48

ANSWER

a 10% of 60 = 6 1% of 60 = 0.6

12% = 10% + 1% + 1%

12% of 60 = 6 + 0.6 + 0.6 = 7.2

b 50% of 48 = $\frac{1}{2}$ of 48 = 24 1% of 48 = 0.01 × 48 = 0.48

49% = 50% − 1%

49% of 48 = 24 − 0.48 = 23.52

12% = 10% + 1% + 1%

49% = 50% − 1%

★ ACTIVITY ★

In pairs or groups of three see how many of these amounts you can make.

Choose a percentage:

50%	20%	22%	100%
1%	75%	5%	15%
25%	2%	10%	200%

Choose an amount:

800	40	8	60
20	10	200	6000
2000	50	400	500

Work out the percentage of the amount – if your answer is on this board, score one point.

400	2.5	100	0.08	0.25
0.6	25	800	600	2500
4	1000	0.8	250	20 000
900	200	40	60	400

Work out your percentage without a calculator.
(Your opponent can use a calculator to check.)

Exercise B9.4

LOW

This exercise is to be completed without a calculator.

Unit B

1 Work out

a 50% of 106 b 25% of 84
c 10% of 80 d 25% of 90
e 10% of 42 f 25% of 200
g 1% of 500 h 1% of 320

2 Work out

a 2% of 900 b 40% of 2000
c 20% of 54 d 30% of 85
e 20% of 42 f 25% of 200
g 5% of 400 h 5% of 60

3 Work out

a 21% of 320 b 15% of 300
c 55% of 120 d 52% of 20
e 3% of 42 f 49% of 200
g 12% of 500 h 9% of 320

4 Which is larger, 25% of 60 or 20% of 70?
Show your working.

5 Which is smaller, 50% of 30 or 40% of 40?
Show your working.

*6 a Work out **i** 10% of 250 **ii** 5% of 500
b Explain why the answers are the same.

7 ☆ **PROBLEM** ☆

a 25% of a number is 100. What is the number?
b 10% of a number is 35. What is the number?

Adding and subtracting fractions

B9.5

- Add and subtract fractions

☆ ACTIVITY ☆

Look at this rectangle.
How many squares is

a $\frac{1}{2}$ **b** $\frac{1}{3}$ **c** $\frac{1}{6}$ of the rectangle?

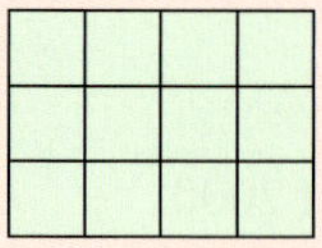

Explain why $\frac{1}{2} + \frac{1}{3} + \frac{1}{6} = 1$

→ **To add or subtract fractions, they must have the same denominator. If they don't, change them so that they do.**

EXAMPLE

Work out **a** $\frac{2}{5} + \frac{1}{4}$ **b** $\frac{5}{6} - \frac{3}{8}$

ANSWER

a $\frac{2}{5} = \frac{8}{20}$ $\frac{1}{4} = \frac{5}{20}$

$\frac{2}{5} + \frac{1}{4} = \frac{8}{20} + \frac{5}{20}$

$= \frac{13}{20}$

b $\frac{5}{6} = \frac{20}{24}$ $\frac{3}{8} = \frac{9}{24}$

$\frac{5}{6} - \frac{3}{8} = \frac{20}{24} - \frac{9}{24}$

$= \frac{11}{24}$

Exercise B9.5

MEDIUM

This exercise is to be completed without a calculator.

1 Work out these fractions of amounts.
Give your answers in their simplest form.

a $\frac{1}{4}$ of £90 b $\frac{3}{5}$ of £35 c $\frac{5}{8}$ of £16 d $\frac{2}{3}$ of £24

2 What fraction of £500 is left after deducting a $\frac{1}{5}$ of £500 and b $\frac{1}{2}$ of £500?

3 Add these fractions. Give your answers in their simplest form.

a $\frac{3}{8} + \frac{1}{8}$ b $\frac{3}{8} + \frac{5}{16}$ c $\frac{2}{7} + \frac{1}{10}$ d $\frac{2}{5} + \frac{7}{12}$

4 Subtract these fractions. Give your answers in their simplest form.

a $\frac{5}{8} - \frac{3}{8}$ b $\frac{8}{9} - \frac{8}{27}$ c $\frac{3}{10} - \frac{1}{5}$ d $\frac{2}{3} - \frac{1}{4}$

Multiplying and dividing fractions

B9.6

- Multiply and divide with fractions

Jenny has $\frac{1}{3}$ of a cake. She gives Joel $\frac{1}{2}$ of her $\frac{1}{3}$.

What fraction does Joel get?

$\frac{1}{2}$ of $\frac{1}{3} = \frac{1}{2} \times \frac{1}{3}$

$= \frac{1}{6}$ ⇨ Joel gets $\frac{1}{6}$ of the cake.

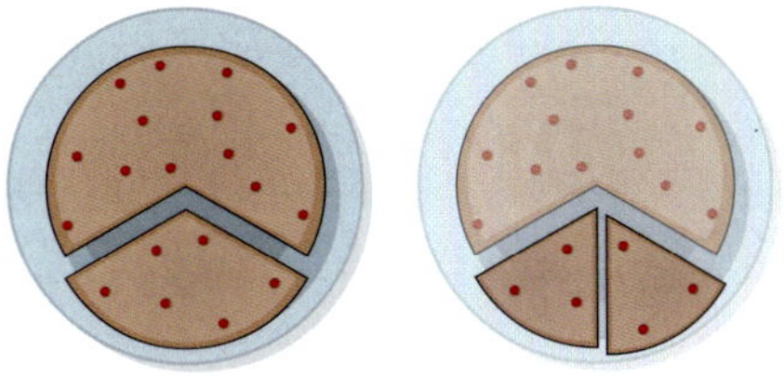

→ **To multiply fractions, multiply the numerators and multiply the denominators.**

EXAMPLE

Work out **a** $\frac{2}{7} \times \frac{4}{5}$ **b** $\frac{3}{4}$ of 84 **c** $\frac{2}{15} \div \frac{1}{3}$ **d** $\frac{1}{8} \div \frac{2}{3}$

ANSWER

a $\frac{2}{7} \times \frac{4}{5} = \frac{2 \times 4}{7 \times 5} = \frac{8}{35}$

b $\frac{3}{4}$ of $84 = 3 \times 84 \div 4 = 63$

c $\frac{2}{15} \div \frac{1}{3} = \frac{2}{15} \times \frac{3}{1} = \frac{6}{15} = \frac{2}{5}$

d $\frac{1}{8} \div \frac{2}{3} = \frac{1}{8} \times \frac{3}{2} = \frac{3}{16}$

→ **Dividing by 2 is the same as multiplying by $\frac{1}{2}$.**

→ **To divide by a fraction, change from ÷ to × and use the reciprocal fraction.**

Exercise B9.6

MEDIUM

1 Multiply these fractions. Give your answers in their simplest form.

a $\frac{2}{3} \times \frac{1}{4}$ b $\frac{2}{7} \times \frac{3}{10}$ c $\frac{4}{5} \times \frac{6}{7}$ d $\frac{4}{9} \times \frac{3}{8}$

2 Divide these fractions. Give your answers in their simplest form.

a $\frac{2}{5} \div 4$ b $\frac{1}{2} \div 2$ c $\frac{3}{4} \div 5$ d $\frac{9}{10} \div 6$ e $\frac{1}{5} \div \frac{1}{2}$ f $\frac{2}{9} \div \frac{1}{4}$ g $\frac{5}{16} \div \frac{1}{3}$ h $\frac{3}{10} \div \frac{2}{5}$

3 ☆ **PROBLEM** ☆

a Choose from this list: $\frac{1}{2}$ $\frac{3}{8}$ $\frac{2}{5}$ $\frac{4}{7}$ $\frac{1}{10}$

i the two fractions that have the lowest product

ii three fractions that add up to 1.

b Choose from this list: $\frac{5}{8}$ $\frac{1}{2}$ $\frac{1}{5}$ $\frac{2}{7}$ $\frac{3}{10}$

i the two fractions that have the lowest product

ii three fractions that add up to 1.

Mixed numbers

B9.7

- Use efficient methods to calculate with fractions

- A mixed number contains a whole number part and a fraction part – for example, $2\frac{1}{5}$ or $5\frac{3}{8}$.
- To change a **mixed number** to a fraction, use the denominator to find how many of the fraction there are in one whole.

$2\frac{1}{3} = (6 \text{ lots of } \frac{1}{3}) + \frac{1}{3} = \frac{7}{3}$

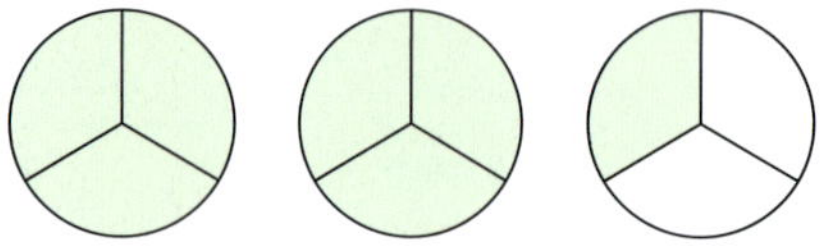

$\frac{7}{3}$ is an **improper fraction** – the numerator is bigger than the denominator.

- To change an improper fraction to a mixed number, divide the numerator by the denominator.

$\frac{11}{5} = 2\frac{1}{5}$ ← This is the remainder.

- To add and subtract mixed numbers, change the fractions to improper fractions first.

EXAMPLE

Work out these fraction sums. Give your answer as a mixed number.

a $4\frac{1}{2} + 2\frac{3}{5}$ **b** $5\frac{1}{6} - 2\frac{3}{4}$ **c** $\frac{3}{5} \times 9$

ANSWER

a $4\frac{1}{2} = \frac{9}{2}$ $4 \times 2 + 1 = 9$

$2\frac{3}{5} = \frac{13}{5}$ $2 \times 5 + 3 = 13$

$4\frac{1}{2} + 2\frac{3}{5} = \frac{9}{2} + \frac{13}{5}$

$= \frac{45}{10} + \frac{26}{10}$

$= \frac{71}{10} = 7\frac{1}{10}$

b $5\frac{1}{6} - 2\frac{3}{4} = \frac{31}{6} - \frac{11}{4}$

$= \frac{62}{12} - \frac{33}{12}$

$= \frac{29}{12} = 2\frac{5}{12}$

c $\frac{3}{5} \times 9$

$= \frac{3}{5} \times \frac{9}{1}$

$= \frac{3 \times 9}{5 \times 1}$

$= \frac{27}{5}$

$= 5\frac{2}{5}$

5 goes into 27 5 times remainder 2.

Exercise B9.7

MEDIUM LOW

This exercise is to be completed without a calculator.

1 Multiply these fractions. Give your answers as a mixed number.

a $\frac{2}{5} \times 4$ b $\frac{3}{4} \times 10$

c $\frac{5}{6} \times 9$ d $\frac{1}{4} \times 3$

e $\frac{5}{6} \times 11$ f $\frac{2}{3} \times 4$

g $2\frac{1}{2} \times 5$ h $\frac{2}{7} \times 3$

2 Gareth's cat eats $\frac{3}{4}$ of a tin of cat food each day.
What is the least number of tins Gareth needs to buy for 7 days?

3 a How many halves are there in $2\frac{1}{2}$?

b How many fifths are there in $1\frac{1}{5}$?

c How many quarters are there in $3\frac{1}{4}$?

4 Work out these fraction sums. Give your answers in their simplest form.

a $1\frac{2}{5} + 2\frac{3}{8}$ b $9\frac{1}{2} + 2\frac{1}{4}$

c $2\frac{3}{4} + 5\frac{1}{3}$ d $1\frac{9}{10} + 3\frac{1}{2}$

e $2\frac{1}{2} - \frac{4}{13}$ f $2\frac{1}{4} - \frac{3}{5}$

g $1\frac{5}{6} - 2\frac{1}{4}$ h $1\frac{3}{4} - 2\frac{2}{3}$

Recurring decimals and exact fractions

B9.8

- Recognize equivalence between recurring decimals and exact fractions

*

RICH TASK

On a calculator, work out the fraction family of 11ths as decimals:

$\frac{1}{11}$ $\frac{2}{11}$ $\frac{3}{11}$ $\frac{4}{11}$ $\frac{5}{11}$ $\frac{6}{11}$ $\frac{7}{11}$ $\frac{8}{11}$ $\frac{9}{11}$ $\frac{10}{11}$

What do you notice about these decimals?

Divide the numerator by the denominator.

Not all fractions have an exact decimal equivalent.

You should know that $\frac{1}{3} = 0.333\,333\ldots$ and $\frac{2}{3} = 0.666\,666\ldots$

When you get a repeating string of digits on your calculator, the chances are it's a recurring decimal.

These fractions have recurring decimal equivalents.

→ **In recurring decimals, a digit or digits repeat over and over again.**

→ **Recurring decimals can be written as exact fractions.**

EXAMPLE

Write these fractions as decimals.

a $\frac{7}{9}$ **b** $\frac{1}{15}$ **c** $\frac{16}{33}$ **d** $\frac{8}{27}$

ANSWER

a $\frac{7}{9} = 7 \div 9 = 0.777\,77\ldots = 0.\dot{7}$ Only 7 repeats.

b $\frac{1}{15} = 1 \div 15 = 0.066\,66\ldots = 0.0\dot{6}$ 6 repeats but not 0.

c $\frac{16}{33} = 16 \div 33 = 0.484\,84\ldots = 0.\dot{4}\dot{8}$ 4 and 8 both repeat.

d $\frac{8}{27} = 8 \div 27 = 0.296\,296\,2\ldots = 0.\dot{2}9\dot{6}$ 296 repeats as a group of 3 numbers.

Use dots above the recurring digits to show how the decimal recurs.

Dots go above the first and last number.

Exercise B9.8

MEDIUM

1 a Work out the fraction family of 9ths as decimals:

$\frac{1}{9}$	$\frac{2}{9}$	$\frac{3}{9}$	$\frac{4}{9}$	$\frac{5}{9}$	$\frac{6}{9}$	$\frac{7}{9}$	$\frac{8}{9}$

b Write down a quick way to recognise these decimals as fractions.

2 Work out $\frac{22}{7}$. Give your answer to 2 decimal places. You should recognise this number: write down the Greek symbol for which it is an approximation.

3 a Work out as a recurring decimal

i $\frac{1}{6}$ **ii** $\frac{5}{6}$

b Write your answers to 3 decimal places.

4 a Work out

i $\frac{1}{3}$ of 12 **ii** 0.333×12 **iii** 0.3×12

b $\frac{1}{3} = 0.\dot{3}$.

Write down $\frac{1}{3}$ correct to

i 3 decimal places **ii** 1 decimal place.

c Use your answers to parts a and b to explain why it is sometimes more accurate to use a fraction than a decimal in calculations.

5 Copy this number line.

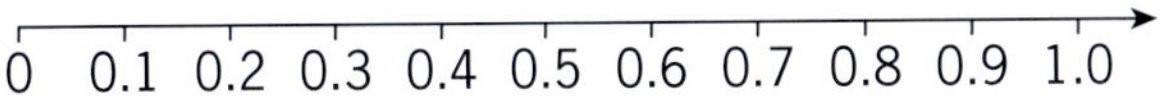

Mark on your number line the position of these recurring decimals

a $0.\dot{3}$ b $0.\dot{5}$ c $0.\dot{7}$

Unit B

Summary and assessment

B9

CHECK OUT

You should now be able to

- find equivalent fractions, decimals and percentages
- perform calculations using fractions, decimals and percentages without a calculator
- understand that in some cases a fraction is exact where a decimal may not be.

Exam-style question

Nickels, dimes and quarters are all American coins.
A nickel is worth 5 cents, a dime 10 cents and a quarter is worth 25 cents.
There are 100 cents in one dollar.

a What percentage of a ten dollar bill is a quarter?

b What percentage of a quarter is **i** a nickel **ii** a dime?

When lots of information is given in a question, it can be helpful to write down the numbers you need for each question part.

Claire's answer

a A quarter is 25 cents,
10 dollars is 1000 cents
$\frac{25}{1000} \times 100 = 2.5\%$

b i a nickel is 5 cents,
a quarter is 25 cents
$\frac{5}{25} \times 100 = 20\%$

ii a dime is 10 cents,
a quarter is 25 cents
$\frac{10}{25} \times 100 = 40\%$

Always show your working clearly.

Exam practice

1 a How many halves are there in $6\frac{1}{2}$?

b Work out 20% of £400. *(3 marks)*

2 Which of these fractions is closest to $\frac{1}{4}$?

$\frac{4}{20}$ $\frac{7}{30}$ $\frac{9}{40}$

Show how you decide. *(2 marks)*

3 A brand of shampoo is made in three sizes, holding different amounts.

Travel size:	50 ml
Standard size:	250 ml
Large size:	400 ml

a What fraction of the large size is
i travel size **ii** standard size?

b Write your answers to part a as decimals. *(4 marks)*

Case study 3: Business

One out of every two small businesses goes bust within its first two years of trading. Mathematics can be applied to reduce the risk of failure for a business as well as to maximise its profits.

A manager needs to know how much cash is coming into and going out of the business. Accountants must set a suitable budget that includes realistic performance targets, and limits expenditure to what the business can afford.

Example

Annie sells hand made cards at a monthly craft fair.

The production costs and selling price per card are:

Cost of materials used	Production time	Wages paid	Selling price	Profit
£0.30	15 minutes	£1.00	£2.55	£1.25

This is Annie's cash flow budget for her first three craft fairs (some of the information is missing):

	January (£)	February (£)	March (£)
TOTAL SALES INCOME	56.10	71.40	63.75
Materials used	6.60	8.40	7.50
Wages			
Craft fair fees	10.00	10.00	10.00
Advertising	5.00	5.00	5.00
TOTAL EXPENDITURE	43.60		
NET CASH SURPLUS/DEFICIT	12.50		
CASH BALANCE BROUGHT FORWARD	-	12.50	
CASH BALANCE TO CARRY FORWARD	12.50		

How many cards did Annie sell in each of the three months?

Use this information to calculate the wages paid for each month.

Calculate the total expenditure for each month. During which month were Annie's expenses highest?

The net surplus (profit) or net deficit (loss) is calculated using the formula Balance = Income − Expenditure

Copy the table and complete the missing values.

On separate copies of the table template, show how the cash flow could change if

a) the craft fair fees were increased to £15

b) the cost of the materials used per card increased to £0.40

c) the selling price per card was increased to £2.75.

Investigate how other changes to costs/income might affect Annie's cash flow.

The breakeven point is when a company's expenditure is equal to its income. If the company can operate at levels above the breakeven point, it will make a profit. If sales fall below this point, the company will make a loss.

For Annie's cards:

Fixed costs = craft fair fees (£10) + advertising (£5) = £15;

Variable cost per card = material costs (£0.30) + wages (£1) = £1.30 per card;

Total costs = fixed costs + variable costs

To calculate total variable cost, multiply: variable cost per card × number of cards

To calculate revenue, multiply: sales price per card × number of cards

If 20 cards are sold, revenue = £2.55 × 20 = £51

Variable costs and revenue increase in direct proportion with the number of cards produced.

Fixed costs: £15, Variable costs: £1.30 per card, Sales price: £2.55 per card

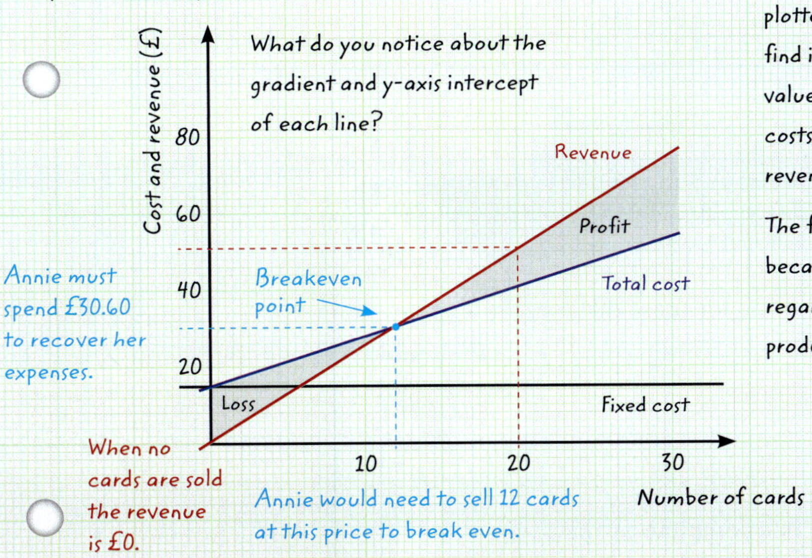

Make sure you understand where the plotted values come from. You may find it useful to draw up a table of values (showing number of cards, fixed costs, variable costs, total costs and revenue).

The fixed cost line is horizontal because the fixed costs do not change regardless of the number of cards produced.

Plot your own charts to show the breakeven point if

a) the craft fair fees were increased to £15

b) the cost of the materials used per card increased to £0.40

c) the selling price per card was increased to £2.75.

Investigate how other changes to costs/income could affect Annie's profit/loss.

10 Circles, angles and lines

INTRODUCTION

A fractal is a geometrical shape that can be split into parts so that each part is a smaller copy of the whole shape. Here is an example of the von Koch Snowflake. The von Koch Snowflake fractal is constructed from an equilateral triangle. On the middle third of each side is built another equilateral triangle and the process is repeated over and over again.

What's the point?

Fractal geometry is a relatively new branch of mathematics that is finding applications all the time. The von Koch snowflake can be used to model the reception of an antenna.

CHECK IN

1 Give the mathematical name for these distances in a circle.

a

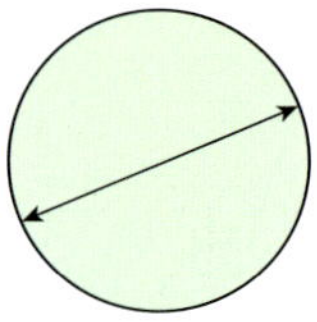

b

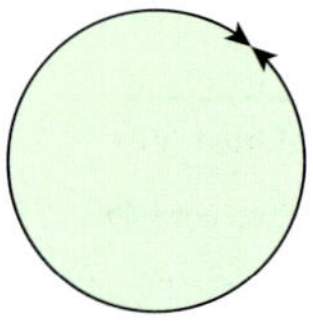

c

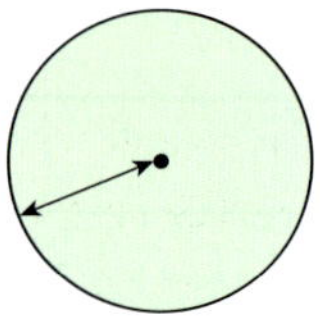

2 How many degrees are in a complete turn?

Orientation

What I should know	What I will learn	What this leads to
A3	■ Recognise different angles ■ Using mathematical language to talk about circles ■ Find angles in parallel lines	C18

RICH TASK

In this six-sided polygon every vertex has been joined to every other vertex by a line. There are nine diagonals.

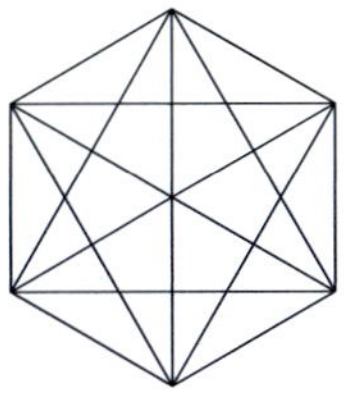

If the polygon had 12 sides, how many diagonals would you need to draw?

What if it had 100 sides?

Circles

- Know and use terms associated with lines and circles

p. 263

A **line** is straight, infinitely long and infinitely thin. It has no ends.
A **line segment** is that part of a line which lies between two points.
It has two ends. The ends are usually labelled with capital letters.

However, the word *line* is commonly used to mean *line segment*, as in 'the line *PQ*'.

A **circle** is the locus of a point which moves so that it is always the same distance from a fixed point (the **centre**, O).
This distance is the **radius** of the circle.

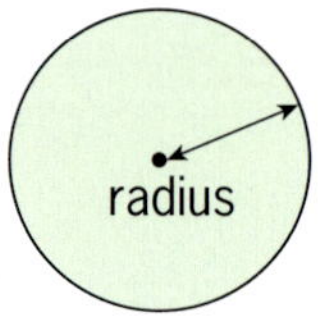

A **chord** of a circle is a line segment which has its two ends on the circumference.

The perimeter of the circle is called its **circumference**.

The **diameter** is the longest chord and it passes through the centre, O. So its length is twice the length of the **radius**.

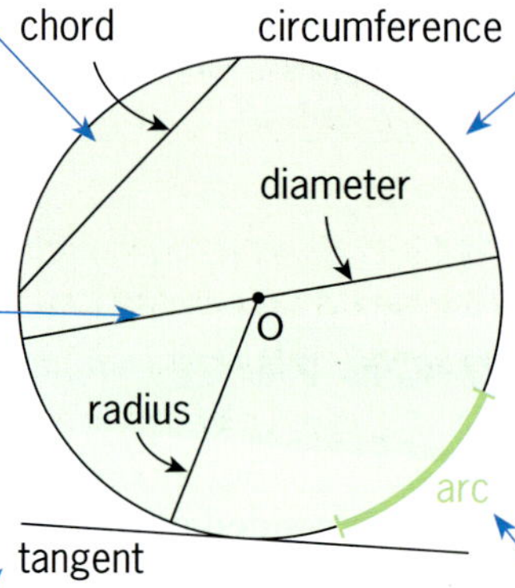

A **tangent** to a circle is a straight line which touches the circle at just one point.

An **arc** is part of the circumference.

A **sector** of a circle is a region inside a circle which is cut off by two radii.

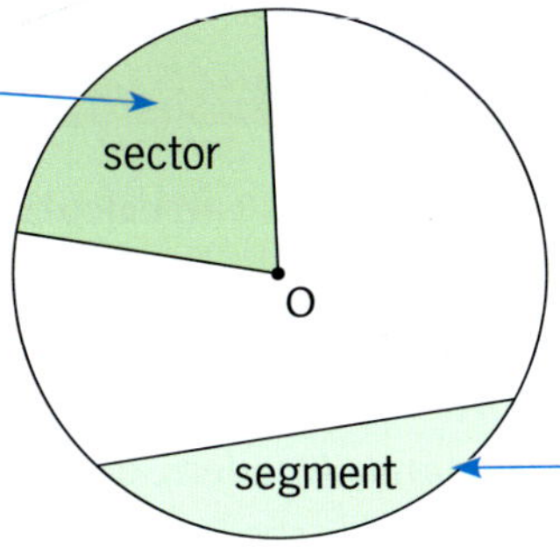

A *sector* includes the centre **O** and has a letter **O**.

The word *segment* does not have the letter **O**.

A **segment** of a circle is a region inside a circle which is cut off by a chord.

You can have *major* and *minor* sectors and segments.

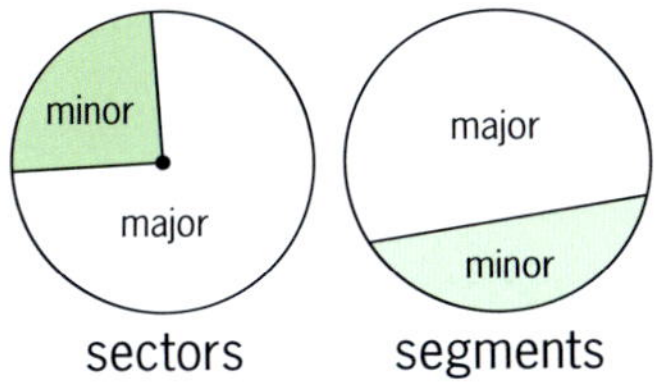

The shaded sector is a semi circle.

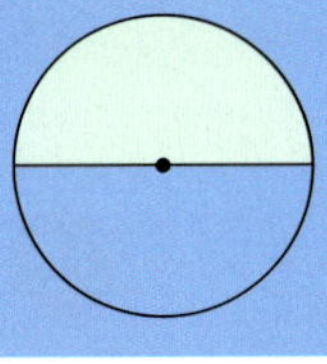

Exercise B10.1

MEDIUM

1 What do you call a sector which has an arc equal to

a half the circumference b a quarter of the circumference?

2 Point X is outside a circle. Draw two tangents from X so that they touch the circle at points A and B. Use a ruler to measure AX and BX. What do you notice? Repeat for other positions of X. Can you give a reason for your results?

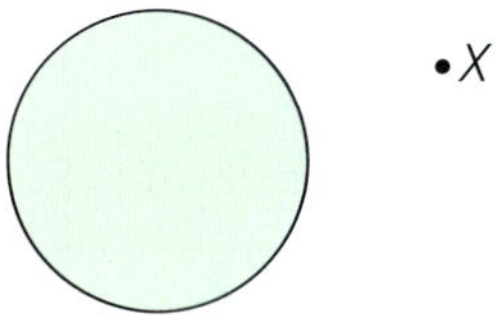

3 Draw any diameter AB of any circle. Draw the chords AX and BX where X is anywhere on the circle's circumference. What size is angle AXB? Do you get the same answer for all possible positions of X?

4 If you cut out a major sector OAB of a circle and bring the two radii OA and OB together, what 3-D shape do you make? Do you get the same result if you start with a minor sector?

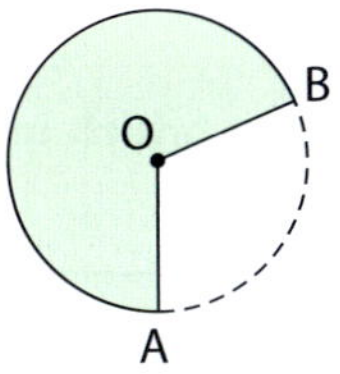

5 a You draw n chords in the same circle. What is the maximum number of regions of a circle that you can make when

i $n = 1$ **ii** $n = 2$ **iii** $n = 3$ **iv** $n = 4$?

b Repeat for values of $n > 4$.

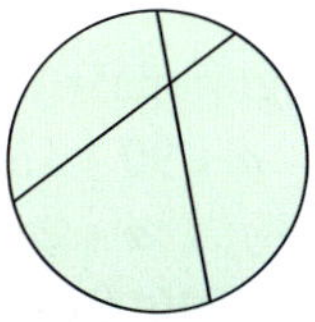

$n = 2$

6 ☆ **RESEARCH** ☆

What have Woodhenge, Callanish and Avebury in common with Long Meg and her daughters? Which famous site could be added to this list? Find how old and how big they all are.

Angles and degrees

- Recognise and estimate different kinds of angles and calculate angles which make a full turn

About 4000 years ago, the people of Mesopotamia (modern-day Iraq) thought that the Earth took about 360 days to circle the Sun and so they divided the circle into 360 degrees.

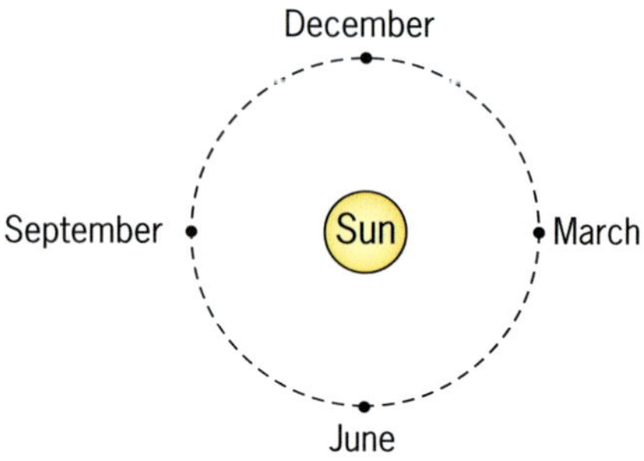

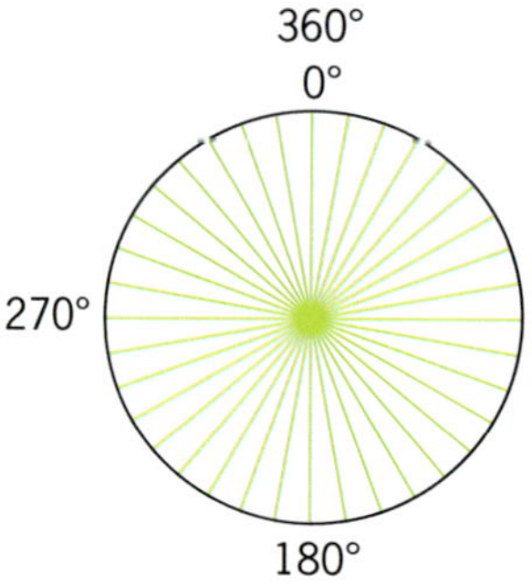

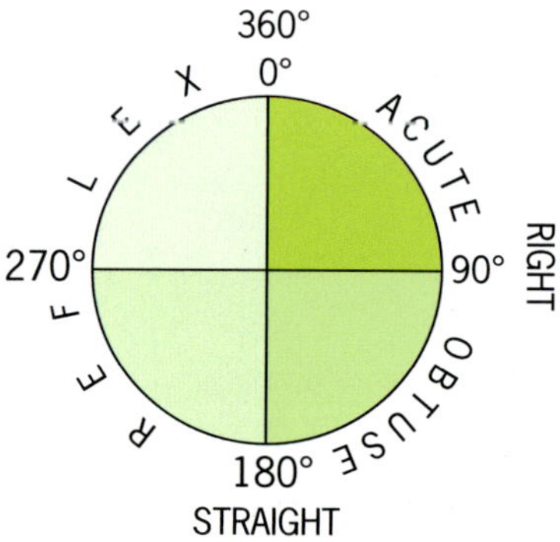

Angles have different names which depend on their size.

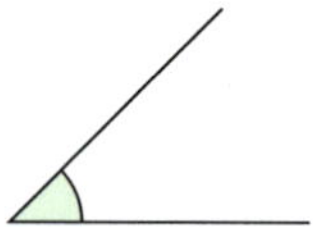

Acute angles are between 0° and 90°.

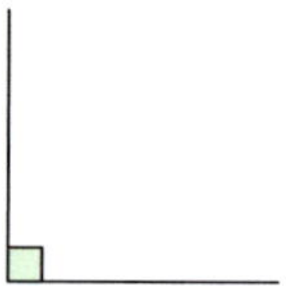

Right angles are exactly 90°.

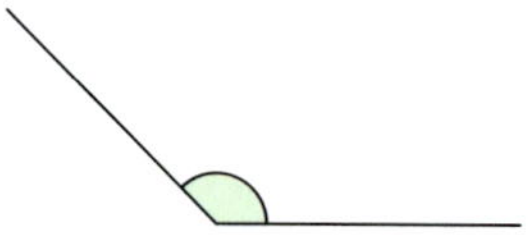

Obtuse angles are between 90° and 180°.

Straight angles are exactly 180°.

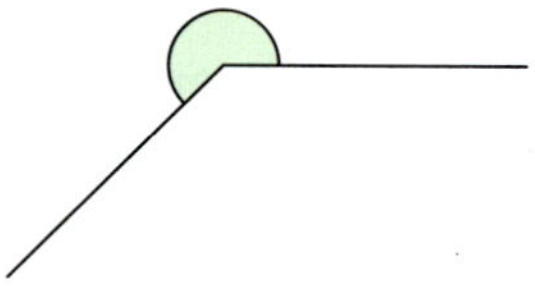

Reflex angles are between 180° and 360°.

→ **Angles which make a full turn at a point add up to 360°.**

EXAMPLE

Calculate the size of angle a in this diagram.

ANSWER

The angles make a full turn of 360°.

$$a + 90 + 150 = 360$$

$$a + 240 = 360$$

Subtract 240° from both sides.

$$a = 360 - 240 = 120$$

So angle a is 120°.

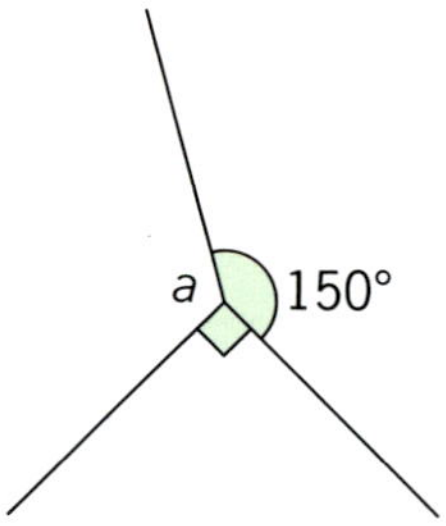

Exercise B10.2

LOW

1 i Give a name to each of these angles.
ii Estimate the size of each angle in degrees.

a 
b 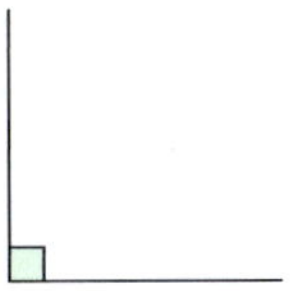
c 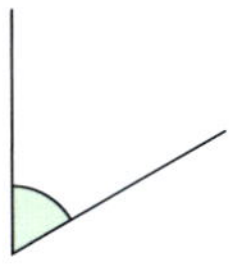
d 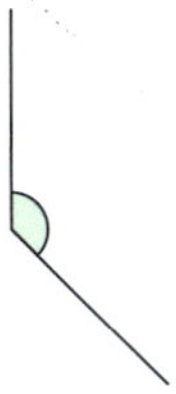
e 
f 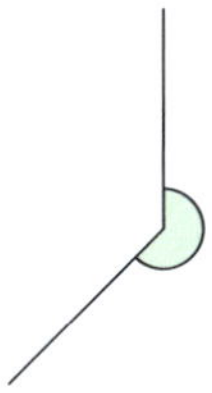

2 For each lettered angle in these quadrilaterals
a say what kind of angle it is
b estimate its size.

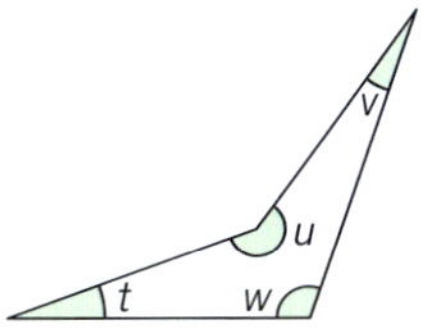

3 You face due north. You turn
a to face due south
b clockwise to face due east
c anticlockwise to face due east
d clockwise to face south-east.
In each case, give the type of angle that you turn through and its size.

4 Calculate these lettered angles.

a 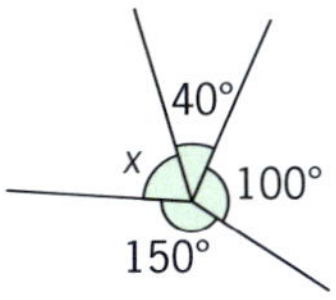

b 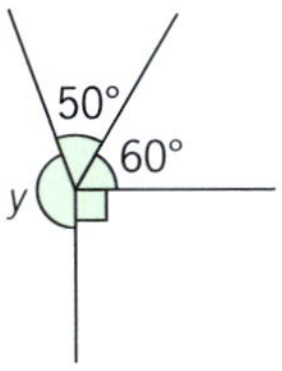

c 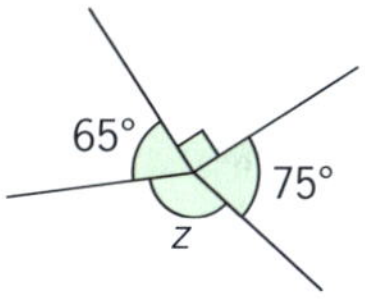

d 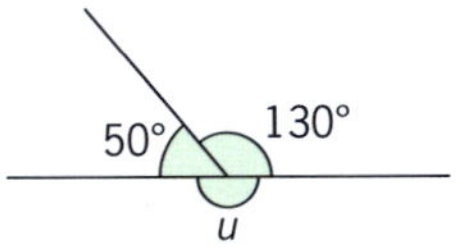

e 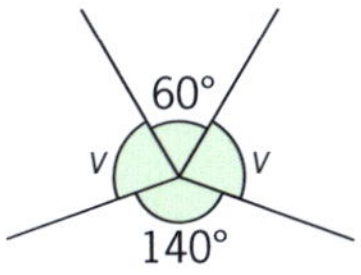

f 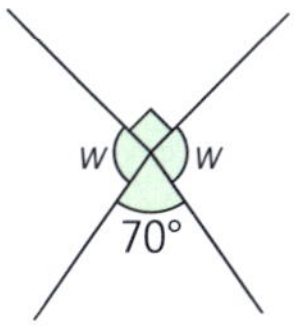

5 Through what angle does the Earth turn around the Sun between 1st March and 1st July?

6 **RICH TASK**

You draw two straight lines from the central coloured dot to any two of the outer black dots. How many different sizes of angle can you make? Find the sizes.

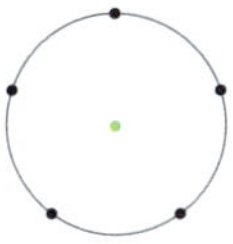

Unit B

Angles at a point

B10.3

- Explore angles at a point which make full, half and quarter turns

A **half turn** makes a straight line of $\frac{1}{2}$ of 360° = 180°.
Angles which make a half turn add up to 180°.

A **quarter turn** makes a right angle of $\frac{1}{4}$ of 360 = 90°.
Angles which make a quarter turn add up to 90°.

You already know that angles making a full turn add up to 360°.

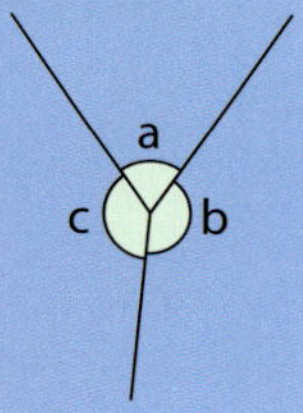

For example,
$a + b + c = 360°$.

EXAMPLE

a Find angle d in this diagram.

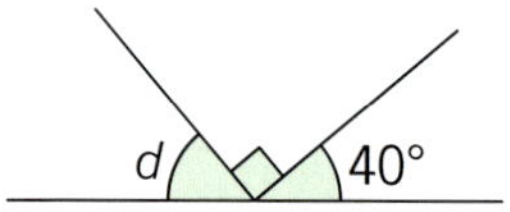

b This diagram shows steel girders of a bridge meeting at point P. Calculate angle g.

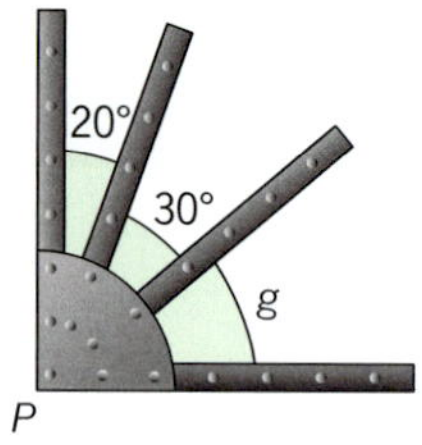

ANSWER

a The angles make a straight line of 180°.

$d + 90 + 40 = 180$

$d + 130 = 180$

$d = 180 - 130 = 50$

So angle d is 50°.

b The angles make a right angle of 90° at P.

$g + 20 + 30 = 90$

$g + 50 = 90$

$g = 90 - 50 = 40$

So angle $g = 40°$.

When two straight lines cross, they make two pairs of **vertically opposite** angles.
Vertically opposite angles are equal,
so $w = y$ and $x = z$.

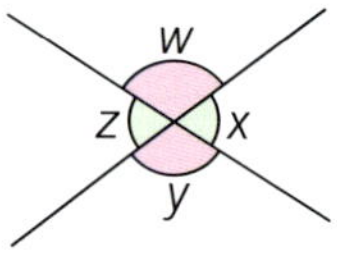

Imagine these scissors opening. The two angles that open up are equal to each other.

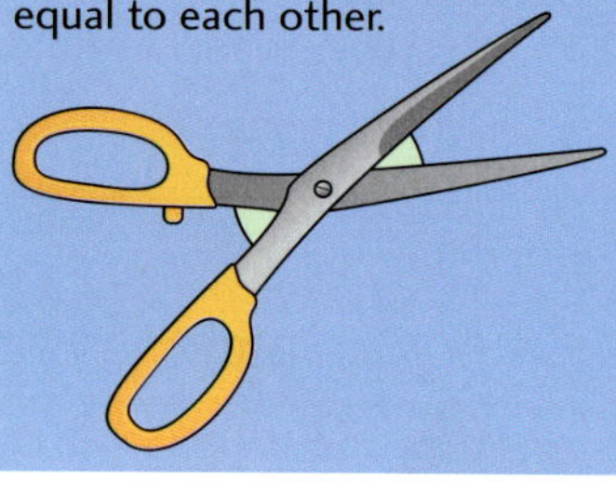

EXAMPLE

Calculate angles p, q and r where these two rail tracks cross.

ANSWER

Angle p is vertically opposite 130°. So $p = 130°$.
Angle q and 130° make a straight line of 180°. So $q = 180 - 130 = 50°$.
Angles q and r are vertically opposite. So $r = q = 50°$.

The angles p, q and r are 130°, 50° and 50° respectively.

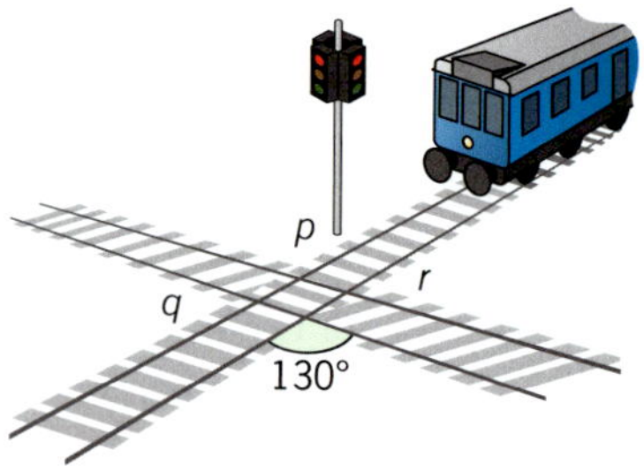

Exercise B10.3

LOW

1 Calculate these lettered angles.

a
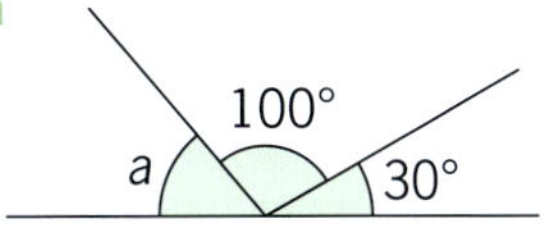

b
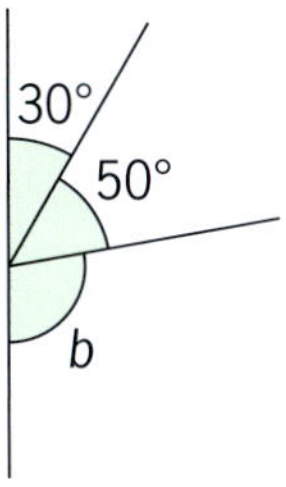

c
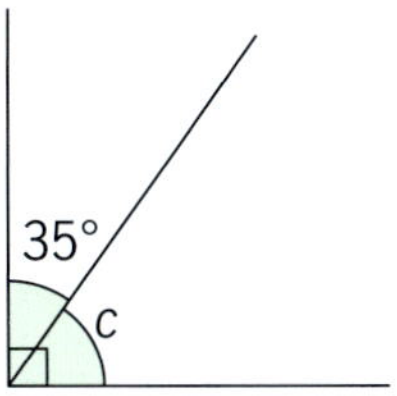

d
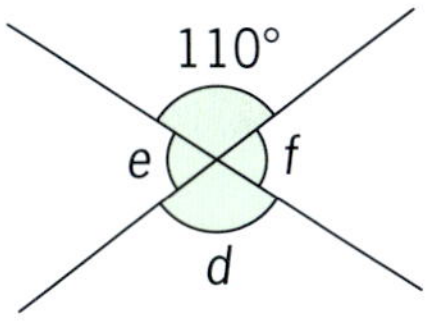

e
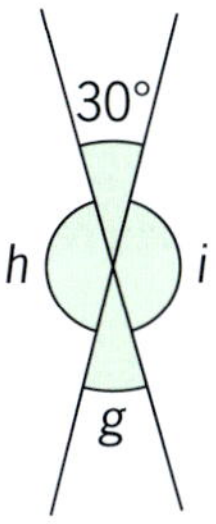

2 Calculate these lettered angles. Angles with the same letter are equal.

a
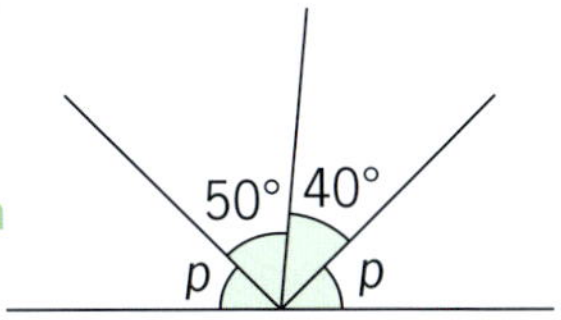

b
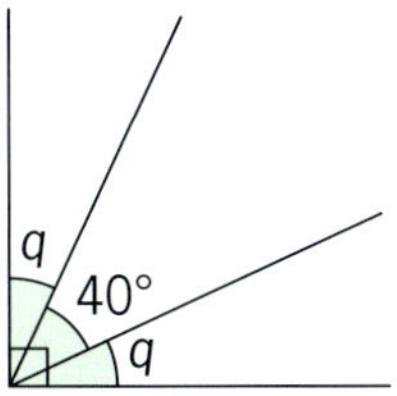

c
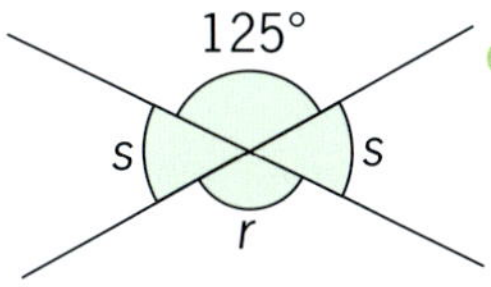

d
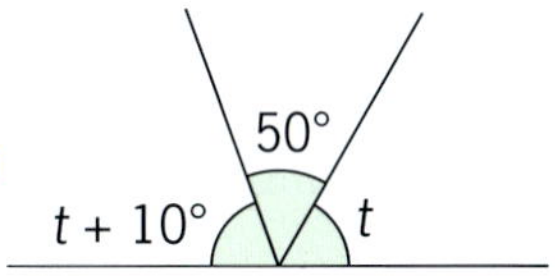

e
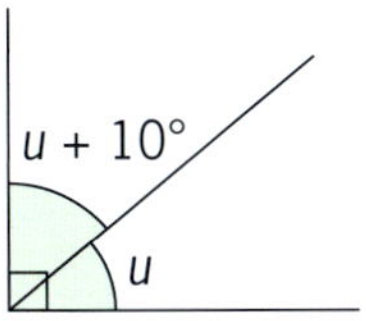

f
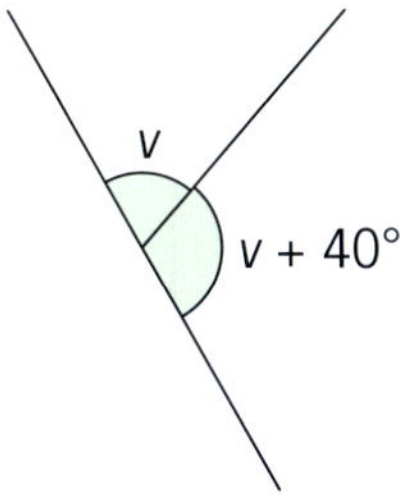

g
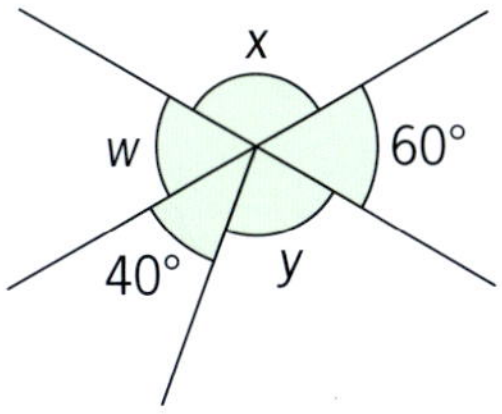

3 Steel girders are used to make this bridge.
Calculate the lettered angles.

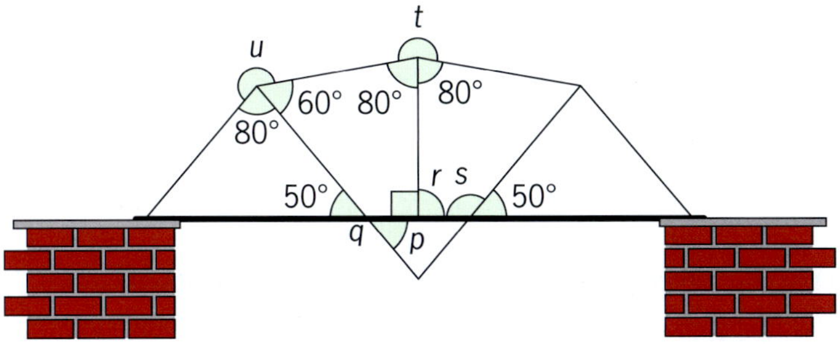

Unit B

Parallel lines and parallelograms

B10.4

- Know about angles on parallel lines and in parallelograms

Parallel lines run in the same direction and so they never meet.
A straight line crossing two parallel lines makes alternate angles and corresponding angles.

Angles a and b are **corresponding** angles.
Angle a can slide down without turning and fit exactly onto angle b.
So corresponding angles a and b are equal.

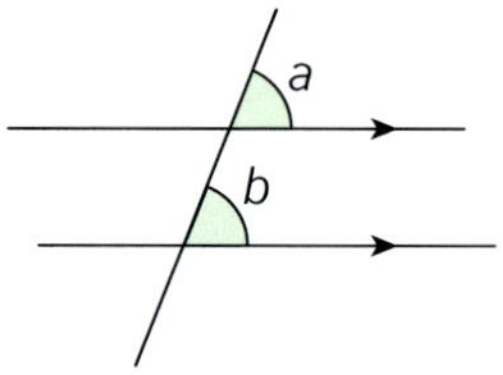

Corresponding angles are the angles inside a letter *F*.

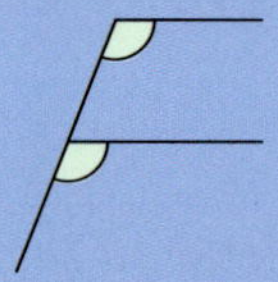

Angles c and d are **alternate** angles.
Angle c can slide down without turning to become vertically opposite angle d.
So alternate angles c and d are equal.

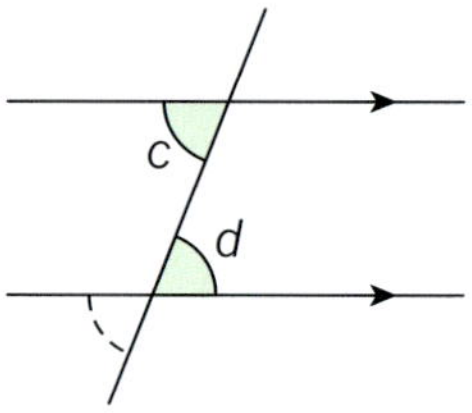

Alternate angles are the angles inside a letter *Z*.

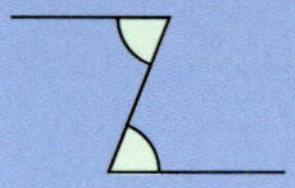

EXAMPLE

Find angles p, q, r and s.

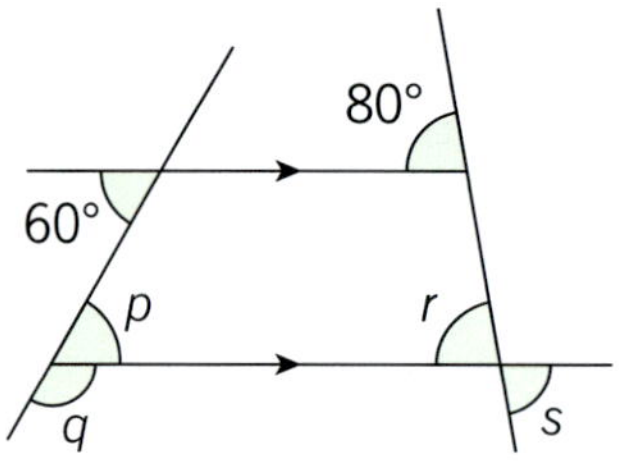

ANSWER

$p = 60°$ (alternate angles in a **Z**)
$q = 180 - 60 = 120°$ (angles on a straight line)
$r = 80°$ (corresponding angles in an **F**)
$s = 80°$ (vertically opposite angle r)

A **parallelogram** is a quadrilateral with opposite sides parallel.
You use arrowheads to show the parallel sides.

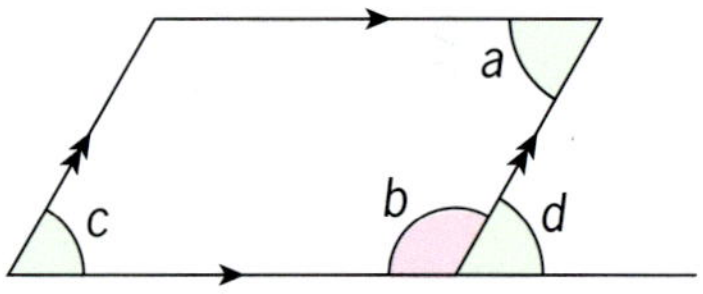

This parallelogram has one side extended to make equal alternate angles $a = d$ and equal corresponding angles $d = c$.
Angles d and b make a straight line, so $d + b = 180°$.

So opposite angles of a parallelogram are equal and adjacent angles a and b add to make 180°.

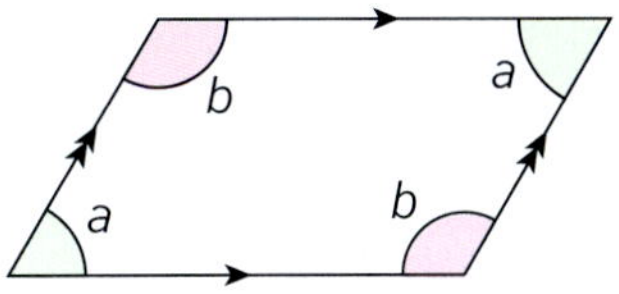

EXAMPLE

Find angles x, y and z.

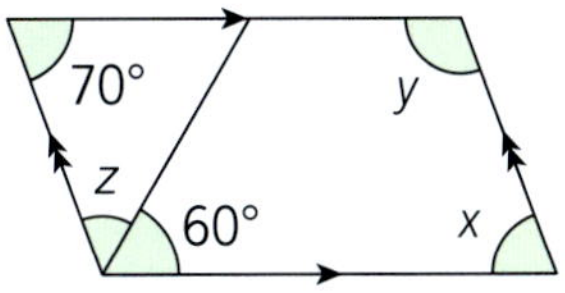

ANSWER

$x = 70°$ (opposite angles in parallelogram)

$y = 180 - 70 = 110°$ (adjacent angles in parallelogram)

$z + 60° = 110°$ (opposite angles in parallelogram)

so $z = 50°$.

Exercise B10.4

MEDIUM

1 Draw any two parallel lines. Draw a third line to cross the parallel lines. How many angles have you made in your diagram?
Measure all the angles with your protractor.
Find pairs of corresponding and alternate angles and colour them differently.

2 Find the lettered angles in these parallelograms.

a
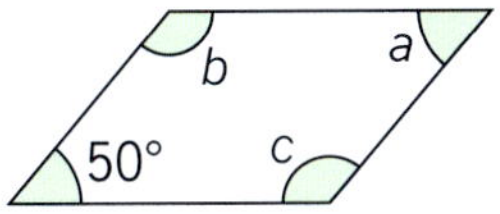

b
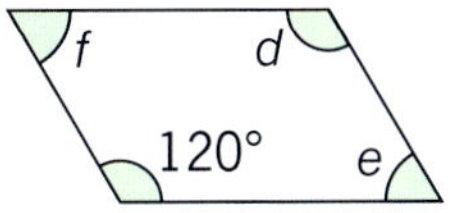

c
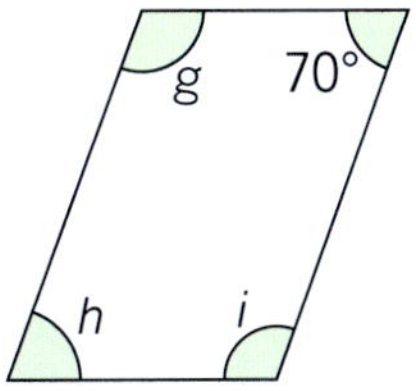

d
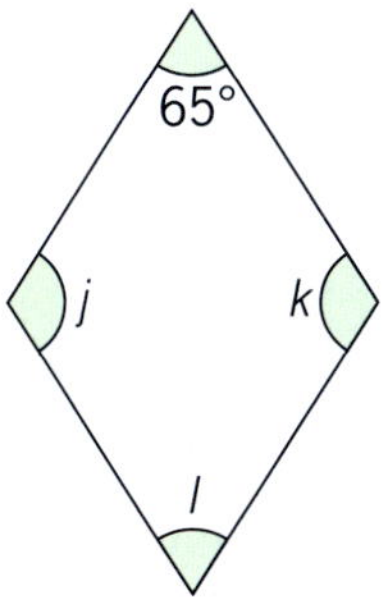

3 Find the lettered angles in these shapes.

a
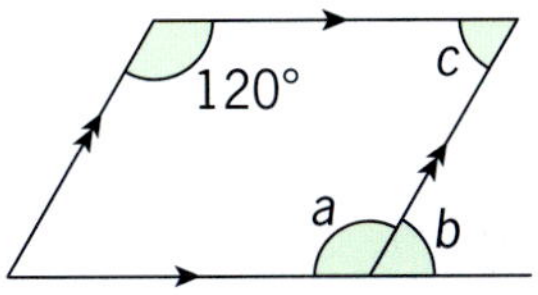

b
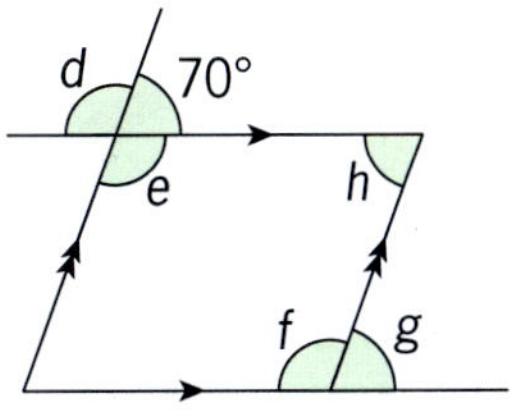

c
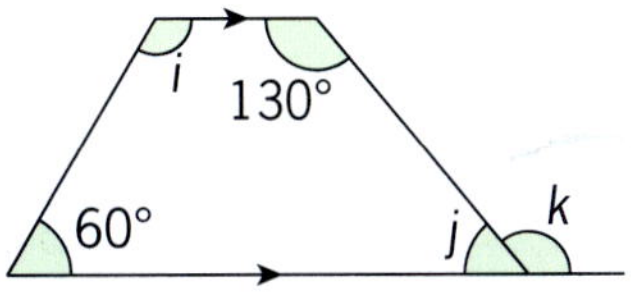

Continued ➤

Unit B

4 Find the lettered angles on these parallel lines.

a
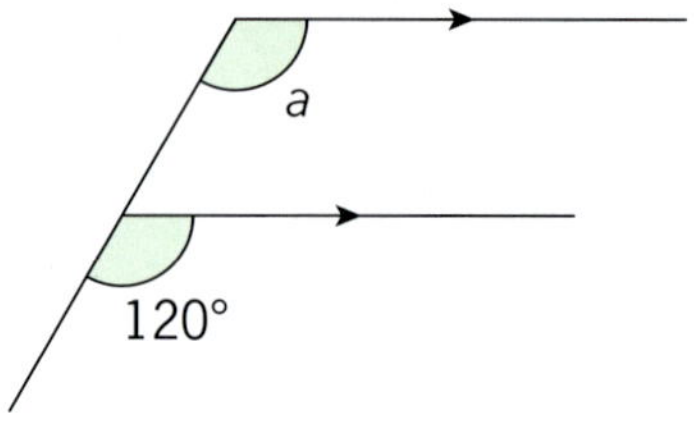

b
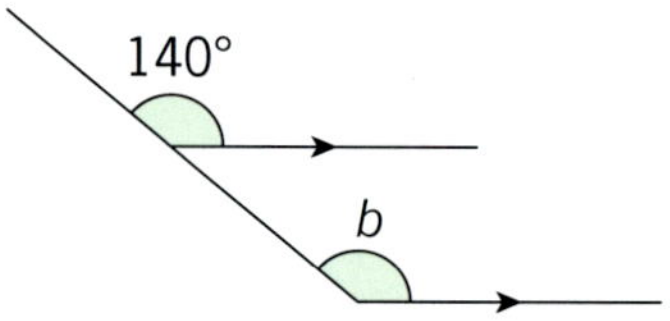

c
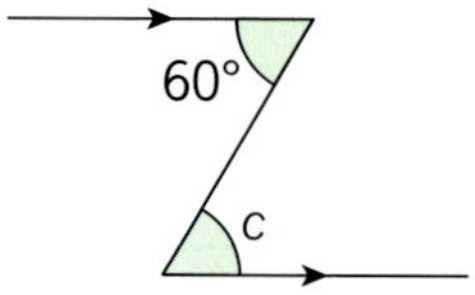

d
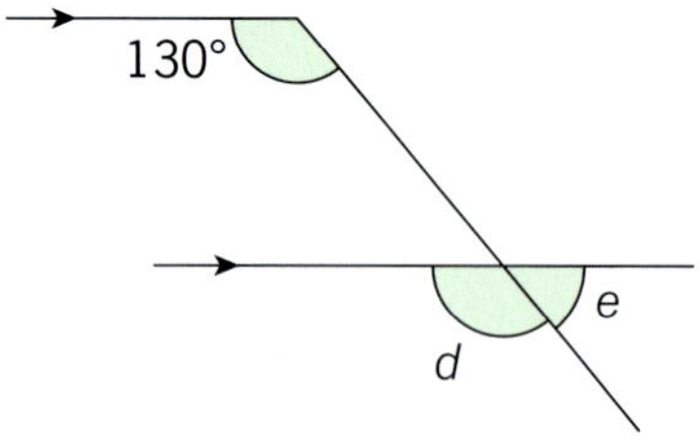

e
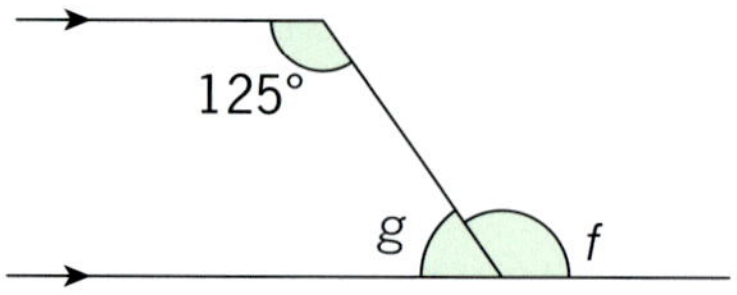

f
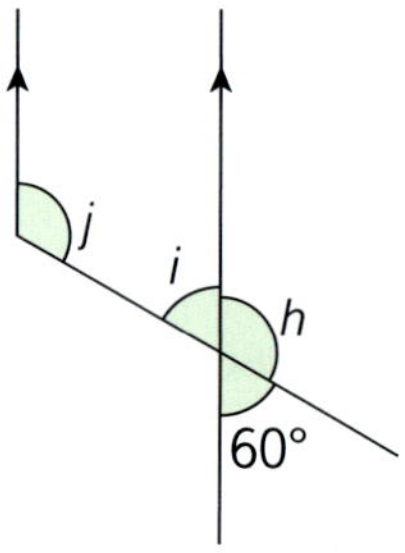

g
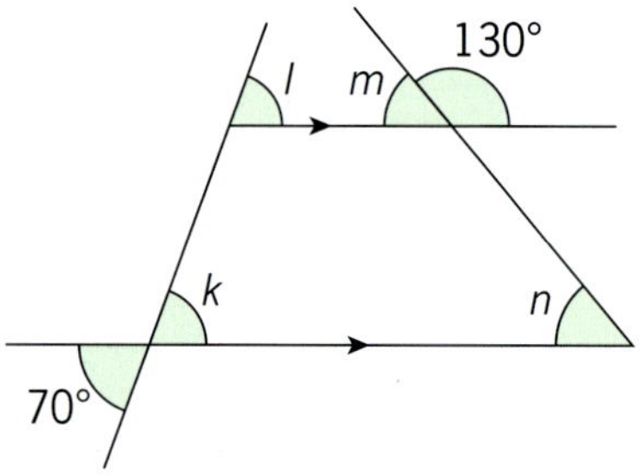

h
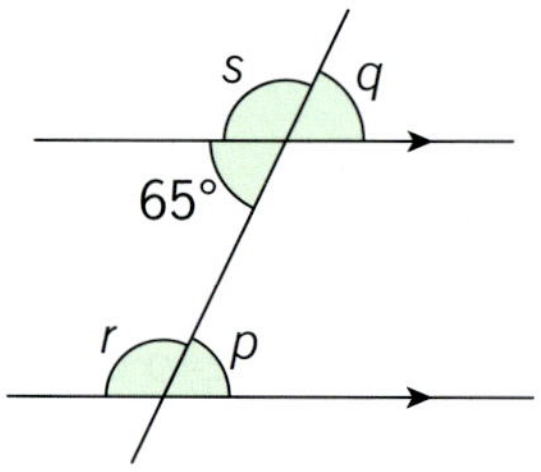

5 A telegraph pole falls across a railway line.
Find the angles p and q.

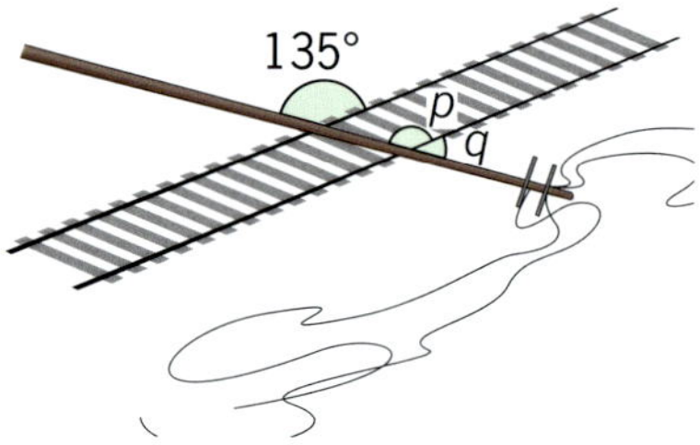

6 A pencil lies on a sheet of lined writing paper.
Find the lettered angles.

Angles in triangles and quadrilaterals

- Revise various facts about triangles and quadrilaterals and prove some of them

RICH TASK

A baby bouncer is attached in a doorframe with sides AB and CD elastic and BC rigid. The baby bounces vertically. Find a relationship between angles x and y and draw its graph. Which part of the graph gives the only practical possibilities for the baby?

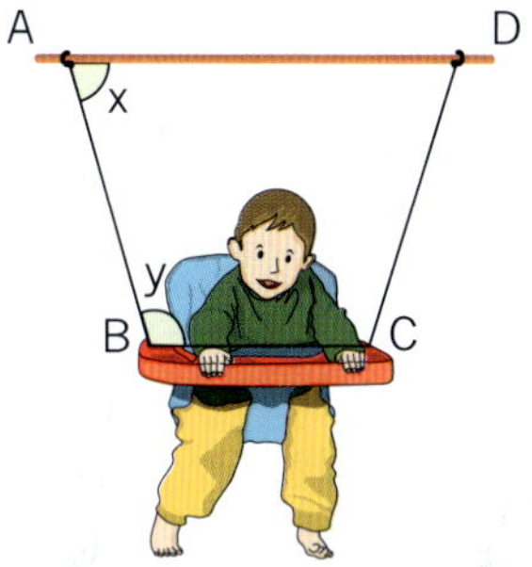

You already know this theorem, but you may not know how to prove it.

A theorem is a statement of a fact which has been proved.

The sum of the angles of any triangle is 180°.

Proof

Let the angles of triangle ABC be a, b and c.
Draw a line DE through C parallel to AB.
So $D\hat{C}A = a$ (alternate angles)
and $E\hat{C}B = b$ (alternate angles).
At point C on the straight line DE, $a + c + b = 180°$.
So the three angles of any triangle ABC add up to 180°.

Can you use this theorem to prove two more theorems?

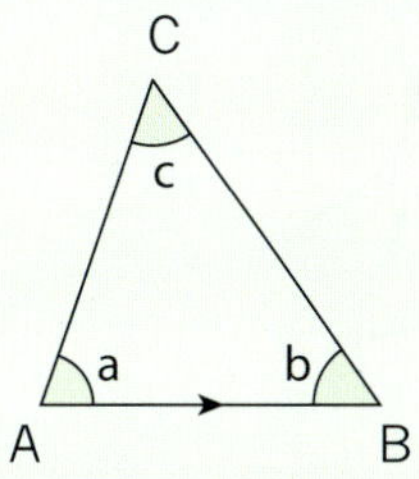

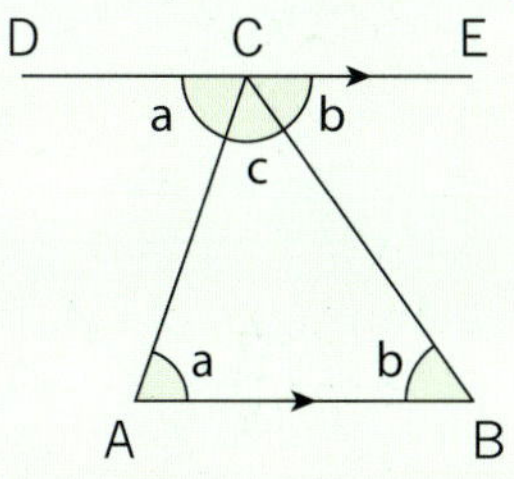

An exterior angle of a triangle is equal to the sum of the two opposite interior angles.

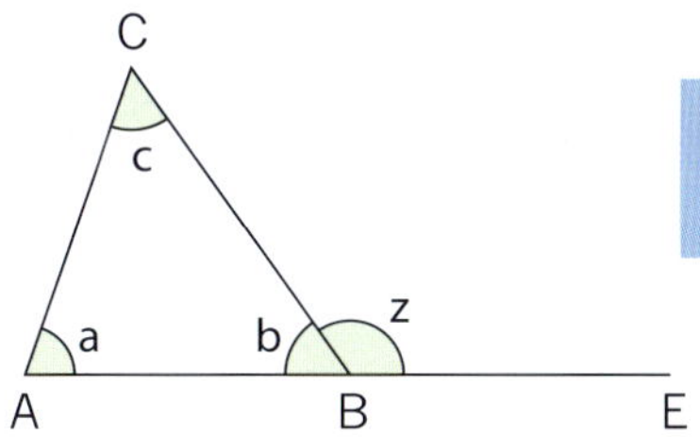

This theorem states that the exterior angle $z = a + c$.

The sum of the interior angles of any quadrilateral is 360°.

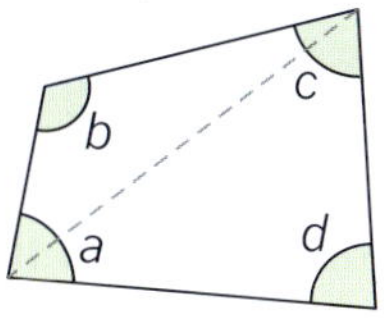

This theorem states that $a + b + c + d = 360°$.

EXAMPLE

This symmetrical footbridge is made from steel girders.
Find the size of angle x.

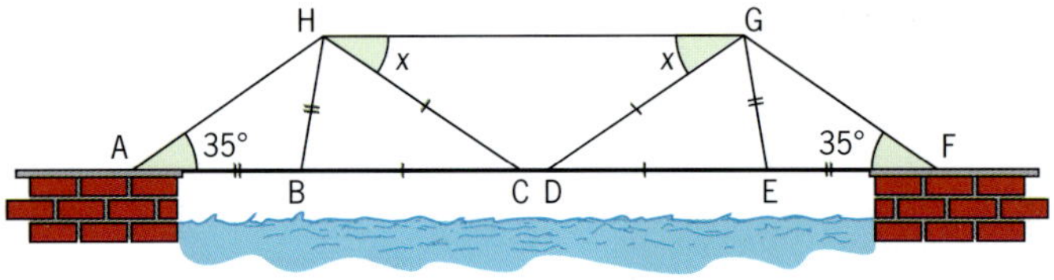

ANSWER

$\triangle$ ABH is isosceles, so $A\hat{B}H = 180 - 2 \times 35 = 110°$
On straight line ABC, $H\hat{B}C = 180 - 110 = 70°$
$\triangle$ HBC is isosceles, so $B\hat{C}H = 180 - 2 \times 70 = 40°$
On straight line BCD, $D\hat{C}H = 180 - 40 = 140°$
In quadrilateral HCDG, $x + x + 140 + 140 = 360°$
Angle $x = \frac{1}{2}(360 - 280) = 40°$

Exercise B10.2

MEDIUM LOW

Find the lettered angles.

1

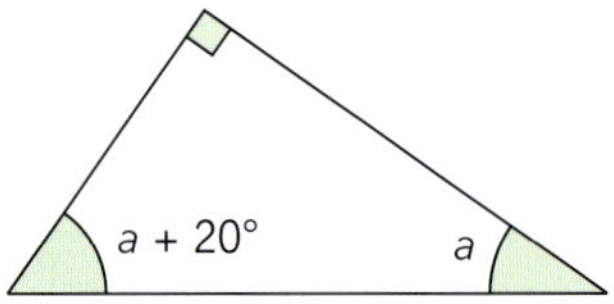

2

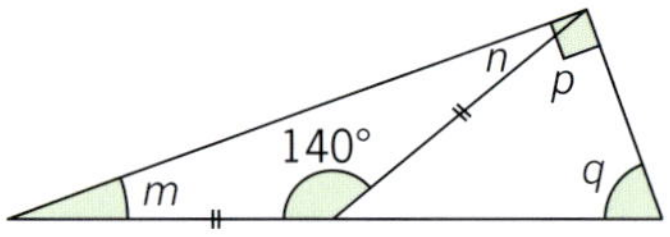

3

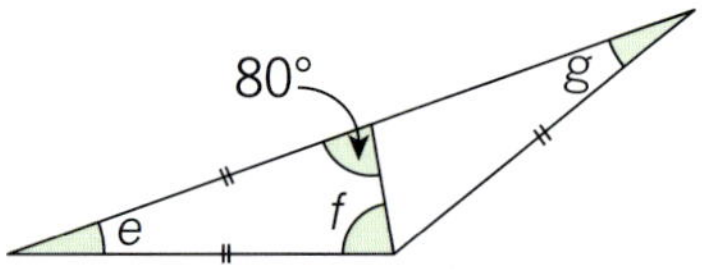

4

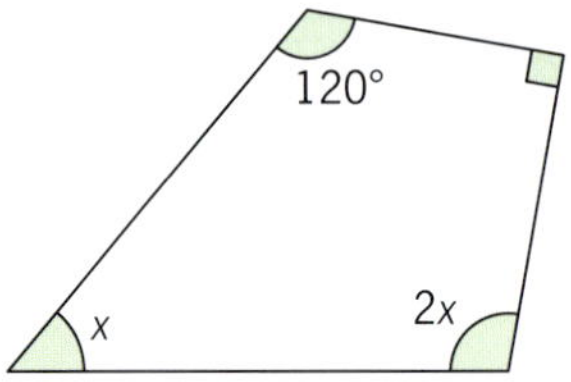

5

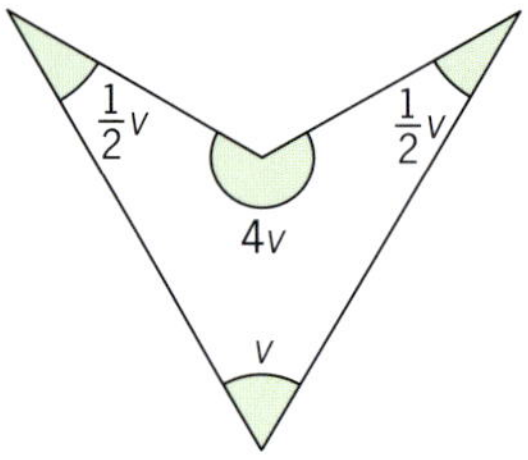

6

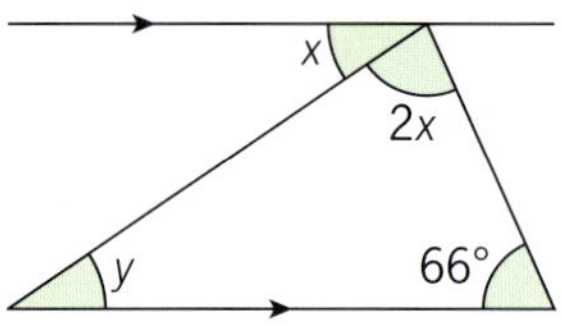

7

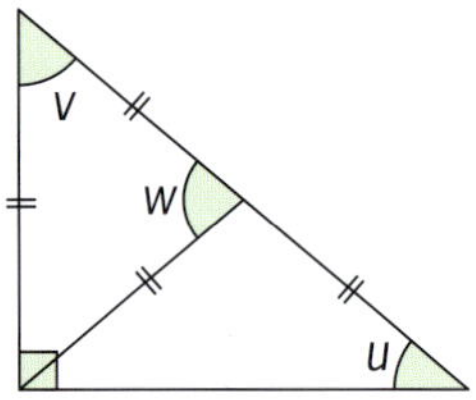

8

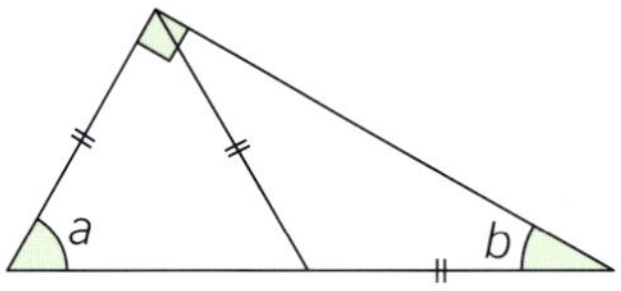

9

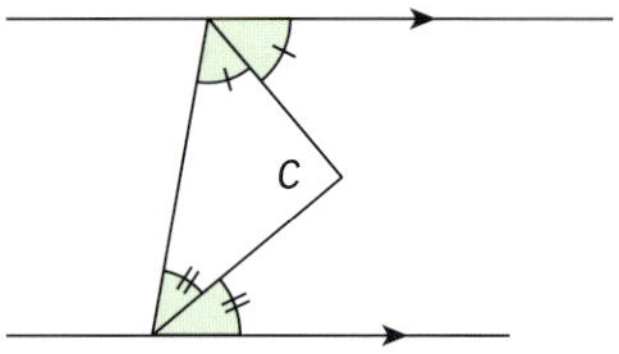

10 Triangle ABC has angle B which is 10° greater than angle A and angle C which is 20° less than angle A. Find angle C.

11 As you visit the four angles of a quadrilateral in turn, each angle is double the previous angle. Find all the angles.

Angles in polygons

B10.6

- Revise the properties and angles of polygons (including quadrilaterals)

ACTIVITY

Here are the names of the common **quadrilaterals**.

Square	Rectangle	Parallelogram
Trapezium	Rhombus	Kite

The **properties** of quadrilaterals tell you about their sides and angles. Draw an example of each of these quadrilaterals and say which of them **must** have

- all sides equal
- all angles equal
- both pairs of opposite sides equal
- both pairs of opposite sides parallel
- both pairs of opposite angles equal
- both diagonals equal
- diagonals which are perpendicular
- diagonals which bisect each other.

What other properties can you list?

'Quad' means 'four' in words such as quad-bike, quadruplets, quadrangle.

Polygons are 2-D shapes with straight edges.
Polygons can be split into triangles by joining one corner to all other corners, as for this hexagon.
The sum of the angles of a polygon can be found from the angles of these triangles.
Sum of angles of a hexagon = $4 \times 180° = 720°$

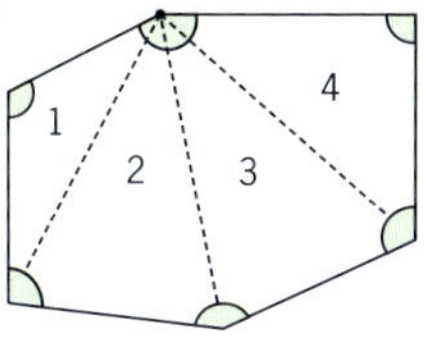

Poly- comes from the Greek word for 'many'.

EXAMPLE

The design of an arrow head is based on a hexagon with two angles of 100°, an angle of 70° and three other equal angles, α.
Calculate the value of the interior angles α and the exterior angles β.

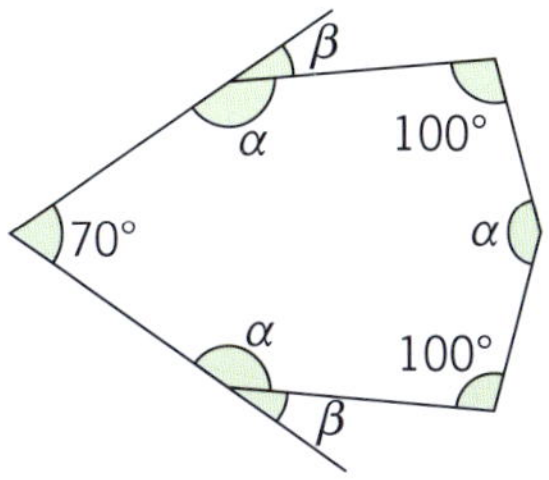

ANSWER

The sum of the six interior angles =
$2 \times 100° + 70° + 3 \times \alpha = 720°$
So $3\alpha = 720 - 200 - 70 = 450°$
Interior angle $\alpha = 450 \div 3 = 150°$
and exterior angle $\beta = 180 - \alpha = 180 - 150 = 30°$

RICH TASK

You can find the sum of the interior angles of any polygon.

For an octagon 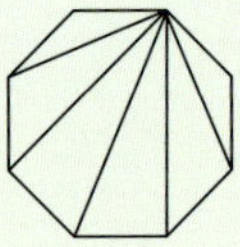 For a pentagon 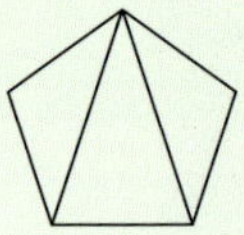

6 triangles, and 6 × 180° = 1080°   3 triangles, and 3 × 180° = 540°

Can you find a general rule?

Exercise B10.3

MEDIUM

1 Copy and complete for **regular polygons**.

Polygon	Number of sides	Each interior angle	Each exterior angle	Sum of all interior angles
Triangle	3			180°
Quadrilateral	4			
Pentagon	5			
Hexagon	6			
Octagon	8			
Decagon	10			

Can you find a general rule?

RICH TASK

Construct a circle of radius 5 cm.
How can you use this to construct a regular hexagon?
Try other regular polygons.

2 Find the lettered interior and exterior angles.

a

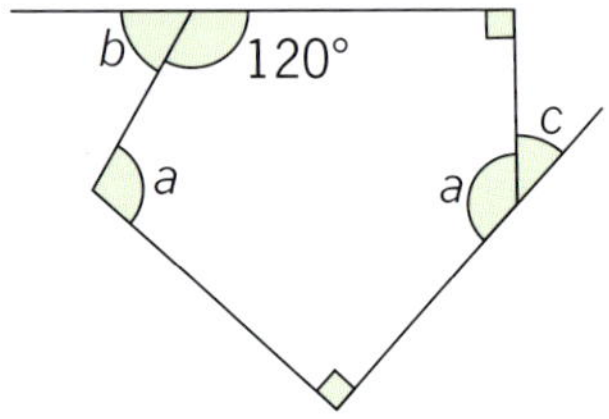

b

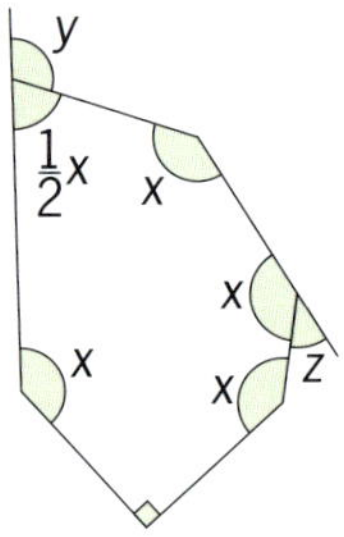

c

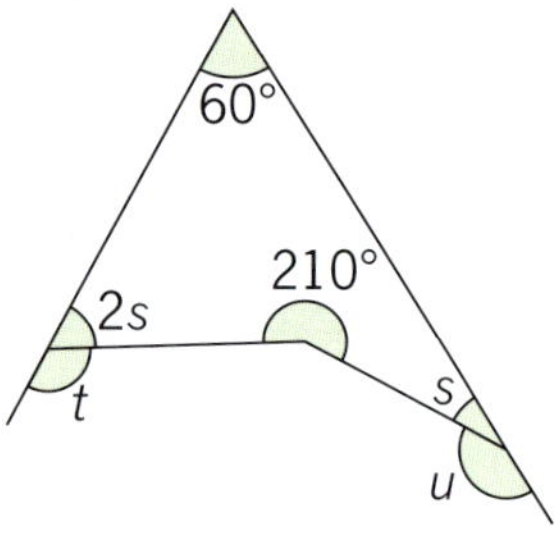

3 **a** A hexagon has three angles of $x°$ each, an angle of $x + 29°$, an angle of $x - 10°$ and a further angle which is double the smallest angle. Find all the angles.

b A pentagon has two angles the same size and three equal angles which are each 10° bigger. Find all the angles.

4 A pentagon has four equal angles and a fifth angle of 100°.
It is used in the design of a star.
Calculate the angles at the points of the star.

7 This is an empty clothes drier.
Find the lettered angles.

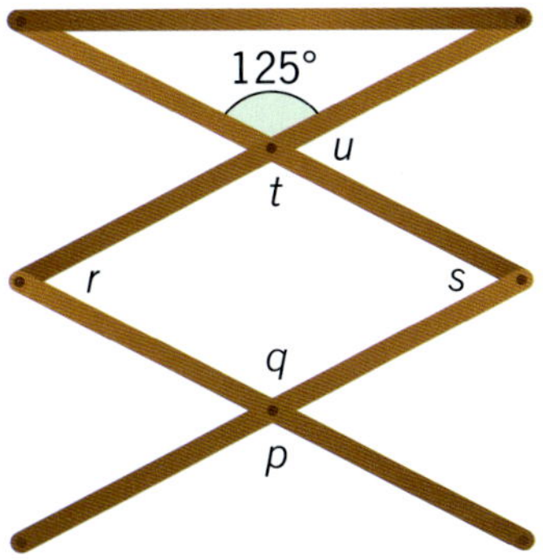

8 Paving slabs like this one are laid across a lawn.
They must touch each other *either* along a side *or* at a corner.
Show several different ways that they might be positioned.
Find any angles that your designs involve.

9 ☆ **CHALLENGE** ☆

A kitchen is fitted with Venetian blinds.
As the blinds open and close, angles x and y change.
Find an expression for angle y in terms of angle x.
Can you draw a graph of the relationship between y and x?
What are the values of x and y when the blind is closed?

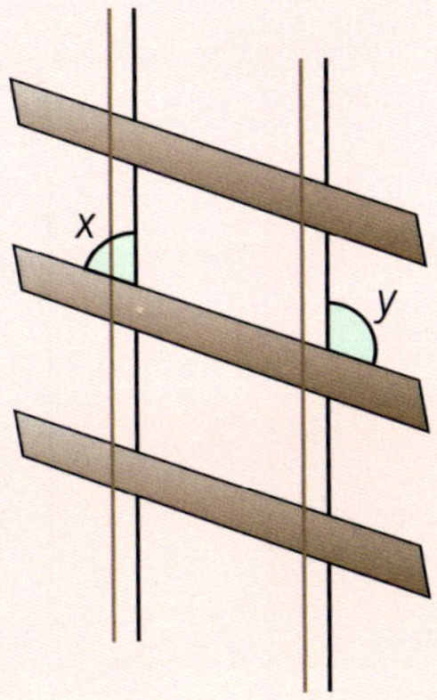

This picture shows, a 'sideways on' view of the blinds.

Unit B

Summary and assessment

CHECK OUT

You should now be able to

- recall and use properties of angles
- solve problems using angles in parallel lines and in polygons.

Exam-style question

The diagram shows an isosceles triangle XYZ drawn on a grid.
X and Y are fixed, but Z can change position.

a Write down the coordinates of Z if XYZ were to be an isosceles right-angled triangle.

b Where could Z be if the triangle were isosceles, but not right-angled?

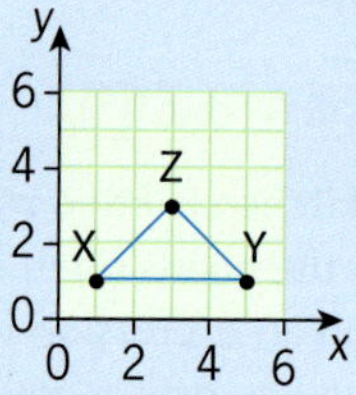

Pierre's answer:

a (5, 5) or (1, 5) or (5, -3) or (1, -3) or (3, 3) or (3, -1)

b could be anywhere on the line $x = 3$, except (3, 3) or (3, -1)

Use the diagram to help.
Add points for Z in pencil and draw in new triangles.

XY is 4 squares long.

Draw in Z so that XZ or YZ is 4 squares long.

Don't forget the right-angle.

The diagram given is isosceles.

Imagine Z moving along the line $x = 3$.

Exam practice

*1 These shapes fit together.

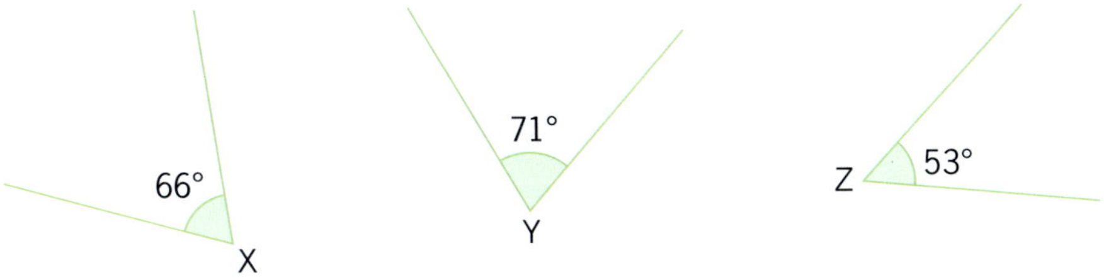

If you put XYZ together will they form a straight line?
Explain how you arrived at your answer. *(2 marks)*

2 ABC and ABD are right-angled triangles.
ACD is an isosceles triangle.
ACDE is a parallelogram.

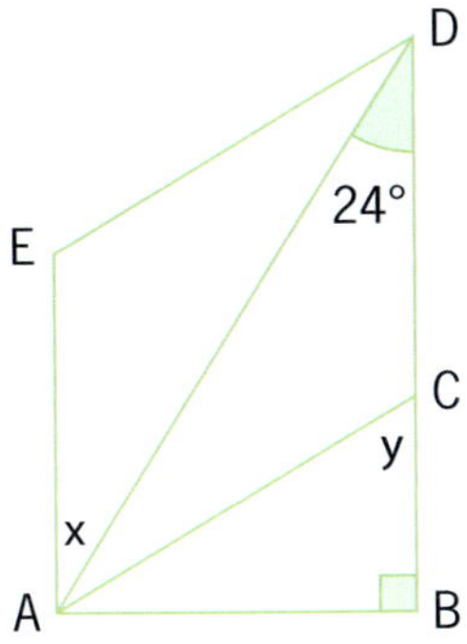

Work out x and y. *(2 marks)*

3 This shape is made from five isosceles triangles with a regular pentagon at its centre.

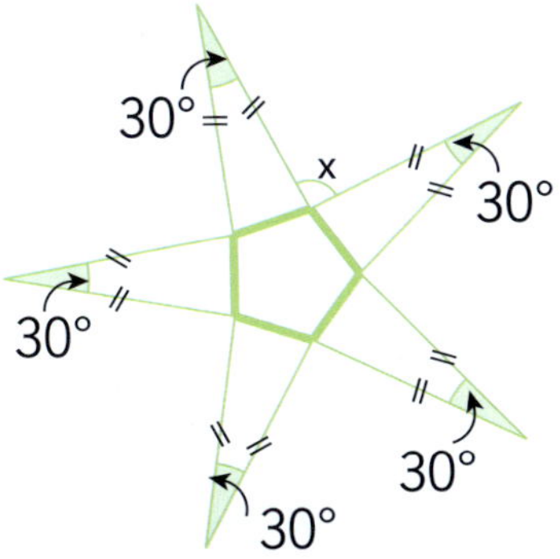

Find the size of angle x. *(4 marks)*

11 Straight lines and inequalities

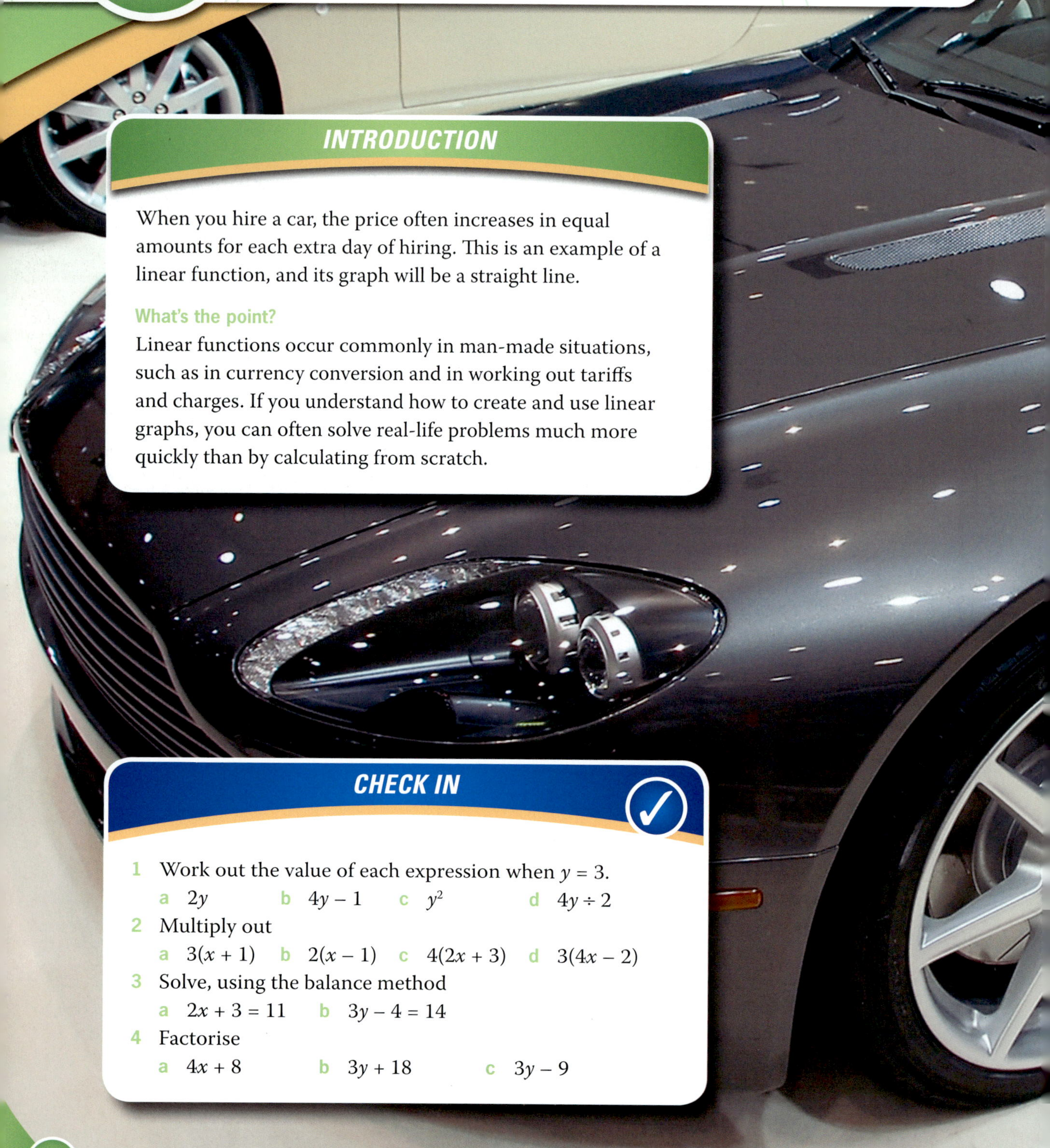

INTRODUCTION

When you hire a car, the price often increases in equal amounts for each extra day of hiring. This is an example of a linear function, and its graph will be a straight line.

What's the point?

Linear functions occur commonly in man-made situations, such as in currency conversion and in working out tariffs and charges. If you understand how to create and use linear graphs, you can often solve real-life problems much more quickly than by calculating from scratch.

CHECK IN

1 Work out the value of each expression when $y = 3$.

a $2y$ b $4y - 1$ c y^2 d $4y \div 2$

2 Multiply out

a $3(x + 1)$ b $2(x - 1)$ c $4(2x + 3)$ d $3(4x - 2)$

3 Solve, using the balance method

a $2x + 3 = 11$ b $3y - 4 = 14$

4 Factorise

a $4x + 8$ b $3y + 18$ c $3y - 9$

Orientation

What I shoud know	What I will learn	What this leads to
A7	■ Construct functions from real life problems ■ Interpret graphs ■ Solve inequalities	B14

RICH TASK

To win a game of 'four-in-a-line,' a player needs to have placed four of their pieces in a line. In this game the winning line of four is clearly marked.

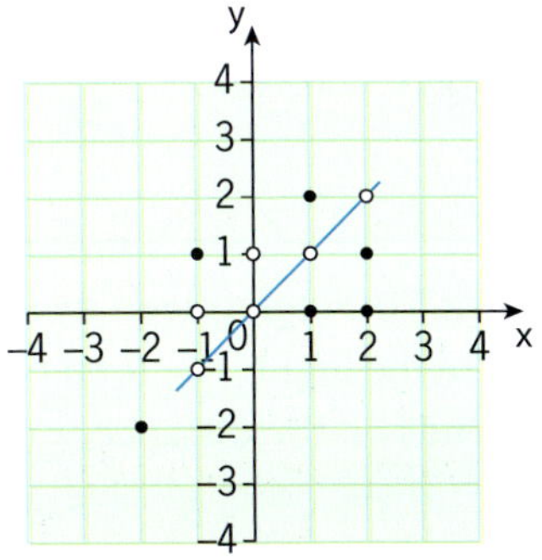

What is the connection between the x and y coordinates of the points on the winning line? Investigate other winning lines of four that are possible on the grid.

Real-life functions

- Construct functions from real-life problems and find their graphs

When scientists, engineers and economists measure **variables**, they often find that one variable depends on another. The relationship between them is often a **function**. The function can be written as a **formula** or as a **flow diagram**.

ACTIVITY

Weather forecasters in the UK measure temperatures in °C (Celsius). In the USA, they are measured in °F (Fahrenheit). This flow diagram changes °C to °F.

C → [×9] → [÷5] → [+32] → F

Use the flow diagram to check that Celsius can be changed to Fahrenheit using this formula $F = \frac{9C}{5} + 32$.

Water freezes at 0 °C.

Use the flow diagram or formula to check that this temperature is the same as 32 °F.

Water boils at 100 °C.

Show that this temperature is the same as 212 °F.

In the activity, a table of values and a graph can be constructed.

°C	0	100
°F	32	212

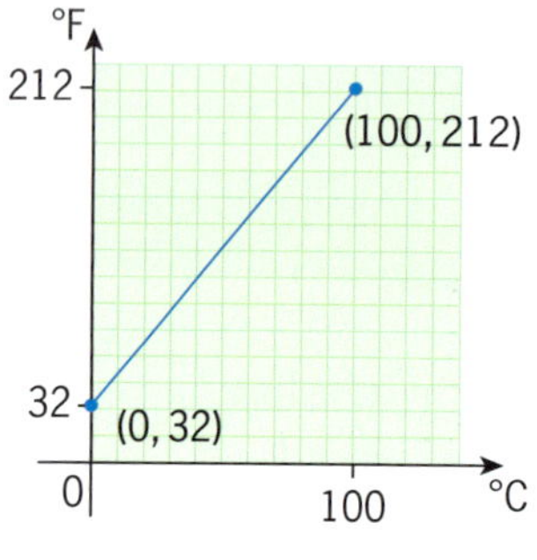

What is the least number of points you need to draw a straight-line graph?
Do you think it is wise to use only the least number of points?

EXAMPLE

When Mr Singh travels by air, he arrives at the airport 3 hours before the flight. The total time that his journey takes, T hours, depends on the distance, x miles, that he flies, where

$$T = \frac{x}{500} + 3.$$

Construct a table of values and draw a graph of the results.
How far does he travel if the total journey takes 8 hours?

EXAMPLE (CONT.)

Unit B

ANSWER

When $x = 500$, $T = \frac{500}{500} + 3 = 1 + 3 = 4$.

When $x = 1000$, $T = \frac{1000}{500} + 3 = 2 + 3 = 5$.

Similarly, when $x = 2000$, $T = 7$ and when $x = 3000$, $T = 9$.

The table and graph are:

x, miles	500	1000	2000	3000
T, hours	4	5	7	9

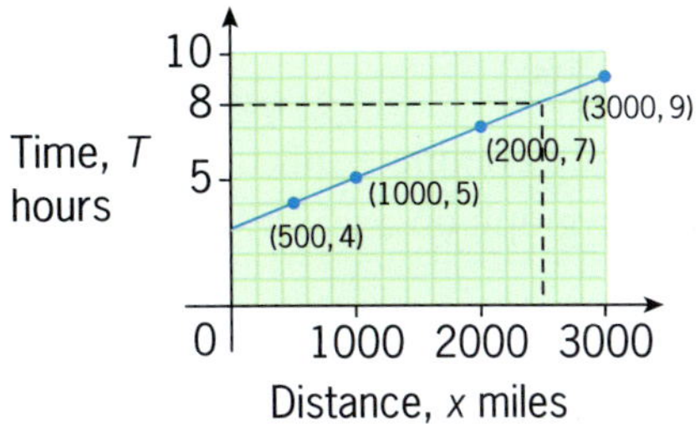

From the graph, when $T = 8$, $x = 2500$.
So a journey of 8 hours means that Mr Singh has flown 2500 miles.

Exercise B11.1

MEDIUM LOW

1 During a storm, the distance between you and the storm, y miles, can be found by counting the number of seconds, x, between the thunder and the lightning.
Copy and complete this table, given that $y = \frac{x}{5}$.

No. of seconds, x	5	10	15	20	25	30
No. of miles away, y						

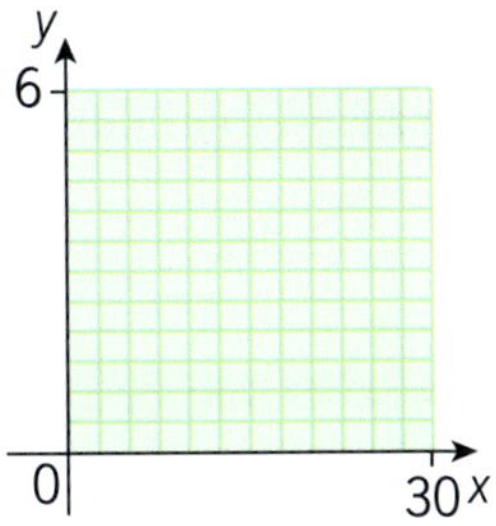

Draw a graph of y against x on axes as shown.

2 A youth club hires a hall for a disco. The cost of electricity, £C, depends on the number of hours, x, that the hall is needed.
Copy and complete this table using the formula $C = 2x + 4$.

No. of hours, x	1	2	3	4	5
Cost, £C					

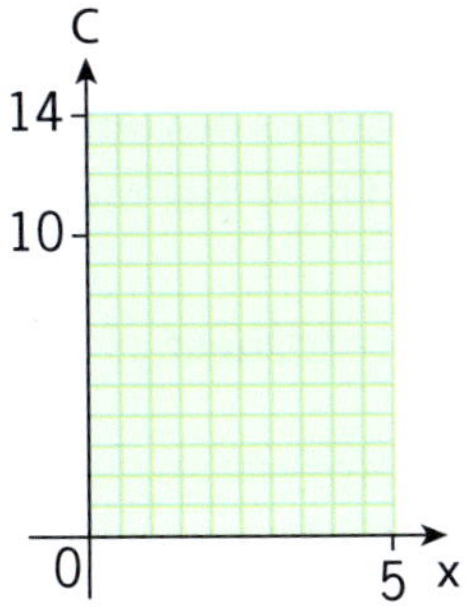

Draw the graph of C against x on axes as shown here.
For how many hours is the hall used, if the charge for electricity is

a £11 b £8?

Continued

3 Alan hires a van. The hire company charges him a basic £20 and then 2 pence for every mile that he travels. If he travels x miles, the cost £C is given by

$$C = \frac{2x}{100} + 20$$

Copy and complete this table.
Draw a graph on axes as shown here.

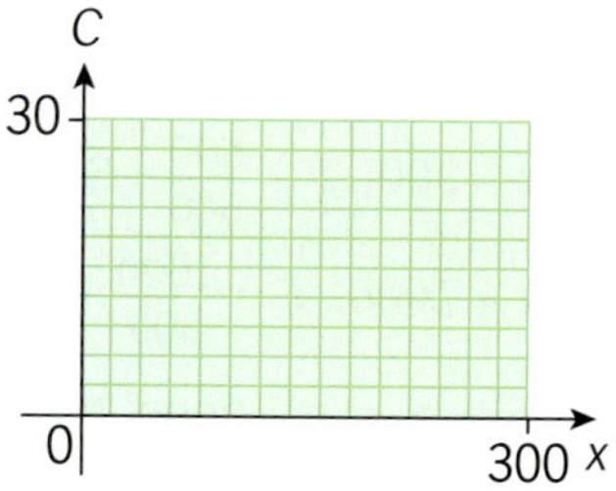

x, miles	50	100	200	300
C, £				

If Alan uses the van for 150 miles, how much is he charged?

*4 **RICH TASK**

As children grow, they need less sleep. A rough guide to the hours of sleep needed, h, is given by their age, a years, where $h = 17 - \frac{a}{2}$.

Construct a table of values of your choice and draw the graph of h against a.

Does this model seem reasonable? Give your reasons for or against.

5 **RICH TASK**

a The mass of the average man, m kg, can be found from his height, h cm, where

$$m = \frac{h - 100}{10} \times 9.$$

Construct a table of values of your choice and draw a graph of m against h. Does this model of the average man seem reasonable?

b Set up a computer spreadsheet as a table of values for m and h.

Real-life graphs

B11.2

- Construct linear functions from real-life scenarios and draw their graphs

RICH TASK

4 A farmer makes a rectangular sheep pen so that it always has a perimeter of 24 metres, but its length x and width y can vary.
Find the possible integer values of x and y.
Draw a graph of all possibilities and find its equation.
Which values of x and y give the maximum possible value of the area of the sheep pen? What kind of rectangle is it in this case?

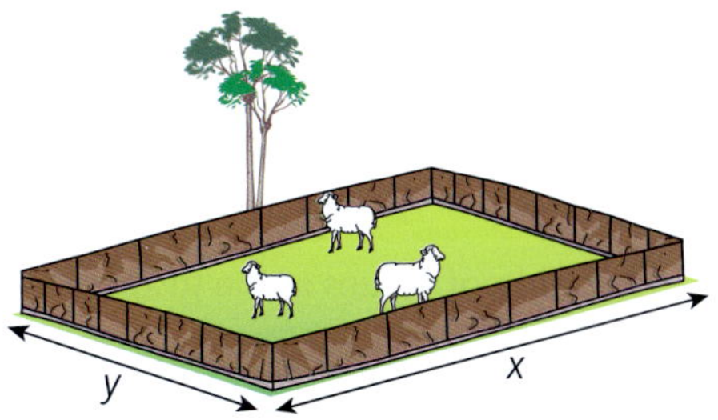

ACTIVITY

A car's windscreen wiper makes angles of x and y with the bottom of the windscreen as shown.

36 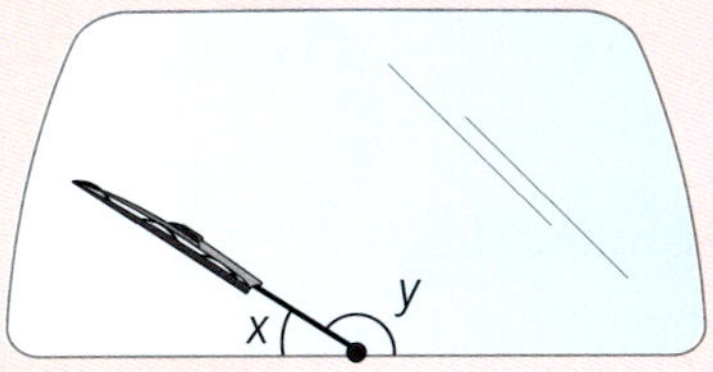

If $x = 60°$, what is the value of y?
What is the equation which links x and y?
Are the values in this table correct? Why?

$x°$	10	60	90	150
$y°$	170	120	90	30

These values can be used to draw the graph of y against x.
Why cannot x have values greater than 180°?

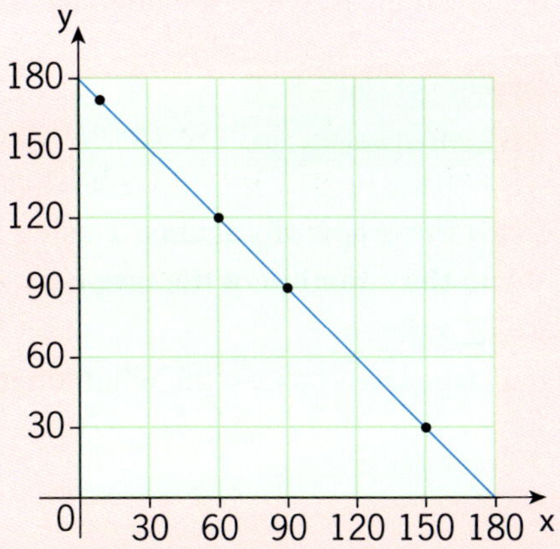

EXAMPLE

A cylindrical water tank has 6 cm of water in it.
A tap is then left running into the tank so that the water level rises 2 cm every minute.
Construct a table to show the depth of water, y cm, against a time of x minutes.
Draw a graph of y against x and write its equation.
How deep is the water after an hour?

ANSWER

At the start, $x = 0$ and $y = 6$
After 1 minute, $x = 1$ and $y = 6 + 2 = 8$
After 2 minutes, $x = 2$ and $y = 6 + 2 \times 2 = 6 + 4 = 10$
After 3 minutes, $x = 3$ and $y = 6 + 2 \times 3 = 6 + 6 = 12$ and so on.

Here are a table of values and a graph.

Time, x minutes	0	1	2	3
Depth, y cm	6	8	10	12

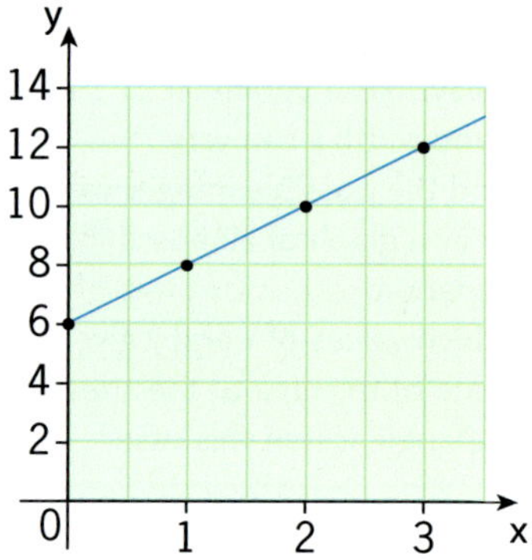

The equation of the graph is $y = 6 + 2x$.
After an hour, $x = 60$ so the depth of water, $y = 6 + 2 \times 60 = 126$ cm.

Exercise B11.2

MEDIUM

1 A square lawn is laid with turf. Plastic edging is fitted round the edge of the lawn.

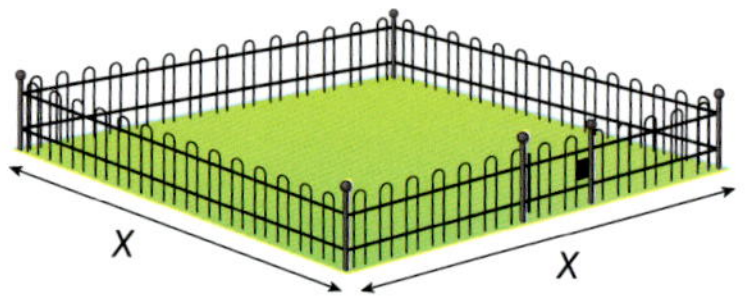

a Copy and complete this table for the length of the plastic edge, y metres, when the lawn has sides x metres long.

Length of sides, x	5	6	7	8	9	10
Length of edging, y						

b Draw the graph of y against x using axes as shown here.
Write the equation of the graph by copying and completing $y = \square \times x$.

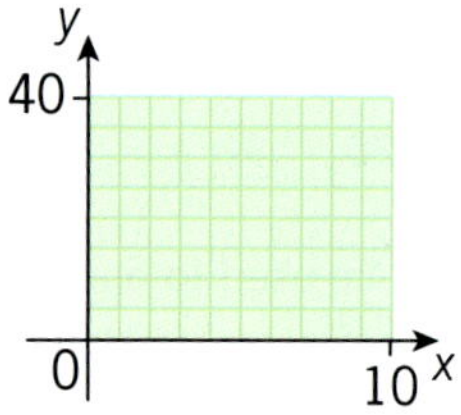

c If a large lawn has $x = 25$, what length of edging is needed?

Unit B

2 A car's petrol tank will hold 40 litres.

a Copy and complete this table of values when the car starts with a full tank.

Amount used, x litres	0	10	20	30	40
Amount left in tank, y litres	40				

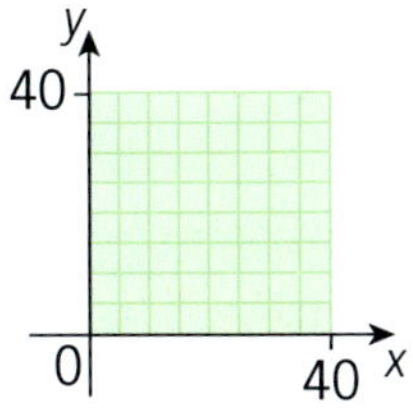

b Draw a graph of y against x on axes labelled as shown here. Write the equation of the graph.

c If $x = 35$, what is the value of y? If the car travels 12 miles on each litre, how far will it have travelled in this case?

3 A taxi cab driver charges you a basic amount of £2 and then £3 for each kilometre of the journey.

a Copy and complete this table of values.

Length of journey, x km	1	2	3	4	5
Cost, £y					

b Draw a graph of y against x. Copy and complete the equation of the graph $y = \square \times x + \square$.

c If $x = 7$, what is the value of y?
Will a £20 note be enough to pay for this journey?

4 **RICH TASK**

This triangle is isosceles.
Make a table of values of y for different values of x.
Draw a graph of y against x and write down its equation.

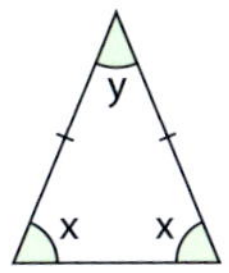

5 **RICH TASK**

How many degrees, y, does the minute hand of a clock turn through in x minutes when x is

a 60 minutes b 30 minutes c 1 minute?

Make a table of values of y for different values of x.
Draw a graph of y against x and write down its equation.

Unit B

Interpreting straight-line graphs

B11.3

- Interpret straight-line graphs in a variety of situations

RICH TASK

Two business men A and B leave their headquarters and travel to the same branch 100 miles away as shown on the graph.

Describe their journeys.

At what time do they leave? How long do they take? Who sets off last? Who gets there first? Who travels faster? Where and when do they cross?

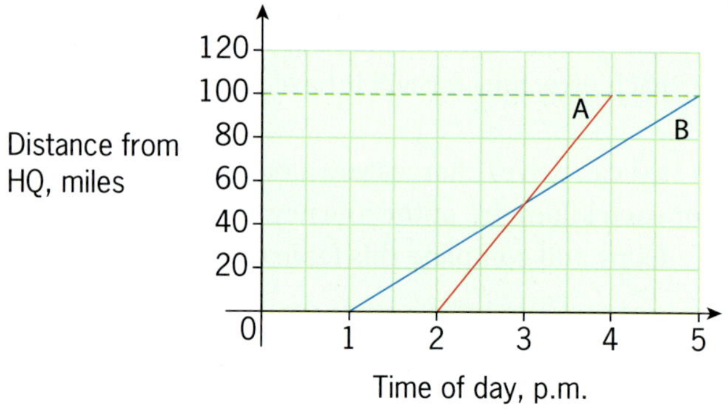

Graphs can tell stories.
You need to be able to interpret information from a graph.

EXAMPLE

Gabby invests a sum of money in a bank account for 4 years at a constant rate of **interest**.
This graph shows how her money grows in value.
Find

a how much she invested at the start

b how long it took to increase to £1300

c how much it increases by each year.

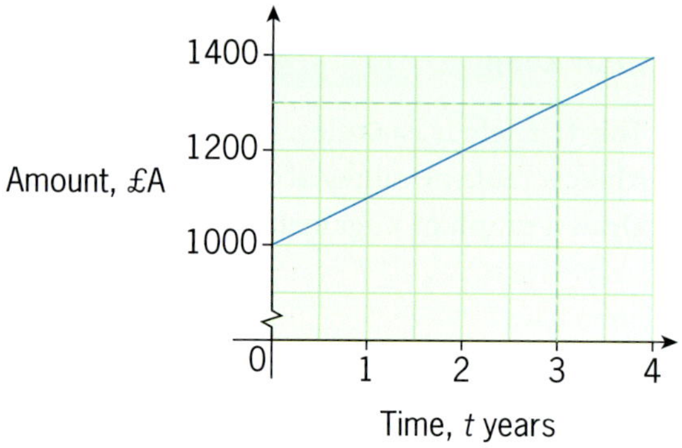

ANSWER

a When $t = 0$, $A = 1000$. So she invested £1000 at the start.

b When $A = 1300$, $t = 3$. It took 3 years to reach £1300.

c Each year, the graph rises by £100.

Notice the jagged line on the A-axis to show that the scale does not start at 0.

Exercise B11.3

MEDIUM LOW

1 The temperature of a liquid in a science experiment rises steadily over 10 minutes, as shown in this graph. Find

a the temperature at the start

b the rise in temperature over the whole 10 minutes

c its temperature after 4 minutes

d how long it takes for its temperature to reach 60°C

e how many degrees the temperature rises each minute.

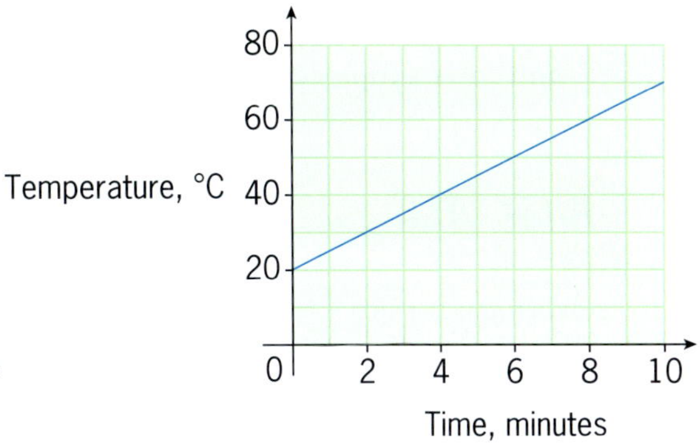

2 A tank of heating oil holds 1000 litres when full. This graph shows the amount of heating oil left in the tank over time. Find

a how much oil is left after 4 weeks

b how much oil was used in the first 4 weeks

c how long it will take for the tank to become half full

d how long it will take for the tank to become empty

e when more oil should be ordered if the order is placed when the tank is three-quarters empty.

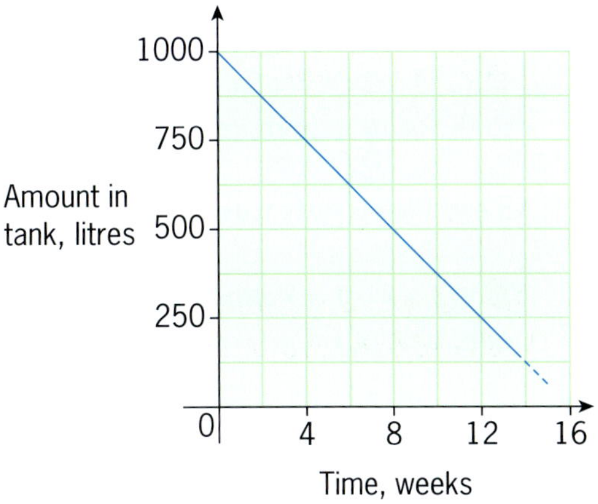

3 These two lines are conversion graphs for British pounds, £, and New Zealand dollars, $.
Line *A* is used when you change pounds into dollars.
Line *B* is used when you change dollars back into pounds.

a Find **i** how many dollars you get for £20
ii how many pounds you get for $100.

b Mrs Weeks changes £40 into dollars but she does not spend them. She then changes all her dollars back into pounds. Does she get back her original £40? If not, why not?

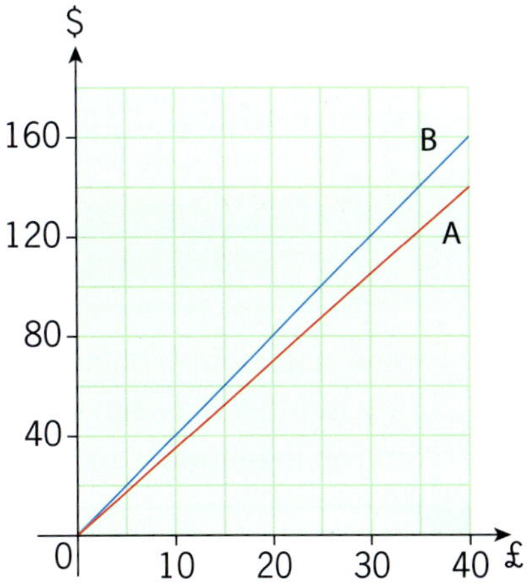

4 If 8 km = 5 miles, how many miles are 80 km?

a Use this information to draw a conversion graph for miles and kilometres up to 160 km.

b Leeds to Skipton is 30 miles and Leeds to Derby is 80 miles. Use your graph to find these distances in kilometres.

Unit B

Inequalities

B11.4

- Know the difference between an equality and an inequality and solve simple inequalities

RICH TASK

When this balance is left alone, the left-hand side always drops. What can you find out about the value of x?

In an **identity**, two expressions are equal to each other regardless of the value of the unknown.

$11x - 4 - 2x + 7 \equiv 3x + 6x + 5 - 2$, because both sides simplify to the same expression. What is this expression?

The sign $\equiv$ means 'is identical to'.

An **equation** is only true for a specific value (or values) of x.

$2x + 3 + x - 1 = 8$ simplifies to $3x + 2 = 8$, which is true for only one value of x. Can you find this value? Do you get $x = 2$?

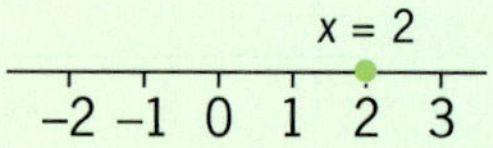

You can solve an equation to find the value of x.

An **inequality** is true for many values of x.

If $x > 4$, then x can have any value greater than 4. These values cover a range of the number line.

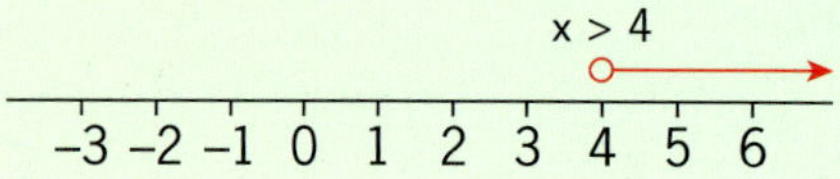

$>$ means 'greater than'
$<$ means 'less than,
$\geq$ means 'greater than or equal to'
$\leq$ means 'less than or equal to'

Some statements contain **two** inequalities. $4 \leq x \leq 10$ means that x is a number between (and including) 4 and 10.

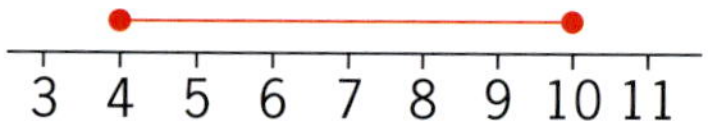

You can manipulate an inequality in the same way as an equation.

EXAMPLE

Find all possible values of x when $5x - 1 \leq 3x + 7$ and show them on a number line.

ANSWER

	$5x - 1 \leq 3x + 7$
Subtract $3x$ from both sides	$2x - 1 \leq 7$
Add 1 to both sides	$2x \leq 8$
Divide both sides by 2	$x \leq 4$

Note that a **solid** dot includes the end value of $x = 4$. However, an **empty** dot would **not** include the end value.

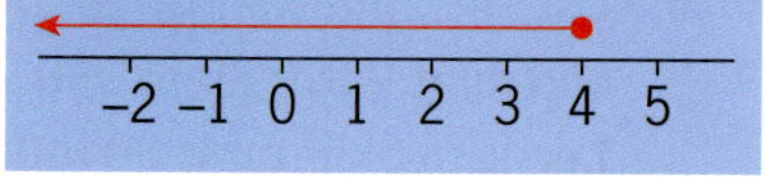

Exercise B11.4

MEDIUM

1 Decide whether each of these statements is an identity, an equation or an inequality.

a $9x + 4 = 8x + 14$ b $9x + 4 + 8x + 14 \equiv 17x + 18$

c $9x + 4 \geq 8x + 14$ d $9x + 4 + 2(3x - 17) < 0$

2 Match the four statements with these four inequalities.

$x \geq 7$ $x < 7$ $x > 7$ $x \leq 7$

a x is less than 7 b x is greater than 7

c x is greater than or equal to 7 d x is less than or equal to 7

3 Write an inequality for each of the coloured regions on these number lines.

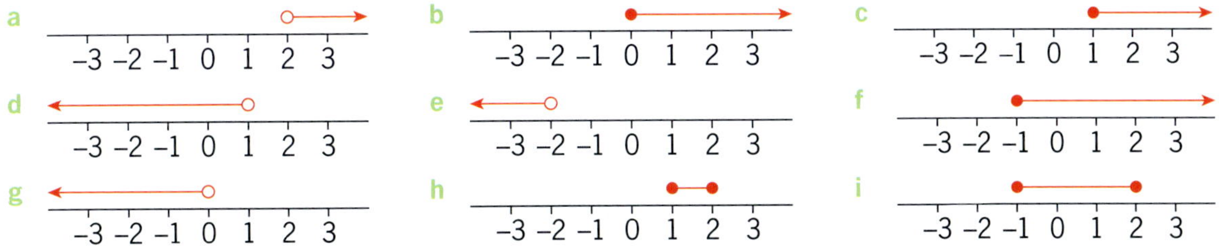

4 Draw and label the x-axis from −4 to 4 .
Colour part of the x-axis to illustrate each of these inequalities.
Remember to use empty dots and solid dots.

a $x > 1$ b $x \geq 1$ c $x \leq 3$ d $x < -2$

e $x \geq -2$ f $1 \leq x \leq 2$ g $-1 < x < 2$ h $-3 \leq x \leq 0$

5 Find the solutions of these inequalities and illustrate them on number lines.

a $5x + 3 > 18$ b $6x - 4 < 8$ c $4x + 9 \geq 25$ d $3x - 7 \leq 2$

e $2(3x - 4) > 22$ f $4(x + 2) < 28$ g $8(2x - 5) > 40$ h $5(x + 5) < 15$

i $6x + 3 < 4x + 9$ j $5x - 7 \geq 2x + 5$ k $2(x - 2) < x + 1$ l $6x + 8 \leq 2(2x + 5)$

m $2x + 5 < 11 - x$ n $4x + 8 \geq 26 - 2x$ o $3(x + 5) \leq 11 - x$ p $6(x + 4) > 3(2 - x)$

6 Find the solutions of these inequalities and illustrate them on number lines.

a $6 < x + 3 < 8$ b $1 \leq x - 2 \leq 3$ c $4 < 2x - 6 < 8$

7 Write down all the possible integer values of x if

a $5 < x < 9$ b $5 \leq x \leq 9$ c $-2 < x < 2$

8 Write down all the possible odd values of x if

a $6 < x < 10$ b $1 \leq x \leq 6$ c $-4 < x < 2$

Summary and assessment

B11

CHECK OUT

You should now be able to

- construct linear functions from real life and draw their graphs
- interpret straight lines
- solve inequalities.

Exam-style question

A plumber uses this graph to work out how much to charge for a job.
The charge for a job is made up of a fixed fee plus an amount for the time that the job lasts.

a How much is the fixed fee?

b How much would the plumber charge for a job lasting 4 hours?

c Write down a formula for the charge £C in terms of the fixed fee, the rate per hour and the number of hours, h, that the job takes.

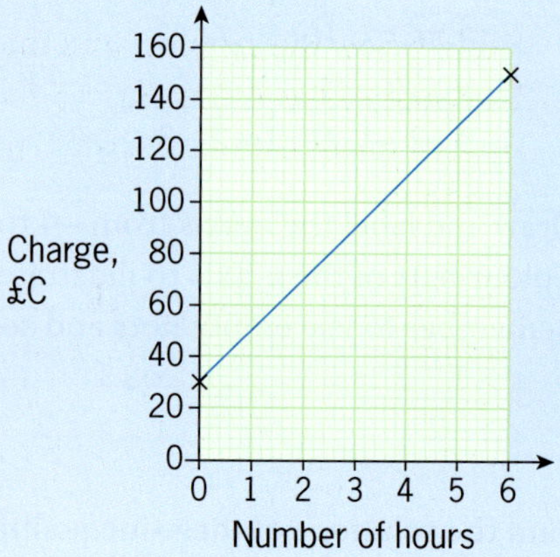

Mia's answer

a The graph shows hours worked and charge.
Zero hours shows a charge £30
Fixed fee = £30

b 4 hours has charge £110

c Rate per hour is the amount charged for each hour
From 0 hours to 1 hour the charge increases from £30 to £50, so the rate per hour is $50 - 30 = £20$

$C = 30 + 20 \times h$

You need to read information on the graph as well as the actual question.

Take care with the scale when reading values from a graph.

You may have to work out some extra information.

Write the final answer as the question asks.

Exam practice

1 a Solve the inequality $2x - 3 \geq 12$.

b i Represent your solution on a number line.

ii Write down all the integer values satisfied by $2x - 3 \geq 12$ that are less than 10. *(4 marks)*

2 The graph shows the cost of making a phone call to a landline charged by Myfone, a mobile phone company.

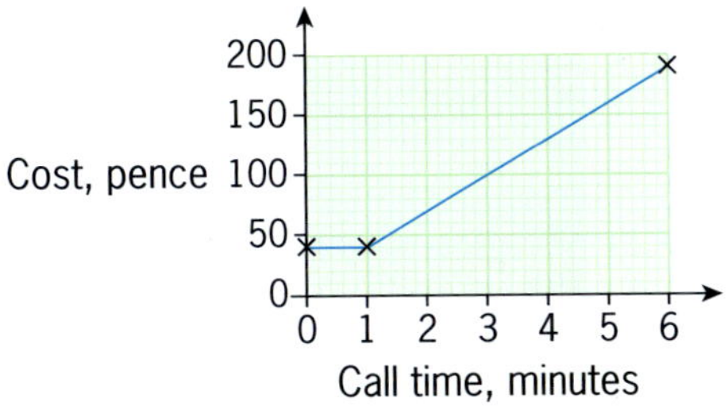

a Explain how the graph shows that it costs the same for a 1-second call as it does for a 59-second call.

b Write a formula for the cost (C pence) of making a call lasting m minutes. *(3 marks)*

3 Jeff likes to go swimming at a local swim club.
He can pay to swim as a member or as a non-member.

Non-member	£3·60 per swim
Member	£24 per year plus £2 per swim

The graph shows the cost for non-members to swim.

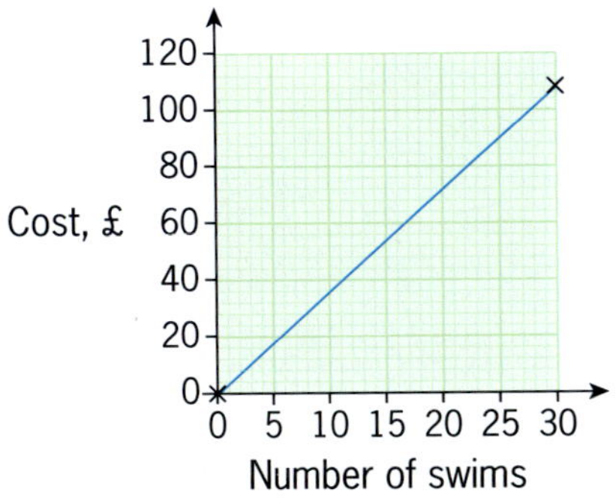

a Copy the graph and on the same graph draw the line to show the cost for members to swim.

b How many times would Jeff need to swim in one year for it to be better for him to be a member of the local swim club?

(3 marks)

12 Transformations

INTRODUCTION

Transformations change points and shapes by moving them from one place to another. When you play a computer game, your character is able to move around the screen because of mathematical transformations. Combinations of these transformations, often taking place in 3-D worlds, allow the characters to move in many different ways.

What's the point?

Mathematicians use transformations not just to change shapes but also to change graphs and statistics. This helps them match the mathematics to real-life situations.

CHECK IN

1 Give the value of these angles.

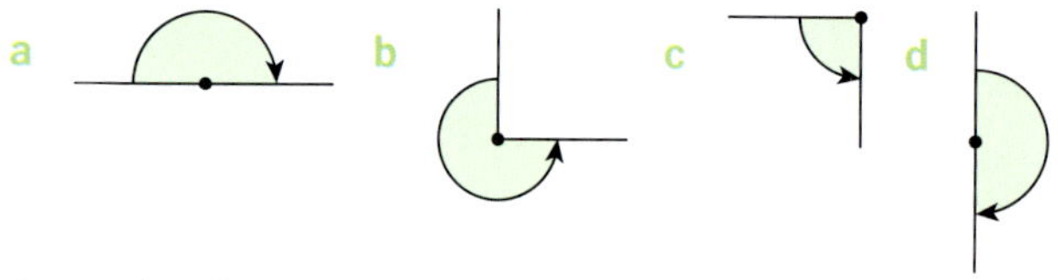

2 State the direction of the turn.

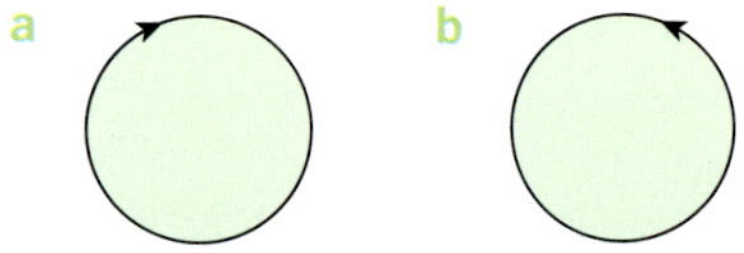

Orientation

What I should know	What I will learn	What this leads to
Key stage 3	■ Find images of reflected shapes ■ Find the images of rotated shapes ■ Translate shapes	B16

RICH TASK

A shape (object) is reflected in the line $y = x$.

To find the new coordinates of each vertex of the shape (image), simply swap the x and y coordinates around.

Investigate rules for finding the image coordinates for this and other transformations.

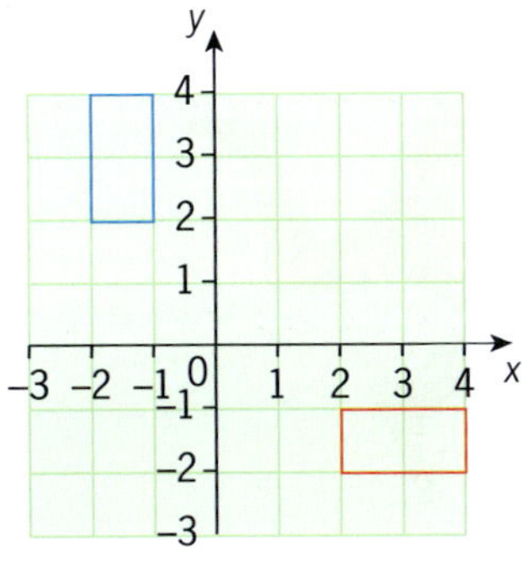

Symmetry

B12.1

- Recognise reflection and rotational symmetry

→ **A shape has reflection symmetry if the shape has two identical parts when folded in half, or if when you hold a mirror on the shape its two halves are identical. The mirror line is called the line of symmetry.**

The badge of the Prince of Wales has only one line of symmetry.

The Flag of Scotland has two lines of symmetry.

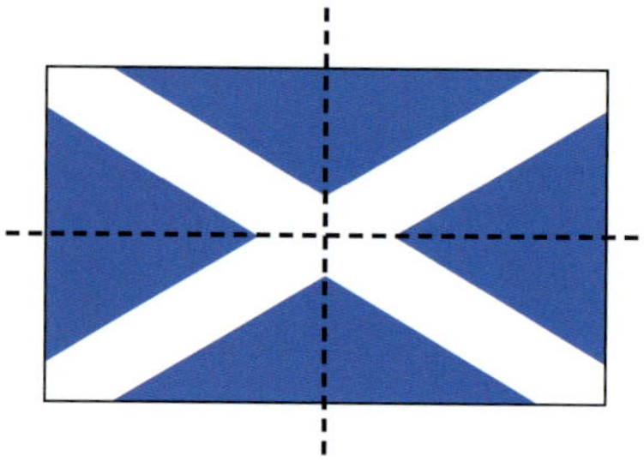

→ **A shape has rotational symmetry if it fits on top of itself exactly when rotated about its centre.**

The order of rotational symmetry is the number of times that it fits on top of itself in a full turn.

All shapes have rotational symmetry of at least order 1 as they fit at least once in a full turn.

Isle of Man flag
order 3

Pentagram
order 5

A spade
order 1

A cog-wheel
order 12

Exercise B12.1

MEDIUM LOW

1 a Copy these shapes and draw their lines of symmetry.

b These diagrams are incomplete.
Use the lines of symmetry to draw the complete shapes.

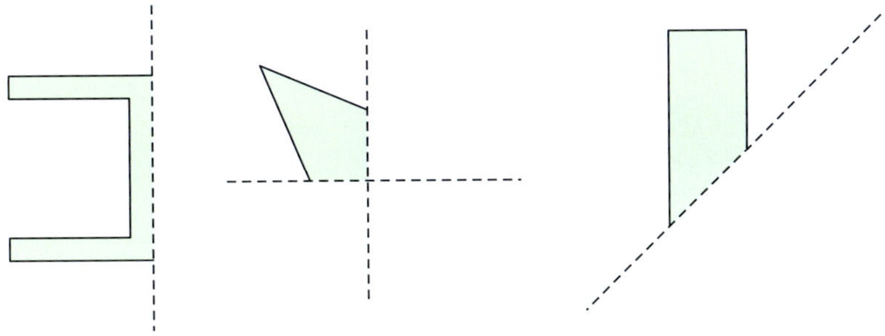

2 Find the order of rotational symmetry of these shapes.

a

b

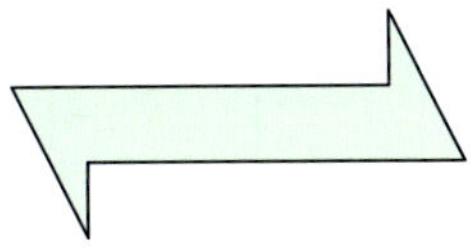

c

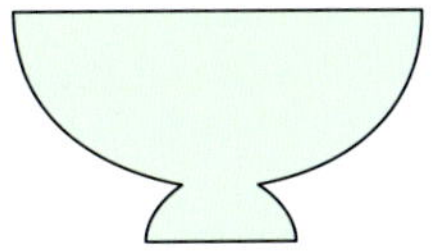

d

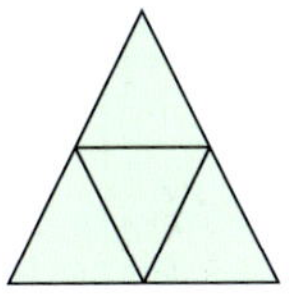

3 How many lines of symmetry does each capital letter of the alphabet have?
Do your answers depend on the type of font?

Unit B

Reflections

- Find images of reflected shapes

When a shape is reflected in a mirror line, every point maps onto its image by moving at right-angles to the mirror line for the same distance beyond the mirror.

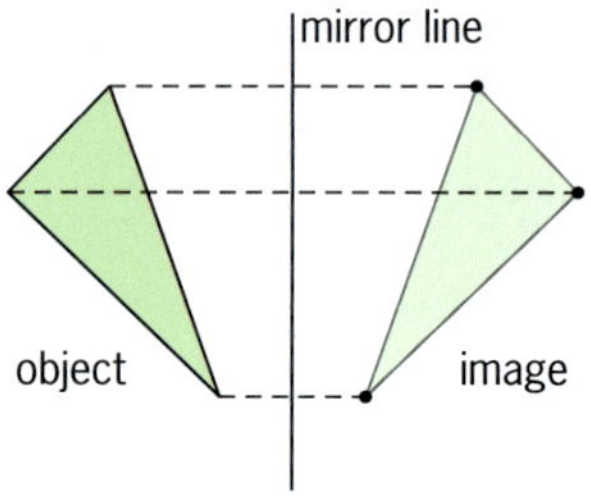

Mirror lines can slope at an angle. To find a mirror line, join at least two object points to their image points and find the midpoints. These midpoints are on the mirror line.

You can check that your work is correct by holding a mirror on the mirror line.

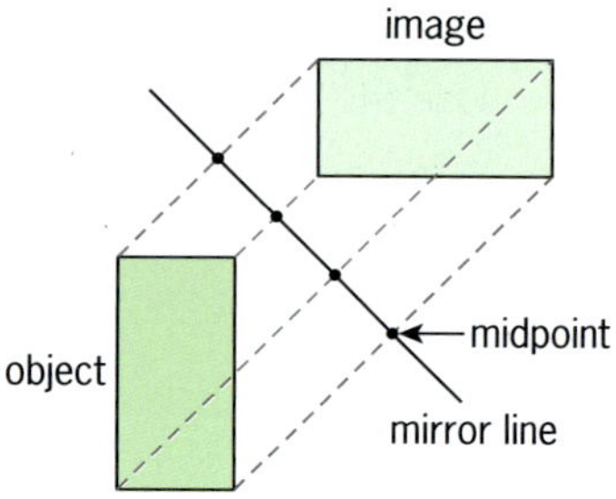

EXAMPLE

a Reflect the triangle with vertices at (1, 2), (2, 7) and (4, 7) in the line $x = 1$.

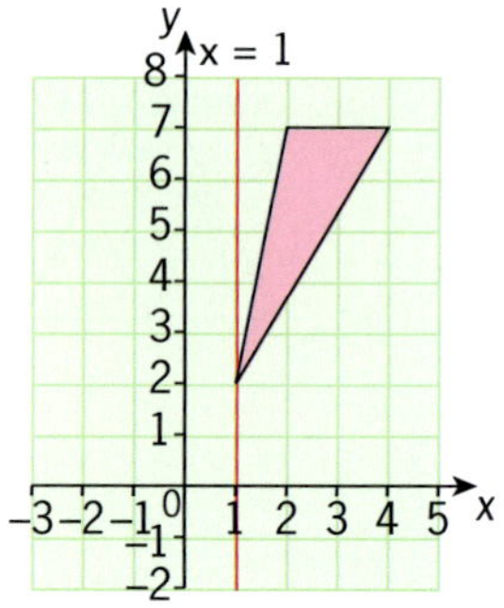

b Reflect triangle ABC in line L.

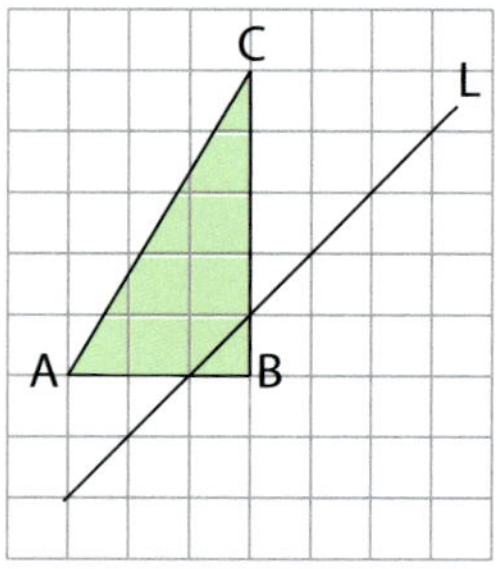

ANSWER

a

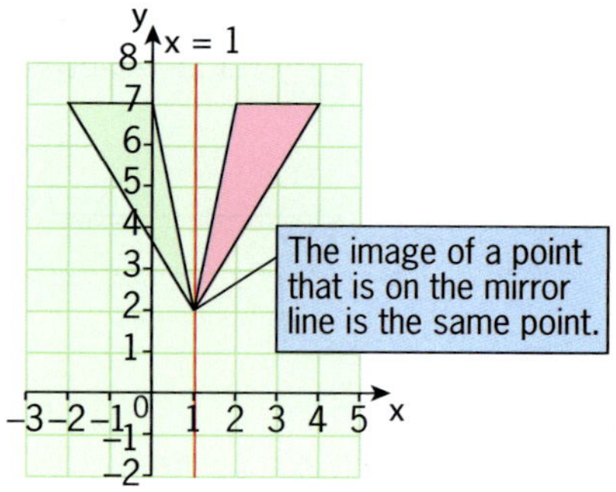

b

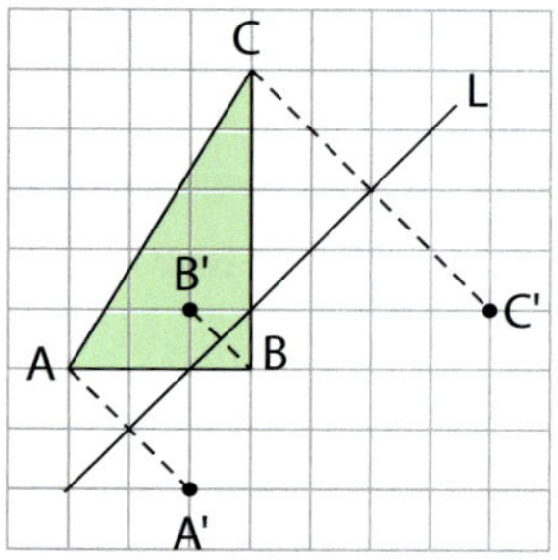

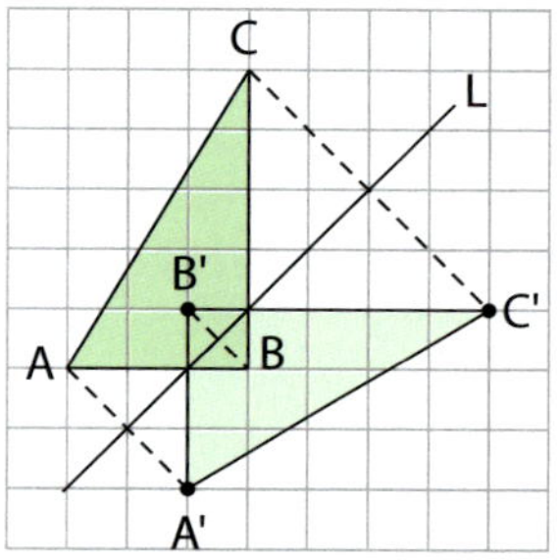

Exercise B12.2

LOW

1 These rectangles are reflected in the given lines. Copy them and find their images.

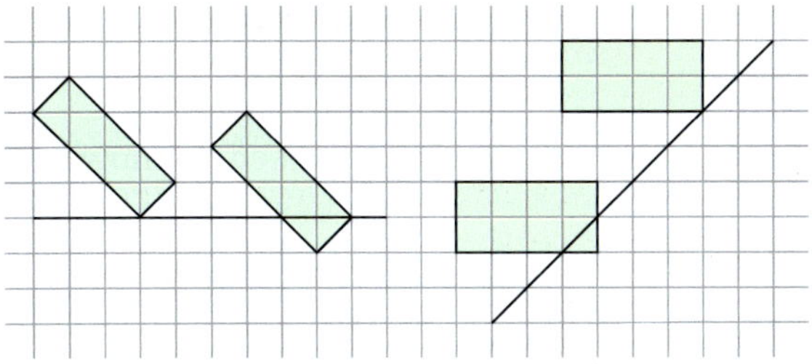

2 Copy this diagram.

a Reflect the triangle T in the line $x = 2$. Label the image U.

b Reflect the triangle T in the line $y = 1$. Label the image V.

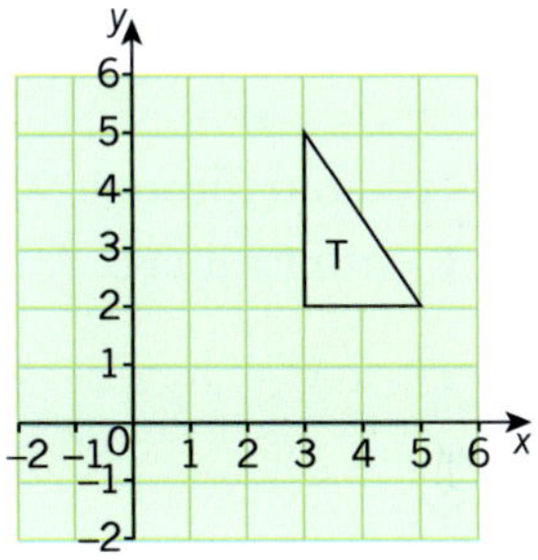

3 Copy this diagram.

a Reflect the kite K in the line $x = -1$. Label the image L.

b Reflect L in the line $y = 1$. Label the image M.

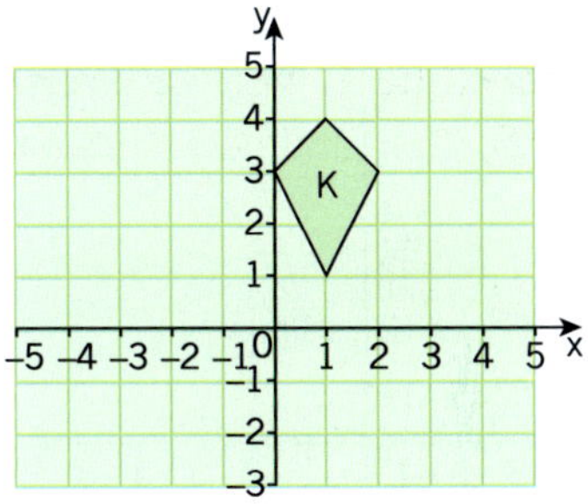

4 Draw triangles A and B onto axes labelled from −10 to 10. Reflect them in the x-axis. Draw triangles C and D and reflect them in the y-axis.

Object triangle
A(3, 1)(3, 4)(1, 4)
B(3, 8)(5, 10)(7, 9)
C(2, 0)(6, 4)(8, 3)
D(2, 10)(6, 10)(6, 8)

Unit B

Rotations

- Find objective images of rotated shapes

When a shape is rotated, you need to know:

the **centre** of rotation

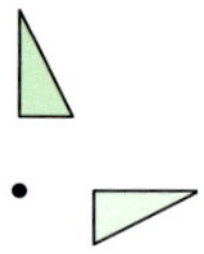

the **angle** of rotation

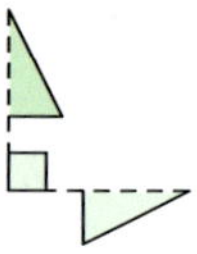

the **direction** of rotation.

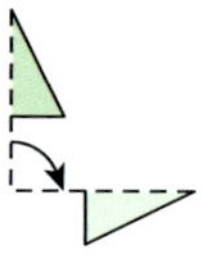

An anticlockwise angle of rotation is positive; clockwise is negative.

Triangle A is rotated through 90° clockwise about point P onto its image A′.

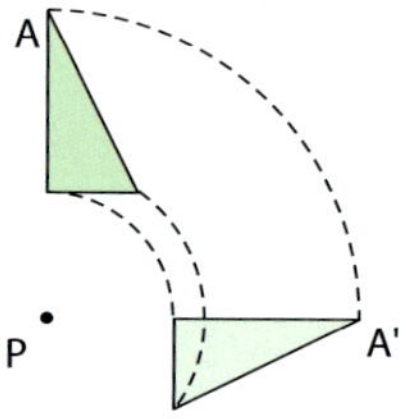

You can check a rotation by using compasses to draw some of the paths traced out.

EXAMPLE

Rotate the triangle through 90° clockwise about centre (1, 2).

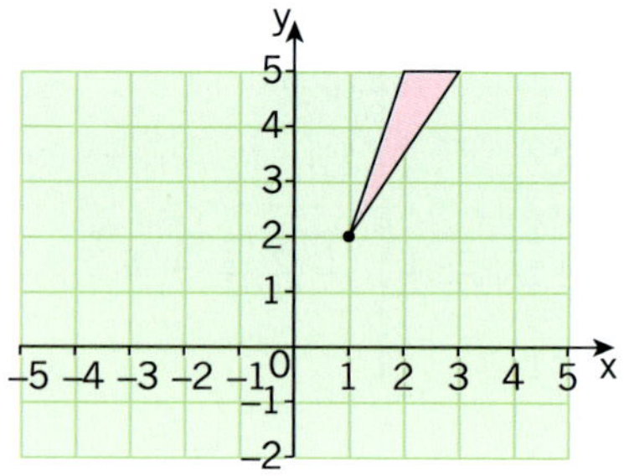

ANSWER

Rotate each vertex through 90° clockwise.

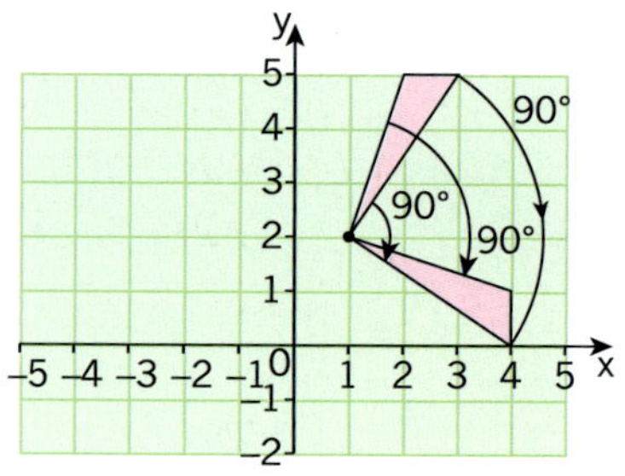

You can use tracing paper to help you do rotations.

EXAMPLE

Rotate the pink triangle through 180° about centre O.

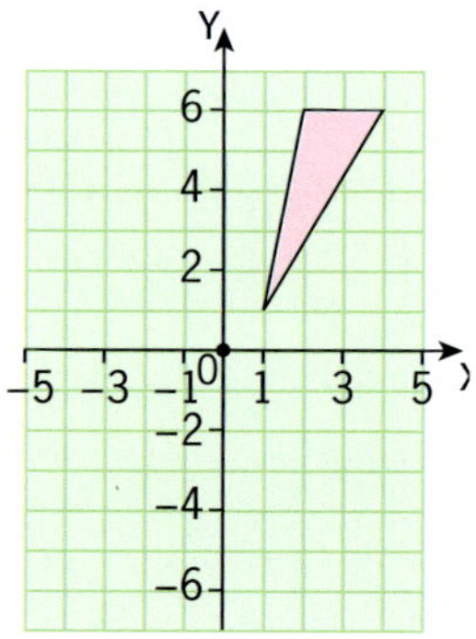

ANSWER

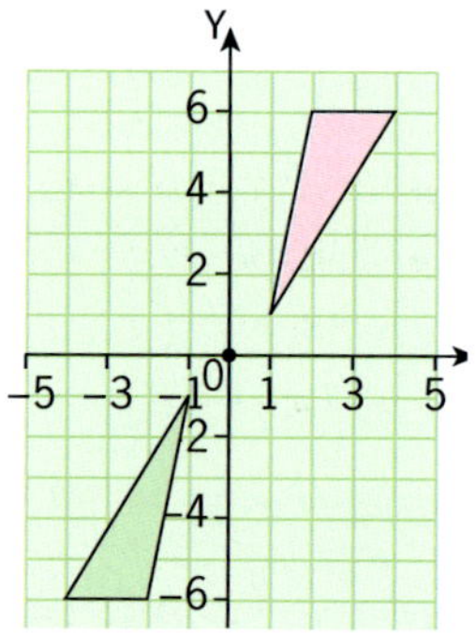

Exercise B12.3

MEDIUM

Unit B

1 Draw these diagrams accurately.
Each coloured shape rotates about point O onto its image.
Use compasses to draw the paths of the lettered points.
Find the angle and direction of rotation in each case.

a

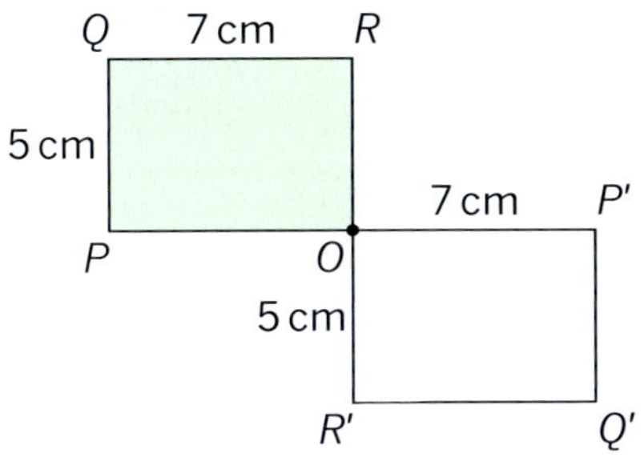

b

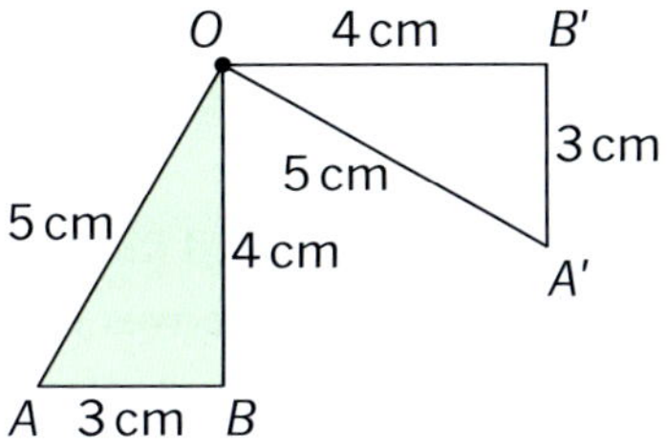

2 Copy these letters onto square grid paper.
Rotate each letter clockwise through 90° about point *O*.
Use tracing paper to help you.

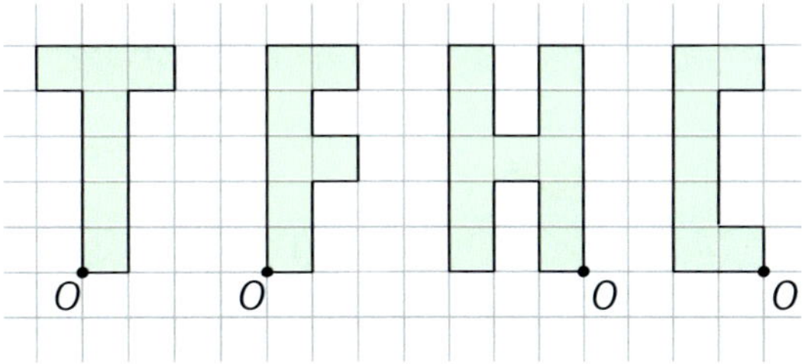

3 Copy this diagram.

a Rotate the triangle T through 180° about (0, 0). Label the image U.

b Rotate the triangle T through 180° about (2, 2). Label the image V.

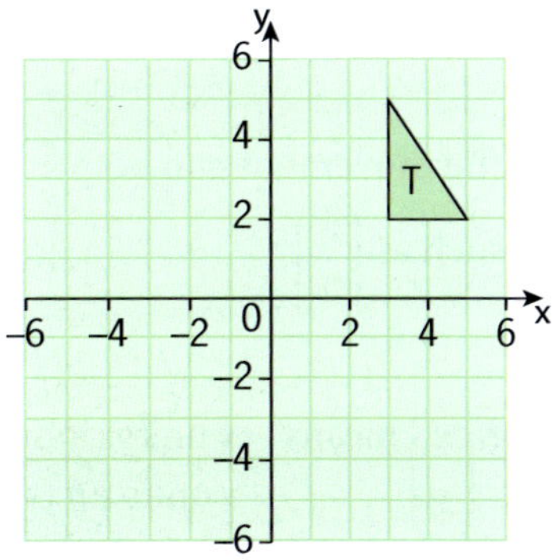

4 Copy this diagram.

a Rotate the kite K through 180° about (1, 1). Label the image L.

b Rotate the kite K through 90° clockwise about (2, 0). Label the image M.

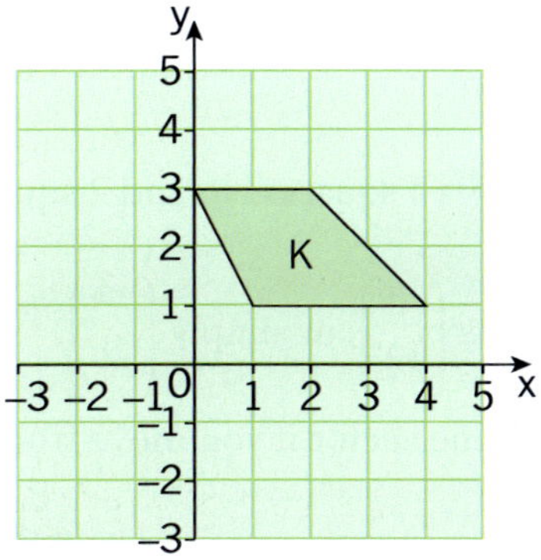

Translations

- Know how to translate a shape using a vector

When a shape is translated, it moves from one position to another without turning.

You can describe a translation
- → ***either*** **by giving the distance it moves and its direction**
- → ***or*** **by using a grid and giving how far it moves up or down and left or right.**

Aircraft A translates onto aircraft B; the translation is described by:

either 5 km on a bearing of 053°

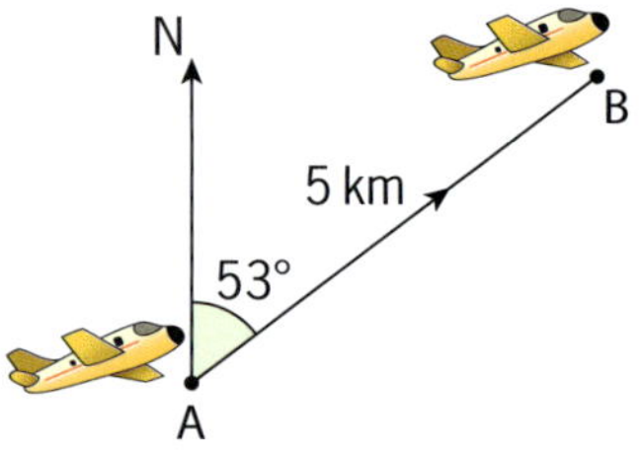

or 4 km right and 3 km up the grid.

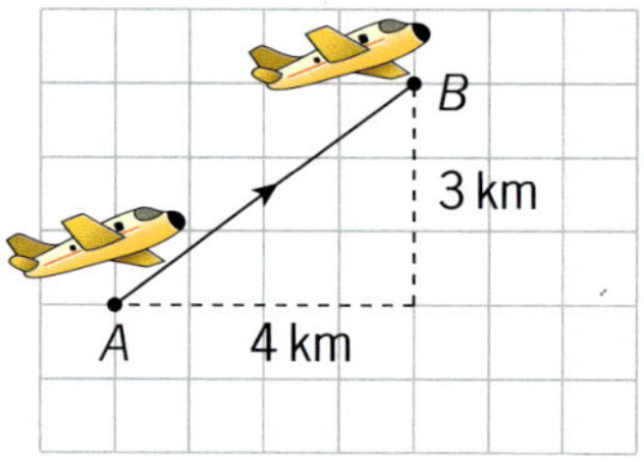

p. 120

When using a grid, the movement can be written as a vector, such as $\begin{pmatrix}4\\3\end{pmatrix}$

$\begin{pmatrix}4\\3\end{pmatrix}$ top number means 4 squares **right**; bottom number means 3 squares **up**

In general, the vector $\begin{pmatrix}x\\y\end{pmatrix}$ means x units right (+) or x units left (–)
y units up (+) or y units down (–).

EXAMPLE

Triangle A′B′C′ is the image of triangle ABC under a translation. Find the vector of the translation.

ANSWER

Corner A moves 4 squares left and 2 squares up to reach its image A′.

So the vector is $\begin{pmatrix}-4\\+2\end{pmatrix}$, or simply $\begin{pmatrix}-4\\2\end{pmatrix}$.

You get the same vector if you move from B to B′ or from C to C′.

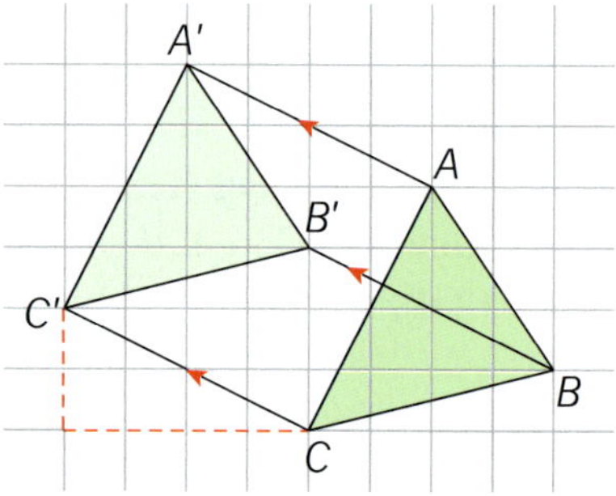

Exercise B12.4

MEDIUM

1 The points A to F are on a grid.

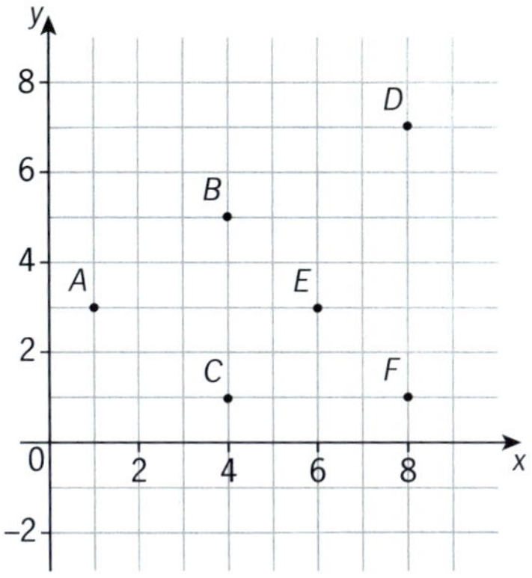

a Find the image points when these object points are translated by these vectors.

Object point	A	B	C	D	E	F
Translation	$\begin{pmatrix}3\\2\end{pmatrix}$	$\begin{pmatrix}3\\-2\end{pmatrix}$	$\begin{pmatrix}4\\1\end{pmatrix}$	$\begin{pmatrix}-3\\2\end{pmatrix}$	$\begin{pmatrix}0\\-5\end{pmatrix}$	$\begin{pmatrix}-2\\-2\end{pmatrix}$

b Find the vectors which translate these object points onto their images.

Object point	B	B	C	D	E	F
Image point	D	E	F	A	B	C

2 Write the vectors of the translations which map

a P onto Q b P onto R c Q onto R d Q onto P.

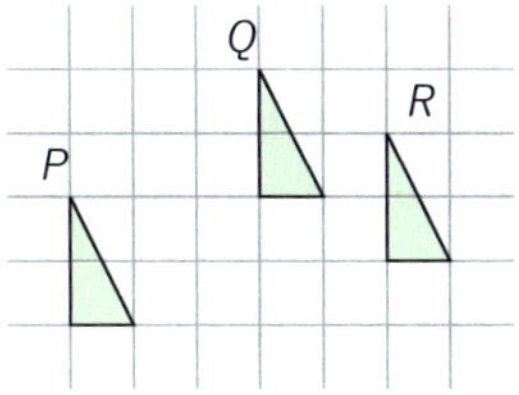

3 Copy this diagram.

a Translate the kite K by the vector $\begin{pmatrix}-4\\3\end{pmatrix}$. Label the image L.

b Translate the kite K by the vector $\begin{pmatrix}2\\-4\end{pmatrix}$. Label the image M.

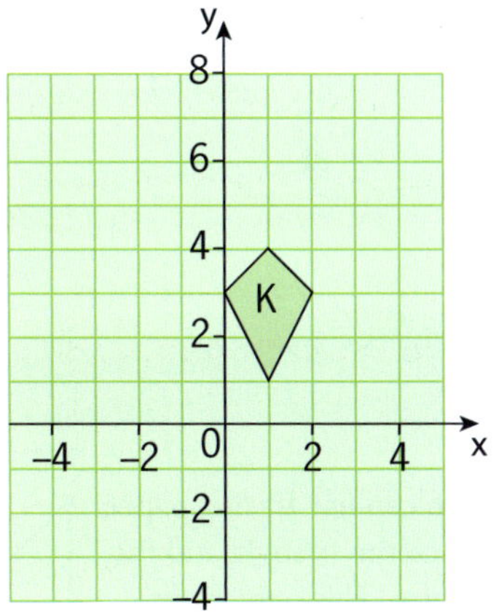

4 ★ CHALLENGE ★

Find how a knight moves in chess.
Use vectors to describe all possible moves from a square.
Find the least number of moves needed to land on other squares.
(In this diagram, to reach the numbered squares needs 1 and 2 moves.)
Is there any pattern to your answers?

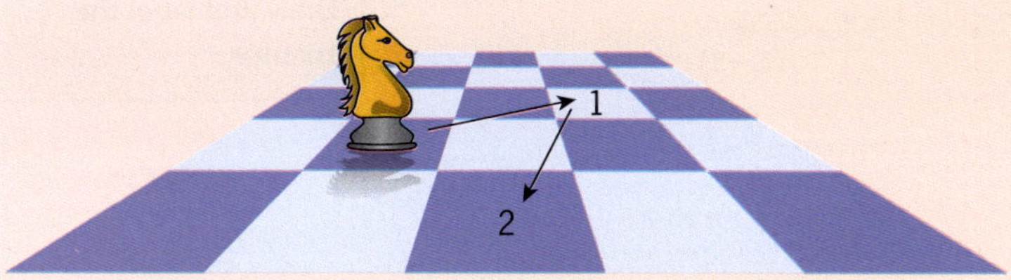

Summary and assessment

CHECK OUT

You should now be able to

- recognise reflection symmetry and rotation symmetry of 2-D shapes
- transform 2-D shapes using
  - reflection
  - rotation
  - translation.

Exam-style question

Rotate the triangle 90° anti-clockwise about centre (3, 0).

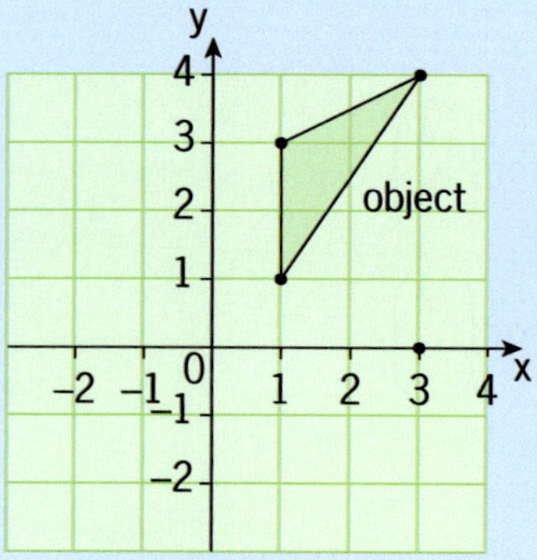

Tina's answer:

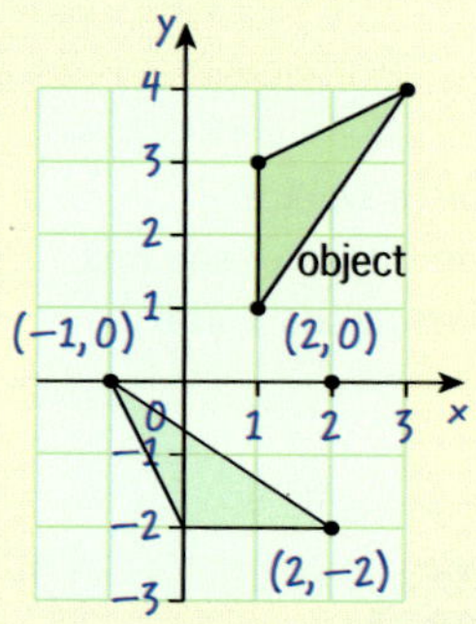

You can use tracing paper. Trace the triangle and the centre. This only works when the x and y scales are the same.

The corners of the triangle are on the grid lines, so the corners of the image will also be on grid lines.

Use a pencil at the centre so it does not move and rotate the tracing paper.

Remember that a vertical line will be horizontal after a 90° turn.
Draw and label the triangle.

Exam practice

1 Describe **i** the line symmetry and **ii** any rotational symmetry of these shapes.

a

b

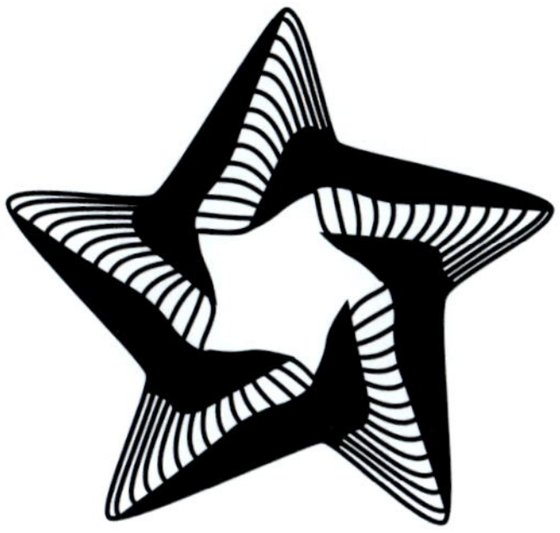

(4 marks)

2 The diagram shows a triangle T.

a Rotate triangle T 90° anticlockwise about centre (0, 0). Label the image R.

b Reflect triangle T in the x-axis. Label the image S.

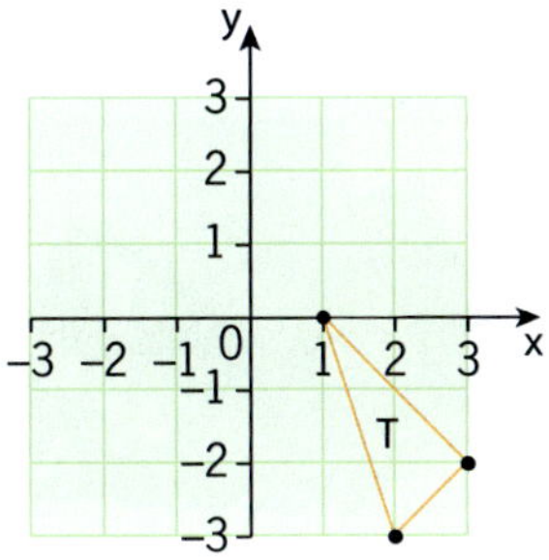

(4 marks)

3 Copy this diagram

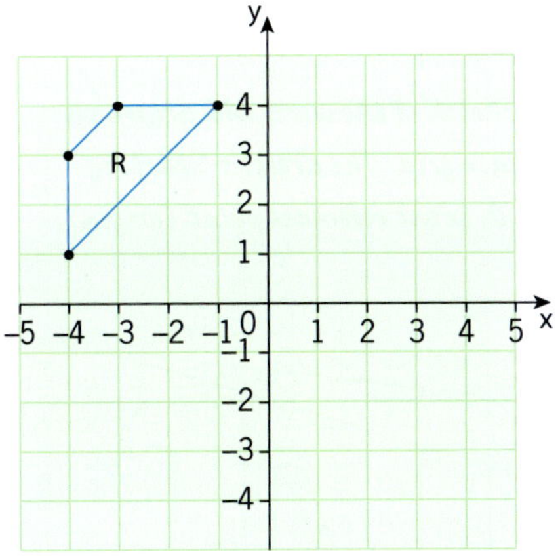

a Translate trapezium R using vector $\begin{pmatrix} 5 \\ -3 \end{pmatrix}$. Label it S.

b Translate trapezium S using vector $\begin{pmatrix} -1 \\ 2 \end{pmatrix}$. Label it T. *(4 marks)*

Case study 4: Art

Graffiti artists often sketch their designs before projecting them onto the surface. They sometimes use grids or parts of their body as measuring tools to help them copy the proportions accurately.

1. A graffiti artist projects an image from a sketchpad of length 20cm and height 14.8cm onto a wall of length 6m.
   a. What scale factor is being used?
   b. What is the height of the graffiti wall?

   The artist's hand-span is 150mm.

   c. What are the dimensions of the wall in terms of hands?

Grid method

The T is made up of a trapezium and a scalene triangle. The coordinates of its vertices are

(-4, -1), (-4, 4.6), (-6, 4), (-4, 6), (4, 7), (-2.6, 5) and (-2.6, -1).

2. Describe
   a. the shapes used to make up i. the A ii. the G
   b. the coordinates of the vertices of i. the A ii. the G.
3. The dimensions of the sketch are 12cm × 10cm.
   a. What scale factor would you use to project this image onto a surface of dimensions 24cm × 20cm?
   b. What would the effect be on the area of the image?
   c. Calculate the area of
      i. the original sketch ii. the enlarged image.
   d. Use a 2cm square grid to draw the enlarged image.

When a grid is being used, it is the shapes that make up the design and their borders that are important.

Here is a sketch of the word TAG drawn on a 1cm square grid. The artist's 'starting position', O, is the reference point (origin):

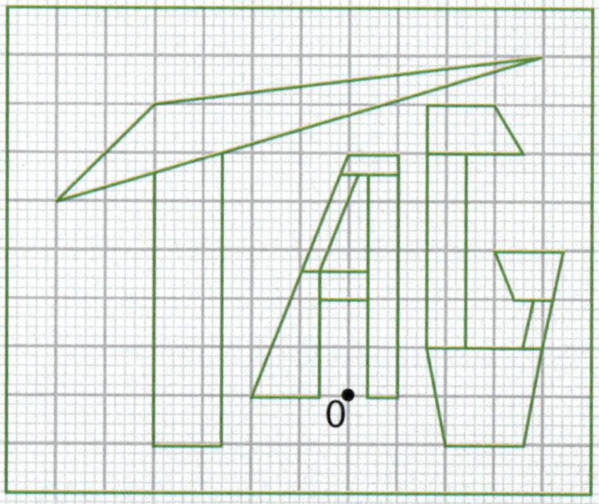

Sketch your own graffiti tags using geometric shapes. Use the grid method to create enlarged copies of your images.

Crop circles are geometric patterns that are displayed in crop fields.

CROP CIRCLES

This crop circle was found in Wiltshire in 2008. The design is based on an equilateral triangle.

You can recreate the pattern using these steps:

1. *Use your compasses to draw a circle*
2. *Choose a point on your circle. Use your compasses (do not alter them) to step round the circle. Mark at each point your pencil touches the circle*
3. *Join three of the marks to form an equilateral triangle*
4. *Join the other three marks to form an equilateral triangle in the reverse direction*
5. *Alter your compasses. Draw a circle, with the same centre, passing through the six intersections of the two triangles*
6. *Alter your compasses again. Draw a circle, centred at a point of one of the triangles, so that it touches the second circle you drew*
7. *Construct two more circles as in step 6 at the other two points of that triangle*
8. *Erase your construction lines.*

Geometric construction steps 1 to 4

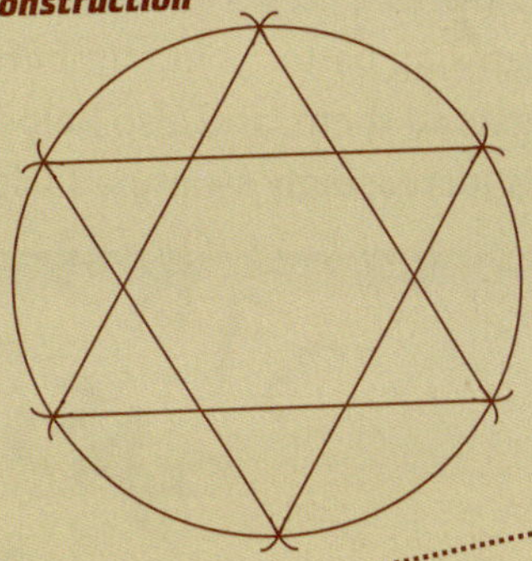

Geometric construction steps 5 to 7

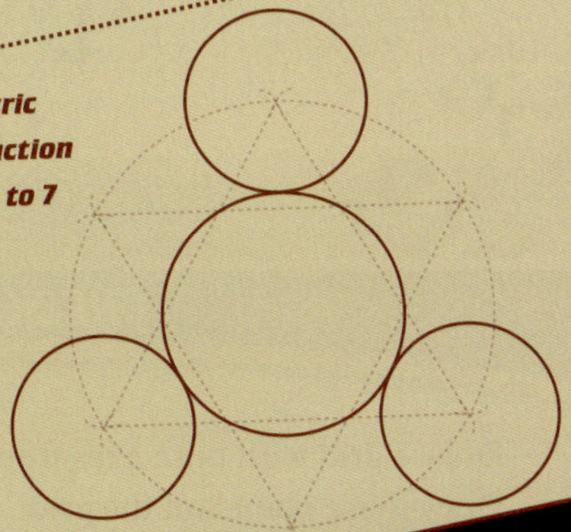

Create your own crop circle designs using a ruler and compasses. Write the steps to instruct someone how to recreate your pattern.

13 Bivariate data and time series

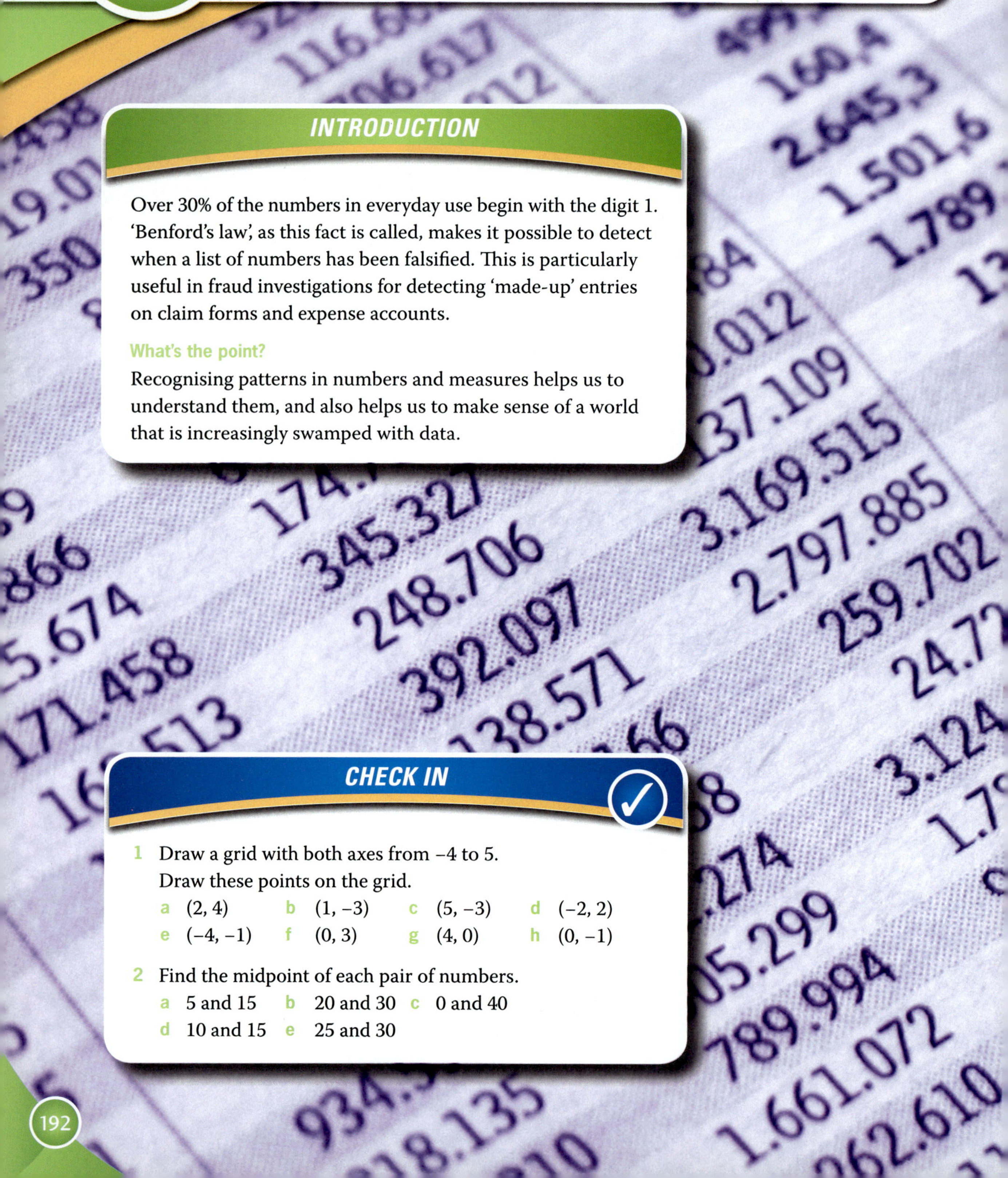

INTRODUCTION

Over 30% of the numbers in everyday use begin with the digit 1. 'Benford's law', as this fact is called, makes it possible to detect when a list of numbers has been falsified. This is particularly useful in fraud investigations for detecting 'made-up' entries on claim forms and expense accounts.

What's the point?

Recognising patterns in numbers and measures helps us to understand them, and also helps us to make sense of a world that is increasingly swamped with data.

CHECK IN

1 Draw a grid with both axes from −4 to 5.
Draw these points on the grid.

a (2, 4) b (1, −3) c (5, −3) d (−2, 2)
e (−4, −1) f (0, 3) g (4, 0) h (0, −1)

2 Find the midpoint of each pair of numbers.

a 5 and 15 b 20 and 30 c 0 and 40
d 10 and 15 e 25 and 30

Orientation

What I should know	What I will learn	What this leads to
A6	■ Draw and use scatter graphs and understand correlation ■ Draw and use time series graphs	Careers involving statistics include environmental statistics, medical statistics, social policy, actuarial science…

RICH TASK

The taller you are, the heavier you are.
Investigate this statement.

Scatter graphs

- Draw and interpret scatter graphs

→ **A scatter graph can be used to represent paired data.**

Paired data is data that is collected for two variables that may be linked

RICH TASK

A survey was conducted on a sample of people who live at different altitudes, to measure their lung capacity. The **scatter graph** shows the results.

Does where you live affect your lung capacity? Is there any other data that could have been collected?

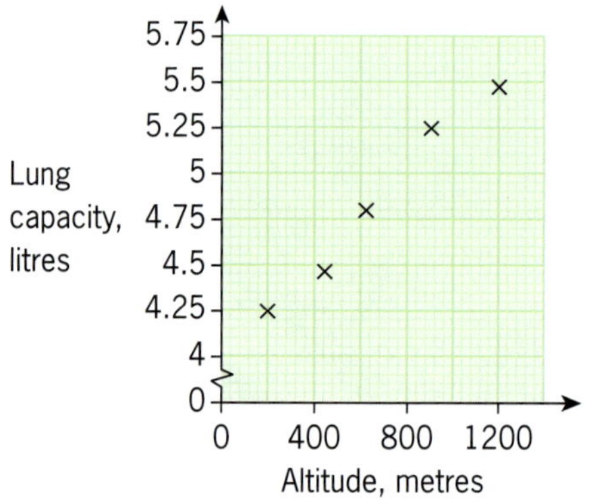

EXAMPLE

Freddie carried out an experiment to find how high his power ball would bounce.

Drop height, cm	30	40	50	60	70	80	90	100
Bounce height, cm	27	48	53	62	69	89	102	110

Draw a scatter graph to represent these data.

ANSWER

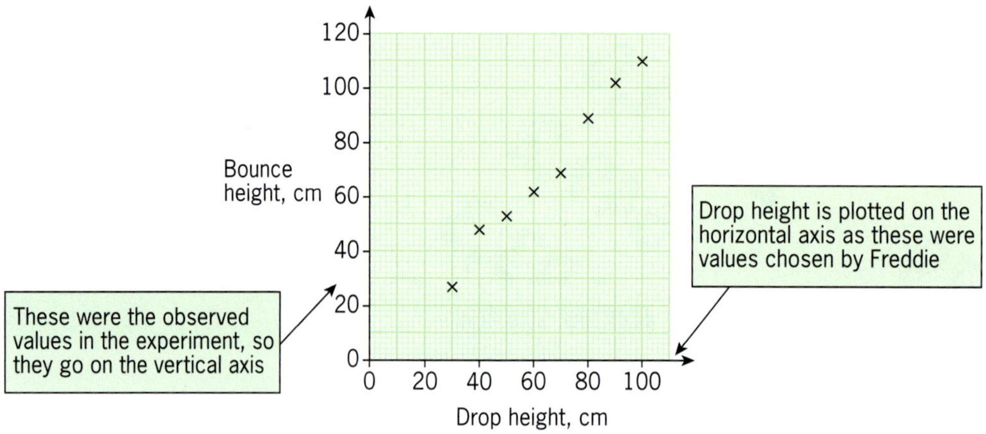

Exercise B13.1

MEDIUM

1 Draw a scatter diagram to represent these data recorded by Iain on the percentages achieved in both statistics and mathematics exams by 10 students.

Statistics %	78	82	74	75	93	70	66	62	77	89
Mathematics %	70	76	61	70	89	65	59	58	73	82

2 Carrie conducted an experiment to find out whether height and arm span were related. She measured the heights and arm spans of eight of her friends.

Height (cm)	136	158	126	131	149	157	143	152
Arm span (cm)	134	146	122	130	147	150	134	145

a Draw a scatter diagram to represent these data.

b Anthropometric data is the name given to data collected about body measurements. What paired data might it be useful to collect for bicycle design?

*3 Read the information and answer the questions that follow.
Louise recorded information about the average number of minutes per day spent playing computer games, and reaction time.

Minutes per day spent playing computer games	40	60	75	40	35	20	80	50	45
Reaction time, seconds	5.2	4.3	3.9	5.5	6.0	7.2	3.6	4.8	5.0

Barry carried out an experiment to find the reaction time of volunteers after they had consumed alcohol.

Units of alcohol	2.5	9.0	5.0	3.5	4.5	8.5	7.5	6.0	2.0	7.0	8.0
Reaction time (sec)	3.6	9.2	4.7	4.8	3.4	8.8	7.5	4.7	1.4	7.3	9.2

Jordan recorded the reaction times, in seconds, of a group of students when woken at 3 a.m. one morning and their reaction time at mid-morning on another day.

3 a.m. reaction time	5.3	4.6	7.8	7.9	8.4	7.0	5.8	6.6	5.0	9.5
Mid-morning reaction time	3.7	3.1	6.2	6.5	7.0	5.7	4.5	4.7	3.3	7.8

a Draw three scatter diagrams to represent the information collected by Louise, Barry and Jordan.

b What can you say about the effects of computer games, alcohol and sleep on reaction time?

Unit B

Correlation

- Appreciate correlation
- Distinguish between positive, negative and zero correlation

➜ **A scatter diagram shows you whether there is correlation between two variables.**

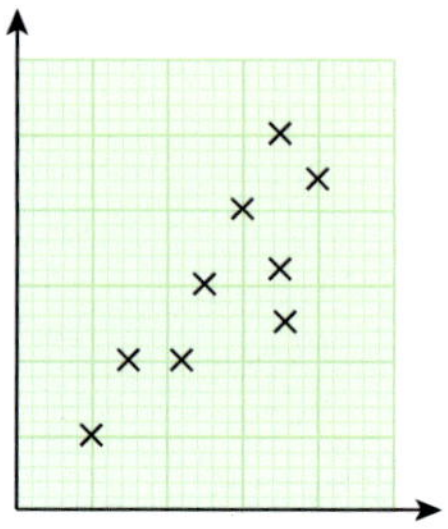

Positive correlation

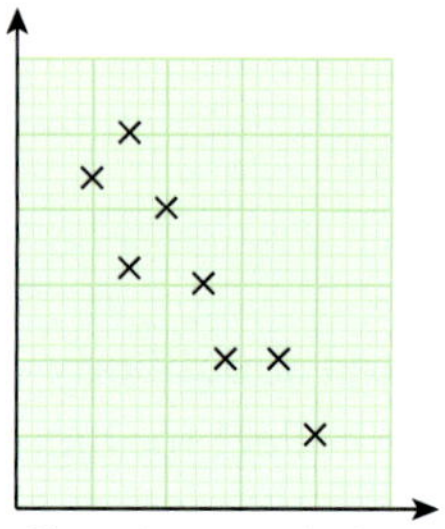

Negative correlation

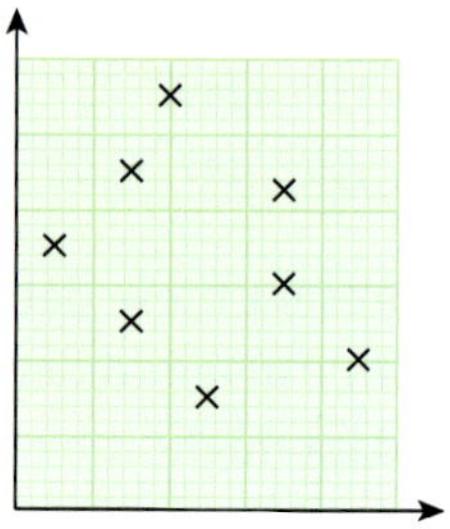

No correlation

There is correlation if the graph appears to show a line. If there is no line then there is no correlation, correlation shows that two variables may be linked.

EXAMPLE

Jon collected these data from a beach cafe on eight consecutive days one summer.

Midday temperature, °C	15	22	21	18	20	19	20	17
Ice-cream sales, £	100	154	139	122	140	132	135	115
Cold drinks sales, £	82	112	105	94	109	96	107	92

Comment on the relationship between the midday temperature and sales at the cafe, using appropriate scatter graphs.

ANSWER

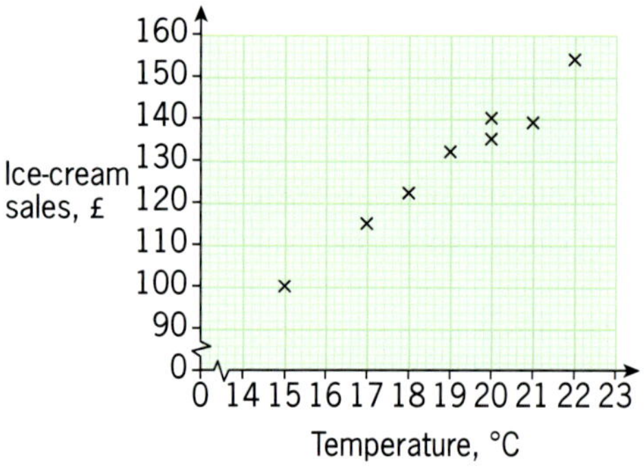

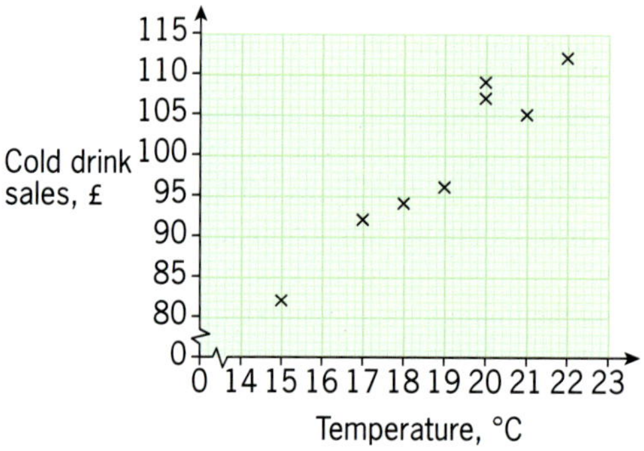

Both graphs show positive correlation.
The correlation is stronger between temperature and ice-cream sales than between temperature and cold drink sales.
The graphs show that as the temperature increases, the sales of ice-cream increase and the sales of cold drinks also increase.

What would a scatter graph of cold drink sales vs ice-cream sales look like?

Exercise B13.2

MEDIUM

1 The graph shows the results of the survey on altitude and lung capacity described on page 194. Write down the correlation shown by the graph, and describe the relationship shown by the variables.

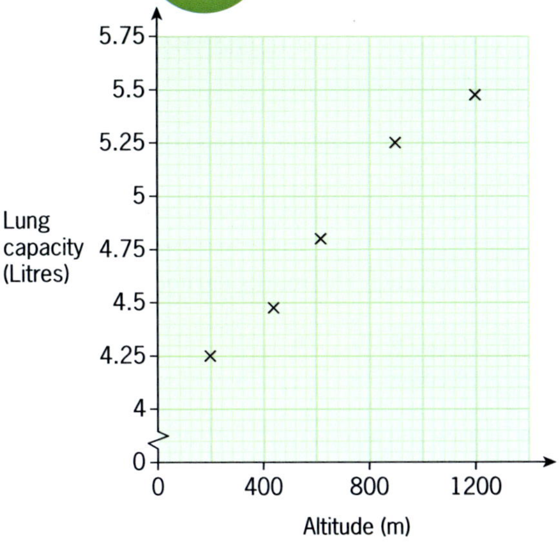

▲ Question 1

*2 For each of the graphs you drew in exercise B13.1 questions 1 and 2

a describe any correlation shown by the graph

b describe the relationship shown between the variables.

*3 Emma recorded how long it took a group of friends to write their name with their right hand and with their left hand.

Right-hand time taken, seconds	4.6	6.5	7.2	4.8	5.8	6.2	4.8	9.8	4.6	4.0	6.8	7.0
Left-hand time taken, seconds	6.2	8.4	9.6	6.4	7.0	7.5	6.6	6.5	6.6	6.0	8.5	9.2

a Draw a scatter diagram of these data.

b Describe the correlation shown.

c Describe the relationship between the times taken to write your name with each hand.

*4 The table shows the place, year, magnitude (measured on the Richter scale) and number of deaths (to the nearest 500) of major earthquakes between 1900 and 1940.

Place	USA	Chile	Italy	Italy	China	Japan	China	China	India	Chile	Turkey
Year	1906	1906	1908	1915	1920	1923	1927	1932	1935	1939	1939
Magnitude	8.3	8.8	7.5	7.5	8.6	8.3	8.3	7.6	7.5	7.8	7.9
Number of deaths (000s)	0.5	20	120	30	180	143	200	70	60	30	23

Draw an appropriate scatter graph to investigate whether the number of deaths is related to the magnitude of the earthquake as measured on the Richter scale. Explain what you find.

Unit B

Lines of best fit

B13.3

- Draw lines of best fit and understand what these represent

When a scatter diagram shows correlation you can often draw a **line of best fit**.

EXAMPLE

The table gives the marks earned in two exams by 10 students.

Maths %	70	76	61	70	89	65	59	58	73	82
Statistics%	78	82	74	75	93	70	66	62	77	89

a Draw a scatter graph for the data.
b Describe the correlation and the relationship shown.
c Draw a line of best fit.
d Predict the statistics mark for a student who scored 62% in Maths.

One way to draw a line of best fit is 'by eye':
- Find the mean of each variable
- Plot the means as a single point M (maths 70.3, statistics 76.6)
- Draw a straight line through M, with half the data points above and half below the line.

ANSWER

a, **c** See graph.
b Positive correlation. Students who scored higher in Maths also tended to score higher in Statistics.
c Maths mean: $\frac{703}{10} = 70.3$

Statistice mean: $\frac{766}{10} = 76.6$

d Predicted Statistics mark =70%

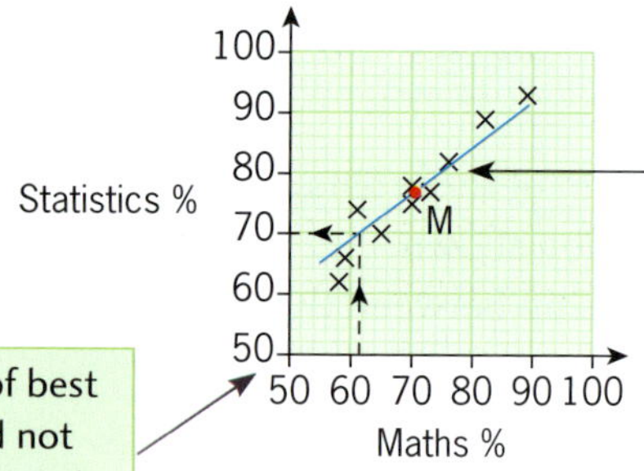

A line of best fit need not pass through the origin.

The line of best fit may not pass though any of the points, but will be close to them.

Sometimes data do not appear to follow a straight line, but do follow a curve. The graph shows the annual heating costs and the number of rooms in a survey of hotels.

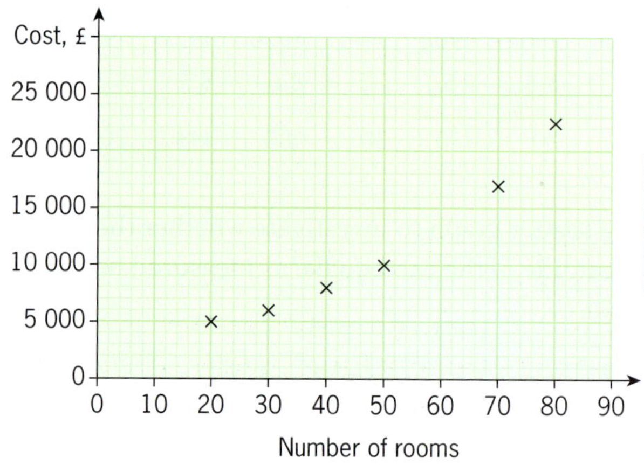

This is an example of **non-linear correlation**.

Exercise B13.3

MEDIUM

1 The graphs show the information collected by Jon on midday temperature and sales of cold drinks and ice-cream at a beach cafe.

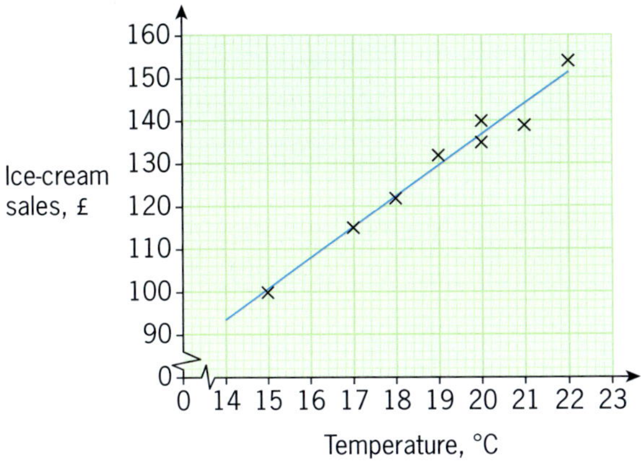

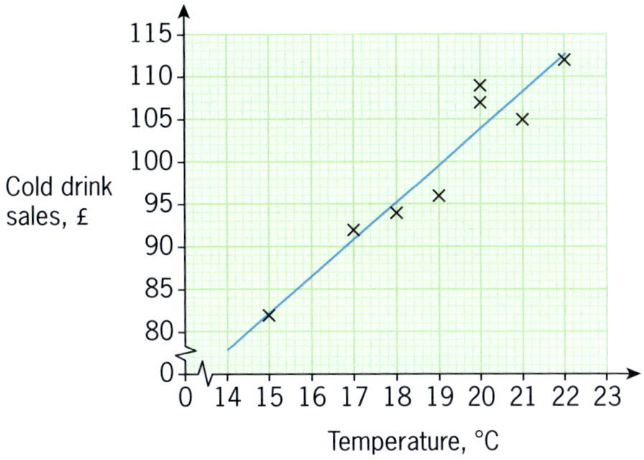

Use the graphs to estimate

a ice-cream sales when the midday temperature is 16 °C.

b cold drink sales when the midday temperature is 18.5 °C.

* 2 The graph shows the altitude of the location where a sample of men live and their lung capacity.

a Estimate the lung capacity for a man living at an altitude of 720 m.

b Explain why it may not be sensible to use the graph to estimate the lung capacity of

i a man living at an altitude of 1500 m,

ii a woman.

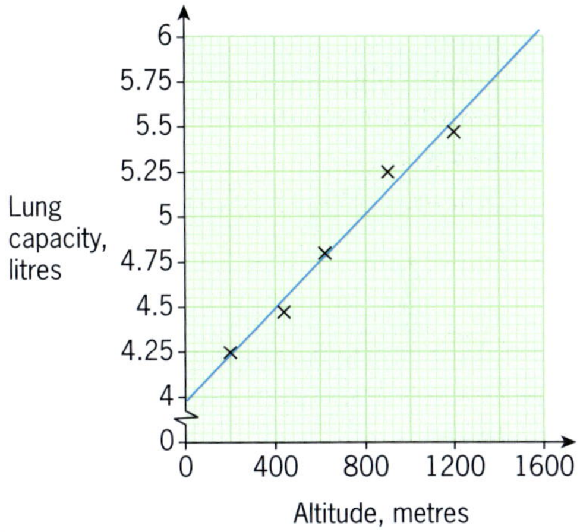

* 3 The table shows the average life span, in years, and average gestation period, in days, for some mammals.

Mammal	Black bear	Cat	Goat	Hamster	Horse	Lion	Mouse	Pig	Rabbit	Red fox	Sheep	Squirrel
Life span, years	18	15	8	4	24	15	3	10	5	3	12	10
Gestation, days	210	63	145	16	336	106	21	114	31	52	150	44

a Use the data to draw a scatter diagram and estimate

i the life span of a guinea pig that has a gestation period of 68 days

ii the gestation period of a tiger that has a life span of 16 years.

b Discuss the reliability of your results with reference to animal size.

Unit B

Time series

B13.4

- Draw and interpret line graphs for time series

Time series data is collected over a period of time at regular intervals. It is often displayed as a line graph.

Common real-life examples of time series data are utility bills, unemployment rates, and average house prices.

EXAMPLE

Jenny drew this table for her quarterly gas bills over a period of two years.

	Jan–March	April–June	July–Sept	Oct–Dec
2008	£65	£38	£24	£60
2009	£68	£42	£30	£68

a Draw a time series graph to represent these data.

b Describe the annual changes in Jenny's gas bill shown by the graph.

c Roughly what would you expect the 2010 April–June gas bill to be?
Explain your answer.

ANSWER

a

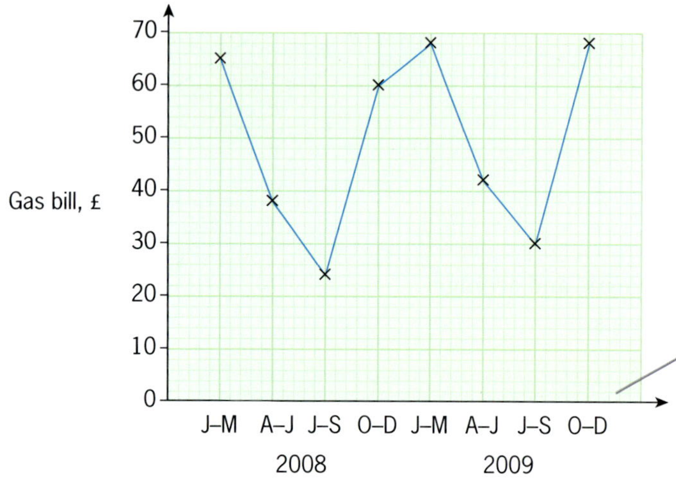

Time is on the horizontal axis.

b The bills are highest in the January–March quarter in each year and are lowest in the July–September quarter each year.

c The bill for April–June 2010 may be in the range £45–£50, because the 2009 prices are all higher than in 2008, so we would expect the 2010 prices to be higher than in 2009.

Exercise B13.4

MEDIUM

Draw a time series graph to represent the information given in each table.
Describe what is shown in each time series graph.

1 The table shows information about Richard's monthly mobile phone bill.

Jan	Feb	Mar	Apr	May	Jun	Jul	Aug	Sep	Oct	Nov	Dec
£12	£14	£10	£18	£16	£22	£13	£10	£15	£15	£17	£20

2 The table shows information about Luigi's quarterly electricity bills over a two-year period.

	Jan–Mar	Apr–Jun	Jul–Sep	Oct–Dec
2004	£55	£24	£21	£46
2005	£59	£28	£23	£50

3 The table shows information about monthly ice-cream sales at Haroun's shop during one year.

Jan	Feb	Mar	Apr	May	Jun	Jul	Aug	Sep	Oct	Nov	Dec
£16	£10	£8	£16	£22	£36	£48	£50	£12	£15	£16	£39

4 Tasmin kept a record of how much money she had earned from odd jobs during three years.

	Jan–Apr	May–Aug	Sep–Dec
2007	£15	£39	£27
2008	£21	£45	£30
2009	£24	£54	£42

5 Alec kept a record of his quarterly expenses over a period of two years.

	Jan–Mar	Apr–Jun	Jul–Sep	Oct–Dec
2004	£47	£56	£33	£18
2005	£39	£48	£29	£16

Unit B

Summary and assessment

CHECK OUT

You should now be able to draw and interpret

- scatter diagrams
- time series graphs.

Exam-style question

Two samples of students with similar alcohol, smoking and exercise patterns took part in a survey on pulse rates, in beats per minute, before and after exercise.

Males

Before	82	70	96	86	74	75	72	85
After	140	96	152	150	98	130	119	120

Females

Before	78	88	76	80	78	60	65	70
After	130	140	100	135	120	90	104	116

Draw scatter graphs to represent these data.
Comment on any differences and similarities shown between male and female pulse rate

Use the same scales on both graphs to make comparisons easier.

Only draw a line of best fit if appropriate.

Yumi's answer:

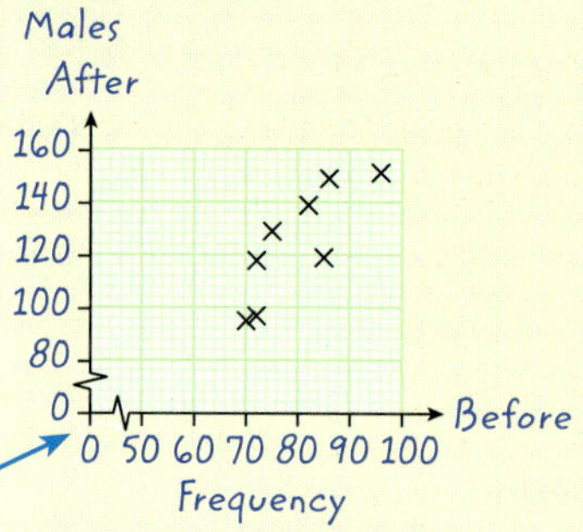

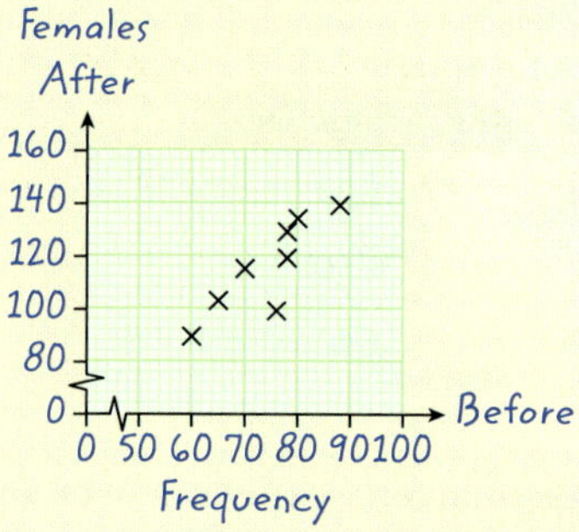

Both show positive correlation
Females have lower pulse rate before exercise and it doesn't increase as much as their line of best fit would be less steep.

Draw separate diagrams next to each other.

Give reasons for your comments.

Exam practice

*1 The table gives the mean average monthly temperature in °C for July in Northern Ireland from 1996 to 2007.

Draw a time series graph to show these data.

Describe what the graph is showing.

1996	1997	1998	1999	2000	2001	2002	2003	2004	2005	2006	2007
14·2	14·9	13·7	15·5	14·4	14·3	13·6	15·6	13·8	15·0	16·7	13·9

(*3 marks*)

*2 The table gives the number of marriages, to the nearest hundred, in the UK for each month during 1992 and 1993.

	January	February	March	April	May	June
1992	8500	15900*	15700	22700	33900*	32500
1993	9200*	11900	15400	22300	32800*	33100
	July	**August**	**September**	**October**	**November**	**December**
1992	37800	49100*	41600	26200*	13900	13800
1993	41900*	40200	40000	25200*	13500	14200

The numbers with an asterisk are the months where there were five Saturdays.
Draw a time series graph to display this data and comment on any trends shown.

(*4 marks*)

*3 A simulation was carried out of two different types of queuing in a bank.
Queuing method A: a single queue line forms and people wait for next free counter.
Queuing method B: queue lines form for individual counters.
Each person taking part in the simulation had these waiting times in each queuing method.

A	1	7	3	0	0	6	8	6	7	1	2	5	2	9	3
B	1	6	0	1	0	4	6	3	4	5	1	6	4	5	3

Work out summary statistics.
Comment on which method you think is best for queuing. (*6 marks*)

14 More straight-line graphs

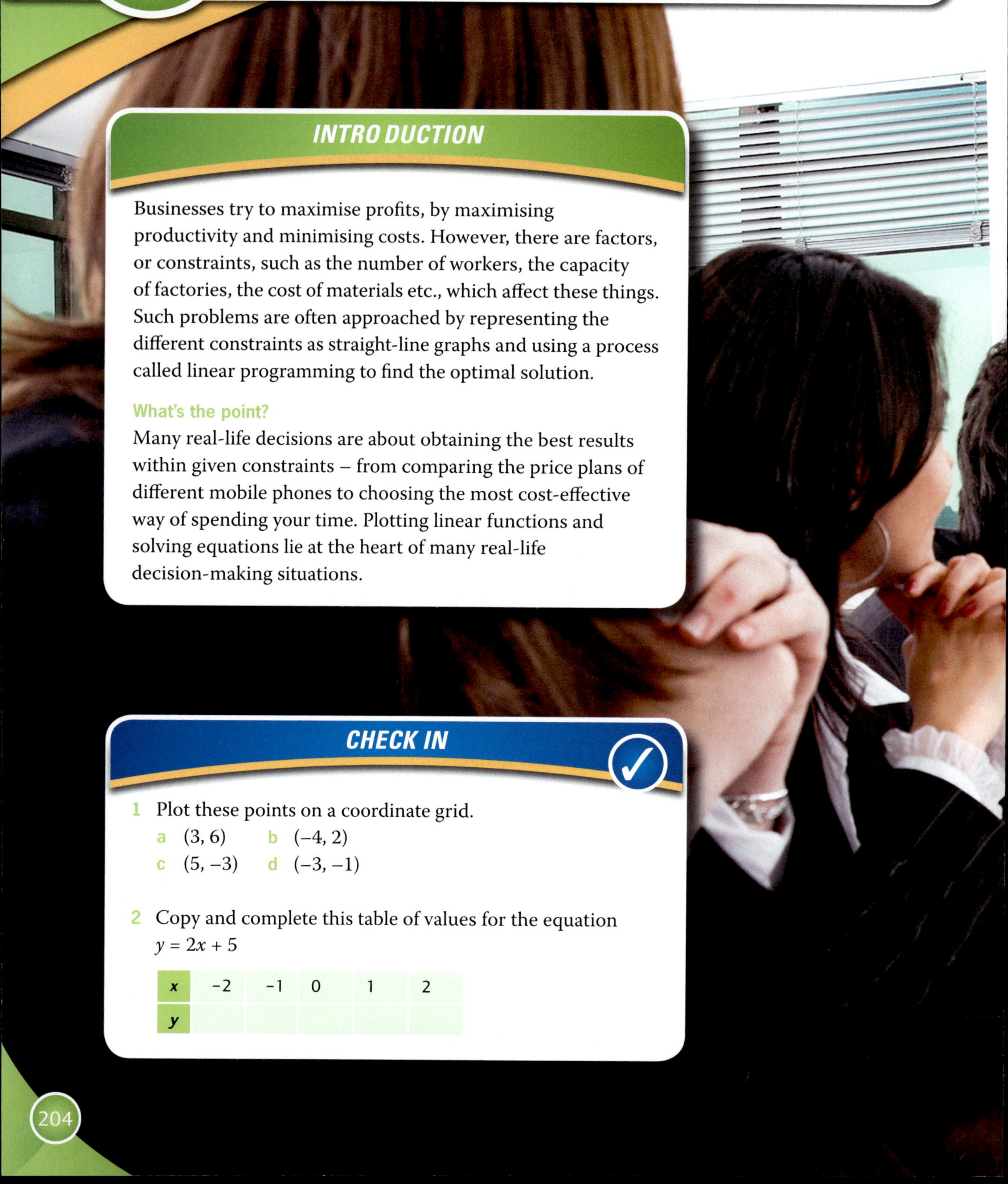

INTRODUCTION

Businesses try to maximise profits, by maximising productivity and minimising costs. However, there are factors, or constraints, such as the number of workers, the capacity of factories, the cost of materials etc., which affect these things. Such problems are often approached by representing the different constraints as straight-line graphs and using a process called linear programming to find the optimal solution.

What's the point?

Many real-life decisions are about obtaining the best results within given constraints – from comparing the price plans of different mobile phones to choosing the most cost-effective way of spending your time. Plotting linear functions and solving equations lie at the heart of many real-life decision-making situations.

CHECK IN

1 Plot these points on a coordinate grid.

a (3, 6) b (−4, 2)

c (5, −3) d (−3, −1)

2 Copy and complete this table of values for the equation $y = 2x + 5$

x	−2	−1	0	1	2
y					

Orientation

What I should know	What I will learn	What this leads to
B11	■ Plot straight line graphs and find the gradients of straight lines ■ Solve simultaneous equations	C19 and C21

RICH TASK

This L-shape is drawn on a 10×10 grid numbered from 1 to 100.

It has five numbers inside it.

We can call the shape L_{35} because the largest number inside it is 35.

The total of the numbers inside L_{35} is 138.

Find a connection between the L-number and the total of the numbers inside the L-shape

What happens if you work on a 9×9 grid? An 8×8 grid?

What about an $n \times n$ grid?

1	2	3	4	5	6	7	8	9	10
11	12	13	14	15	16	17	18	19	20
21	22	23	24	25	26	27	28	29	30
31	32	33	34	35	36	37	38	39	40
41	42	43	44	45	46	47	48	49	50
51	52	53	54	55	56	57	58	59	60
61	62	63	64	65	66	67	68	69	70
71	72	73	74	75	76	77	78	79	80
81	82	83	84	85	86	87	88	89	90
91	92	93	94	95	96	97	98	99	100

Straight-line graphs

- Draw straight-line graphs from their equations

You can use a function machine or an equation to complete a table of values.

ACTIVITY

This function machine shows the rule 'double and add 4'.
It can be written as the equation $y = 2x + 4$.

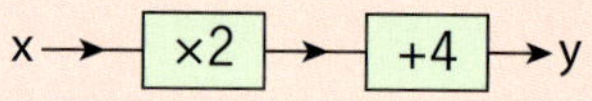

When $x = 0$, $y = 2 \times 0 + 4 = 0 + 4 = 4$.
When $x = 1$, $y = 2 \times 1 + 4 = 2 + 4 = 6$.
What are the values of y when $x = 2$ and when $x = 3$?

You can complete a table of values.

x	0	1	2	3
y	4	6	8	10

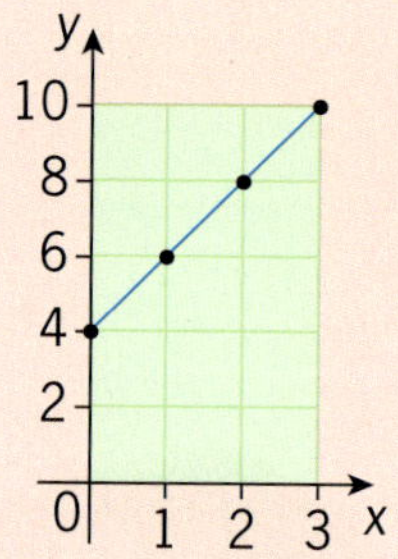

What are the coordinates of the points used to draw this graph?

EXAMPLE

Draw the graph of $y = 12 - 2x$ by constructing a table of values using x-values from 0 to 6.

ANSWER

When $x = 0$, $y = 12 - 2 \times 0 = 12 - 0 = 12$
When $x = 2$, $y = 12 - 2 \times 2 = 12 - 4 = 8$
When $x = 4$, $y = 12 - 8 = 4$
When $x = 6$, $y = 12 - 12 = 0$

The table of values is

x	0	2	4	6
y	12	8	4	0

The points (0, 12), (2, 8), (4, 4) and (6, 0) are used to draw the graph.

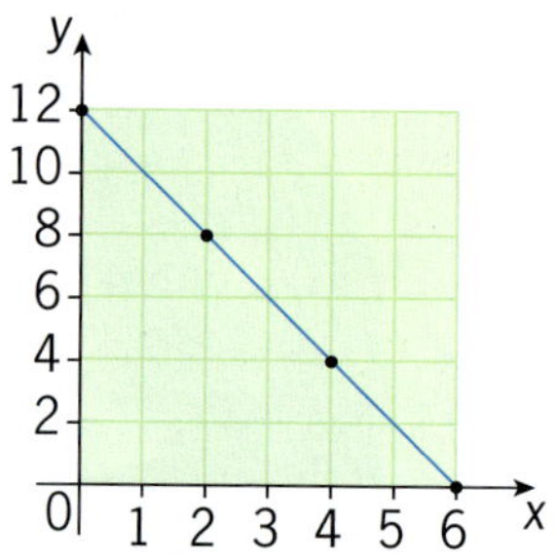

Do you remember the least number of points required to draw a straight-line graph?

Exercise B14.1

MEDIUM

1 Use this function machine to complete this table.
Plot points on axes as labelled here.
Draw a straight-line graph and write its equation.

x → ×2 → +1 → y

x	0	1	2	3
y				

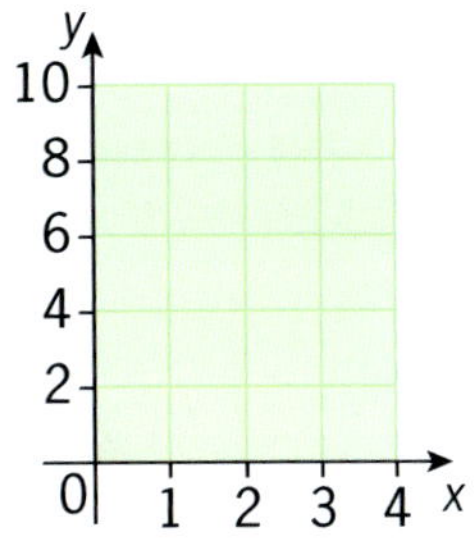

2 Copy and complete this table for each of these equations.
Draw the graph of each equation on axes as labelled here.

a $y = x + 3$ b $y = x - 3$ c $y = 2x + 1$
d $y = 2x - 2$ e $y = 12 - x$ f $y = 3 - x$

x	0	1	2	3	4
y					

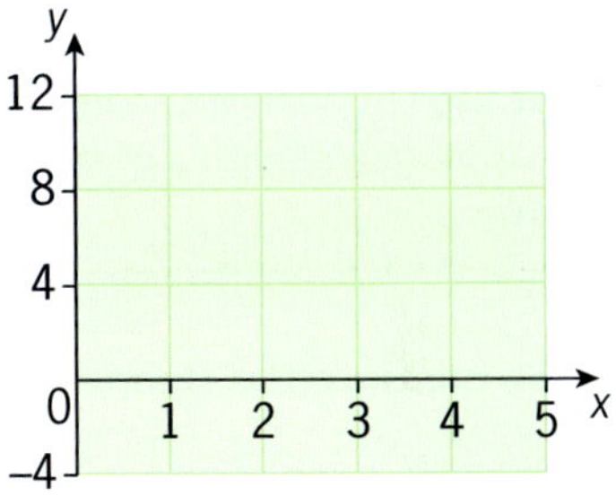

3 Decide whether each point is on the given line.
You do not need to draw any graphs.

a (3, 10), $y = x + 7$ b (5, 16), $y = 2x + 6$ c (3, 11), $y = 3x - 2$
d (4, 0), $y = 8 - 2x$ e (5, –4), $y = 1 - x$ f (–6, 1), $y = \frac{1}{2}x - 2$

4 Two roads run due north and due east. Taking the roads as the x- and y-axes, a fence lies along the line $y = 10 - x$.
a With both axes labelled from 0 to 12, draw the roads and the fence.
b Find the area enclosed by the two roads and the fence.

5 A footpath runs along the line of the x-axis. Two walls lie along the lines $y = 2x$ and $y = 9 - x$.
a Draw a diagram of the footpath and walls.
b Find the area of the triangle which lies between the walls and path.

6 Grubshaw Silver Band hires a hall for a concert. The cost of the hire, £C, depends on the number of hours, x, that the hall is needed.
Given that $C = 2x + 30$, draw a graph of C against x for $0 \le x \le 5$.
For how many hours is the hall used if the club is charged £37?

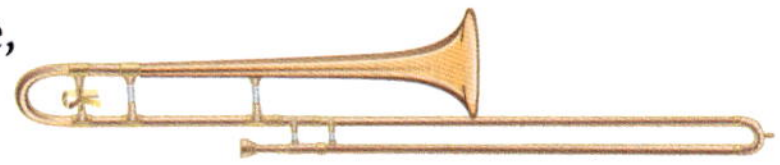

7 ✮ **ICT CHALLENGE** ✮

Use a computer spreadsheet to create tables of values for $y = 3x - 8$ and also $y = 2x + 12$ for positive values of x.

By extending the spreadsheet downwards, can you find the coordinates of the point where the graphs of these two equations intersect?

Gradients of straight lines

B14.2

- Find the gradient of a straight line and recognise parallel lines

RICH TASK

Investigate what these road signs mean.
The modern signs use percentages, but the older signs are still in use.
Can a 4 × 4 vehicle climb a 1 : 1 hill?
Are these gradients the same as those used by mathematicians when they work with gradients of straight lines?

→ **You measure the steepness of a sloping line by finding its gradient.**

To find the gradient of a straight line:

→ **choose any two points on the line**

→ **draw a right-angled triangle including one vertical and one horizontal side**

→ **gradient =** $\dfrac{\textbf{amount up}}{\textbf{amount across}}$

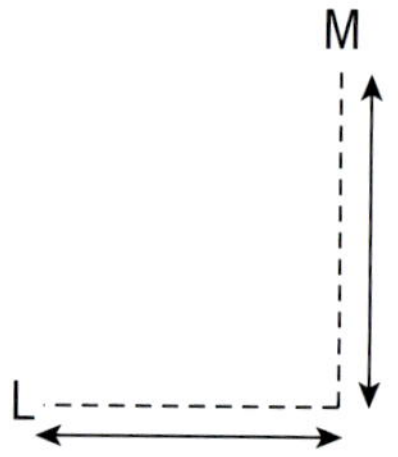

EXAMPLE

The lines $y = 2x + 4$, $y = x + 4$, $y = -2x + 4$ are drawn here. Find their gradients. What is the gradient of the line $y = mx + 4$?

ANSWER

The gradient of $y = 2x + 4$ is 2.
The gradient of $y = x + 4$ is 1
The gradient of $y = -2x + 4$ is -2.
The gradient is the number which multiples the x.
The gradient of $y = mx + 4$ is m.

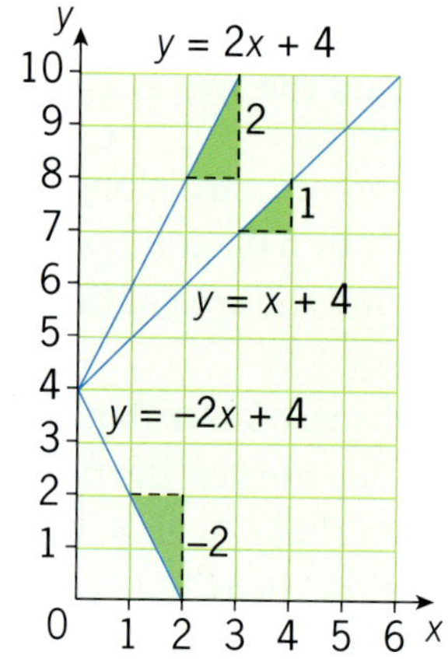

→ **The line $y = mx + c$ (where m and c are numbers) has a gradient of m.**
It cuts the y-axis at the point $(0, c)$.
The line $y = 5x + 2$ has a gradient of 5. It is a rising line.
The line $y = -4x + 2$ has a gradient of -4. It is a falling line.

→ **Parallel lines have equal gradients.**

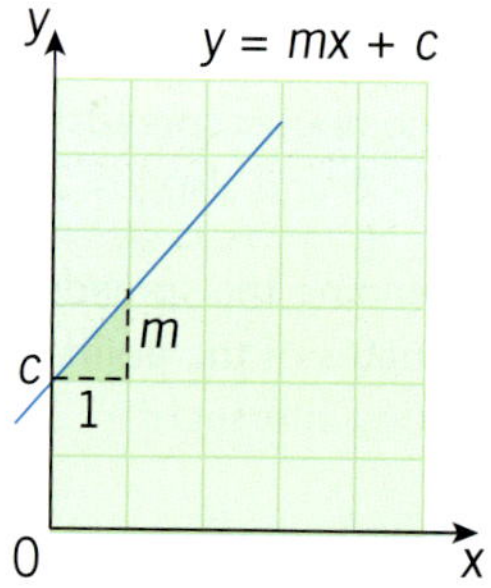

Exercise B14.2

MEDIUM

1 Write the gradients of the straight lines P to V.

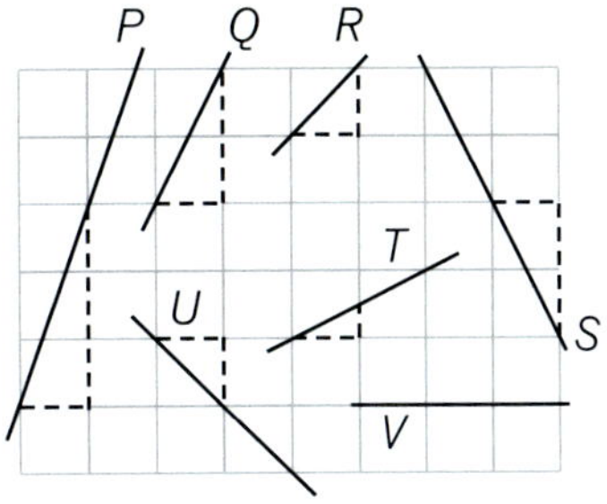

2 Draw and label both axes from 0 to 8.
Join these pairs of points with straight lines.
Find the gradient of each straight line.

a (1, 4) and (3, 8)
b (1, 4) and (2, 7)
c (1, 4) and (7, 7)
d (1, 4) and (6, 4)
e (1, 4) and (5, 0)
f (1, 4) and (3, 0)

3 Write down the gradient of the straight lines with these equations.
Do not draw any graphs.

a $y = 5x + 2$
b $y = 4x - 7$
c $y = x + 3$
d $y = \frac{1}{2}x + 6$
e $y = -6x + 4$
f $y = -3x + 1$
g $y = -8x + 3$
h $y = 3 - 8x$
i $y = 2 - 4x$
j $y = 5 - x$
k $y = 1 - \frac{1}{2}x$
l $y = 2 + 3x$

4 Without drawing any graphs, decide which pairs of equations give parallel lines.

$y = 2x + 1$	$y = 4 - x$	$y = \frac{1}{2}x + 3$	$y = -2x + 7$
$y = -x + 5$	$y = 2x - 1$	$y = 6 - 2x$	$y = 1 + \frac{1}{2}x$

Find a rule for identifying parallel lines from their equation.

5 **RICH TASK**

Use a computer's graphical software to draw the graphs of $y = mx + 3$ where m has

a whole-number values from 1 to 5
b whole-number values from −1 to −5
c a value of zero
d any value greater than 100.

Describe your results.
What happens if you draw the graphs of $y = 2x + c$ where c has a similar range of values?

Unit B

Implicit functions

B14.3

- Draw straight-line graphs where y is given implicitly in the equations

RICH TASK

Mr Riley is a market gardener.
He grows apples and pears.
He wants to plant x apple trees and y pear trees.
Each apple tree seedling costs £5 and each pear tree seedling costs £10.
He wants to spend exactly £200.
How many different combinations of apple and pear tree seedlings can he buy?
If you graph these possible choices, what does the graph look like?

When $y = mx + c$, you can find y directly from the equation when you know the value of x.
But with $5x + 2y = 10$, the connection between x and y is implied.
This type of equation is called an implicit equation.

You can draw the graph of a line such as $5x + 2y = 10$ by finding where it cuts both axes.

EXAMPLE

Draw the graph of the straight line $4y + 3x = 12$.

ANSWER

All points on the y-axis have $x = 0$.
Substitute $x = 0$ to get $4y + 0 = 12$ which gives $4y = 12$, so $y = 3$.
The line cuts the y-axis at the point (0, 3).

All points on the x-axis have $y = 0$.
Substitute $y = 0$ to get $0 + 3x = 12$ which gives $x = 4$.
The line cuts the x-axis at the point (4, 0).

Draw the graph using the two points (0, 3) and (4, 0).

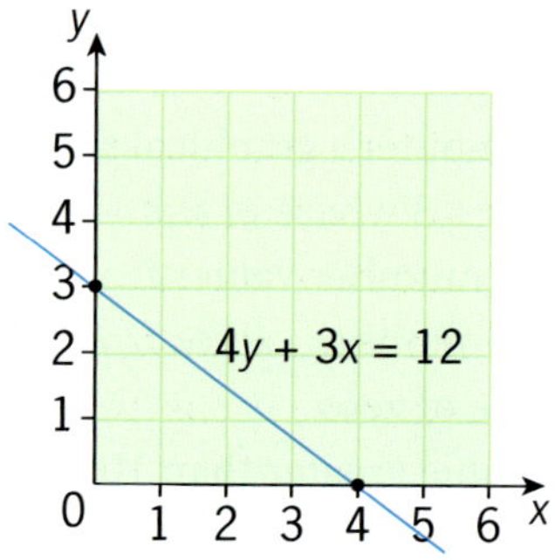

How many points have you used? Is this a disadvantage?

Exercise B14.3

MEDIUM

1 Find the points where the lines with these equations cut the two axes.
Draw the lines on axes labelled from 0 to 6.

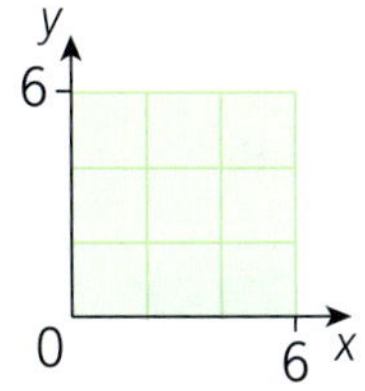

a $3x + 2y = 12$ b $2x + 5y = 10$ c $3x + 5y = 15$
d $4x + 3y = 12$ e $2x + 3y = 9$ f $x + 3y = 6$

2 Find where the lines with these equations cut the two axes.
Draw the lines on axes labelled from –6 to 6.

a $2x - 4y = 8$ b $6x - 5y = 30$ c $3x + 4y = -12$

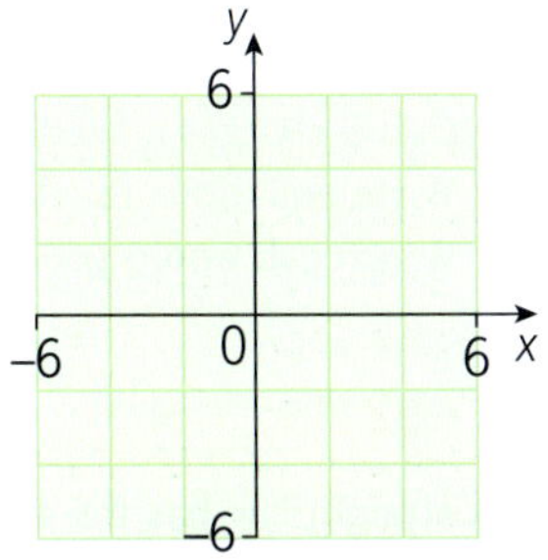

3 Draw the graphs of the lines with these equations.
Label your axes from 0 to 8.
Use your graphs to find the gradient of each line.

a $2x + y = 8$ b $3x + y = 6$ c $x + 2y = 4$

4 Draw the graphs of the lines with these equations.
Find the gradient of each line from your graphs.

a $4x + 2y = 8$ b $3x + y = 6$ c $3x - y = 3$
d $6x + 2y = -12$ e $x + 2y = -4$ f $2x - y = 4$

5 Draw the graph of the equation $2x + y = 6$.
Find all possible values of x and y which are positive integers.

6 **RICH TASK**

Molly buys y birthday cards at £2 each and x Christmas cards at £1 each. Altogether she spends exactly £8.
Draw a graph and find all the possible combinations of cards that she could have bought.
If she bought twice as many Christmas cards as birthday cards, find how many she bought of each type.

Simultaneous equations

B14.4

- Solve simultaneous equations, including in real-life contexts

RICH TASK

Two taxicabs compete with each other.
Cab A simply charges a cost in £ equal to three times the mileage, x.
Cab B charges a fixed cost of £6 plus a rate of £1 per mile.
Write equations for their fares and explore them graphically.
Which cab would you decide to take?

☆ ACTIVITY ☆

A straight line has the equation $x + y = 6$.
Another straight line has the equation $x + 2y = 8$.
Is there a point that lies on both lines at the same time?
What is this point?

p. 110

→ **Simultaneous equations are solved at the same time to find the value of more than one variable**

EXAMPLE

Draw the graphs of $y = 2x + 1$ and $x + y = 7$.
Solve these equations simultaneously.

'Simultaneously' means that the solution should work for both equations

ANSWER

Choose suitable values of x, make tables of values and draw the graphs of the equations.

for $y = 2x + 1$

x	0	1	3
y	1	3	7

for $x + y = 7$

x	0	1	3
y	7	6	4

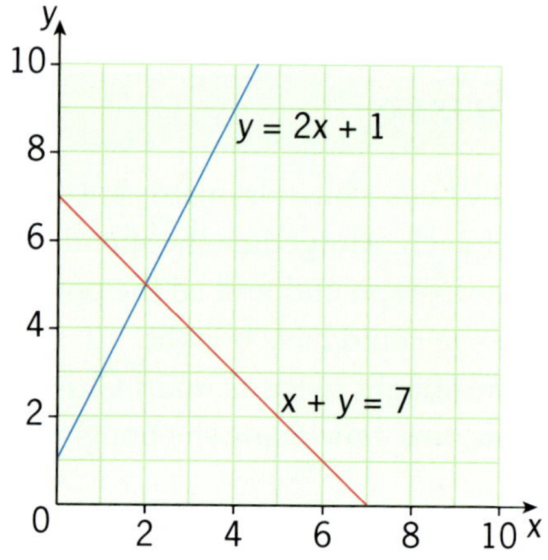

From the graphs, the point of intersection is (2, 5).
So the solution of the simultaneous equations is $x = 2$, $y = 5$.

As a check,
$y = 2x + 1 = 2 \times 2 + 1 = 5$
and $x + y = 2 + 5 = 7$. ✓

Exercise B14.4

MEDIUM HIGH

1 For each pair of simultaneous equations, copy and complete the table of values, draw their graphs on axes labelled from –2 to 10, and write down the solution.

a

x	0	1	2	3
$y = x + 3$				
$y = 2x + 1$				

b

x	0	1	2	3
$y = x + 5$				
$y = 2x + 3$				

c

x	0	1	2	3	4
$y = x + 1$					
$y = 2x - 2$					

d

x	0	1	2	3	4
$y = x + 4$					
$y = 3x - 2$					

e

x	0	1	2	3	4
$y = x + 2$					
$y = 6 - x$					

f

x	0	1	2	3	4
$y = x - 1$					
$y = 7 - x$					

2 For each pair of simultaneous equations, find the points where the graphs cut the axes, draw their graphs, and write down the solution. Check your answers.

a $x + y = 5$
  $x + 2y = 8$

b $2x + y = 8$
  $x + 3y = 9$

c $2x + y = 10$
  $3x + 4y = 25$

d $3x + 2y = 12$
  $x + 6y = 12$

e $2y + x = 10$
  $3y + 2x = 18$

f $2y + x = 6$
  $2y - x = 2$

*3 **RICH TASK**

Gary buys x spanners.
Firm A charges £3 per spanner and no postage, so their total cost £$C = 3x$.
Firm B charges £2 per spanner plus £4 postage, so their total cost £$C = 2x + 4$.
Draw graphs of these costs and advise Gary on his purchase.

Unit B

Summary and assessment

B14

CHECK OUT

You should now be able to

- draw straight lines
- find gradients of straight lines and recognise parallel lines
- recognise that the point of intersection of two straight lines is the solution of simultaneous equations.

Exam-style question

Look at this diagram

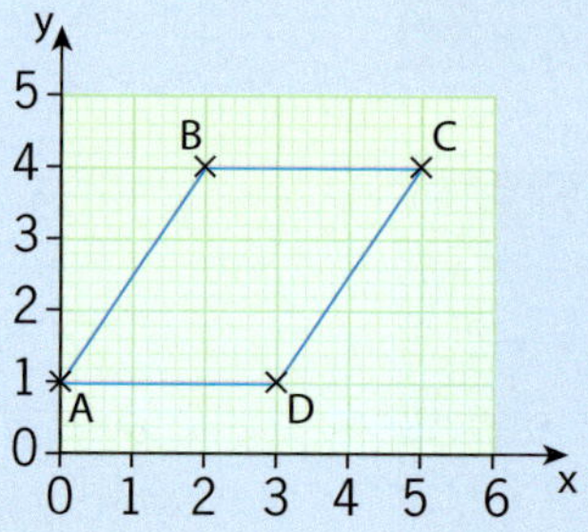

a Explain why the gradient of the lines joining AD and BC are 0.

b What is the gradient of the line joining

i AB **ii** CD?

***c** Explain how this tells you that the shape ABCD is a parallelogram.

Clive's answer:

a AD and BC are both horizontal lines and the gradient of horizontal lines is zero

b i) A(0, 1) B(2, 4)

Gradient AB $= \frac{1-4}{0-2} = \frac{-3}{-2} = \frac{3}{2}$

ii) C(5, 4) D(3, 1)

Gradient CD $= \frac{4-1}{5-3} = \frac{3}{2}$

c Gradient of AB = gradient of CD
So AB is parallel to CD & AB is opposite CD
We were told that gradients of AD and BC are the same and they are opposite sides
A four- sided shape with two pairs o parallel sides is a parallelogram, so ABCD is aparallelogram

All the information you need will be on the diagram.

Write explanations that are complete, clear and concise.

Explain means that you have to use words as well as numbers.

Explain what your results show in your explanation.

Exam practice

1 a Copy and complete the table of values for $y = 3x - 1$

x	0	1		3
y	−1	2		8

b Draw the line $y = 3x - 1$. Use axes from 0 to 8. *(3 marks)*

2 a Work out the gradient of this straight line.

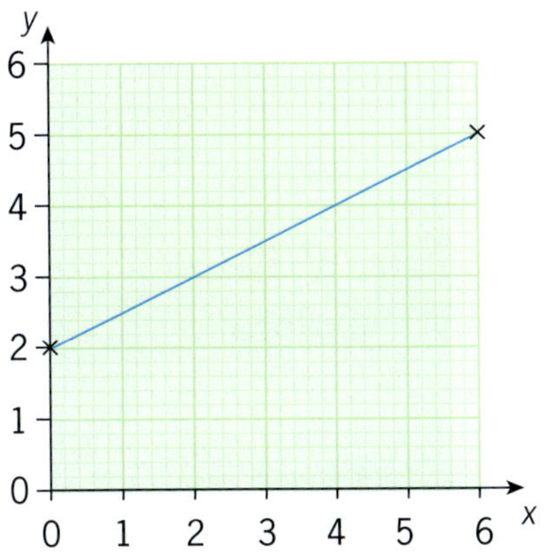

b Copy the grid and on it draw the straight line that passes through the point (2, 2) and is parallel to the line already drawn.

(3 marks)

3 The grid shows the lines $x + y = 5$ and $y = 2x + 1$.

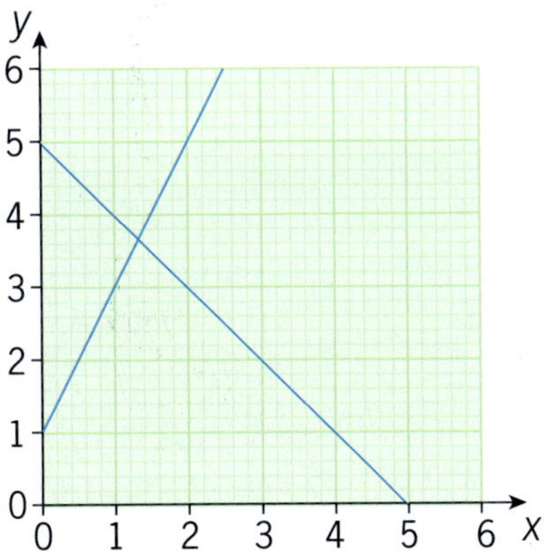

Write down an estimate for the solution of the simultaneous equations

$x + y = 5$

$y = 2x + 1$ *(1 mark)*

15 Estimation and indices

INTRODUCTION

When you use the internet to pay for goods, you need to know that your financial details are safe. To make these details secure they are turned into a secret code (encrypted). The message can be encrypted using the product of two very large prime numbers – the person receiving the message has to know both of these prime numbers so that they can decrypt the message.

What's the point?

The problems involved in identifying very large prime numbers make it very difficult for someone intercepting an encrypted message to crack the code.

CHECK IN

1 Calculate

a 7×7 b $2 \times 2 \times 2$

c $10 \times 10 \times 10$ d $32 \div 100$

2 Find the missing number in this calculation.

$\_\_\_\_\_\_ \times \underline{\quad 5 \quad} = 25$

3 Does 244 divide by 4?

Explain how you know.

Orientation

What I should know	What I will learn	What this leads to
A1	■ Find powers and roots ■ Use index laws ■ Estimate answers to problems	C22

RICH TASK

The number 1000 is made by adding up lots of smaller whole numbers in such a way that when these numbers are multiplied together they produce the maximum product possible.

What are the numbers?

Mental arithmetic

- Develop a range of strategies for mental calculation

You can derive unknown facts from facts that you do know.
Place value can often help.

EXAMPLE

Given that $430 \times 5.8 = 2494$ find

a $43\,000 \times 58$ **b** 0.43×58 **c** $24\,940 \div 43$

ANSWER

a 100 times bigger × 10 times bigger → answer is 1000 times bigger

$43\,000 \times 58 = 2\,494\,000$

b 1000 times smaller × 10 times bigger → answer is 100 times smaller

$0.43 \times 58 = 24.94$

c $2494 \div 430 = 5.8$
10 times bigger ÷ 10 times smaller → answer is 100 × bigger
$24\,940 \div 43 = 580$

Use estimation to check that your answer is correct.

$\div \frac{1}{10}$ is the same as $\times 10$

($\frac{1}{10}$ and 10 are **reciprocals**)

Doubling and halving is also a useful strategy:

$4.68 \times 50 = 2.34 \times 100 = 234$

Using **reciprocals** changes division into multiplication:

$$20.04 \div 0.8 = 200.4 \div 8$$
$$= 200.4 \times \frac{1}{8}$$
$$= 100.2 \times \frac{1}{4}$$
$$= 50.1 \times \frac{1}{2} = 25.05$$

There are lots of other mental strategies you can use.

EXAMPLE

a Find the time lapse between 9.47 a.m. and 1.14 p.m.
b Work out **i** $12.71 - 8.86$ **ii** 49×372

ANSWER

a Sketch a time line: Total time 3 hours 27 minutes

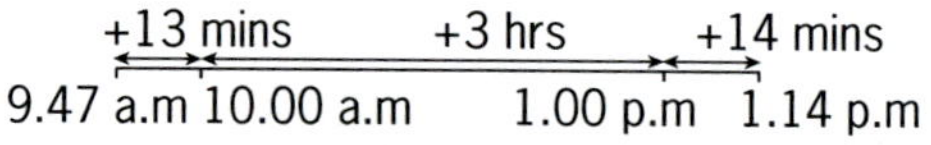

b **i** Round up and compensate: $12.71 - 9 + 0.14 = 3.71 + 0.14$
$= 3.85$

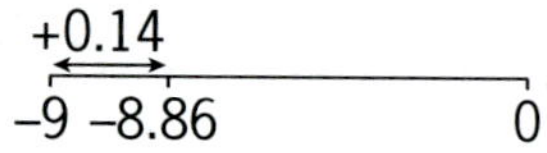

ii Round up to 50: $50 \times 372 = \frac{1}{2}$ of $100 \times 372 = 37\,200 \div 2$
$= 18\,600$

$49 = 50 - 1$ so subtract 372: $49 \times 372 = 18\,600 - 372$
$= 18\,228$

Exercise B15.1

LOW

This exercise is to be completed without a calculator.

Unit B

1 Work out

a 238 + 54	b 429 + 172	c 1057 + 682
d 600 – 327	e 507 – 349	f 891 – 656
g 375 × 54	h 16 × 6	i 48 × 52
j 1.32 ÷ 0.4	k 5.17 ÷ 0.06	l 29 ÷ 0.8

2 a Given 87 × 132 = 11 484, find

 i $\frac{11484}{870}$ ii $\frac{114.84}{132}$

b Given 93 × 214 = 19 902, find

 i $\frac{199.02}{9.3}$ ii 9300 × 0.00214

c Given 76 × 45 = 3420, find

 i $\frac{342}{0.045}$ ii $\frac{342}{7600}$

d Given 32 × 225 = 7200, find

 i $\frac{72}{3.2}$ ii 0.0032 × 2250

3 Work out the change Antonio will receive

a from £5 when buying four chocolate bars costing 56p each

b from £10 when buying a drink for £1.15 and a doughnut for £1.08.

4 a Maisie catches a bus to work at 7:39 a.m. The bus arrives at 8:12 a.m.
How long is Maisie's bus journey?

b Maisie decides to take the train home.
The 5:48 p.m. train leaves the station as Maisie arrives.
The next train she can catch leaves at 6:23 p.m.
How long does Maisie have to wait for the next train?

5 ☆ **PROBLEM** ☆

Use numbers from this list

49 73 77 98 17 84 27 51 34

a Choose two numbers with a total of 100.

b Choose two numbers with a difference of 50.

c Which two numbers would give the highest product?
Work out that product.

Powers and roots

B15.2

- Use the terms 'square', 'square root', 'cube' and 'cube root'

☆ ACTIVITY ☆

There are 16 dots in this square arrangement.
How many dots are there on one side?
What other numbers will allow you to make a square arrangement like this?
List a few and find the number of dots on one side for each.

→ **A square number is a number multiplied by itself.**
5 squared = 5 × 5 = 25
A quick way of writing this is $5^2 = 25$

The opposite of squared is square root.
$\sqrt{\ }$ is the symbol for square root.
$5^2 = 25$ so $\sqrt{25} = 5$

☆ ACTIVITY ☆

Write down a value for $\sqrt{36}$.
Now find the value of -6×-6.
Suggest **another** value for $\sqrt{36}$.

You are expected to know the square numbers from 1^2 up to 15^2 and their square roots.

→ **Square roots can have two answers, one positive and one negative.**

This large cube contains two small cubes along each edge.
There are 8 small cubes altogether.
$2 \times 2 \times 2 = 2$ cubed $= 8$ or $2^3 = 8$

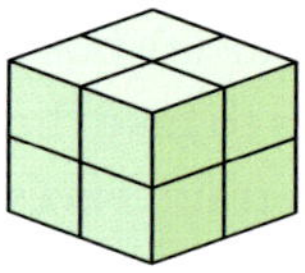

→ **A cube number is a number multiplied by itself and then multiplied by itself again $4^3 = 4 \times 4 \times 4 = 64$**
→ **The opposite of cube is cube root.**
$2^3 = 8$ so $\sqrt[3]{8} = 2$

You should know these cube numbers
$2^3 = 8$
$3^3 = 27$
$4^3 = 64$
$5^3 = 125$
$10^3 = 1000$

EXAMPLE

Work out **a** $3^2 + 4 \times 10$ **b** $\sqrt{9} + \sqrt{4}$ **c** $5^2 - \sqrt{16}$

ANSWER

a $3^2 + 4 \times 10 = 9 + 40 = 49$ Work out power separately from bracket

b $\sqrt{9} + \sqrt{4} = 3 + 2 = 5$ Work out each square root separately

c $5^2 - \sqrt{16} = 25 - 4 = 21$

Exercise B15.2

MEDIUM LOW

Questions 1 to 7 are to be completed without a calculator

1 Find a 7^3 b 20^2 c 15^2 d 6^3 e 5^3

2 Find a $\sqrt{49}$ b $\sqrt{10000}$ c $\sqrt[3]{125}$ d $\sqrt[3]{64}$ e $\sqrt{81}$

3 Write down both the positive and negative answers to

a $\sqrt{100}$ b $\sqrt{121}$ c $\sqrt{9}$ d $\sqrt{196}$

4 a Work out i $\sqrt{169}$ ii $\sqrt{100}$

b Use your answers to a to work out $\sqrt{1.69}$

5 a Work out i $\sqrt{144}$ ii $\sqrt{1.44}$

b Work out i $\sqrt{225}$ ii $\sqrt{2.25}$

6 Write these numbers as a **power of 10**

a 1000 b 1 000 000 c 1 million

7 Work out

a $\sqrt{4}+5^2$ b 3^3-5^2 c $\sqrt{100}+\sqrt{25}$

d $4^3-\sqrt{64}$ e $\sqrt{(25+24)}$

8 Copy this number line.

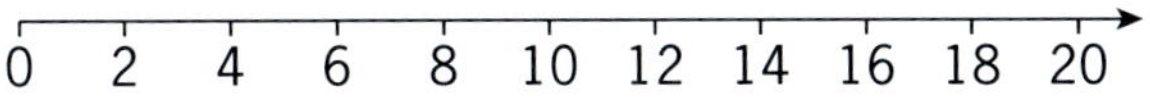

Mark on the number line

a $\sqrt{30}$ b $\sqrt{111}$ c $\sqrt{200}$ d $\sqrt{3}$ e $\sqrt{300}$

9 ☆ **PROBLEM** ☆

Make nine square numbers by adding pairs of numbers from 1 to 18.
Each number can only be used once.
Use the table of numbers to help.

1	2	3	4	5	6
7	8	9	10	11	12
13	14	15	16	17	18

10 ***RICH TASK***

Look at this pattern

1

1 + 3

1 + 3 + 5

1 + 3 + 5 + 7

Write down the sequence of answers.
What would the next number be?
What sort of numbers are the answers?

Estimation and checking procedures

B15.3

- Estimate answers to problems involving decimals

☆ ACTIVITY ☆

Look at these reports from the 2008 Olympics.

By May 2007 almost 85%, \$2.1 billion, of the construction budget for the six main venues had been raised. The Beijing National Stadium, nicknamed "The Bird's Nest", cost \$423 million to build.

The torch relay lasted for 130 days and the torch was carried for 137 000 km. A total of 11 028 athletes from 207 nations competed in 302 events in 28 sports. 4.7 billion viewers worldwide tuned into the television coverage.

On March 6, 2009 Beijing reported that total spending on the games was about \$15 billion, and extra monies raised from the Olympic Games would exceed the original target of \$16 million. Other sources, however, estimated that approximately \$40 billion had been spent on the Games.

Which numbers do you think are accurate and which numbers are estimates?
Why is it okay for some numbers to be reported as estimates?

There are lots of other instances when the exact answer is not the best answer.

EXAMPLE

The latest paperback novels are sold at a supermarket for £3.79 each.
How many can Keith buy for £21?

ANSWER

Keith could calculate $21 \div 3.79$, but does he need the exact answer?
3.79 rounds up to 4, and $21 \div 4 \approx 5$
Keith can buy 5 books.

You would need to be careful if the amount were £23. A sensible estimate might be $24 \div 4 = 6$

You can use estimation to solve problems. Be careful with rounding.

p. 8

➜ **When multiplying, compensate by making one number bigger and one smaller** $3.4 \times 4.4 \rightarrow 3 \times 5 = 15$

➜ **When dividing, compensate by making both numbers bigger or both smaller** $3.4 \div 4.4 \rightarrow 3 \div 4 = 0.75$

Check how accurate these estimates are.

EXAMPLE

Estimate **a** $\frac{3.2 \times 49}{3.8 - 1.9}$ **b** $\frac{6.1^2}{8.7 \times 2.3}$

ANSWER

a $\frac{3.2 \times 49}{3.8 - 1.9} \approx \frac{3 \times 50}{4 - 2} = \frac{150}{2} = 75$ **b** $\frac{6.1^2}{8.7 \times 2.3} \approx \frac{6^2}{9 \times 2} = \frac{36}{18} = 2$

≈ means 'roughly equal to'.

Exercise B15.3

MEDIUM LOW

Unit B

This exercise is to be completed without a calculator.

1 Estimate

a 97×6.2 b 1002×8.4 c 820×18.3

d 22.3×4.45 e 3.8×24.9 f $25.9 \div 2.1$

g $38.4 \div 4.8$ h $405 \div 7.9$

2 Use estimation to work out these questions.
You must show your working.

a How many £2.85 are there in £15?

b How many books for £4.49 can be bought for £50?

3 Estimate a $\sqrt{85}$ b $\sqrt{105}$ c 4.7^2 d 8.1^2 e $\sqrt{50}$

4 Estimate

a $\frac{2.2 \times 3.7}{4.5 - 1.9}$ b $\frac{39 \times 203}{77}$ c $\frac{11.2 + 6.7}{3.25 \times 2.9}$ d $\frac{7.3^2}{9.7 - 4.1}$ e $\sqrt{36 + (4.5^2)}$

f $\frac{19.8}{2.8 + 5.9^2}$ g $\frac{2 \times 623.25}{12.6 + 87.22}$ h $\frac{\sqrt{24.7}}{98.1 - 26}$ i $\frac{\sqrt{40095}}{9.87^2}$ j $\frac{610 \times 4.98}{0.213}$

k $\frac{7.6}{12.5 - 6.8}$ l $\left(\frac{5}{0.4}\right)^2$ m $\sqrt{25.3 - 2.7^3}$ n $\frac{19.4 + 2.3}{0.6 - 0.55}$

5 Sticks of seaside rock cost £1.29 each.
Estimate the number of sticks of rock that can be bought with £10.

6 Chris has £16.72 credit left on his mobile phone.
Estimate the number of games he can download if each game costs £1.35.

7 Dave is making a garden patio.
The patio is 3.2 m by 4.8 m.
The patio slabs are 0.8 m by 0.5 m.

Estimate the number of patio slabs that Dave needs.

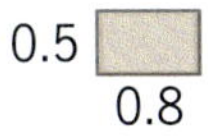

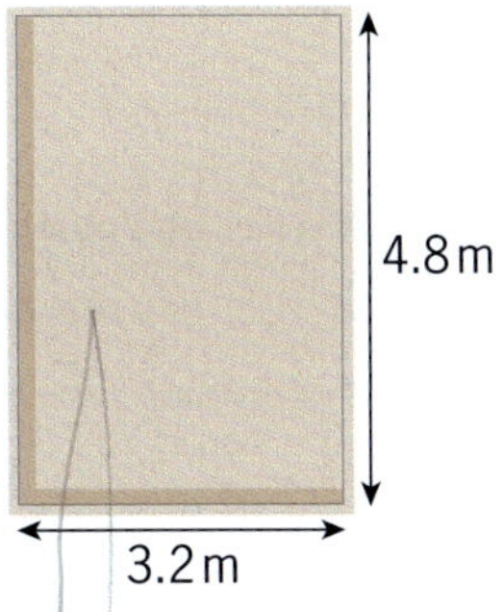

Indices

- Use index notation for simple integer powers
- Use index laws for multiplication and division of integer powers

RICH TASK

Dexter is faced with a dilemma – he wants to buy a new bike costing £500.
In return for doing housework he is offered these options, up to the value of £500:

£20 per week for each week or £2 in week 1, £4 in week 2, £8 in week 3 … the amount doubles each week.

Which option should Dexter accept?

→ **Indices provide a quick way to write a sum where you multiply by the same number repeatedly.**

'Indices' is the plural of 'index'.

EXAMPLE

Simplify **a** $3 \times 3 \times 3 \times 3 \times 3$ **b** $6 \times 6 \times 6 \times 6$

ANSWER

a $3 \times 3 \times 3 \times 3 \times 3 = 3^5$ **b** $6 \times 6 \times 6 \times 6 = 6^4$

p. 56

EXAMPLE

Write 504 as a product of prime factors in index form.

ANSWER

$504 = 2 \times 2 \times 2 \times 3 \times 3 \times 7 = 2^3 \times 3^2 \times 7$

You can multiply numbers using indices.

RICH TASK

a Work out $2^2 \times 2^3$ **i** as a number **ii** as a power of 2.

b Work out $3^6 \div 3^2$ **i** as a number **ii** as a power of 3.

c Try multiplying and dividing other powers. Write down anything you notice.

This leads to two important rules of indices:

→ **Add the indices when multiplying powers of the same number** $a^m \times a^n = a^{m+n}$

→ **Subtract the indices when dividing powers of the same number** $a^m \div a^n = a^{m-n}$

EXAMPLE

Simplify **a** $3^2 \times 3^6$ **b** $5^{12} \div 5^3$ **c** $3^3 \times 5^2$

ANSWER

a $3^2 \times 3^6 = 3^{2+6} = 3^8$

b $5^{12} \div 5^3 = 5^{12-3} = 5^9$

c $3^3 \times 5^2$ cannot be simplified as the numbers are not the same

But you can work out $3^3 \times 5^2 = 27 \times 25 = 675$.

Exercise B15.4

MEDIUM

This exercise is to be completed without a calculator.

1 Work out **a** 2^6 **b** 3^4 **c** 3^5 **d** 5^4 **e** $2^3 \times 3^2$ **f** $4^2 \times 2^4$ **g** $4^3 \div 2^2$

2 What would be the last digit in **a** 22^3 **b** 95^7?

3 Choose from these sets of numbers

a 1^7 2^6 3^4 4^3 5^3

b 1^9 4^6 9^3 16^3 25^2 30^3

i the largest number

ii a number that is not a square number

iii equal numbers.

Explain how you worked out your answers.

4 Write as a product of prime factors in index form

a 48 **b** 42 **c** 600 **d** 1280 **e** 735

5 Simplify, giving your answers as indices.

a $7^4 \times 7^5$ **b** $3^6 \times 3^2$ **c** $8^5 \times 8^4$

d $15^1 \times 15^7$ **e** $4^6 \times 4^6$ **f** $3^8 \div 3^2$

g $8^{12} \div 8^{11}$ **h** $9^7 \div 9^2$ **i** $6^8 \div 6^4$

6 **a** Which of these is not a square number?

3^4 5^3 7^2 9^3

b Which of these is not a cube number?

2^6 3^5 6^3 8^2

Be careful – a square number might not have a power of 2.

7 ☆ **PROBLEM** ☆

Write down the values of x and y if $256 = 16^x$ and $256 = 4^y$.

Summary and assessment

B15

CHECK OUT

You should now be able to

- use mental arithmetic for calculations
- use estimation and checking procedures
- work out powers and roots
- use indices in calculations.

Exam-style question

A room measures 2.25 m by 3.6 m.

Josh chooses carpet for the room that costs £5.99 per square metre and can be bought in a 4m width.

Estimate the cost of buying this carpet.

Takao's answer:

Carpet 4m wide will go across the 3.6 m room side

This means I need to buy a length of carpet 2.25 m long.
This is about 2.5 (rounding down may mean the carpet is too short)

The area of carpet I need is
$4 \times 2.5 = 10\ m^2$

Carpet costs £5.99 / m^2

This is almost £6

$10 \times 6 = 60$
An estimate of the cost of the carpet is £60

Think about the context of the question and what is sensible to use.

Read the question again to make sure you have answered what is asked.

Make estimates that you can work out without doing long complicated sums.

You may have to do more than one calculation.

Exam practice

1 Estimate $\frac{12\cdot8 + 14\cdot09}{32 \times 0\cdot1}$ *(2 marks)*

*2 Is 3^{30} odd or even?
Explain how you know. *(2 marks)*

3 A school librarian is given £400 to spend one year.
She wants to take out two magazine subscriptions for one year.
One subscription costs £11·99 per month and the other £7·99 per month.
She also wants to buy a daily newspaper each day the school is open. The newspaper costs 80p per day and the school opens on 195 days.
Estimate whether or not she has enough money to pay for the magazine subscriptions and the newspaper. *(4 marks)*

Case study 5: Recycling

Recycling waste materials has become an important part of everyday life now that our focus on protecting the environment is stronger than ever.

This time-series chart shows the total waste per person produced by households in the UK and the proportion of this waste that was recycled between 1983/4 and 2007/8.

What can you say in general about the amount of waste produced and recycled by households in the UK during this time? Justify your response by referring to the data.

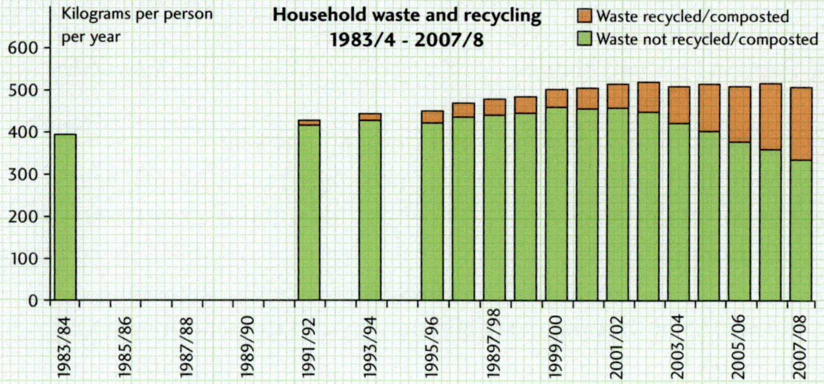

Copy and complete this table giving the values shown in the time-series chart.

Year	1983/84	1991/92	1993/94	1995/96	1996/97	1997/98	1998/99	1999/00	2000/01	2001/02	2002/03	2003/04	2004/05	2005/06	2006/07	2007/08
Waste not recycled/ composted	394	417		423	438	441	443	457	455	456	449	425		376	359	
Waste recycled/ composted	3	11	15	27	32	36			52	60	71	87	113	135	157	173
Total waste	397	427	445		469	477	483	505	507		521	512	517	511		507

The data shows that less than 0.8% of UK household waste was recycled in the year 1983/4. Work out what percentage of UK household waste (to the nearest 0.01%) was recycled each year until 2007/8. Add this information to your table.

Can you think of any reason for the trend shown by this data?

By what percentage has the amount of

a) household waste

b) waste recycled by households

changed during this time in the UK?

What realistic predictions do you think the government could have made about household waste for the year 2008/09?

Manufacturers are responsible for designing packaging that is as environmentally friendly as possible whilst also protecting the product.

A drinks company sells its brand of Cola in 500ml plastic (PET) bottles. The company has reduced the weight of these bottles by a third since the 1970s. The bottles weighed 39g in the 1970s.
What is the weight of a new bottle?
The company plans to start using bottles that weigh 24g.
What further reduction in weight (%) would this be?

Glass milk bottles are 50% lighter than they were 50 years ago.

As well as reducing the consumption of raw materials, lighter packaging also saves money in other ways such as transport costs.

A supermarket sells tomatoes in packs of six. The packaging consists of a plastic tray with a lid as shown. How much lighter (as a %) would each package be if it were made with no lid?

Do you think that not having lids would risk the quality of the tomatoes?

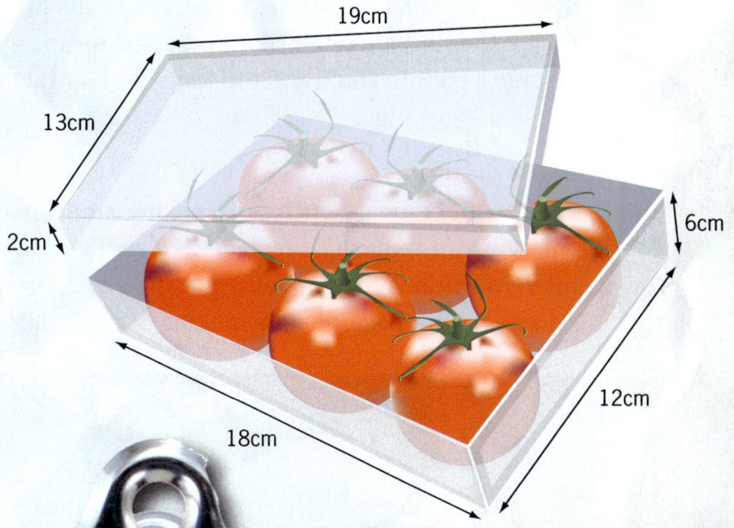

The product/pack ratio compares the weight of the packaging with the weight of the product it contains. Companies use this ratio to assess the suitability of the packaging used for each of their products. They often express it as a percentage to show how much of the overall weight is contributed by the packaging.

Look at some of the packaging you have at home. Could it be adapted to use less material without increasing the risk of damage to the product? If so, how?

How does the packaging used for perishable goods (e.g. food) differ from that used for non-perishable goods (e.g. electrical items)?

Research some well-known manufacturing companies on the Internet to find out about their packaging guidelines. Do they have different rules for different products (e.g. perishable/non-perishable goods)?

16 Similarity and enlargement

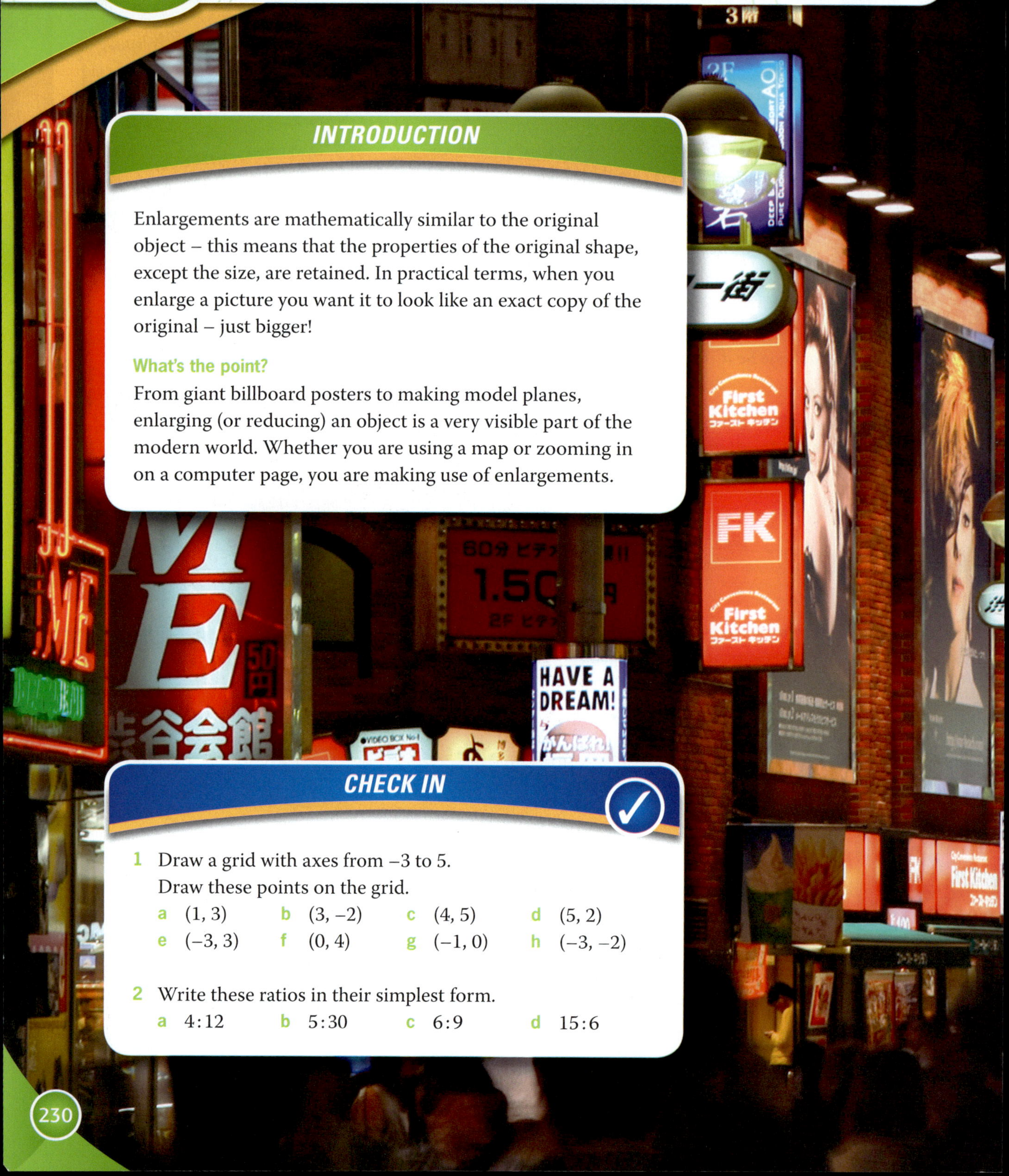

INTRODUCTION

Enlargements are mathematically similar to the original object – this means that the properties of the original shape, except the size, are retained. In practical terms, when you enlarge a picture you want it to look like an exact copy of the original – just bigger!

What's the point?

From giant billboard posters to making model planes, enlarging (or reducing) an object is a very visible part of the modern world. Whether you are using a map or zooming in on a computer page, you are making use of enlargements.

CHECK IN

1 Draw a grid with axes from –3 to 5.
Draw these points on the grid.

a (1, 3) b (3, –2) c (4, 5) d (5, 2)
e (–3, 3) f (0, 4) g (–1, 0) h (–3, –2)

2 Write these ratios in their simplest form.

a 4 : 12 b 5 : 30 c 6 : 9 d 15 : 6

Orientation

What I should know	What I will learn	What this leads to
B12	■ Identify congruence and similarity in shapes ■ Enlarge shapes and describe enlargements of shapes	Careers in graphic design and technical drawing

RICH TASK

You can tessellate combinations of squares, hexagons and equilateral triangles.
Investigate which of the regular polygons tessellate, and which ones don't.

Congruence and similarity

B16.1

- Recognise congruent shapes and similar shapes

RICH TASK

Which shapes in this stencil are identical, and which are the same shape but a different size?

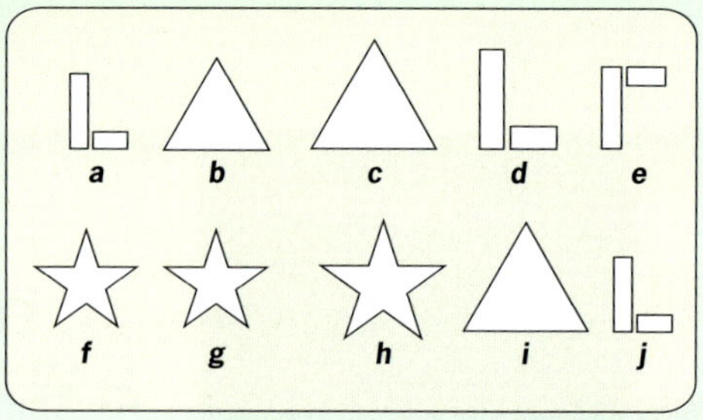

→ **Congruent shapes will fit on top of each other exactly.**
These two quadrilaterals have both sides and angles which are the same size. The quadrilaterals are congruent.

→ **Similar figures have the same shape but different sizes.**
These two triangles have the same angles (so their shape is the same), but the larger triangle has all sides twice as long as the smaller triangle (so their sizes are different). The triangles are similar.

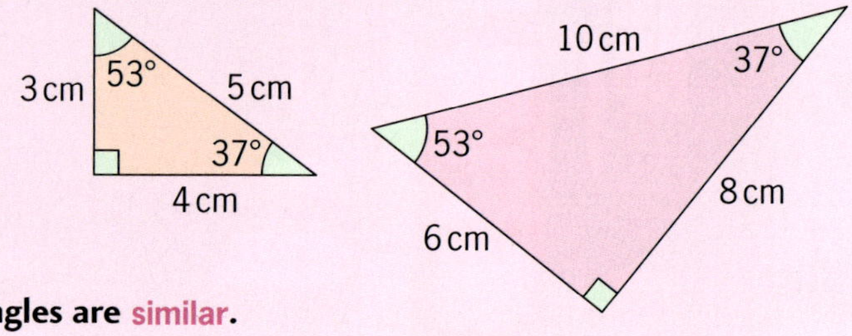

Exercise B16.1

MEDIUM LOW

*1 Explain whether or not these pairs of triangles are congruent.

a

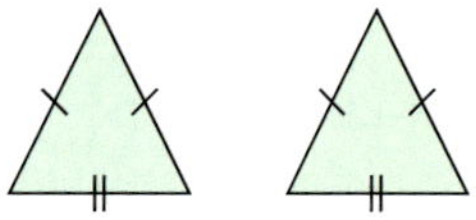

b

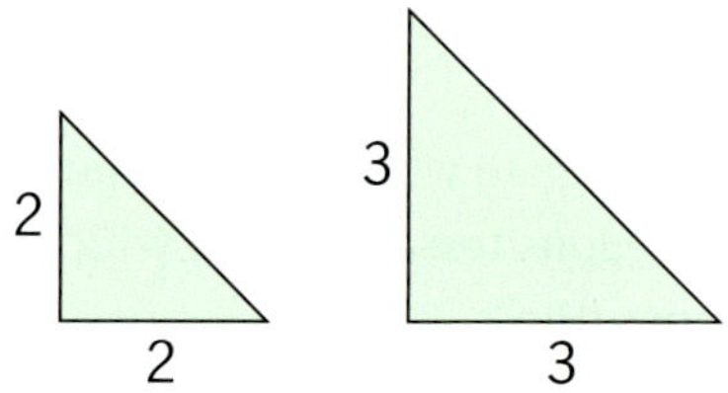

c

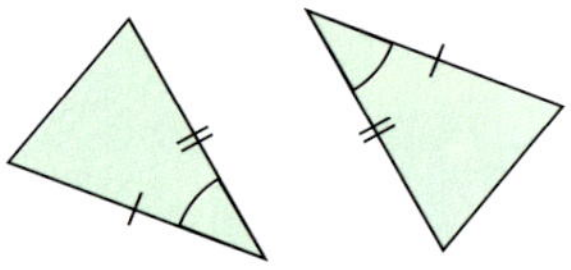

2 The dotted lines show two different ways of splitting a rectangle into two congruent shapes.

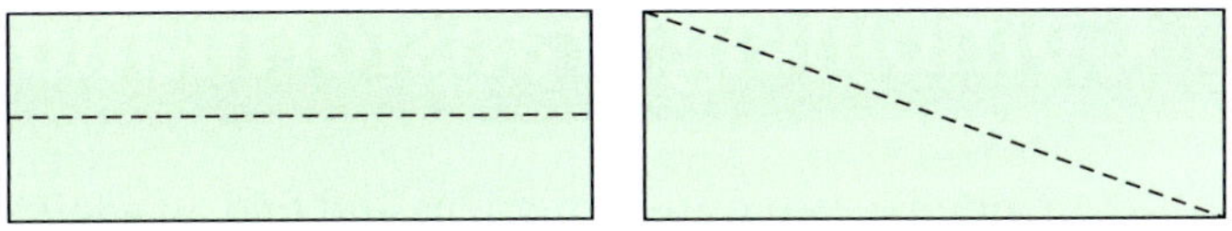

a Draw two copies of the rectangle. Draw dotted lines to show two more ways of splitting the rectangle into congruent shapes.

b For each, state whether the dotted line is also a line of symmetry.

3 Copy these shapes. Draw dotted lines to split them into congruent shapes.

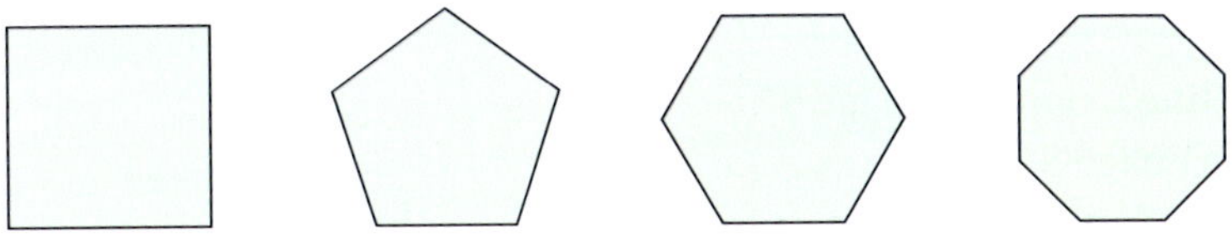

4 Look at the angles and the sides of these shapes. Decide whether the pairs of shapes are similar or not.

a

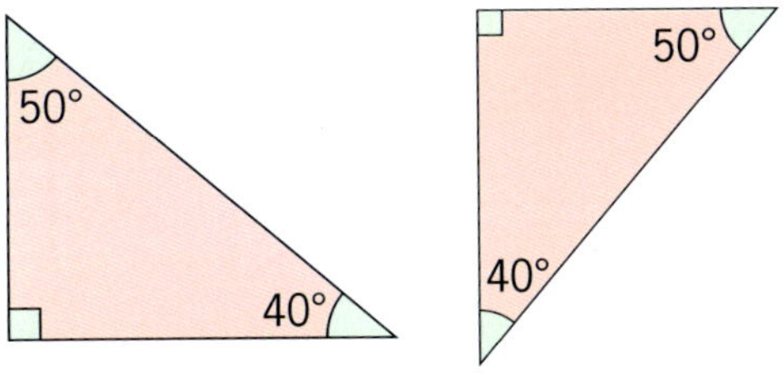

b

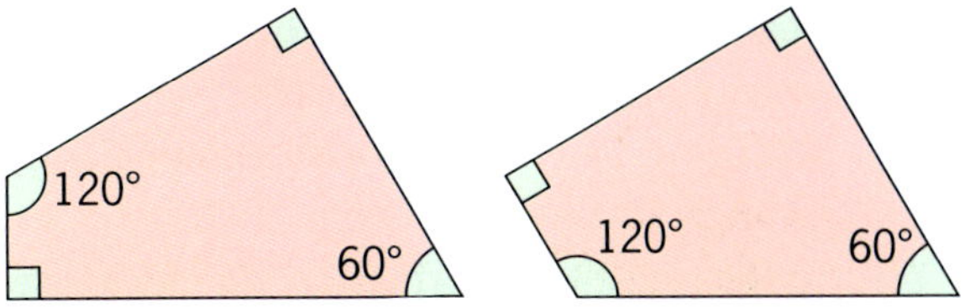

c

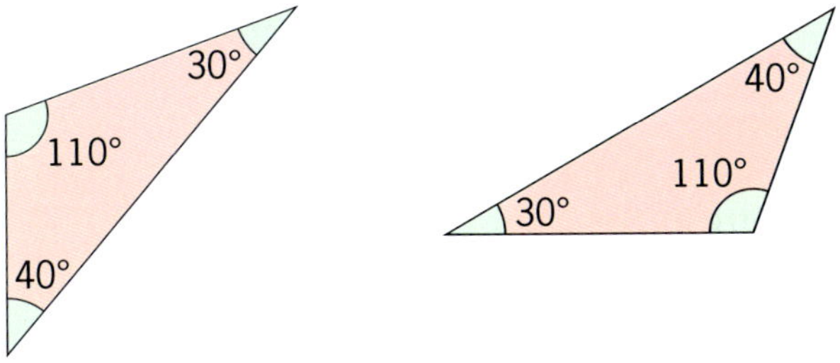

5 On axes labelled from 0 to 12, draw these pairs of triangles.
Use a ruler and protractor to measure sides and angles to decide which of the pairs of triangles are *congruent, similar* or neither.

	First triangle	Second triangle
a	(1, 1)(7, 1)(7, 8)	(9, 5)(9, 11)(2, 11)
b	(1, 8)(4, 1)(10, 5)	(5, 3)(9, 9)(2, 12)
c	(1, 1)(6, 1)(3, 7)	(12, 1)(12, 11)(0, 5)
d	(1, 4)(10, 2)(8, 12)	(1, 9)(7, 1)(12, 7)
e	(4, 1)(4, 7)(0, 5)	(12, 12)(0, 12)(8, 4)
f	(1, 8)(8, 7)(11, 0)	(4, 1)(12, 11)(5, 8)
g	(3, 1)(5, 10)(12, 10)	(9, 1)(1, 12)(1, 5)

6 A model car and the full-sized car are similar. Their lengths are in the ratio 1 : 20. If the model car is 22 cm long and 9 cm high, find the length and height of the full-sized car.

Combining transformations

B16.2

- Combine two transformations and find an equivalent single transformation

An object shape can be mapped onto its final image in two steps using a second transformation after the first transformation.

EXAMPLE

The shape S is firstly reflected in the line L onto its image S′. S′ is then reflected in line M onto the image S″. Describe the single transformation which maps S directly onto S″ in one step.

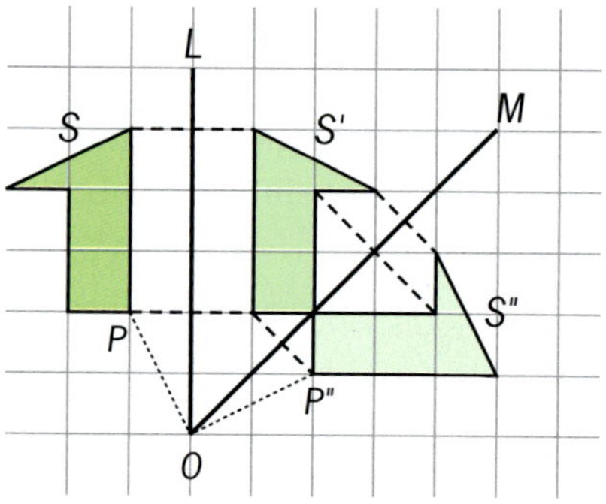

ANSWER

S can be transformed directly onto S″ by a rotation about O.
Draw shape S on tracing paper.
With your pencil point on O, rotate S onto S″. Line OP rotates through 90° onto line OP″.
So the single transformation is a clockwise rotation of 90° about point O.

Exercise B16.2

MEDIUM

p. 182–183

1 Copy this diagram.

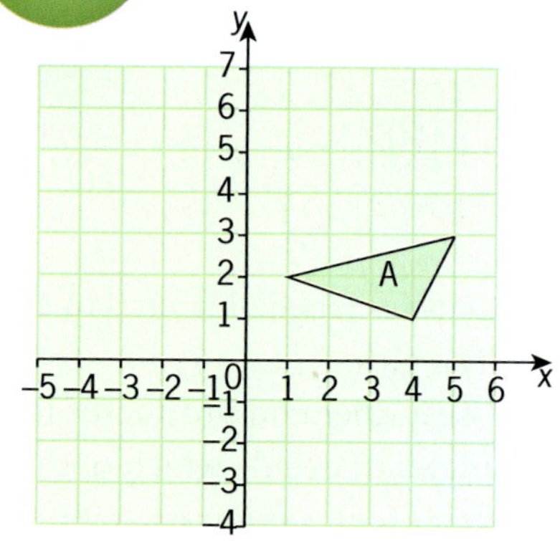

   a Rotate triangle A 90° anticlockwise about centre (0, 0). Label the image B.
   b Rotate triangle B 90° anticlockwise about centre (0, 0). Label the image C.
   c Describe fully the single transformation that takes triangle A to triangle C.
   d Rotate triangle B 90° clockwise about centre (2, 1). Label the image D.
   e Describe fully the single transformation that takes triangle A to triangle D.

2 Copy this diagram.

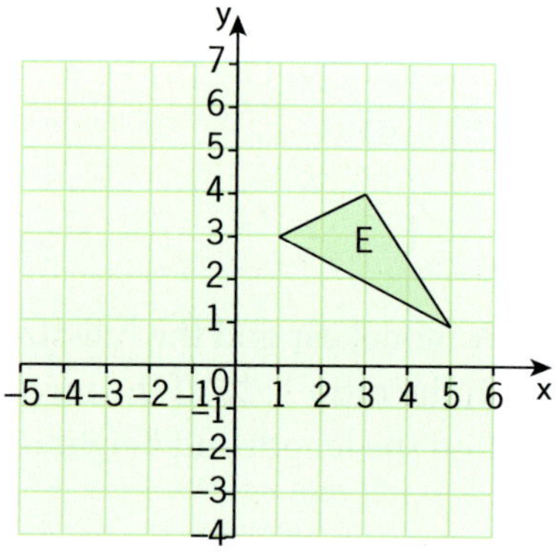

   a Reflect triangle E in the *y*-axis. Label the image F.
   b Reflect triangle F in the *x*-axis. Label the image G.
   c Describe fully the single transformation that takes triangle G to triangle E.
   d Reflect triangle F in the line $x = 2$. Label the image H.
   e Describe fully the single transformation that takes triangle E to triangle H.

3 Triangle T is translated by $\begin{pmatrix} 3 \\ 2 \end{pmatrix}$ onto triangle T′.

Triangle T′ is then translated onto triangle T″ by the vector $\begin{pmatrix} 1 \\ -4 \end{pmatrix}$.

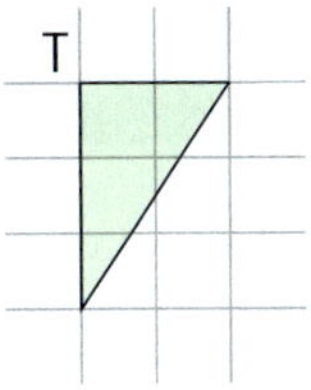

Draw T, T′ and T″ on the same diagram.
Describe fully the *single* transformation which maps T directly onto T″.

*4 a Shape S is reflected in line L onto its image S′.
Shape S′ is reflected in line M onto a second image S″.
Copy this diagram and find the positions of S′ and S″.
What *single* transformation will map S onto S″ in one step?
Give full details.

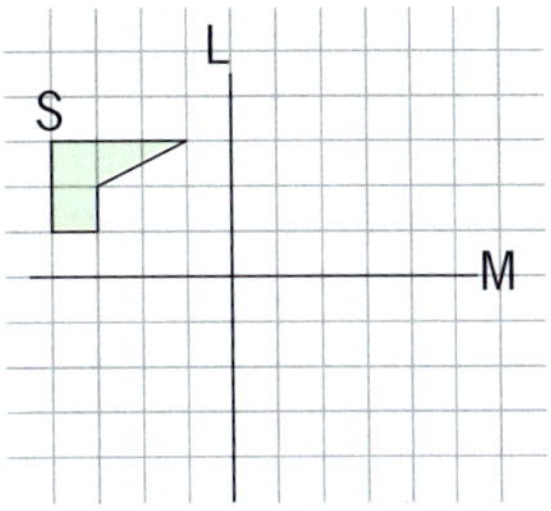

b Repeat part a with these new positions for lines L and M.
Describe the single transformation in full.

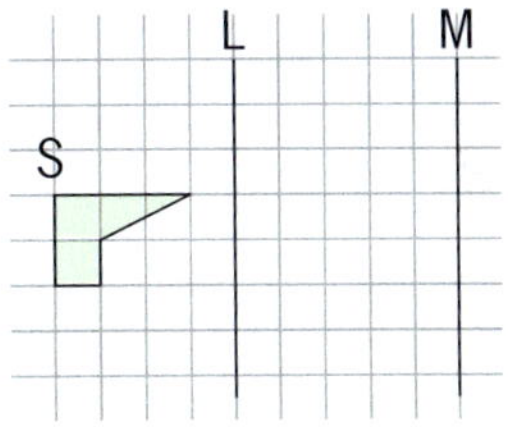

5 Draw the line $y = x$ on axes labelled from −8 to 8.
Draw the trapezium T with corners (1, 7), (4, 7), (4, 4) and (1, 3).
T is reflected in the line $y = x$ onto its image U.
U is reflected in the x-axis onto a new image V.
Draw the positions of T, U and V.
Describe fully the transformation which maps T onto V in one step.

*6 This crossword pattern is not complete. Only a quarter of the total pattern is shown.

a Copy and complete the pattern by reflections into the other three quarters.

b Research crossword patterns and look for the different transformations used.

c Invent a pattern of your own that can be made using four rotations about the centre point.

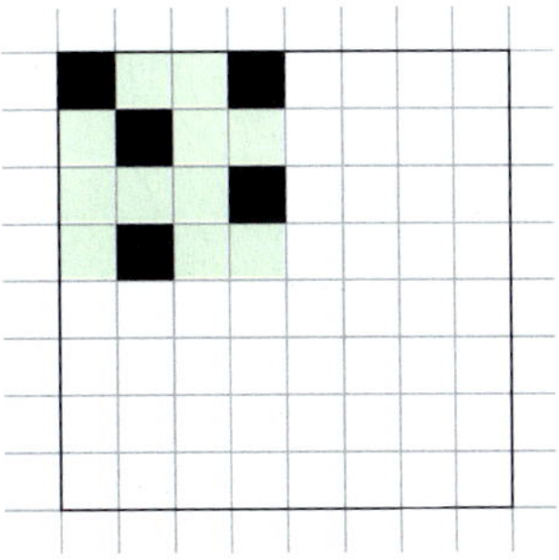

*7 ✯ CHALLENGE ✯

Wallpaper designs involve rotations, reflections and translations.
Analyse wallpaper papers using pattern books or the internet.

Describe the transformations you find.
Invent some patterns of your own.

Enlargements

- Find an image when an object is enlarged

The word *enlargement* is much used in photography when a picture is increased in size. A projector passes light through a small image to make a larger image on a screen.

DISCUSSION

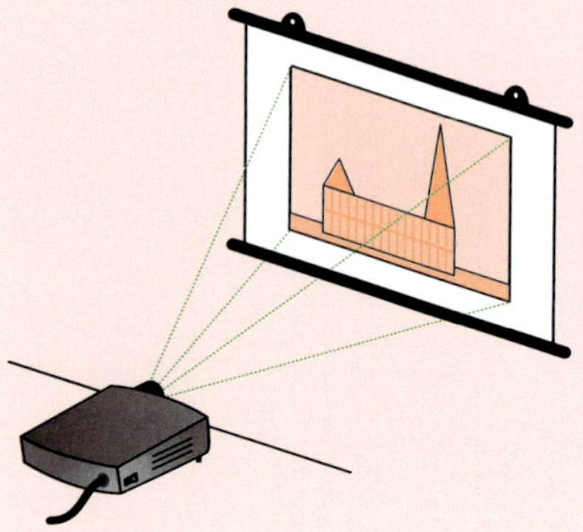

Discuss what difference it makes if the projector is closer to or further from the screen.

A mathematical enlargement has rays from a centre passing through the object to reach the image.
The **scale factor** is the distance between the image and the centre divided by the distance between the corresponding point on the object and the centre.

Point P′ on the image is 6 cm from the centre C. Point P on the object is 2 cm from the centre C. P′ is 3 times further away than P from C because 3 × 2 cm = 6 cm.
So the scale factor of the enlargement is $\frac{6}{2} = 3$.

The object and image are similar.

Measure CQ, CR, CS and their images and check the scale factor of 3.
The **scale factor** is also the multiplier from the image length to the object length.
Side PQ is 1 cm long and its image P′Q′ is 3 cm long.
So the scale factor is 3 because 3 × 1 cm = 3 cm.

→ **An enlargement can have a fractional scale factor.**
ΔABC is enlarged by a scale factor of $\frac{1}{2}$ to the image triangle A′B′C′.

OA′ = $\frac{1}{2}$ OA   OB′ = $\frac{1}{2}$ OB   OC′ = $\frac{1}{2}$ OC

Notice that the image is now smaller.

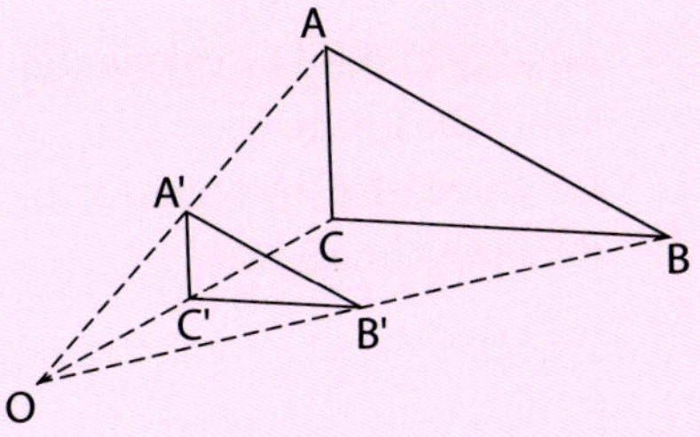

Exercise B16.3

MEDIUM

1 a Copy this diagram, but extend both axes to 16.
  b Enlarge triangle T by scale factor 2, centre (0, 0). Label the image U.
  c Enlarge triangle T by scale factor 3, centre (0, 0). Label the image V.

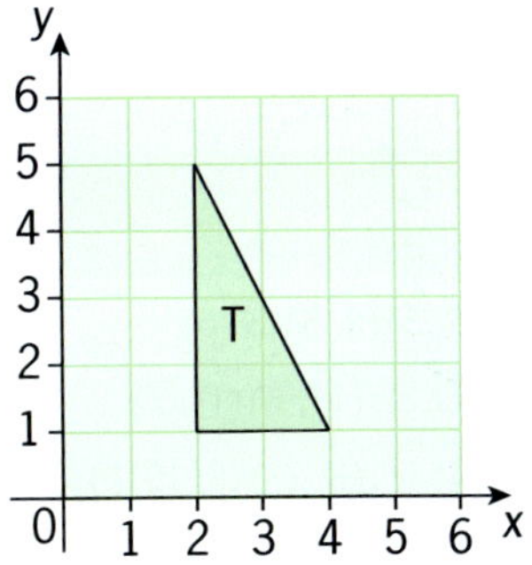

2 a Copy this diagram, but extend both axes from –7 to 7.
  b Enlarge kite K by scale factor 2, centre (1, 3). Label the image L.
  c Enlarge kite K by scale factor 4, centre (1, 3). Label the image M.

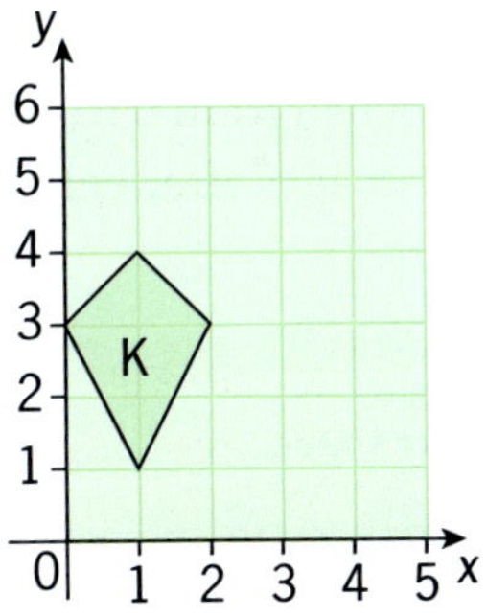

3 a Copy this diagram.
  b Enlarge triangle T by scale factor $\frac{1}{2}$, centre (0, 0). Label the image U.
  c Enlarge triangle T by scale factor $\frac{1}{2}$, centre (2, 2). Label the image V.

Does an enlargement scale factor less than 1 make a shape bigger or smaller?

Continued ➤

4 a Copy this diagram.
b Enlarge kite K by scale factor $\frac{1}{3}$, centre (0, 0). Label the image L.
c Enlarge kite K by scale factor $\frac{1}{3}$, centre (3, 3). Label the image M.

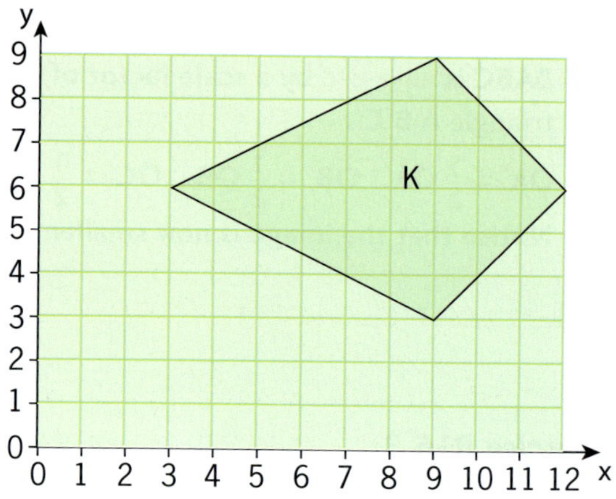

5 On axes labelled from 0 to 24, draw these object shapes. Find their images using the given centres and scale factors.

	Object shape	Centre of enlargement	Scale factor
a	(3, 3)(5, 3)(3, 6)	(2, 0)	2
b	(12, 4)(14, 6)(11, 7)	(9, 6)	3
c	(4, 17)(8, 17)(8, 19)(4, 20)	(5, 18)	2
d	(16, 23)(24, 23)(20, 19)	(20, 11)	$\frac{1}{2}$

On the grid, you can draw 'rays' with your ruler by counting squares across and up like a vector.

6 On axes labelled from 0 to 20, draw each of these object shapes and their three images after enlargements using the given centres and scale factors.

	Object shape	Centre of enlargement	Scale factors
a	(8, 6)(14, 6)(10, 8)	(12, 2)	×2 ×3 $\times\frac{1}{2}$
b	(4, 4)(4, 6)(6, 4)	(2, 2)	×2 ×3 $\times\frac{1}{2}$
c	(6, 12)(6, 16)(10, 20)(10, 2)	(2, 20)	×2 $\times\frac{1}{2}$ $\times\frac{1}{4}$

7 A manufacturer makes trousers based on this shape for the child's size.
Choose your own centre and draw an enlargement (scale factor 2) to create the adult template.

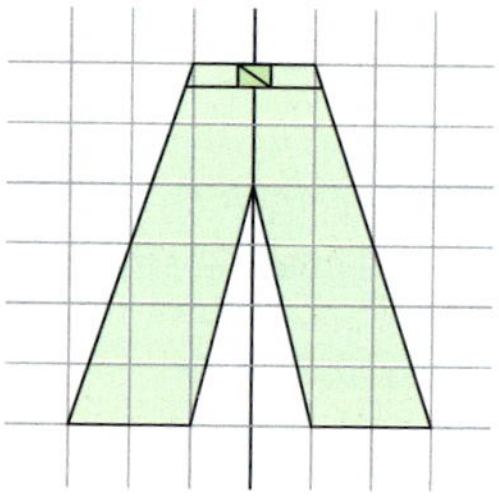

Describing enlargements

B16.4

- Find the scale factor and centre of an enlargement

When you know the centre and scale factor of an enlargement, you can find the image of an object shape.

This triangle T can be enlarged using the centre C with a scale factor of 2. Draw 'rays' from C through the corners of T and extend them.

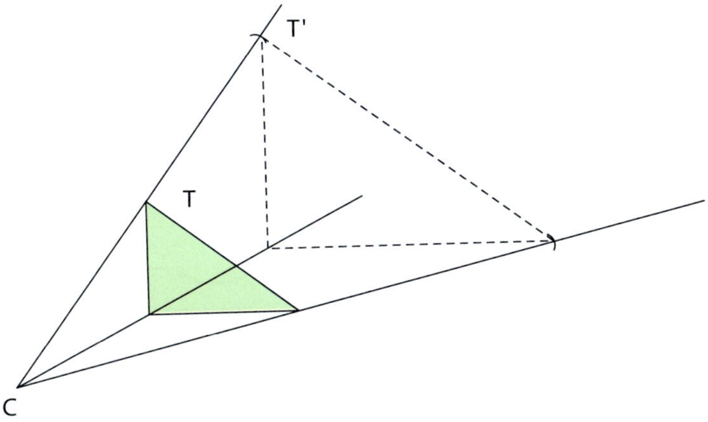

Use a ruler to mark the corners of the image *2 times* further away from C than the object. Join the corners in order to draw the image T′.

EXAMPLE

Triangle T enlarges onto triangle T′.

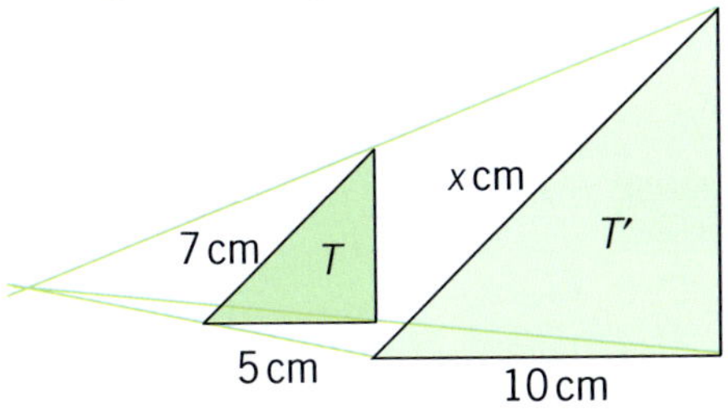

Find **a** the scale factor **b** the value of x.

ANSWER

a The 5 cm side enlarges to give the 10 cm side, so the scale factor is $\frac{10}{5} = 2$.

b The 7 cm side enlarges with the same scale factor.
So $x = 2 \times 7 = 14$ cm.

Exercise B16.4

MEDIUM

1 The quadrilateral Q has been enlarged several times on this diagram to give three images Q_1, Q_2 and Q_3.

a Write the coordinates of the centre of the enlargements.

b Find the scale factor for each enlargement.

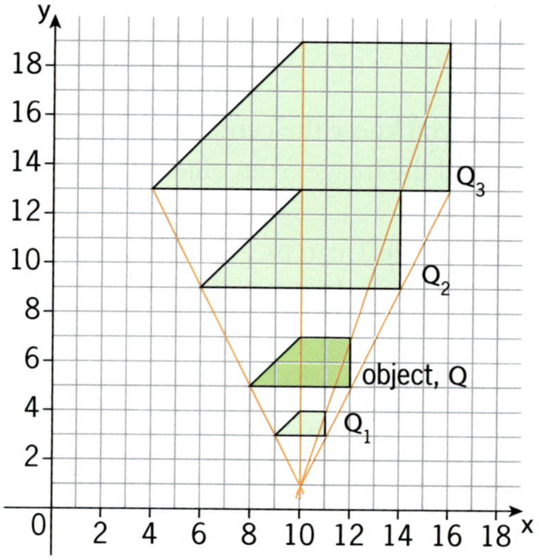

2 A maker of T-shirts has two sizes: one for children and one for adults. An adult's shirt is an enlargement of the child's shirt. What is the scale factor?

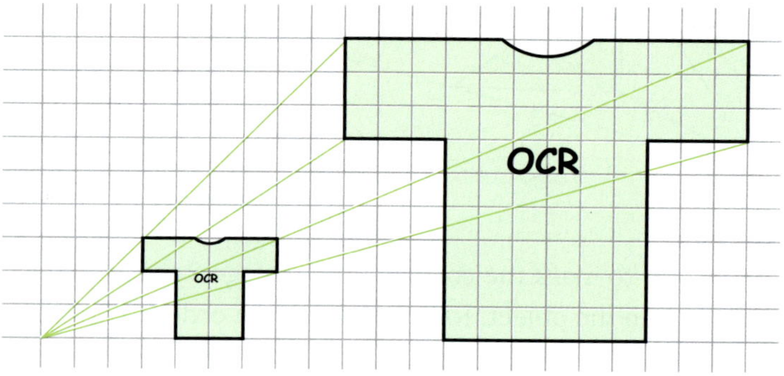

3 On axes labelled from 0 to 24, draw these pairs of objects and images. For each pair, find

i the centre **ii** the scale factor of the enlargement.

	Object	Image
a	(2, 2)(4, 2)(4, 4)	(4, 1)(8, 1)(8, 5)
b	(14, 3)(15, 6)(13, 6)	(18, 7)(21, 16)(15, 16)
c	(4, 8)(6, 8)(5, 10)(4, 10)	(1, 11)(9, 11)(5, 19)(1, 19)

4 An image can be *smaller* than its object. The scale factor is in this case a fraction.
Write the scale factors for these enlargements as a fraction.

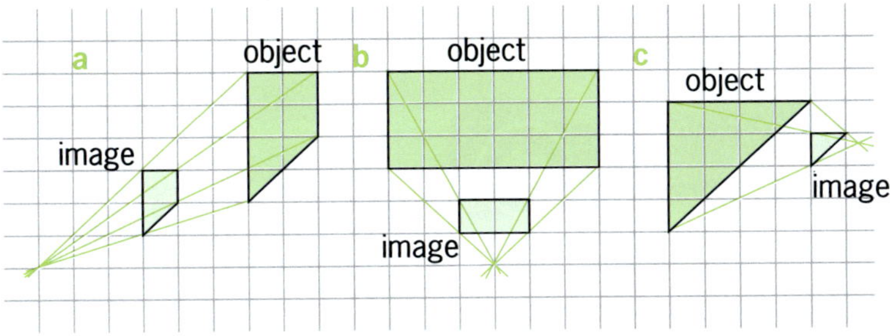

Even though the image is smaller, mathematicians still use the word *enlargement*.

5 For each pair of shapes, draw a diagram to find the centre and the scale factor of the enlargement.
a *Object* (3, 1)(3, 5)(9, 5) *Image* (2, 4)(2, 6)(5, 6)
b *Object* (0, 12)(8, 16)(4, 0) *Image* (4, 6)(3, 9)(5, 10)

6 A film is shown in a community centre on a screen 7 metres wide. The projector uses a roll of film 35 mm wide. What is the scale factor of the enlargement?

*7 ☆ **CHALLENGE** ☆

Explore sizes of paper from the smallest size, A10, to the largest, A0. If an A4 sheet is enlarged to A3 size, what is the scale factor? What scale factors are there when using other sizes of paper? Do you notice anything special?

8 You print a photograph as a rectangle measuring 7″ × 5″. You have two picture frames you can use. Frame A is 180 mm × 130 mm. Frame B is 210 mm × 150 mm. Which of these frames is the better fit?

The symbol ″ is used for *inches*.

9 Photographs are printed to many standard sizes. A popular size is 6″ × 4″. Other popular sizes are 12″ × 8″, 18″ × 12″, 10″ × 8″ and 7″ × 5″. Which of these sizes are true enlargements of the original 6″ × 4″ photograph, and which will need the original photograph to be cropped? Research the size of posters which can be made from prints. Are poster sizes true enlargements?

Summary and assessment

B16

CHECK OUT

You should now be able to

- understand congruence and similarity
- transform 2-D shapes using enlargement
- describe transformations and combinations of transformations.

Exam-style question

An equilateral triangle YPQ is drawn inside a square WXYZ as shown in the diagram.

*Explain why triangles PYZ and QYX are congruent, and show that PWQ is isosceles.

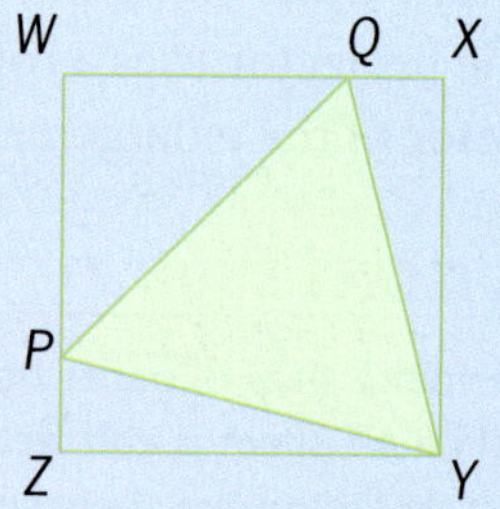

Jay's answer:

PY = QY as the triangle PQY is equilateral

Angle at Z = angle at X = 90° as they are at the corners of a square

ZY = XY as they are sides of a square

So PYZ and QYX must be congruent (two sides and one angle are the same) This means ZP = XQ

So PW = QW as they are the remainder of the sides of the square

PWQ has two sides the same, so it is isosceles

Use facts that you know about angles and side lengths in squares and equilateral triangles.

Use angle and line facts you know about isosceles triangles to finish the question.

Use the facts to give a reason for every statement you make.

Remember to write a statement that answers the question.

Exam practice

1 Enlarge this quadrilateral, scale factor 2, centre (0, 0).

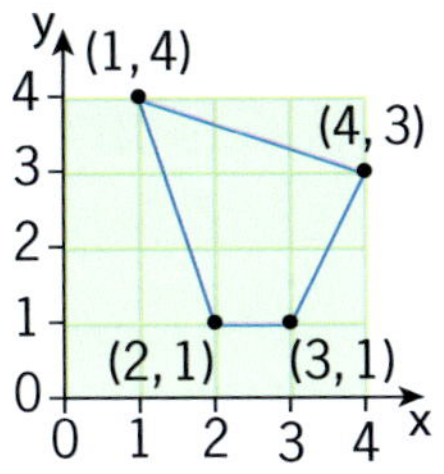

(*3 marks*)

* 2

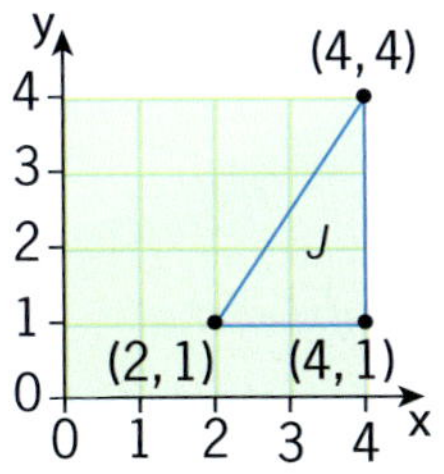

a Reflect triangle J in the x-axis. Label it K.

b Rotate triangle K 180° about centre (0, 0). Label it L.

c Describe fully the single transformation that takes triangle L to triangle J.

(*6 marks*)

* 3 Ben played a round of golf.
On the diagram the dotted lines show the positions of the golf ball after each stroke at the first hole

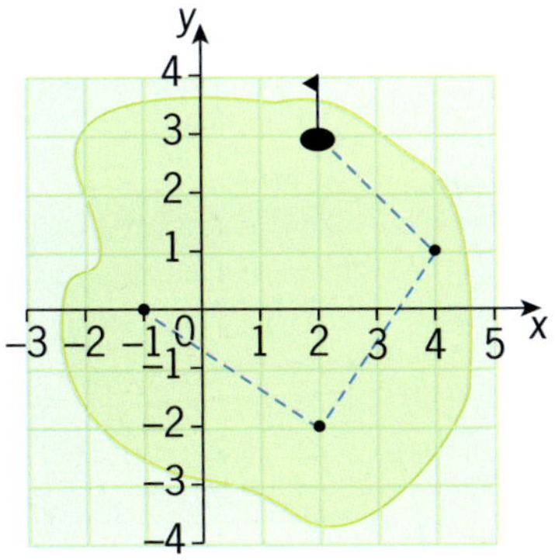

Describe the movement of the golf ball to the end of each stroke relative to its position before that stroke. (*4 marks*)

17 Percentages and proportional change

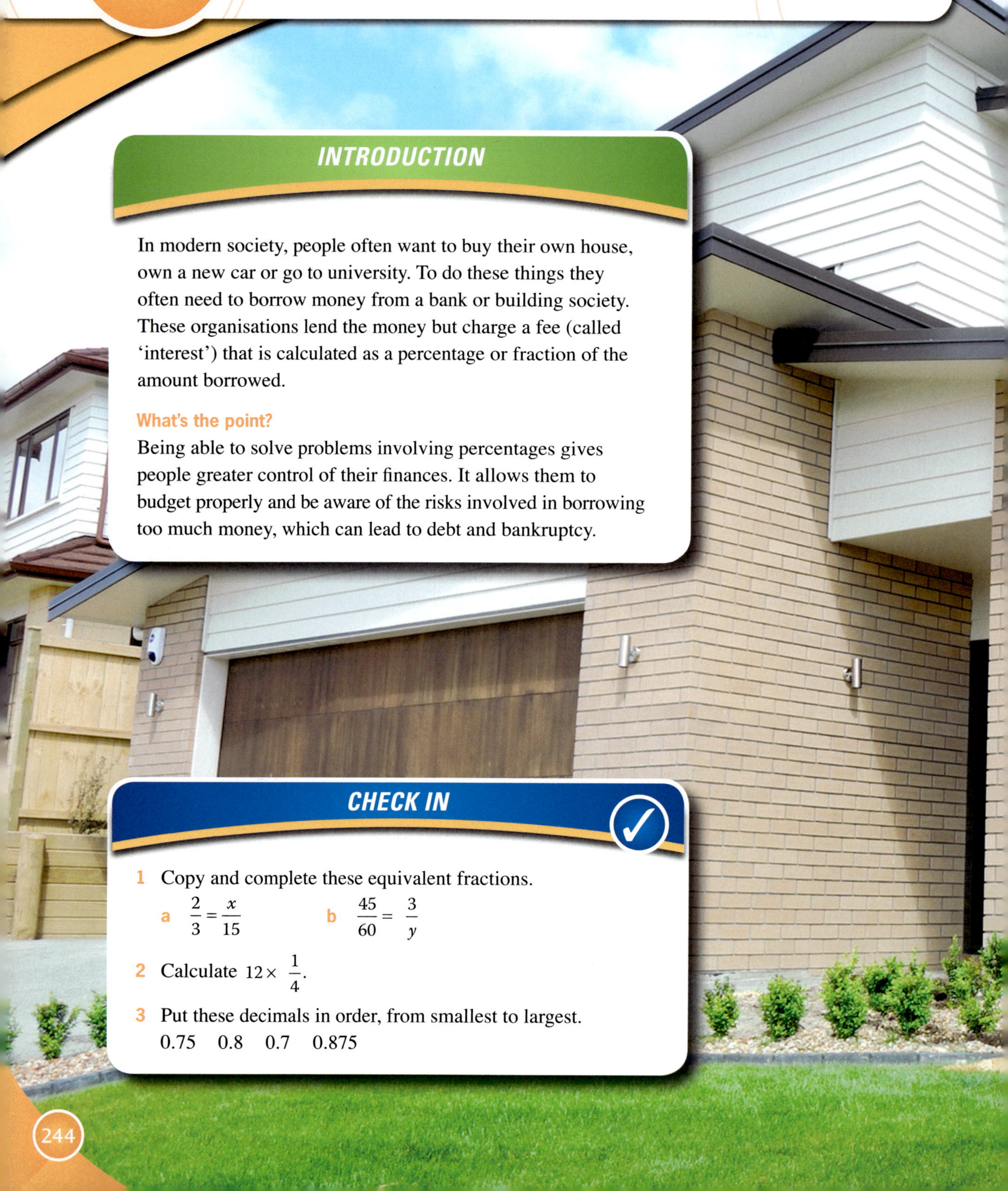

INTRODUCTION

In modern society, people often want to buy their own house, own a new car or go to university. To do these things they often need to borrow money from a bank or building society. These organisations lend the money but charge a fee (called 'interest') that is calculated as a percentage or fraction of the amount borrowed.

What's the point?

Being able to solve problems involving percentages gives people greater control of their finances. It allows them to budget properly and be aware of the risks involved in borrowing too much money, which can lead to debt and bankruptcy.

CHECK IN

1 Copy and complete these equivalent fractions.

a $\frac{2}{3} = \frac{x}{15}$ b $\frac{45}{60} = \frac{3}{y}$

2 Calculate $12 \times \frac{1}{4}$.

3 Put these decimals in order, from smallest to largest.

0.75 0.8 0.7 0.875

Orientation

What I should know	What I will learn	What this leads to
B9	■ Calculate percentage increases and decreases ■ Calculate simple interest	Aspects of everyday life involving financial management

RICH TASK

A pair of trainers costs £80.
They are increased in price by 10%.

By what percentage do they need to be reduced to get back to the original cost of £80?

Percentages

C17.1

- Solve simple percentage problems in real-life situations

☆ PROBLEM ☆

The table shows the amount of stamp duty payable for different purchase prices of property in August 2009.

Purchase price of property	Stamp duty payable
Up to £175 000	0
£175 001 to £250 000	1% of purchase price
£250 001 to £500 000	3% of purchase price
£500 001	4% of purchase price

Work out the difference in stamp duty payable on properties purchased for

a £175 000 and £175 001

b £250 000 and £250 001

c £500 000 and £500 001

What is the real cost of spending the extra £1 to buy a property in these cases?

To properly manage your finances, you should understand percentages.

p. 138

A **percentage** is often charged on goods sold and bought, such as VAT (value added tax), commission, stamp duty and so on.

When you work, a percentage of your pay is paid in tax, national insurance and income tax.

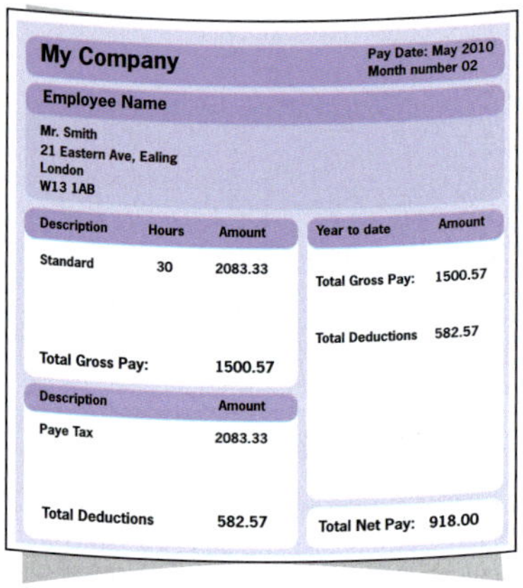

To find a percentage of an amount — **20% of £300**

→ write the percentage as a fraction — $\frac{20}{100} = \frac{1}{5}$

→ and multiply by the amount — $\frac{1}{5} \times £300 = £60$

EXAMPLE

The tables show:

- the amount you are allowed to earn before tax is charged (your income tax allowance) for different age groups in the tax year 2009–10
- the basic and higher percentage rate of tax payable on different amounts earned.

Income tax allowances 2009–10	
Personal Allowance for people aged under 65	£6 475
Personal Allowance for people aged 65–74	£9 490
Personal Allowance for people aged over 74	£9 640

Income tax rates 2009–10	
Basic rate: 20%	£0–£37 400
Higher rate: 40%	Over £37 400

Work out the income tax payable for a 28-year-old earning £50 000.

ANSWER

First deduct personal allowance: 50 000 – 6475 = £43 525

43 525 – 37 400 = £6 125

20% tax on £37 400 $= \frac{20}{100} \times 37\,400$ = £7480

40% tax on £6 125 $= \frac{40}{100} \times 6125$ = £2450 ⇒ Total tax = £9930.

Exercise C17.1

MEDIUM LOW

1 In a sale, a clothes shop displays a notice announcing 20% off everything. How much do you save on articles priced at

a £35 b £90 c £560?

2 An energy company offers 8% discount on both gas and electricity bills if both services are supplied to a household. How much discount is given over a year if the total gas bills are £503 and total electricity bills are £225?

3 Use the tables in the example to work out the income tax for

a a 20-year-old earning £16 000 b a 42-year-old earning £40 000
c a 72-year-old earning £12 000 d a 30-year-old earning £98 000.

4 The table shows the percentage of commission payable to the auction house for items they sell on your behalf. The hammer price is the price the item sells for.

Hammer price	Commission
Up to £500	15%
£501 – £1000	12.5%
£1001 – £59 999	10%
£ 60 000 – £149 999	8%
£ 150 000 – £299 999	7%
£ 300 000 – £599 999	5%
£ 600 000 – £1 499 999	4%

Work out the commission for items sold for

a £495 b £75 000 c £140 000
d £150 000 e £1200 f £22

Unit C

Percentage increase and decrease

C17.2

- Solve simple percentage problems, including increase and decrease

☆ PROBLEM ☆

A trader buys £600 of fresh cut flowers to sell at a market stall.
The stall costs £70 for the day.
The flowers are organised in bunches to sell.
80% of the flowers are sold at a 30% profit.
With an hour left the trader reduces his prices by 50% and sells 75% of the remaining flowers.
Did the trader make a profit or loss, and by how much?

EXAMPLE

Two televisions are priced at £800 and £240 respectively.
They are each reduced by 25%.
Is the saving on each television the same?

ANSWER

25% of £800 = £200 saving ⇨ this television will now cost £600.
25% of £240 = £60 saving ⇨ this television will now cost £180.
No, the saving is different.

25% is the same as a quarter.

EXAMPLE

a Work out the sale price of a necklace bought for £150 and sold at a 64% profit.

b Work out the sale price of a dress costing £80 reduced by 35% in a sale.

ANSWER

a 64% of 150 = $\frac{64}{100} \times 150 = 96$

150 + 96 = 246 The sale price is £246.

b 35% of 80 = $\frac{35}{100} \times 80 = 28$

80 – 28 = 52 The sale price is £52.

RICH TASK

In the previous example, try to think of a single decimal number that you can multiply £80 by to get the answer of £52.
Use your answer to find a quick way to work out percentage increase and decreases.

Exercise C17.2

MEDIUM

1 Work out the sale price of

a a camera that cost £480 and is reduced by 22% in a sale

b a toy that cost £89 and is reduced in a sale by 45%.

2 What is the selling price of

a a bag of sweets bought for 25p and sold for 40% profit

b a necklace bought for £6 and sold for 30% profit?

3 Ali earns £78 per week working part-time.
She is given a 6% pay increase.
How much does Ali now get paid each week?

4 Jules is paid an annual salary of £34 900.
She is given a pay rise of 3.2%.
What is her new annual salary?

5 Donna buys a new sofa for £1760.
Donna pays 15% of the sale price and the remainder over 12 months in equal instalments. How much is each instalment?

6 ☆ PROBLEM ☆

A company wanted to promote 200g chocolate bars with a special offer.
These chocolate bars normally cost 164 pence.
The company could choose to

a increase the amount of chocolate in the bar by 25% or

b reduce the price of the bar by 25%.

Which should they choose to give customers the better deal?

7 **RICH TASK**

Find the effect on price of:

- a 10% increase followed by a 10% decrease
- a 10% decrease followed by a 10% increase
- a 10% increase followed by another 10% increase
- a 10% decrease followed by another 10% decrease.

Take a good guess at first, then use real numbers to arrive at the answers.
Justify your responses clearly.

Unit C

Percentage difference

C17.3

- Solve simple percentage problems in real-life situations, including increase and decrease

☆ PROBLEM ☆

Stan and Ollie each buy and sell a car.
Stan buys a car for £2000 and sells it for £2400.
Stan makes £400 profit.
Ollie buys a car for £4000 and sells it for £4400.
Ollie makes £400 profit.
Who made the better deal?

You can work out percentage increase and decrease using a formula.

→ **Percentage increase** $= \dfrac{\text{increase}}{\text{original amount}} \times 100\%$

→ **Percentage decrease** $= \dfrac{\text{decrease}}{\text{original amount}} \times 100\%$

These formulae can also be used to work out **percentage profit** and **loss**.

EXAMPLE

Find the percentage loss on a sound system bought for £360 and sold for £150.

ANSWER

$$\%\text{ loss} = \frac{\text{loss}}{\text{original amount}} \times 100$$

$$= \frac{360 - 150}{360} \times 100 = \frac{210}{360} \times 100 = 58.3\%.$$

Exercise C17.3

MEDIUM

1 a Jon puts on weight. His weight changes from 51 kg to 58 kg. Work out the percentage gain in Jon's weight.

b Dave loses weight. His weight changes from 98 kg to 84 kg. Work out the percentage loss in Dave's weight.

2 Janine buys a watch at a boot fair for £30 and sells it for £52. Work out her percentage profit.

3 Ricky buys golf clubs for £300 and sells them a year later for £120. Work out his percentage loss.

4 Rob adds 40% for profit to the price he pays for bicycles in his cycle shop. He sells bicycles to friends at a discount of 25% of the selling price. What percentage profit or loss is made on a bicycle bought for £370 and sold to a friend?

Simple interest

C17.4

- Solve simple percentage problems in real-life situations

Money that is invested attracts interest.
The interest paid is given as a percentage, usually a rate per year.
Simple interest is calculated on the amount initially invested.

EXAMPLE

Tula invests £450 at 7% simple interest for two years.

a How much interest does she get?

b How much is the investment worth after two years?

ANSWER

a $7\% \text{ of } £450 = \frac{7}{100} \times 450$

$= £31.50$

Over two years Tula gets $2 \times £31.50 = £63$.

b After two years the investment is worth $£450 + £63 = £513$.

→ **To find simple interest for *y* years, calculate percentage × amount × *y* years**

Exercise C17.4

MEDIUM

1 £15 000 is invested at 8% simple interest.

a How much interest will be gained over four years?

b How much is the investment worth after four years?

2 Nathan has saved £100.
He puts it in a savings account paying 12% simple interest.

a How much interest is added to the account in three years?

b How much are Nathan's savings worth after three years?

3 Work out the value of £900 invested for

a six years at 2% simple interest **b** two years at 6% simple interest.

4 Goldie borrows £250. The interest charged is 6.5% simple interest.
How much does Goldie owe at the end of one year?

5 ☆ **PROBLEM** ☆

Mrs Brown borrows £800.
There are two ways in which she can borrow the money:

a at a simple interest rate of 5% for five years

b at a simple interest rate of 8% for three years.

Which option involves paying less interest and by how much?

If you **invest** money, the bank **pays** you interest.
If you **borrow** money, the bank **charges** you interest.

Repeated percentage change

C17.5

- Solve simple percentage problems in real-life situations

RICH TASK

An ordinary piece of paper is said to be about 0.081 mm thick.

- Take a sheet of ordinary A4 paper and fold it in half.
- Fold it a second time and then a third – it should now be the thickness of a fingernail.
- Continue folding in half for as many times as you can.
- How many folds is it possible to make?
- How thick is the paper at each fold?
- Try to describe the rate at which the thickness of the paper grows.

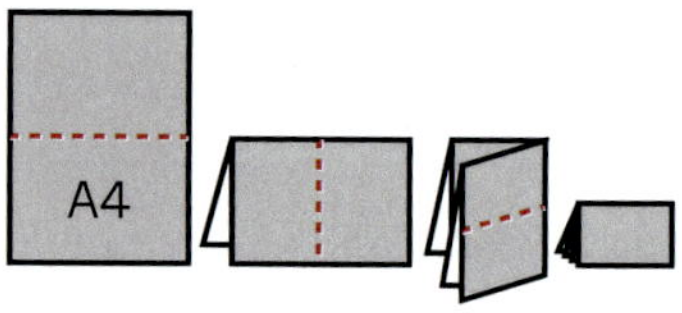

EXAMPLE

£11 000 is invested in a savings account.
7% compound interest is paid annually.
How much money is in the account after

a one year **b** three years?

ANSWER

100% + 7% = 107% = 1.07 as a decimal

a After one year 11 000 × 1.07 = £11 770

b After two years 11 770 × 1.07 = £12 593.90

After three years 12 593.90 × 1.07 = £13 475.47 to the nearest penny.

Compound interest is an example of repeated percentage increase.

1.07 is the **decimal multiplier.** You could write a single calculation with **powers.**

$11\,000 \times 1.07 \times 1.07 \times 1.07$
$= 11\,000 \times 1.07^3 = 13475.47$

→ **Depreciation** **is an example of repeated percentage decrease.**

EXAMPLE

A car bought for £36 000 depreciates in value by 10% each year.
How much is the car worth after

a one year **b** four years?

ANSWER

100% – 10% = 90% = 0.9

a After one year 36 000 × 0.9 = 32 400

b After two years 32 400 × 0.9 = 29 160

After three years 29 160 × 0.9 = 26 244

After four years 26 244 × 0.9 = 23 619.60

The car is worth around £23 600 after four years.

You could write a single calculation:
$36\,000 \times 0.9 \times 0.9 \times 0.9 \times 0.9$
$= 36\,000 \times 0.9^4 = 23\,619.60$

Exercise C17.5

MEDIUM HIGH

1 Five hundred pounds is invested for two years at 8% compound interest.

a How much interest is added at the end of year 1?

b How much interest is added at the end of year 2?

c What is the total investment worth at the end of two years?

2 A car loses 10% of its value each year.
When new, it is valued at £40 000.

a i How much did it lose in value after one year?

ii What is its value at the end of year 1?

b i How much did it lose in value after two years?

ii What is its value at the end of year 2?

c What is the car's value at the end of year 3?

3 A bouncy ball is dropped from a height of 6 m.
After each bounce it rises to 90% of its previous height.
To what height does it rise after a one bounce b two bounces?

4 The population of a species of small mammal on an island increases at a rate of 12% per year. One year the population was 1.5 million.
What size will the population be in

a one year's time b four years' time?

5 The population of a rare species of bird is falling at a rate of 8% per year.
One year the population was 28 000.
What size will the population be in

a one year's time b three years' time?

For questions 5 – 7, try to use a single calculation involving powers. See the blue boxes on page 252.

6 £19 000 is invested in a savings account.
6.95% compound interest is paid annually.
How much is in the account after

a one year b two years c five years?

7 £42 000 is invested in a savings account.
5.2% compound interest is paid annually.
How much is in the account after

a one year b three years c six years?

Unit C

Proportional change

C17.6

- Solve word problems about proportion

☆ PROBLEM ☆

A photographic developing company advertises its prices.
These are the print sizes and prices for developing 35 mm films.

Print size		Up to 27 exposures	Up to 40 exposures
Compact	12 cm × 9 cm	99p	£2.99
30% bigger	15 cm × 10 cm	£2.49	£3.49
100% bigger	17. 2 cm × 12.5 cm	£3.25	£3.75

Are the larger sizes really 30% and 100% bigger than compact sized prints?
Justify your response.
Which print size is the best buy

a for up to 27 exposures

b for up to 40 exposures?

Supermarkets often present bewildering choices, so it's helpful to know which is the best value for money.

EXAMPLE

Two sizes of a brand of shampoo are on the chemist's shelf.
You can buy 400 ml for £3.39 or 250 ml for £2.75. Which is the better value?

ANSWER

You can compare these in different ways:

	Larger size	Smaller size
Working out amount per pence	$\frac{400}{339}$ =1.18 ml/p	$\frac{250}{275}$ = 0.91 ml/p
Working out price per ml	$\frac{339}{400}$ = 0.85 p/ml	$\frac{275}{250}$ = 1.1 p/ml
For 100 ml of each size	0.85 × 100 = 85 p/100ml	1.1 × 100 = 110 p/100ml

Supermarkets often use price per 100 ml.

It is cheaper to buy the larger size.

☆ PROBLEM ☆

The largest pyramid in Egypt is believed to have been built by 4000 slaves over 30 years.
If you doubled the number of slaves to 8000 how long would it take?
How many slaves would be needed if you wanted to build the pyramid in 10 years?

Exercise C17.6

MEDIUM

1 a A 420 g bag of chocolates costs £1.59.
How much would a 200 g bag cost?

b Three litres of oil costs £2.19.
How much would it cost to buy 1200 ml of oil?

2 a It takes five cleaners six hours to clean an office block.
How long would it take if there were
i 10 cleaners **ii** six cleaners **iii** four cleaners?

b It takes six workers 14 days to fix the potholes in a stretch of road.
How long would it take if there were
i eight workers **ii** five workers?

3 For each of these choices, show which is better value.

a five litres of spa water for £1.29 or two litres for 87p

b 75 ml of hair gel for £1.19 or 200 ml for £2.89

4 ☆ **PROBLEM** ☆

Molly makes honey biscuits to sell at country fairs.
Here is the recipe she uses and the costs of each ingredient.

Honey biscuits (makes 6 large biscuits)	Cost of ingredients	
70 g butter	Butter	75p for 250 g
30 g soft brown sugar	Soft brown sugar	£2.00 for 1 kg
150 g self-raising flour	Self-raising flour	80p for 1 kg
60 g clear honey	Honey	£2.70 for 450 g

Molly bakes two batches of biscuits in the oven at a time.
The cost of using the oven to bake two batches is £1.50.
Molly wants to make a minimum profit of 25% when she sells the cakes and biscuits.
How much should Molly sell them for at the country fairs?

5 ☆ **PROBLEM** ☆

The energy in a 25 g bag of crisps is 132 kilocalories.
A child aged 5–10 years has a guideline daily amount of 1800 kilocalories.
How many 25 g bags of crisps would they need to eat in one day to exceed their guideline daily amount of calories?

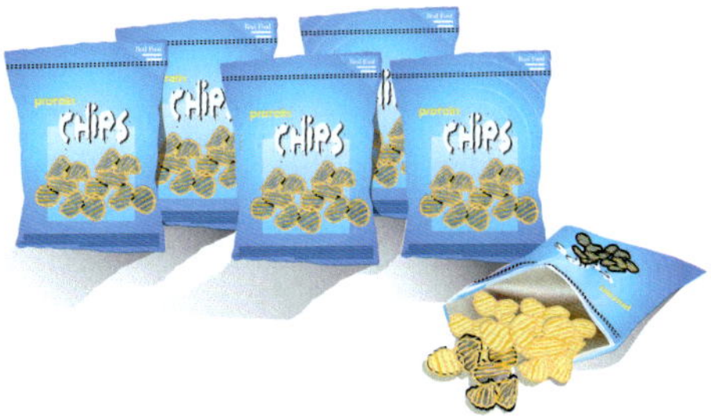

Unit C

Summary and assessment

C17

CHECK OUT

You should now be able to

- calculate with percentages in problem solving
- work out simple interest
- work out compound interest
- use proportional change in problem solving.

Exam-style question

Cheryl invested £2000 at 3·5% simple interest per annum.
Danni invested £2000 at 3·3% compound interest per annum.

Whose investment was worth more after five years, and by how much?

Mike's answer:

Cheryl's simple interest

$2000 \times \frac{3\cdot5}{100} \times 5 = 350$

Cheryl's investment is worth £2350

Danni's compound interest

$2000 \times 1\cdot033 \times 1\cdot033 \times 1\cdot033 \times 1\cdot033 \times 1\cdot033 = 2000 \times (1\cdot033)^5 = £2352\cdot51$

Danni's investment is worth £2352·51

Difference $2352\cdot51 - 2350 = 2\cdot51$

Danni's investment is worth more by £2·51

Work out each part of the question separately.

Take care, the interest is worked out differently for simple and compound interest.

Answer the final part of the question.

You may need to do several different calculations to fully answer the question.

Exam practice

1 Work out

a 32% of 500 ml

b 12·5% of 140 g. *(4 marks)*

2 Anne invested £20 000 in premium bonds.
In the first year she won nine prizes worth £25 each and one £50 prize.
Work out her winnings as a percentage of her investment. *(3 marks)*

3 A recipe for carrot cake uses 100 g of carrot, 80 g of walnuts and two eggs.
Gareth picks a carrot weighing 285 g.
He wants to use all the carrot to make a cake.
How many eggs should he use?
Show how you arrive at your answer. *(3 marks)*

18 Area and perimeter

INTRODUCTION

The United Kingdom is not a very regular shape. However, cartographers have managed to work out its area as 244 820 km^2.

What's the point?

Cartographers use lines of latitude and longitude to divide countries into much smaller regular shapes, such as trapeziums. They can then calculate the area of each of these smaller pieces, and add them together to calculate the area of the country.

CHECK IN

1 Evaluate

a 60 × 10	**b** 71 × 1000	**c** 4.8 × 10
d 26.3 × 100	**e** 4.5 × 1000	**f** 600 ÷ 100
g 750 ÷ 10	**h** 6500 ÷ 1000	**i** 32 ÷ 10

2 Evaluate

a $6 \times \frac{8}{5}$	**b** $9 \times 2\frac{1}{2}$	**c** $6 \times 1\frac{3}{4}$

3 Evaluate

a 5.8 + 2	**b** 14.8 + 0.7	**c** 6.4 + 2.6

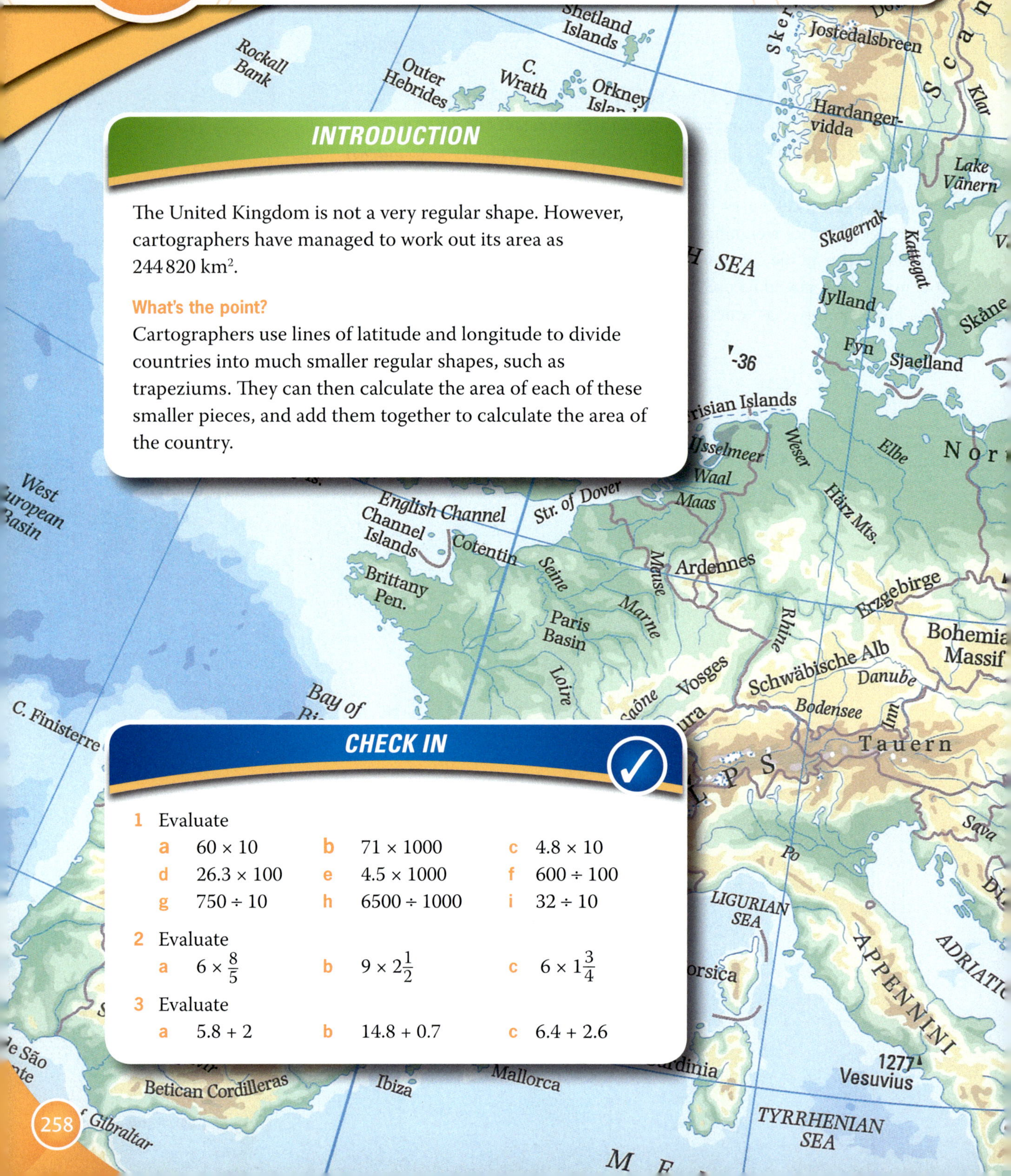

Orientation

What I should know	What I will learn	What this leads to
B10	■ Find the perimeter and area of shapes. ■ Calculate scale factors	Careers in manufacturing and industrial design

RICH TASK

A farmer has 240 m of fencing. He wants to enclose a field with the maximum possible area.

Investigate.

Perimeter and area of basic shapes

C18.1

- Find area of irregular shapes
- Calculate and use the area of rectangles and triangles

→ The perimeter of a shape is the total distance around its edge.

Perimeters of awkward shapes can be measured using a length of string.

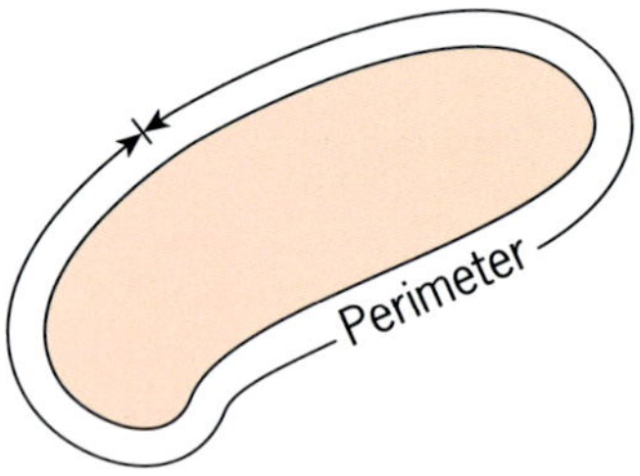

Perimeters of shapes with straight sides can be measured with a ruler.

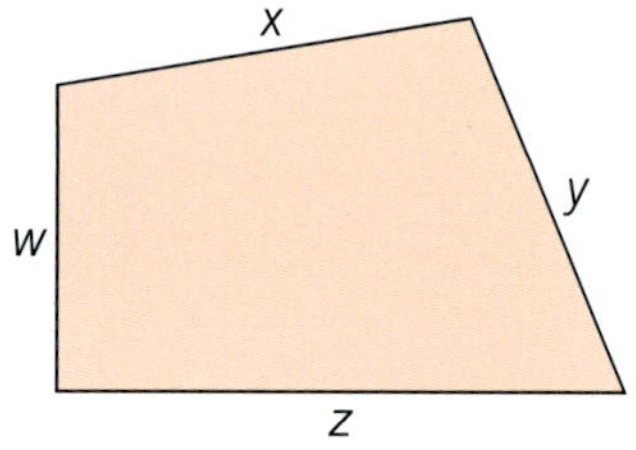

Perimeter = $w + x + y + z$

→ Area is a measure of the amount of space enclosed by a shape.

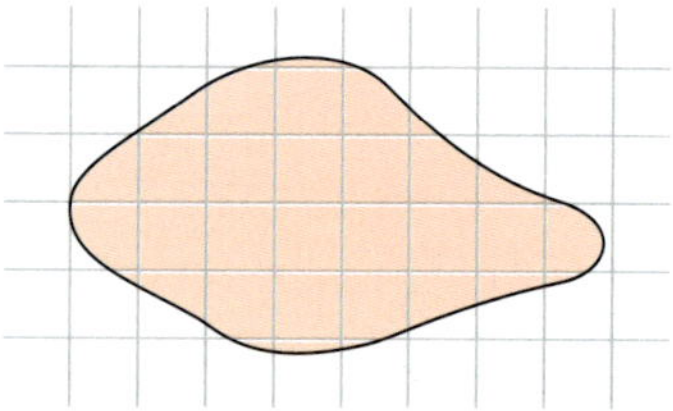

Irregular areas can be found by counting squares on a grid.

Several rows of squares make a rectangle. This rectangle has a length of l squares and a width of w rows. Its area is = 6 × 4 = 24 squares

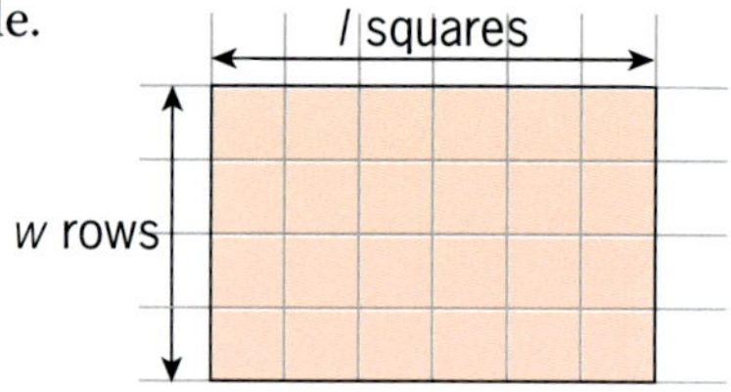

→ The area of a rectangle = $l \times w$.

A triangle can be made by cutting a rectangle in half. The triangle has a base b and a height h.

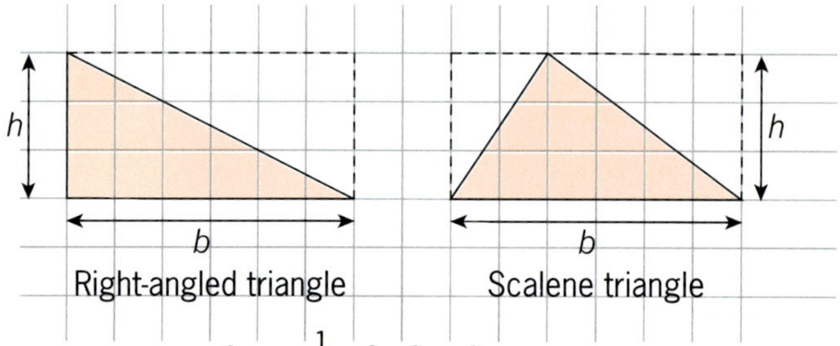

Area = $\frac{1}{2} \times 6 \times 3$ = 9 squares

→ The area of a triangle = $\frac{1}{2}b \times h$.

EXAMPLE

Calculate the area of

a the rectangular top surface of this tray

b the gable-end of this house.

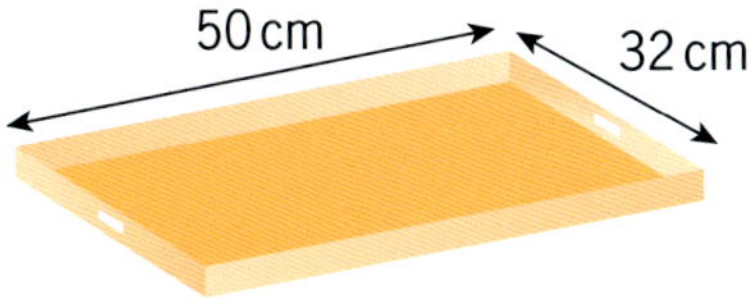

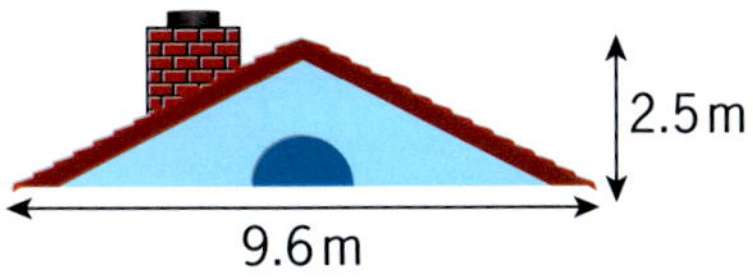

ANSWER

a Area of tray = 50 × 32

= 1600 cm².

b Area of gable-end $= \frac{1}{2} \times 9.6 \times 2.5$

$= \frac{1}{2} \times 24 = 12\ \text{m}^2$.

EXAMPLE

Find the perimeter of this field.

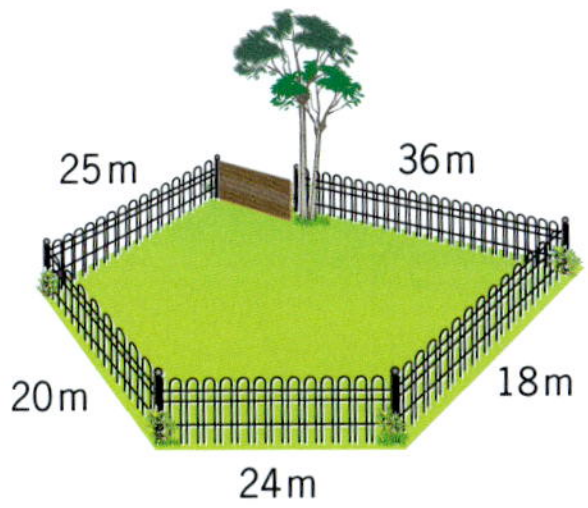

ANSWER

Its perimeter is 20 + 25 + 36 + 18 + 24 = 123 metres.

Exercise C18.1

MEDIUM LOW

1 Use a ruler to find the perimeters of these shapes.

a

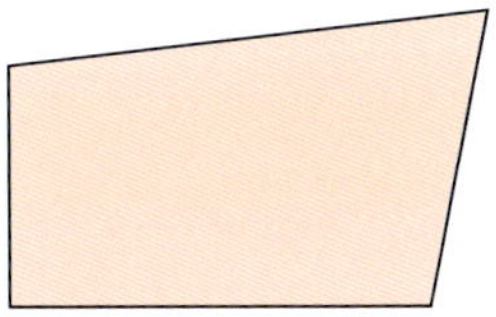

b

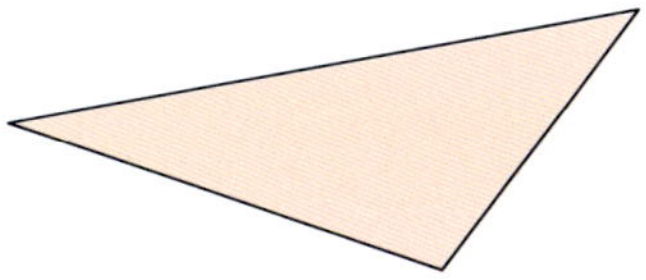

2 On centimetre squared paper, draw shapes by joining these points in order. Find their perimeters.

a (1, 1)(1, 5)(3, 5)(6, 2)   b (1, 1)(2, 5)(3, 3)(6, 4)(6, 1)

3 This plan of a sitting room of a house has a scale of 1 cm = 2 metres. Spiked strips are fitted around the edge of the room before laying a new carpet. Find

a the actual perimeter of the room in metres

b the cost of the strips if a one-metre length costs 65 pence.

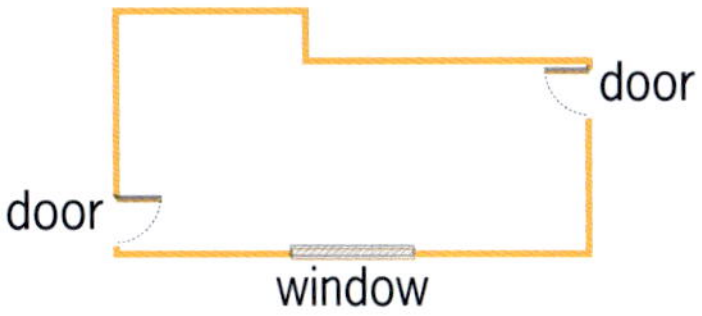

Continued ➤

Unit C

Unit C

4 Use tracing paper with a centimetre grid to find the area (in cm^2) of these shapes.

a
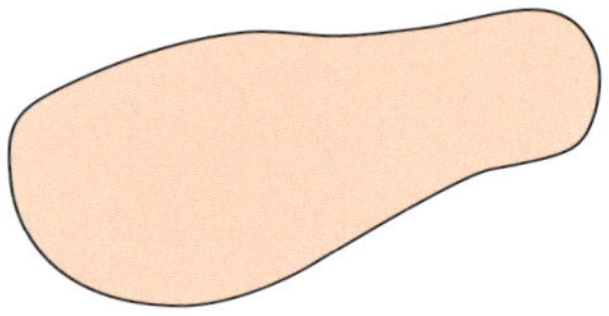

b
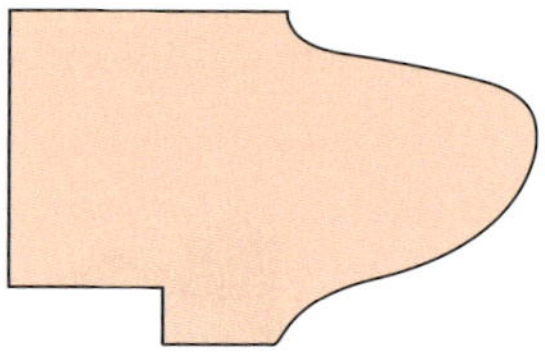

5 This map of the Isle of Man has a scale of 1 cm = 13 km.
Estimate the area of the map and so estimate the area of the island.

6 Calculate the areas of these shapes. All measurements are in centimetres.

a
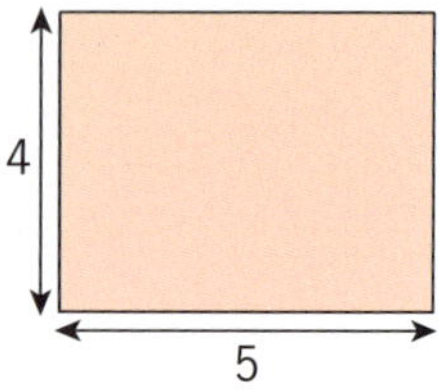

b
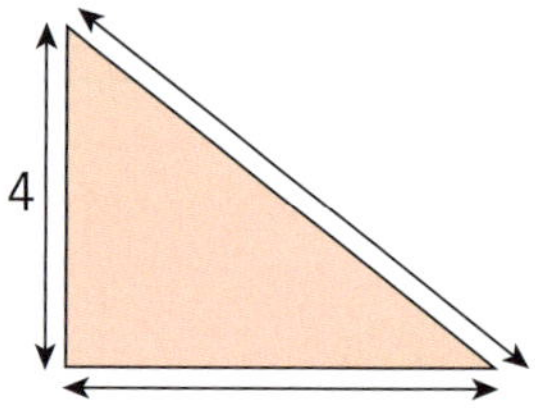

c
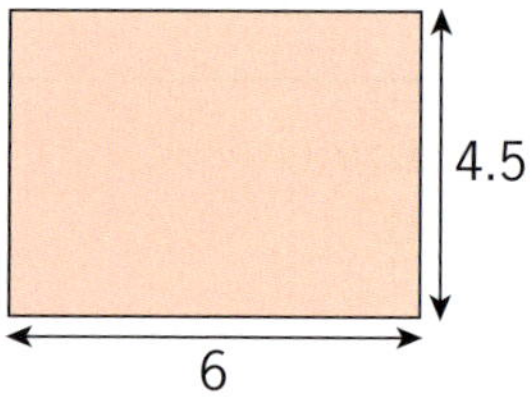

d
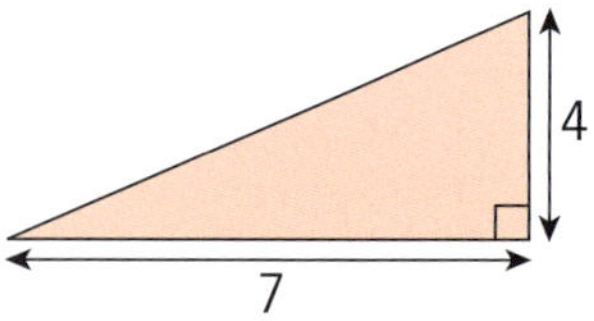

e
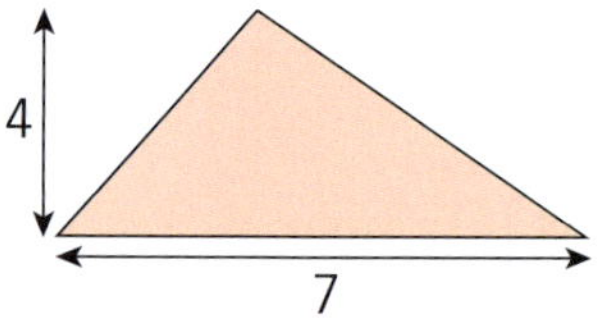

f
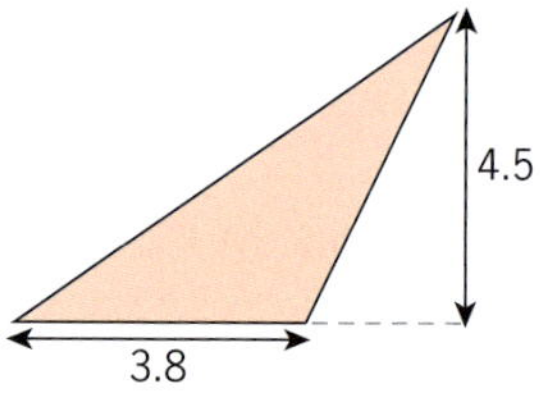

7 A full-size newspaper has a double page 76 cm by 60 cm. Find the area of

a a double page b a single page.

8 **RICH TASK**

Which two states in the USA have roughly rectangular shapes?
Find their lengths, widths and areas from the internet or reference books.
What do you notice about their areas?
Suggest a reason for any anomaly.

Circumference and area of a circle

C18.2

- Calculate and use the circumference and area of a circle

→ **The perimeter of a circle is called its circumference.**

If you cut pieces of string as long as the diameter, you can wrap three of these pieces around the circumference and still have a little gap left.

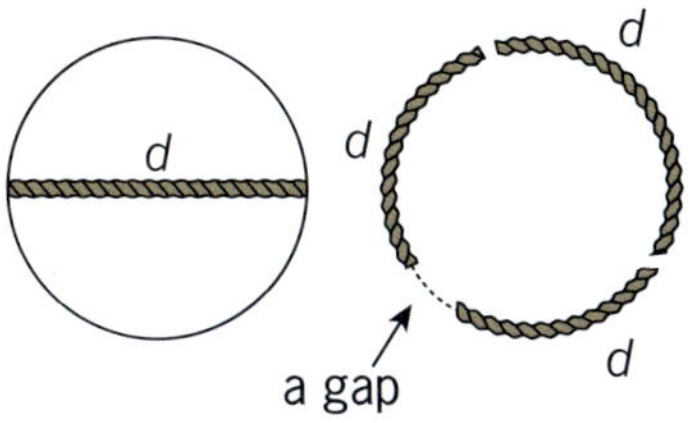

So the circumference, C = (3 and a bit) × the diameter
= 3.14 × the diameter
= $\pi \times d$

The number is 3.141 592 65….. and its exact value cannot be written as a decimal. It is called pi and written as the Greek letter π.

→ **The circumference of a circle $C = \pi d$ or $C = 2\pi r$**

EXAMPLE

A roundabout at a junction has a radius of 6 metres. Find its circumference to the nearest metre.

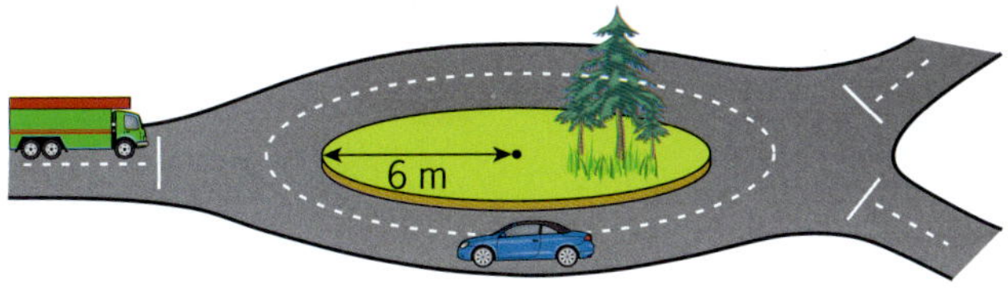

ANSWER

Its diameter is $2 \times 6 = 12$ metres
The circumference is $\pi d = 3.14 \times 12 = 37.68$
= 38 metres (to the nearest metre).

Remember to round your answer to the required accuracy.

★ ACTIVITY ★

The radius of this circle and the sides of this square are all 4 cm long.
Draw and cut out several of these squares. How many squares do you need to cover the circle? You will need to trim the squares and try to fill any gaps.
Repeat with other sizes of circle.
What do you conclude?

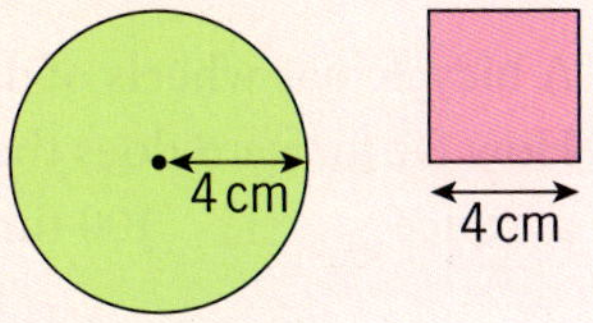

For a radius r cm and square with sides r cm,
the area of one square = $r \times r = r^2$
and the area of the circle = (3 and a bit) × the area of a square
= $3.14\, r^2$

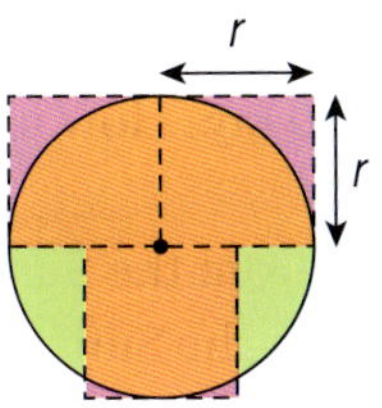

You can use this flow diagram to calculate the area.

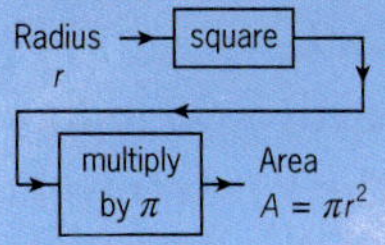

→ **The area of a circle = πr^2.**

★ ACTIVITY ★

This method gives you a value for π.
Cut a circle into 12 pieces. Rearrange the pieces to make a shape which is (almost) a parallelogram. Find its area.
Can you use this area to calculate an approximate value of π?
Explore cutting the circle into more than 12 pieces. Explore other sizes of circle.

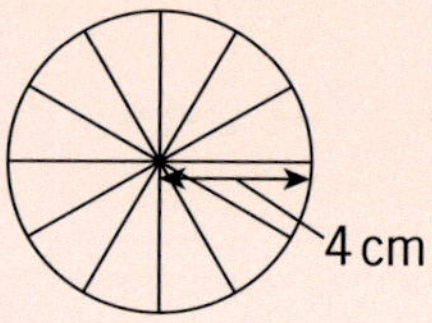

EXAMPLE

An Olympic gold medal has a diameter of 7 cm.
Calculate its area to the nearest cm^2.

ANSWER

Its radius is $\frac{1}{2} \times 7 = 3.5$ cm.
Its area is $\pi \times r^2 = 3.14 \times 3.5^2 = 3.14 \times 12.25 = 38.465 \approx 38\ cm^2$.

Exercise C18.2

MEDIUM

1 Calculate the circumference of a circle (to the nearest whole cm) with
  a a diameter of 5 cm
  b a diameter of 8 cm
  c a radius of 3 cm
  d a radius of 10 cm.

Give your answers in this exercise to 3 significant figures where appropriate.

2 Calculate the circumference (to 3 significant figures) of
  a a £2 coin with a diameter of 28.4 mm
  b a circular pond of diameter 4.5 metres
  c a dinner plate with a radius of 12 cm
  d a jam jar with a radius of 3.4 cm.

3 A bicycle has wheels of diameter 40 cm.
How far forward does the bicycle move if its wheels go round
  a once
  b 100 times?

4 ***RICH TASK***

  a The Earth has a radius of 6370 km. How long is the equator?
  b The Earth is 93 million miles from the Sun. How far does the Earth travel relative to the Sun
    i in one year
    ii in one day
    iii in one hour?

Unit C

5 Calculate the area of the surface of each of these objects.

	Object	Radius
a	Roundabout	6 metres
b	Saucer	5 cm
c	Vinyl record	15 cm
d	2 pence coin	1.3 cm

	Object	Radius
e	Dinner plate	20 cm
f	Dustbin lid	50 cm
g	Frisbee	24 cm
h	10 pence coin	2.8 cm

6 A donkey is tethered to a rope 12 metres long.
What area of grass can it eat?

7 A circular carpet is 4.6 metres across. What is its area?

8 A circular helipad (for landing a helicopter) has a radius of 14 metres.
The cost of building it was £85 per square metre.
Find

a the area of the helipad, to the nearest square metre

b the cost of building it, to the nearest £100.

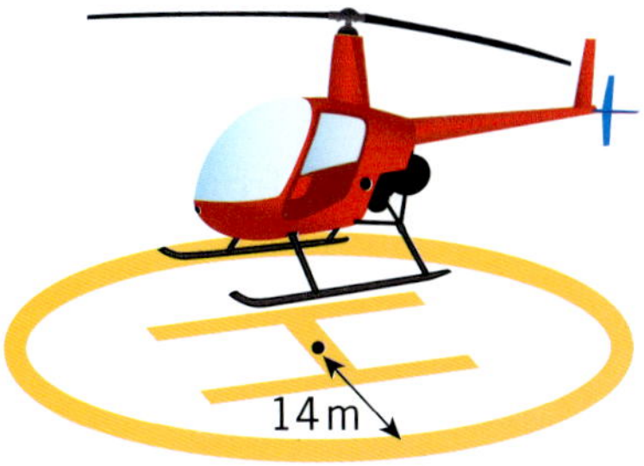

9 A metal washer has the external and internal diameters shown.
Find

a the external and internal radii

b the area of the hole in the washer

c the area of the washer.

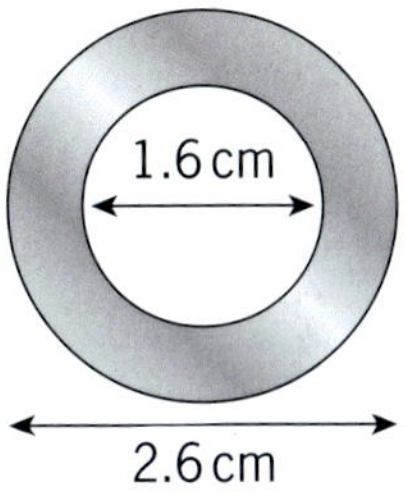

10 A car's disk brake has an outer radius of 9 cm and an inner radius of 6 cm.
Find the area of the brake's surface.

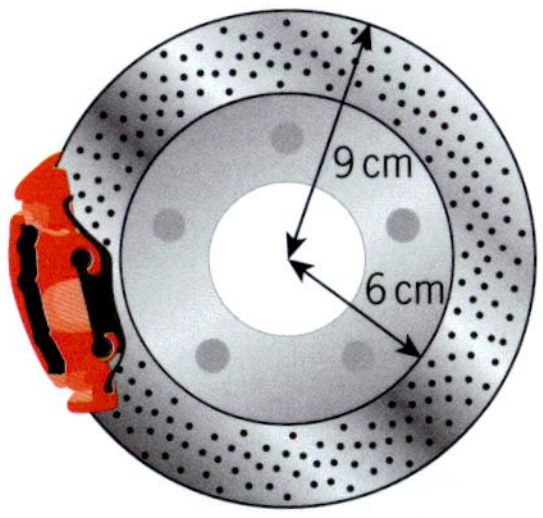

11 Measure the radius of your protractor and find its area.

Area of compound shapes

- Find and use area of shapes made from rectangles and triangles

RICH TASK

An arrow head has this shape. Find the area of

a triangle ABC
b triangle DBC
c the arrow head.

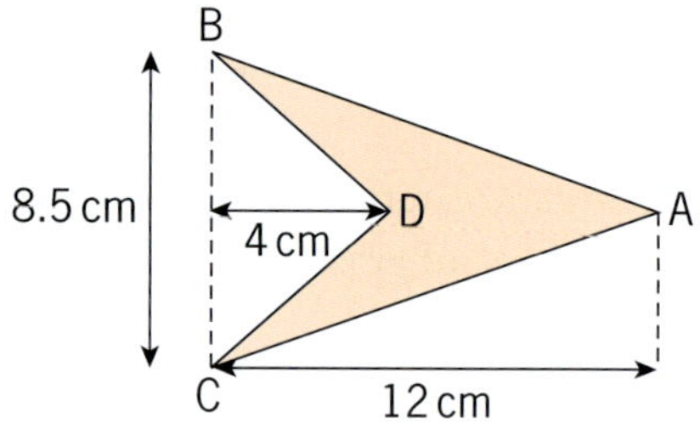

There are many possible shapes that can be made from rectangles and triangles.
This L-shape is made from two rectangles.

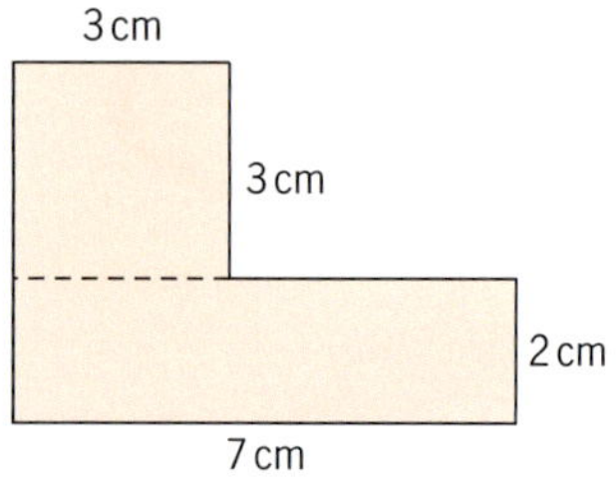

You get the same answer if you split the shape this way

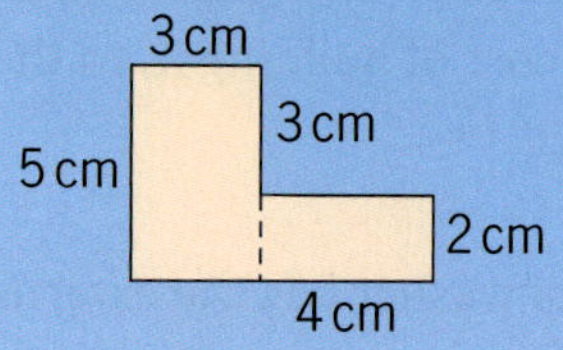

The area of the L-shape = (3 × 3) + (7 × 2) = 9 + 14 = 23 cm^2.

EXAMPLE

A house has a cupboard under the stairs.
The door of the cupboard has this shape.
Find the area of the door.

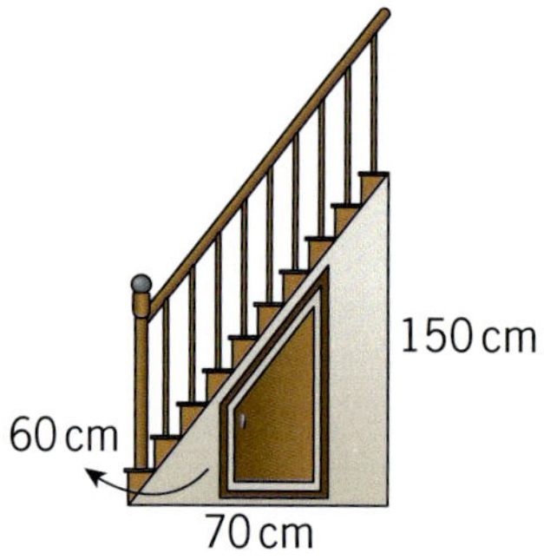

ANSWER

The shape can be split into a rectangle and a triangle. The area of the door is

$60 \times 70 + \frac{1}{2} \times 70 \times 90 = 4200 + 3150$
$= 7350\,\text{cm}^2$.

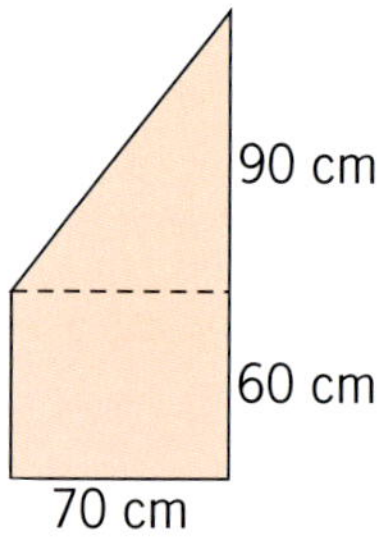

You can also think of the door as a trapezium of area

$\frac{1}{2} \times 70 \times (60 + 150)$
$= \frac{1}{2} \times 210 \times 70$
$= 7350\,\text{cm}^2$.

Exercise C18.3

MEDIUM

1 Sketch each of these shapes. Split them into rectangles and triangles to find their areas. All the measurements are in cm.

a

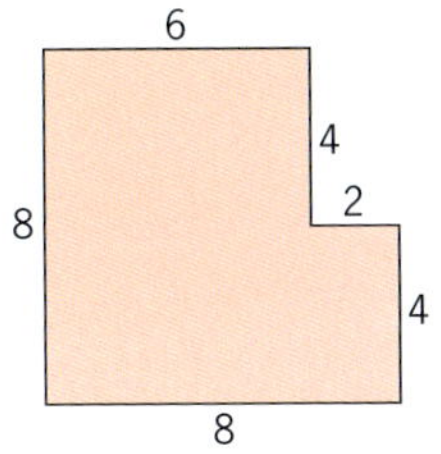

b

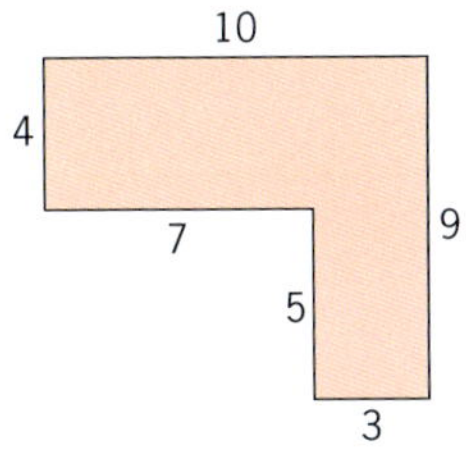

c

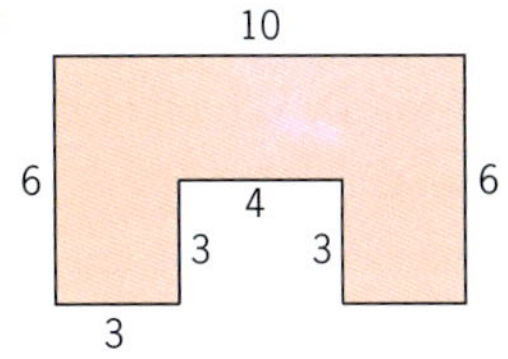

d

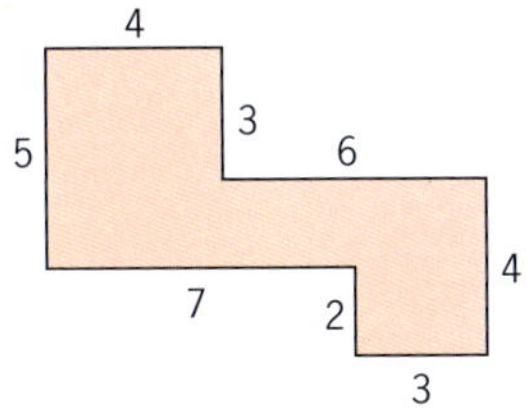

e

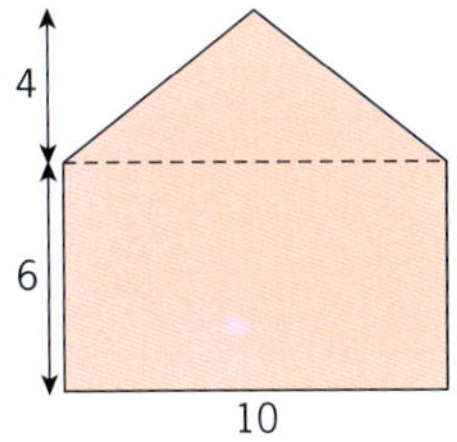

f

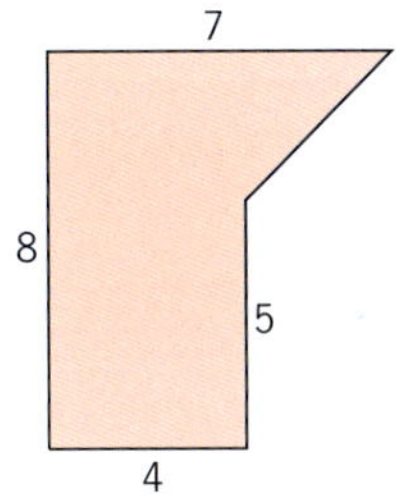

g

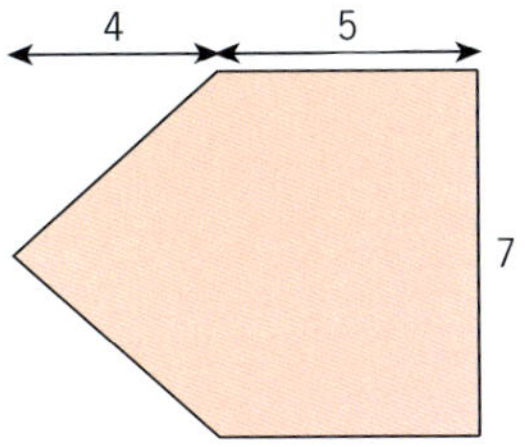

h

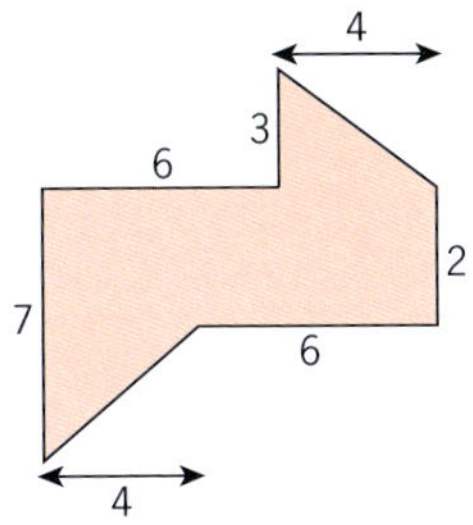

2 Mr Wagstaff is making a lawn for his house.

a Find the area of the lawn.

b Find the cost of laying turf if each square metre costs £8.10.

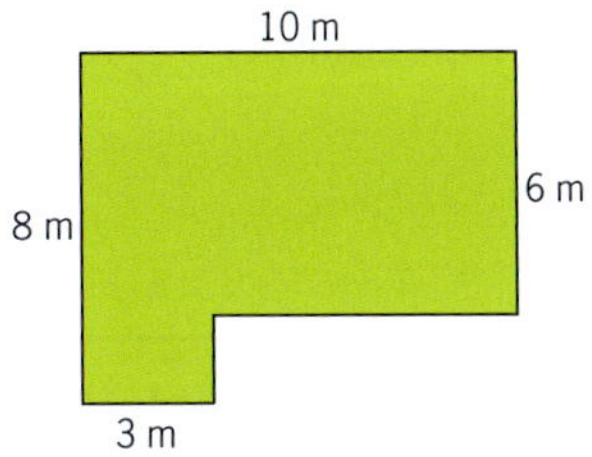

3 Find at least three different ways of working out the area of any L-shape. Try your methods on this L-shape.

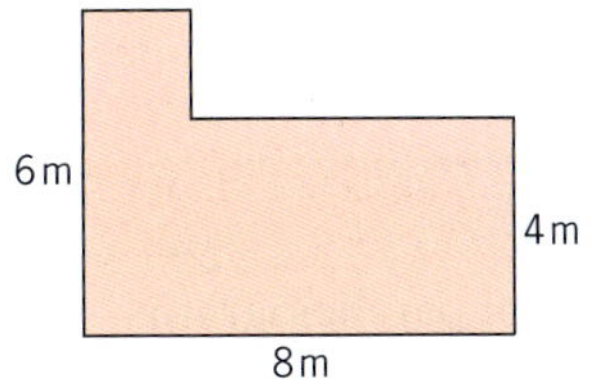

Unit C

Area of parallelograms and trapeziums

C18.4

- Find the area of parallelograms and trapeziums

A parallelogram can be made by rearranging a rectangle.
It has the same area as the original rectangle.

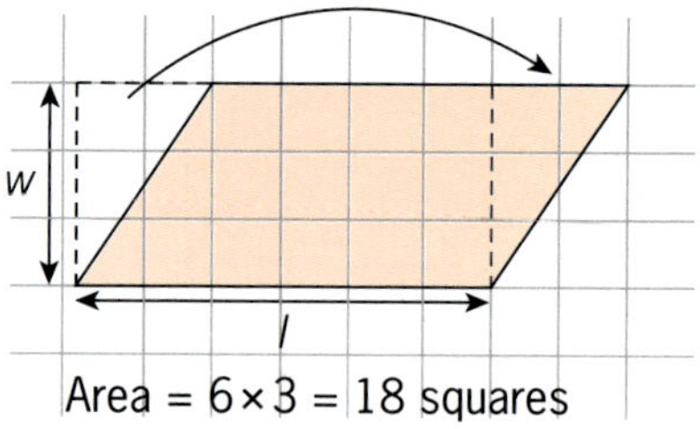

→ **The area of a parallelogram = $l \times w$.**

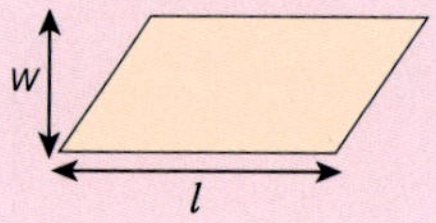

A **trapezium** can be made from two triangles which have the same height, h.

A trapezium is a quadrilateral with (at least) one pair of sides parallel.

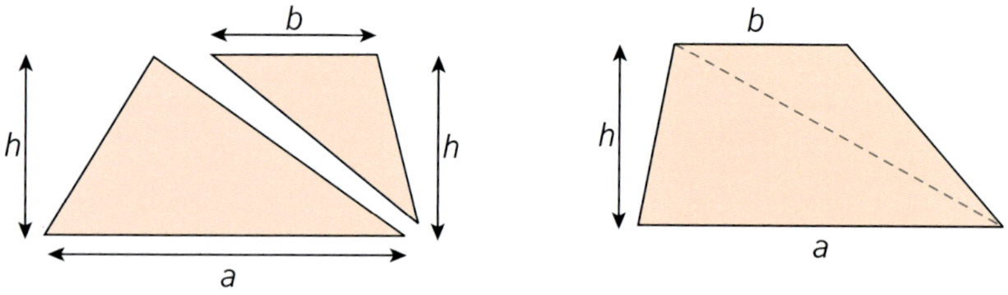

The area of the trapezium, A, $= \frac{1}{2} h \times a + \frac{1}{2} h \times b$.

→ **For a trapezium, $A = \frac{1}{2}h(a + b)$.**

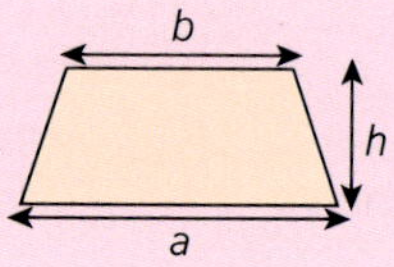

$\frac{1}{2}h$ is a **common factor.**

EXAMPLE

Calculate the area of each shape.

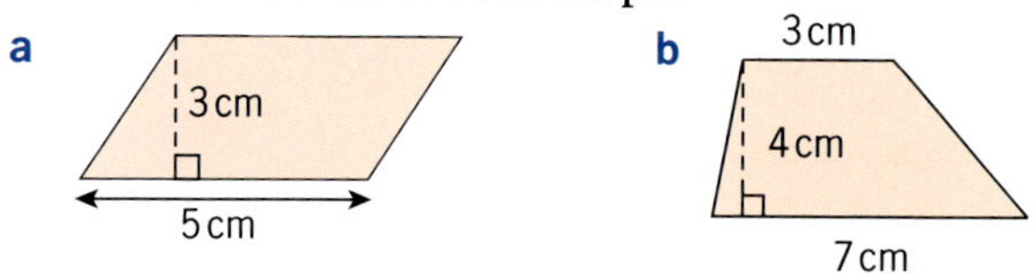

ANSWER

a Area of parallelograms
$= 5 \times 3$
$= 15\ \text{cm}^2$

b Area of trapezium
$= \frac{1}{2}(3 + 7) \times 4$
$= 5 \times 4$
$= 20\ \text{cm}^2$

Always include the units in your answer.

Exercise C18.4

MEDIUM

1 Use the formula $A = \frac{1}{2}h(a + b)$ to find the area of these trapeziums. All the measurements are in cm.

a
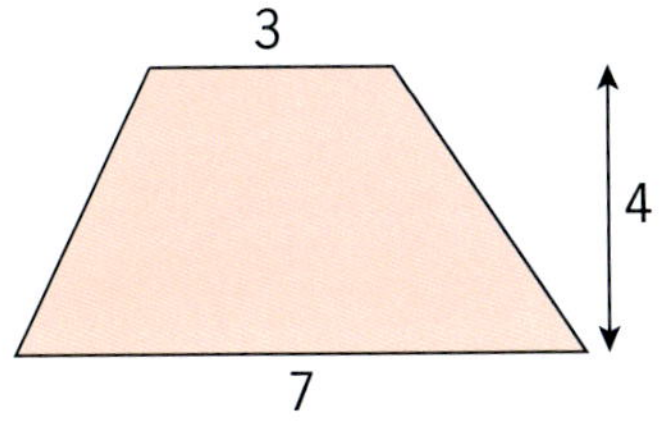

b
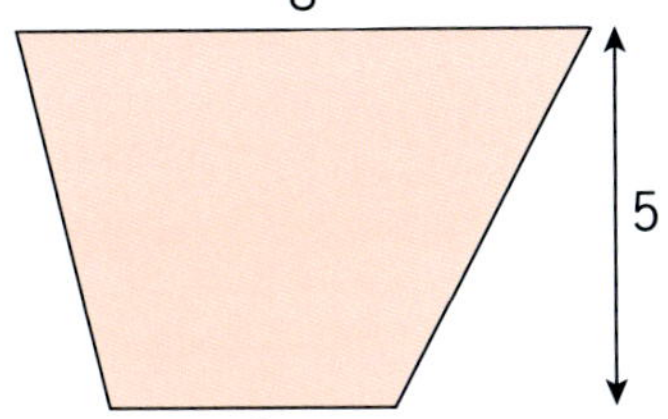

c
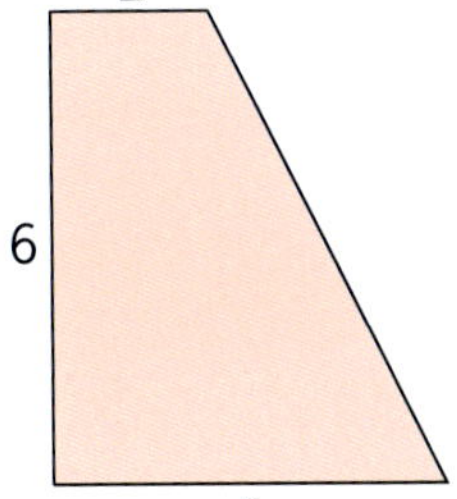

d
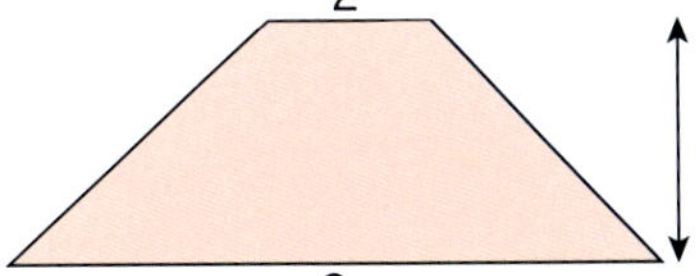

2 Calculate the area of these parallelograms. All the measurements are in cm.

a
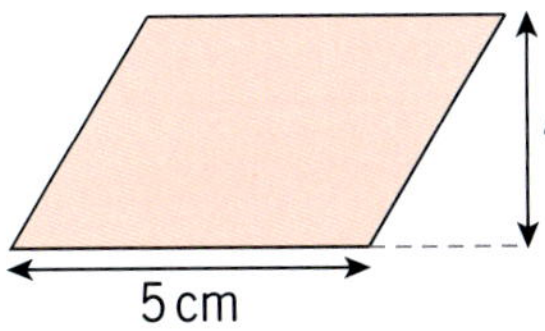

b
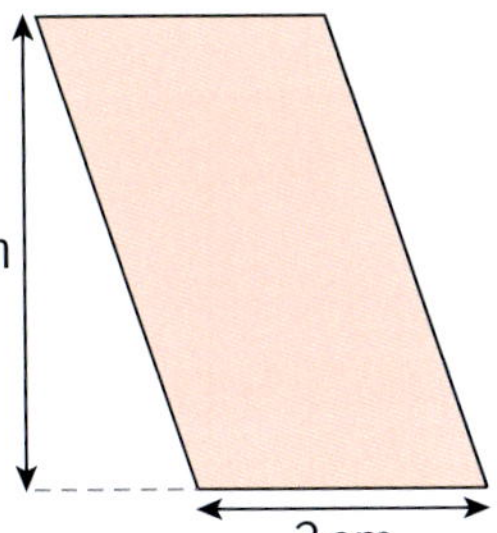

3 A chevron on a roadside warning sign is made from two parallelograms. Find the area of the sign.

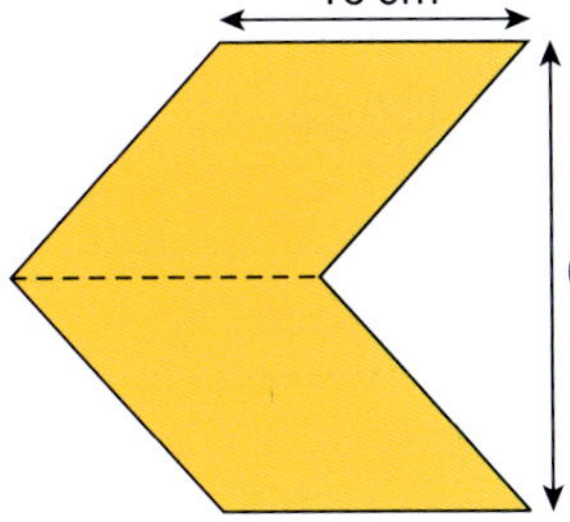

4 **RICH TASK**

Can you find a square with the same area as this trapezium?

Can you find other trapeziums with the same height and same area?

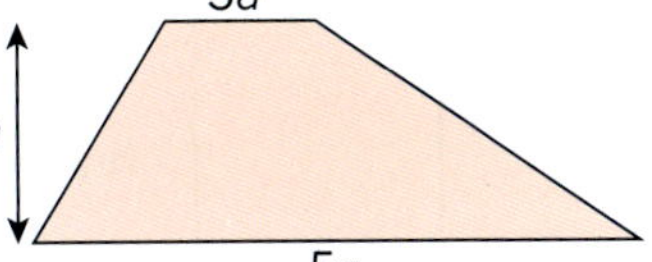

Unit C

Area and perimeter formulae

- Use formulae for area and perimeter

The area formulae that you know are:

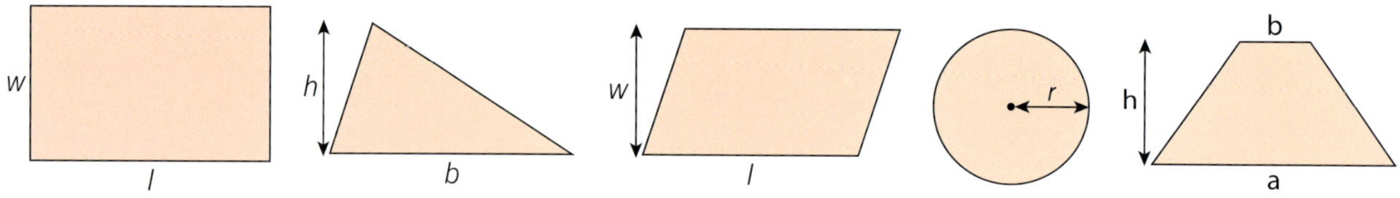

$A = l \times w$
for a rectangle

$A = \frac{1}{2} b \times h$
for a triangle

$A = l \times w$
for a parallelogram

$A = \pi \times r^2$
for a circle

$A = \frac{1}{2} h(a + b)$
for a trapezium

Two formulae for perimeter that you know are:

$P = 2l + 2w$ for a rectangle

$C = \pi \times d$ for a circle

If you know the area or perimeter of a shape, you can often work out one of its dimensions.

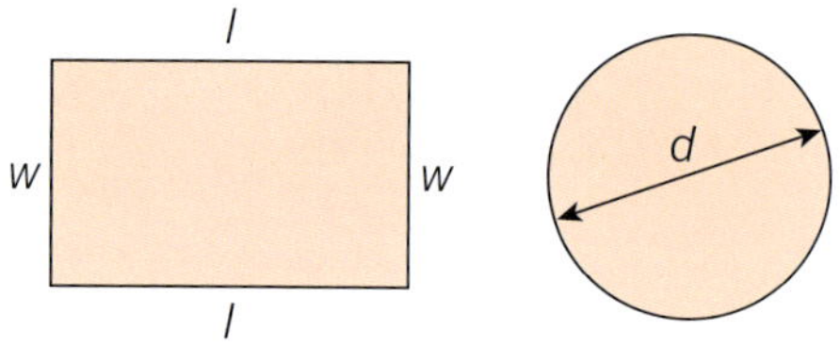

EXAMPLE

A triangle has an area of 120 cm².
Its base is 48 cm long.
What is the height, h, of the triangle?

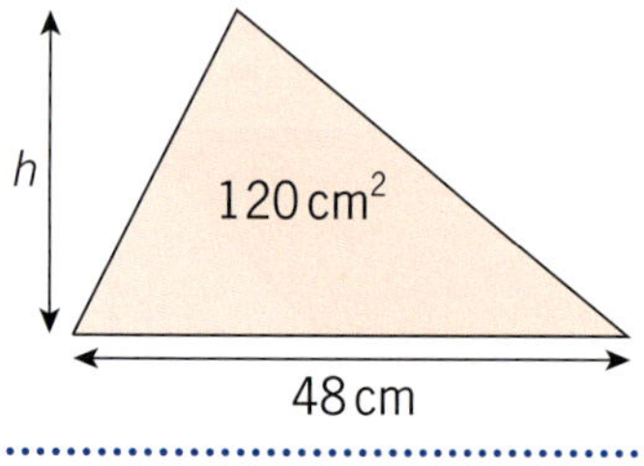

ANSWER

The correct formula is $A = \frac{1}{2} \times b \times h$

Substitute for A and b $120 = \frac{1}{2} \times 48 \times h$

Simplify $120 = 24 \times h$

Divide both sides by 24 $h = \frac{120}{24} = 5$

The height of the triangle, h, is 5 cm.

EXAMPLE

A 2-euro coin has an area of 521 mm².
Calculate its radius.

ANSWER

The formula for a circle is $A = \pi \times r^2$

Substitute for A and π $521 = 3.14 \times r^2$

Divide both sides by 3.14 $r^2 = \frac{521}{3.14}$

$= 165.9$

Square root both sides $r = \sqrt{165.9}$

$= 12.9$

The radius of the coin is 12.9 mm.

Exercise C18.5

MEDIUM

1 Find the lettered dimension in these shapes with the given areas.

a
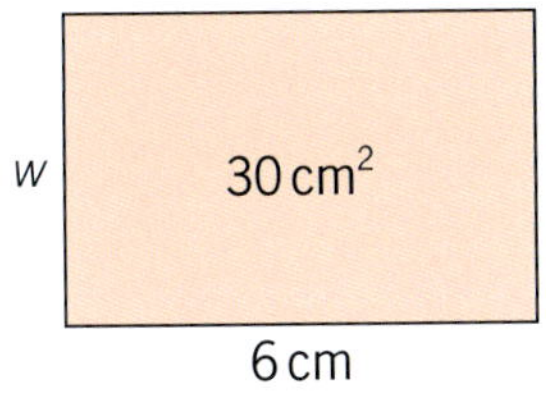

b
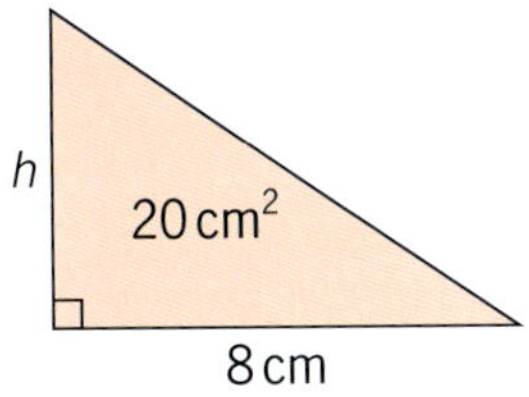

c
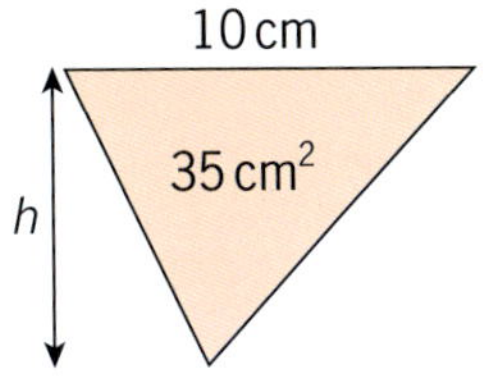

d
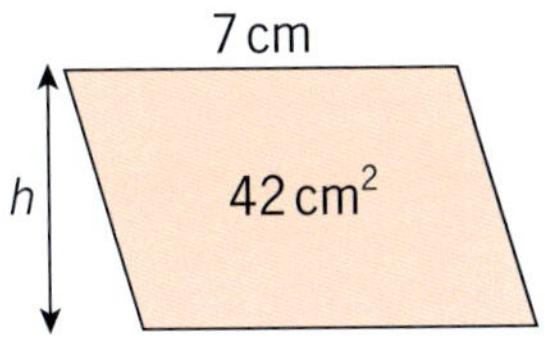

e
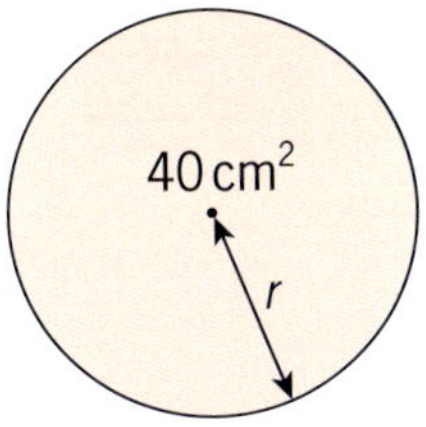

f
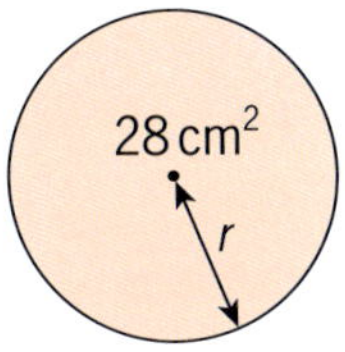

2 A triangular vegetable plot is 24 metres long and has an area of 120 m^2. What is its greatest width?

3 The gable end of a house has an area of 5.75 m^2. The house is 4.6 metres wide. How high is the gable end?

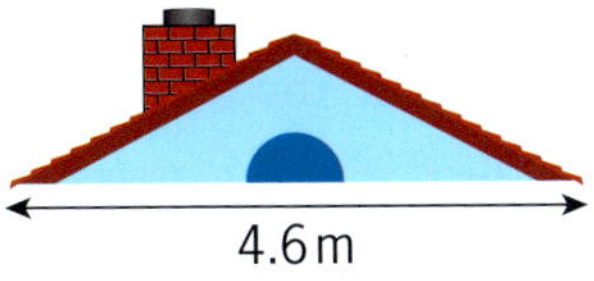

4 A pendant in the shape of a parallelogram is 3.5 cm long. If it has an area of 5 cm^2, how wide is it to the nearest mm?

5 a Find the diameter of a circle which has a circumference of
 i 50 cm ii 100 cm iii 650 mm.

 b Find the radius of a circle which has an area of
 i 20 cm^2 ii 85 m^2 iii 100 mm^2.

6 A circular coffee table has an area of 8820 cm^2. Find its circumference to the nearest cm.

*7 **RICH TASK**

A rectangle has an area of 36 cm^2.
Its length and width (in cm) are both integers.
How many different possible sizes can this rectangle have?
Draw a graph of the perimeters of these possibilities against their lengths.
What can you say about the shape of the rectangle with the least perimeter?

Unit C

Length and area scale factors

- Find and use length and area scale factors

Here is an enlargement.

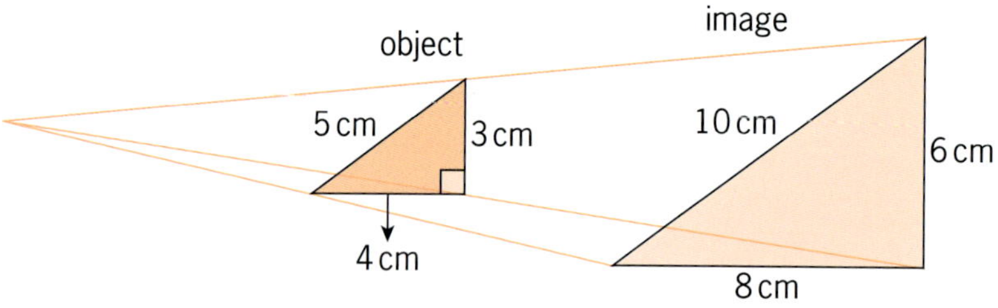

> Angle C = angle C′ = 90°.

Look at the angles. The **shapes** of the two triangles are the same, so **corresponding angles** are equal.
The enlargement has not changed any angles.

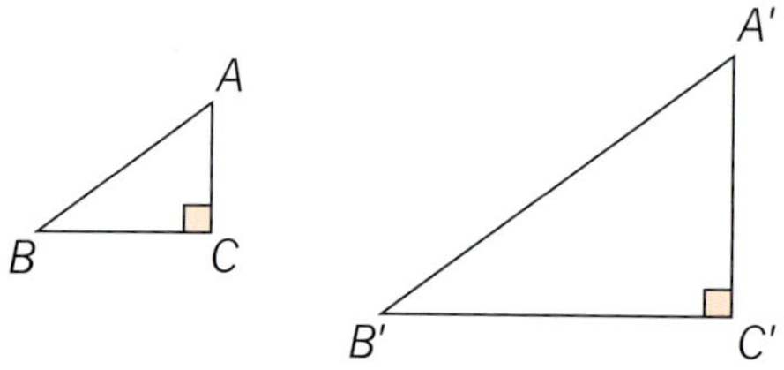

Look at the sides. The **sizes** of the two triangles are different.
Each side of the image = 2 × the equivalent side of the object.
The length **scale factor** is 2.

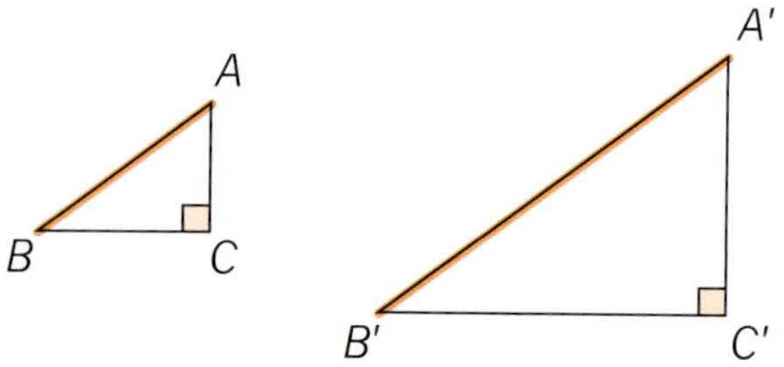

Calculate the **perimeters** of the triangles.

The perimeter of triangle ABC = 5 + 4 + 3
= 12 cm.

The perimeter of triangle A′B′C′ = 10 + 8 + 6
= 24 cm.

The perimeter **scale factor** is 2.

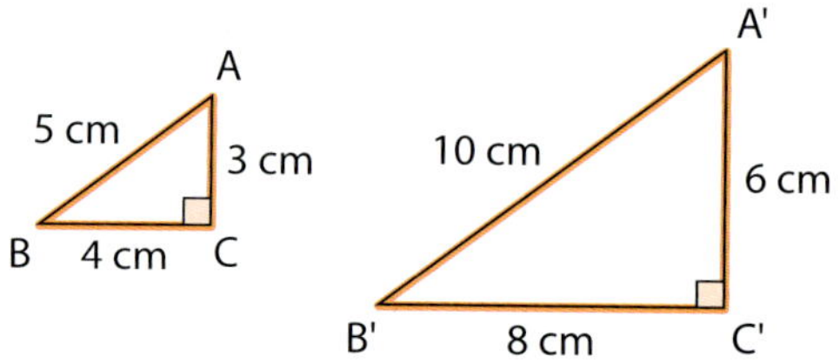

Calculate the **areas** of the triangles.

The area of object triangle

ABC = $\frac{1}{2} \times 3 \times 4 = 6$ cm².

The area of image triangle

A′B′C′ = $\frac{1}{2} \times 6 \times 8$
= 24 cm².

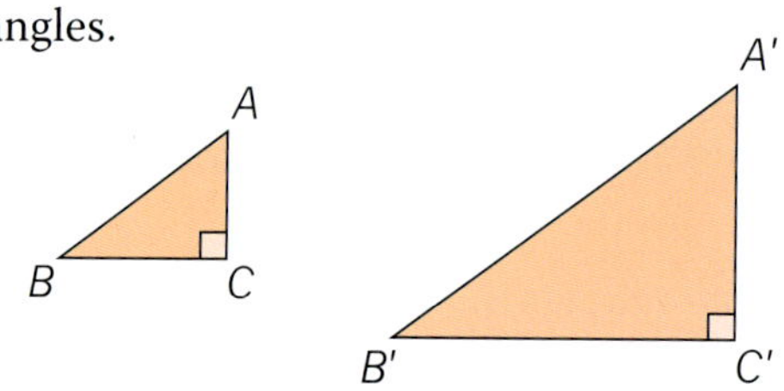

The area of the image = 4 × the area of the object.
The area scale factor is 4.

> You can fit four object triangles into the image triangle.

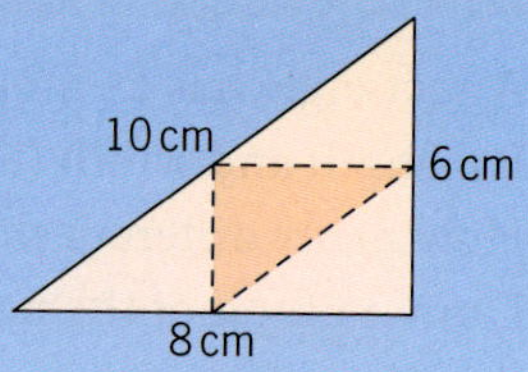

→ **If an enlargement has a length scale factor of m, then the perimeter scale factor is m and the area scale factor is m^2.**

Exercise C18.6

MEDIUM

1 a Draw a graph and label both axes from 0 to 8. Draw the object triangle (2, 3), (4, 3), (3, 2) and image triangle (2, 6), (6, 6), (4, 4). Find the centre of enlargement, the length scale factor and the area scale factor.

b Repeat on new axes labelled from 0 to 10 with the object shape (2, 2), (2, 4), (3, 4), (3, 2) and the image shape (4, 4), (4, 10), (7, 10), (7, 4).

2 **RICH TASK**

For each object and image, find the length scale factor and the area scale factor. Enter your results in a table.

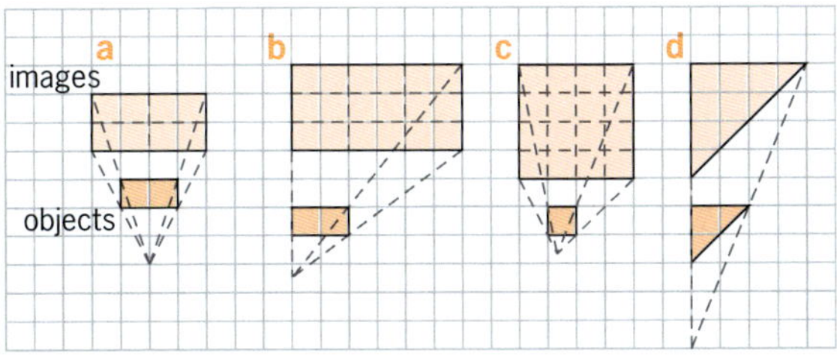

	a	b	c	d
Length scale factor				
Area scale factor				

Can you see the connection between length and area scale factors?

Hint: see the key point above! The important thing is to be able to understand it and explain it to someone else.

3 An advert in a magazine has an area of 15 cm². An enlargement of the advert three times longer and wider appears in a newspaper. Find the area of the newspaper advert.

4 Tinned tomatoes are sold in two sizes. The label on the larger tin has an area of 315 cm². It is an enlargement of the label on the smaller tin which has an area of 35 cm². Find

a the area scale factor

b the length scale factor

c the length of the larger label if the smaller one is 7 cm long.

5 ✯ **CHALLENGE** ✯

The V&A Museum in London sell copies of paintings. You can buy paintings by Toulouse-Lautrec as posters (75 cm by 50 cm) and as prints (30 cm by 20 cm).

Are the posters a mathematical enlargement of the prints? What are the length and area scale factors?

Summary and assessment

CHECK OUT

You should now be able to

- find the perimeter and area of
  - simple shapes made from rectangles and triangles
  - parallelograms and trapeziums
  - circles
- use π in exact calculations.

Exam-style question

Rectangle R has length 12 cm and width 3 cm.

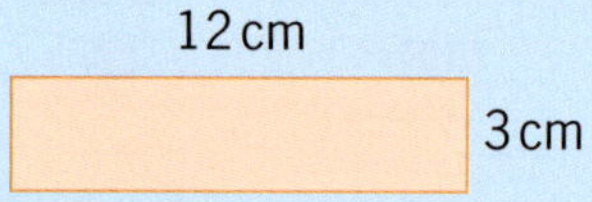

a Square A has the same area as rectangle R. Work out the side length of square A.

b Square B has the same perimeter as rectangle R. Work out the side length of square B.

Andrea's answer:

a Area of R $= 3 \times 12 = 36\,\text{cm}^2$

First work out the area of R.

A square has sides same length so I need to find a number times itself

Use facts you know about side lengths of squares.

$6 \times 6 = 36$

Square A has side length 6 cm

b Perimeter of R $= 12 + 3 + 12 + 3$
$= 30$ cm

First work out the perimeter of R.

A square has 4 sides all the same length

Remember the units in your answers.

$30 \div 4 = 7{\cdot}5$

Square B has side length $7{\cdot}5$ cm

Exam practice

1 Work out the area and perimeter of these shapes.

a

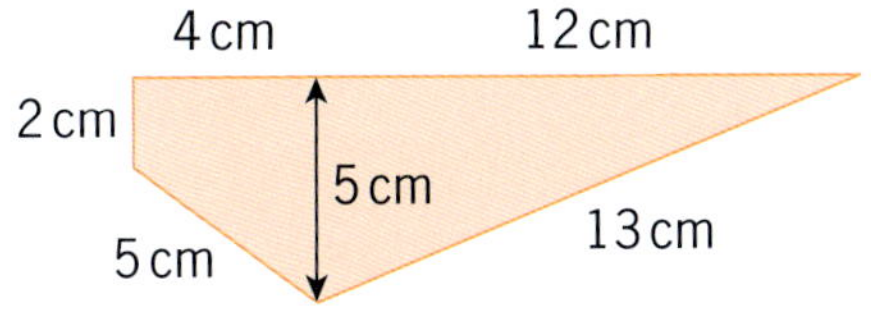

b

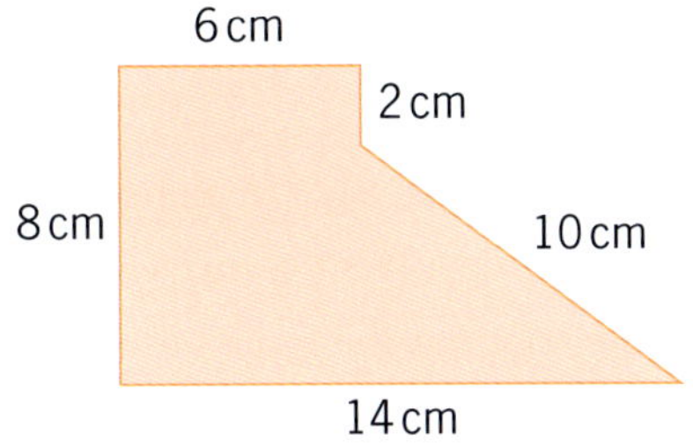

(*8 marks*)

2 The diagram shows the top of a child's shape-sorting toy.

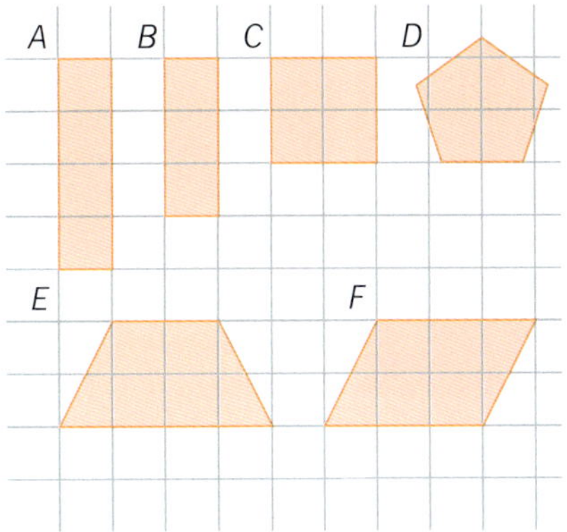

a Which shapes (if any) have the same area as

 i rectangle A ii rectangle B?

b Which shapes have the same perimeter? (*4 marks*)

3 An earring is made from two circles as shown.

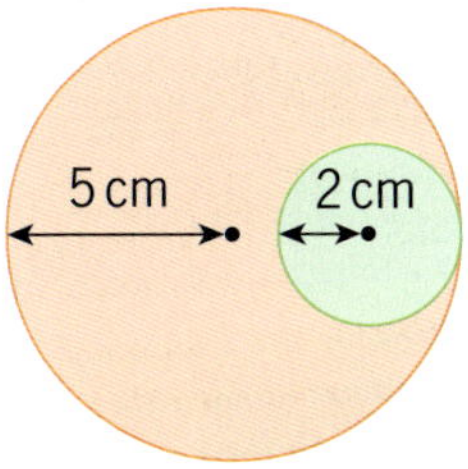

Work out the orange shaded area. (*3 marks*)

Case study 6: Holiday

Mathematics can help you to plan and budget for a holiday, as well as to understand currency, temperature and other units of measure at your destination.

holiday

Paris

LOUISE'S family are planning to go on holiday. Her parents will pay for the trip, but she must raise her own spending money.

1 How much money could Louise save in the three months from March to May if she hired a DVD once a week instead of going to the cinema?

2 A neighbour offers to pay Louise £10 per week if she takes her dog for a 30-minute walk every weekday before school. What hourly rate of pay does this represent? How much would Louise earn if she walked the dog every weekday throughout March, April and May?

3 Louise's brother sold 18 of his CDs for £45. How much did he sell each CD for? He then sold another two CDs and 11 DVDs for £43.50. How much did he charge for each DVD?'

How could you raise money towards a holiday fund or to buy a new item? How long would it take you to reach your target amount?

If you are going on holiday outside of the UK, then you will need to convert your money from £ Sterling to the local currency of your destination. Many European countries now use the Euro, €.

In 2005, the average £ Sterling : Euro exchange rate was 1 : 1.46.

In that year, how many

a) Euros would you receive in exchange for £150

b) £ Sterling would you receive in exchange for 120 EUR?

Suppose that you are charged £70 (with no commission) to buy 91.7 EUR.

What is the exchange rate? Give your answer as a ratio £ Sterling : Euro.

Some companies charge a commission fee to exchange currency. With an added charge of 1%, how many Euros would you now receive (at the same exchange rate) for £70?

What is the £ Sterling/Euro exchange rate today? Research the commission rates that some companies are charging to exchange currency. What is the most/least amount that the companies you have found would charge you to exchange £150 into Euros?

Deciding on your method of transport is an important part of planning a holiday.

Some travel options between Oxford and Paris are shown.

WHICH travel option would you choose? Explain your response with reference to the travel times and costs. All times given are local. Paris is in the time zone GMT + 1 hour.

The foreign travel legs of the same journey options can be paid for in Euros for the following prices:

Return flight BHX to Paris CDG 151.49€; return Eurostar journey 130€, return flight London Heathrow to Paris CDG 162.82€.

How does each of the prices in Euros compare with the corresponding price in GBP?

Explore travel options from your hometown to different destinations. Be careful, there are some times hidden costs such as additional taxes and fees.

■ Different countries often use different units of measure for quantities such as temperature.

An Internet site states that the maximum and minimum temperatures in Rome on a particular day are 99°F and 63°F respectively.

A formula that can be used to convert between °C and °F is temp(°C) = 5(temp(°F) – 32)/9

What are the corresponding maximum and minimum temperatures in Rome?

The maximum and minimum temperatures in London on the same day were 25°C and 16°C. What are these temperatures in °F?

What are the maximum and minimum temperatures in your home town today? Use the formula in the example to convert the temperatures you have found from °C to °F.

19 Algebraic manipulation

INTRODUCTION

Engineers and scientists use algebraic expressions to model and explain the behaviour of real events and activities. Without algebra there would be no aircraft, mobile phones or plasma TVs.

What's the point?

Algebra is an extension of arithmetic, but using letters instead of numbers. By using algebra, we can find and use rules to explain real-life phenomena.

CHECK IN

1 Substitute the following values into the expression $x^3 - 2x$.

a 2 b 2.1 c −3.2

2 Solve the equations

a $2x + 3 = 17$ b $3(x - 4) = 18$

Orientation

What I should know		What I will learn		What this leads to
B14	→	Manipulate algebraic expressions Use trial and improvement to find approximate solutions to equations	→	C21

RICH TASK

Think of a number between 1 and 10.

- Double it.
- Add 4.
- Halve your answer.
- Take away the number you first thought of.

The answer is always 2.

Investigate.

Algebraic manipulation

C19.1

- Revise understanding of algebraic terminology and manipulation

p. 98

→ **An** expression **is a collection of algebraic terms separated by + or – signs.**

$3x + 4y - 2z$ is an expression.

→ **An** equation **is an algebraic statement that is only true for particular values of the** variable**.**

$5x + 2 = 12$ is an equation.

(Its solution is $x = 2$)

→ **An** identity **is an algebraic statement that is true for *all* values of the variable.**

$2x + 3x \equiv 5x$ is an identity (whatever value x takes, it will *always* be true).

An expression may contain brackets.

EXAMPLE

Find the values of a and b in this identity.

$3(2x + 9) + 4(2x - 5) \equiv ax + b.$

ANSWER

Expand the brackets. $3(2x + 9) + 4(2x - 5) = 6x + 27 + 8x - 20$

Collect 'like terms' $= 14x + 7$

Compare $14x + 7$ with $ax + b$. $a = 14$ and $b = 7$.

The inverse of expanding brackets is called **factorising**.

EXAMPLE

Factorise **a** $8x + 12y$ **b** $7x^2 - 4x$ **c** $6z^2 + 15z$

ANSWER

a 8 and 12 have a common factor of 4. $8x + 12y = 4(2x + 3y)$

b $7x^2$ and $4x$ have a common factor of x. $7x^2 - 4x = x(7x - 4)$

c $6z^2$ and $15z$ have two common factors: 3 and z. $6z^2 + 15z = 3z(2z + 5)$

You can check the answers by expanding the brackets.

Exercise C19.1

MEDIUM

1 Write expressions as simply as possible for the length of these arrangements of pipes.

a 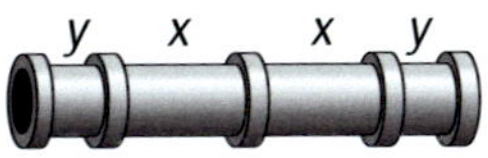

b 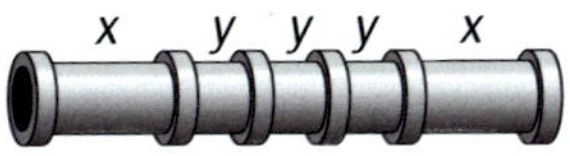

2 Simplify these expressions.

a $3x + 5y + 2z + 2x - 4y - 2z$

b $2p + 3q - 4 + 5p + 7q + 6$

c $3m + 3n + 3 - 2m - 3n - 4$

d $9x - 8y + 7z - 6x + 5y - 4z$

e $s + t + u - s + 4t - u$

f $1 + p - q + 2p - 2q + 3$

g $8 + a - 2b + a + 2b - 8 - 2a$

h $r + 2r - s + 5t - 3r + s$

3 Find the values of a and b to make these statements into identities.

a $8x + 7 + 2x + 3 \equiv ax + b$

b $4y - 2 + y + 5 \equiv ay + b$

c $p - 3 + 6p - 2 \equiv ap + b$

d $3x + 2y - x + 4y \equiv ax + by$

e $2s + 3t - u + s - 4t + u \equiv as + bt$

f $5y - 2z + 4 - 2y + 2z - 6 \equiv ay + b$

g $8 - 3m + n - 2m - n \equiv am + b$

h $3x + y - 4y - 3x + 5y \equiv ax + by$

4 Expand the brackets and simplify as far as possible.

a $3(4x + 2) + 2(6x + 2)$

b $5(2x + 3) + 2(2x - 3)$

c $6(p + 2) + 2(p - 3)$

d $3(2q - 1) + 5(4 + q)$

e $3(r - 4) + 4(2r + 3)$

f $6(2 + 3n) + 3(2n - 4)$

g $x(x + 4) + x(x + 2)$

h $z(z + 7) + z(z - 3)$

i $x(x + 5) + 3(x + 4)$

j $y(y + 1) - 2(y - 3)$

5 Factorise these expressions.

a $4x + 6$

b $6y + 9$

c $12z + 8$

d $3x + 15$

e $5x + 20$

f $18z - 15$

g $5 + 8y + 3 + 4y$

h $14 - 9y + 1 - y$

i $7 - 5t - 3 + 3t$

j $x^2 + 3x$

k $y^2 - 5y$

l $4z - z^2$

m $6p - p^2 + 2p$

n $3b^2 + 8b - 2b$

o $8c^2 - 6c + 2c$

6 **RICH TASK**

Four lengths of pipeline each contain six pipes of length x and three pipes of length y.
Which one of these expressions gives the total length of all this pipeline?

$4 + 6x + 3y$ $\quad$ $4(6 + x + 3 + y)$ $\quad$ $4(6x + 3y)$

Find three other expressions which also give the total length.

Unit C

Trial and improvement

- Find approximate solutions to equations using trial and improvement methods

RICH TASK

Mr Gupta keeps oil for his central heating in a tank which is a cube of side x metres.
When the oil in the tank is 1 metre deep, he orders more oil.
An order of 0.5 m^3 is enough to just fill his tank.

What are the dimensions of his tank?

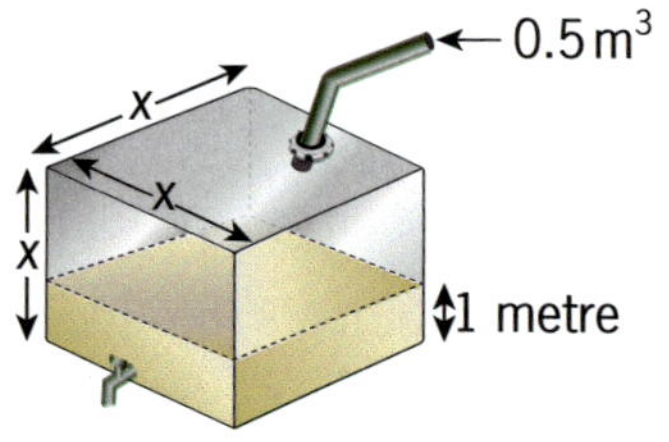

To solve $x^3 = 27$, you find the cube root of 27. What is the value of 3^3? So what is the value of x?

Solving $x^3 = 37$ is not so easy. You can choose a sensible value of x, see if it is close to the correct answer, and then improve on it.

Some calculators can find cube roots.

Set your work out in a table.

x	x^3	
3	27	too small
3.5	42.875	too big
3.4	39.304	still too big
3.3	35.937	too small

Why do you know that the solution lies between 3.3 and 3.4? Which of these two values is nearer to the solution? Try the mid-value 3.35 to find out.

x	x^3	
3.35	37.595	just too big

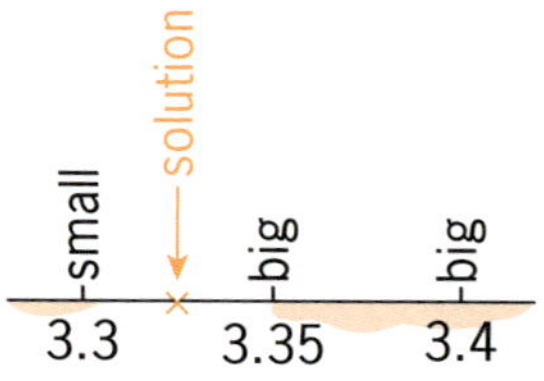

If 3.35 and 3.4 are both too big, then the solution must be nearer to 3.3 than to 3.4.
So the solution of the equation $x^3 = 37$ is $x = 3.3$ to 1 decimal place.

To get a more accurate answer to 2 decimal places, try values between 3.3 and 3.35. The solution to 2 d.p. is 3.33. Work out 3.33^3 and see.

Exercise C19.2

MEDIUM

1 Solve these equations using a trial and improvement method.
Record your results in a table. Take $x = 5$ as your start value in each case.

a $x^2 + x = 72$

x	x^2	$x^2 + x$	Too big or too small?

b $x^3 + x = 738$

x	x^3	$x^3 + x$	Too big or too small?

c $x^3 - x = 336$ d $x^2 + 5x = 104$ e $x^3 + 4x = 240$

f $x^3 - 4x = 480$ g $3x^2 + x = 154$ h $x^3 + x^2 = 810$

2 Find the solution of $x^3 = 10$, correct to 1 decimal place, by copying this table and filling in the blanks. Try $x = 2$ as a start.
The think-bubble gives you some ideas. Remember to check your answer.

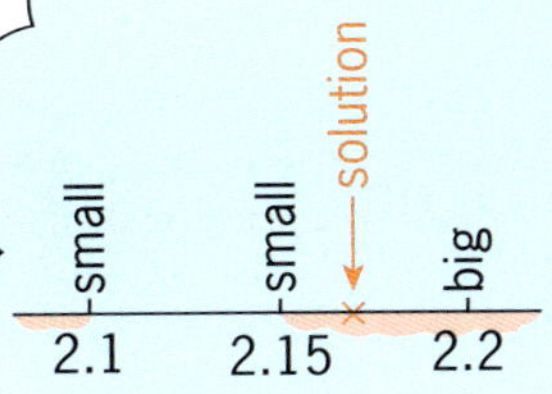

Start with x = 2.
Is the answer too big or too small?
Try x = 3. Is this a better choice?
Choose x = 2.1 which is much nearer to 2.
Try 2.2.
You now know that the solution is between 2.1 and 2.2, so try the mid-value 2.15.

x	x^3	Too big or too small?
2	8	too .....
3	....	far too .....
2.1	....	....
2.2	....	....
2.15	....	....

3 Solve the equation $x^2 + x = 77$, giving your answer correct to 1 decimal place. Use these headings to record your values of $x^2 + x$.
Try $x = 5$ as a start.

x	x^2	$x^2 + x$	Too big or too small?

4 Find a positive solution to these equations, giving answers correct to 1 decimal place.
Some starting values are suggested.

a $x^3 + x^2 = 100$ (try $x = 4$) b $x^3 + 3x^2 = 370$ (try $x = 6$) c $2x^3 - 5x = 280$ (try $x = 5$)

d $x - \sqrt{x} = 16$ e $x^2 + \sqrt{x} = 700$ f $x^3 + x^2 = x$

5 **RICH TASK**

Try using a computer spreadsheet to set up some of the tables of values above.

Further trial and improvement

C19.3

- Extend methods of trial and improvement and apply them to solve real-life problems

The method of trial and improvement can be used to solve more complex problems.
To solve the equation $x^3 = x^2 + 6x$, what headings do you think you may need in your table?

EXAMPLE

Solve the equation $x^2 = x^3 - 10x$, correct to 1 decimal place, taking $x = 2$ as your starting value.

ANSWER

Calculate values of the two sides of the equation, using columns for x^2 and $x^3 - 10x$.

x	x^2	$x^3 - 10x$	Comment
2	4	8 − 20 = −12	x^2 is bigger than $x^3 - 10x$
3	9	27 − 30 = −3	x^2 is still bigger
4	16	64 − 40 = 24	x^2 is now smaller

The solution lies between $x = 3$ and $x = 4$.
Try the mid-value $x = 3.5$ as the first decimal value of x.

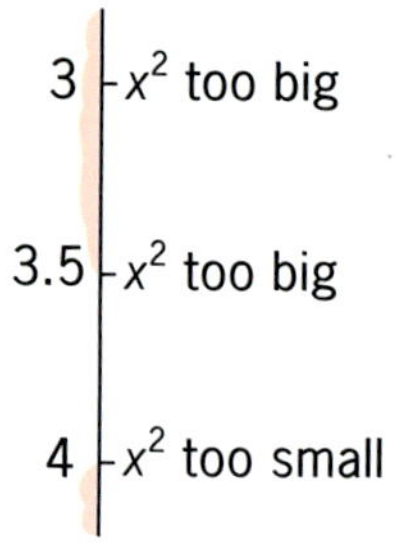

x	x^2	$x^3 - 10x$	Comment
3.5	12.25	42.875 − 35 = 7.875	x^2 is bigger than $x^3 - 10x$
3.6	12.96	46.656 − 36 = 10.656	x^2 is still bigger
3.7	13.69	50.653 − 37 = 13.653	x^2 is only just bigger
3.8	14.44	54.872 − 38 = 16.872	x^2 is now smaller

The solution lies between $x = 3.7$ and $x = 3.8$.
Try the mid-value $x = 3.75$.

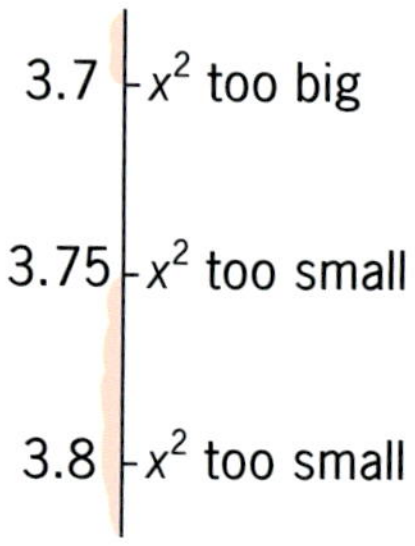

x	x^2	$x^3 - 10x$	Comment
3.75	14.06…	52.7… − 37.5 = 15.2…	x^2 is smaller than $x^3 - 10x$

From the diagram, the solution is between 3.7 and 3.75.
It is closer to 3.7 than to 3.8.
So, to one decimal place, the solution is $x = 3.7$.

Exercise C19.3

MEDIUM

1 a When the integer n, its square n^2, and its cube n^3 are all added together, the total is 584. Find the value of n.

b If the total of n, n^2 and n^3 is 600, find the value of n correct to one decimal place.

2 This patio is 4 metres longer than it is wide. Its area is 70 m^2 and its width is x metres.
Write down an equation in x for its area.
Use a trial and improvement method to find the value of x to 1 decimal place. Take $x = 5$ as your first approximation.

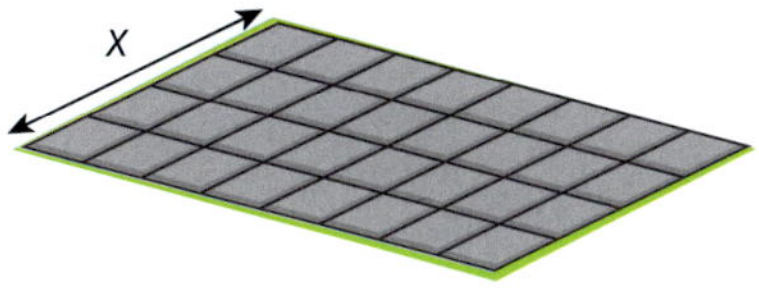

3 Wa Chai has a garden 15 metres by $2x$ metres.
He wants a square pond of side x metres in the middle of the garden.
He has 90 m^2 of turf to cover the rest of the garden to make a lawn.
Write an equation for x and solve it using a trial and improvement method.
Take $x = 4$ as your first approximation.

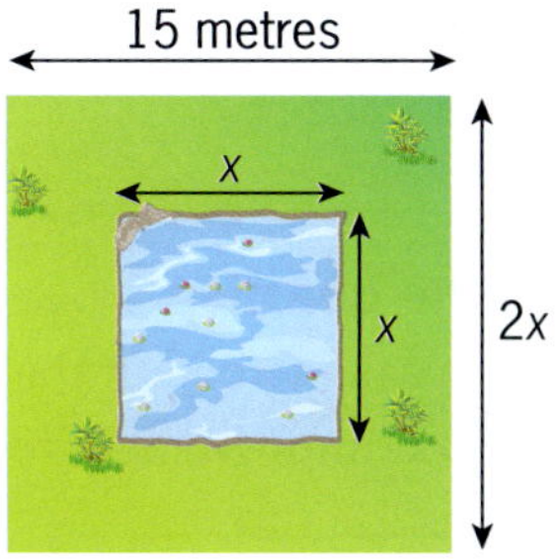

4 The difference between the cube of a certain number, n, and the square of this number is equal to the number itself. Write down an equation for this number and find its value to one decimal place.

5 A plant pot in the shape of a cuboid has a square base of side x cm and four vertical rectangular sides of height 10 cm.
Find the value of x if the total area of these five shapes is

a 500 cm^2 b 400 cm^2.

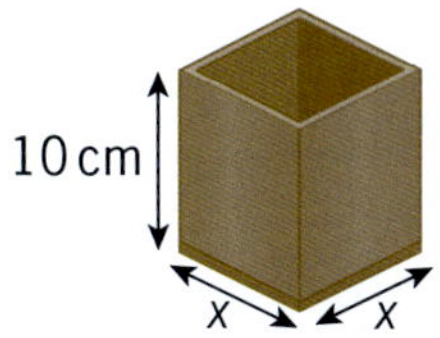

6 Construct your own table of values to solve these equations, correct to 1 decimal place.

a $x^2 = x^3 - 20x$ (try $x = 2$)
b $x^3 = x^2 + 5x + 1$ (try $x = 1$)
c $x^2 = x + 18$ (try $x = 2$)
d $x^3 = x + 200$ (try $x = 3$)
e $x^2 = \sqrt{x} + 18$ (try $x = 3$)
f $x^3 = x - 900$ (try $x = -10$)

7 ***RICH TASK***

n is an odd number. The next odd number after n is $n + 2$.

a Write down the next odd number after $n + 2$.

b If these three odd numbers are multiplied together to give 50 505, find their values. You need a calculator or computer spreadsheet to help you.

Summary and assessment

CHECK OUT

You should now be able to

- use trial and improvement to solve equations
- expand and simplify algebraic expressions multiplying a single term over a bracket.

Exam-style question

You have a calculator, but the calculator cannot work out square roots.
Use trial and improvement to find $\sqrt{14}$ correct to 2 decimal places.

Martha's answer:

$3 \times 3 = 9$
$4 \times 4 = 16$
So try values between 3 and 4
$3{\cdot}5^2 = 12{\cdot}25$ too low
$3{\cdot}6^2 = 12{\cdot}96$ too low
$3{\cdot}7^2 = 13{\cdot}69$ too low, but close
$3{\cdot}75^2 = 14{\cdot}0625$ too high, but very close
$3{\cdot}74^2 = 13{\cdot}9876$ too low, but very close
$3{\cdot}745^2 = 14{\cdot}025025$
Answers below and above 14 came from $3{\cdot}74^2$ and $3{\cdot}745^2$ so the answer lies between 3·74 and 3·745
Rounded to 2d.p. answer is 3·74

Decide on a strategy and explain your thinking.

Show all your working clearly.

Justify your answer.

Give your final answer in the form asked for.

Exam practice

1 Multiply out and simplify.

a $x(4x + 7)$

b $2(3x - 7) + 3(x + 5)$ *(3 marks)*

2 Use trial and improvement to solve this equation

$$x^3 - x = 17$$

Give your answer to one decimal place.
Show all your trials and their outcomes. *(4 marks)*

3 Team Quizzers took part in a quiz.
In round one they scored x points.
Their round two score, also x, was cubed and added to their score from round one.
The total was almost 465 points.
If points are awarded to one decimal place, find how many points were scored in round one. *(4 marks)*

20 3-D shapes

INTRODUCTION

In computer games, 3-D characters and objects are made of thousands of triangles. The triangles are then given different colours to create the illusion of three dimensions. Computer programmers can write a piece of computer code to represent a triangle and then repeat this over and over again to represent much more complex shapes.

What's the point?

Triangles are the most basic polygon. By using triangles in areas as diverse as computer graphics and architecture, we can create sophisticated shapes that we would otherwise not be able to create.

CHECK IN

1 Draw these circles

a radius = 6 cm   b diameter = 9 cm

2 Calculate the area of these shapes. State the units of your answer.

a

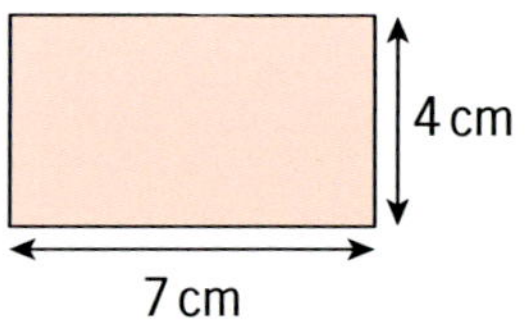

b

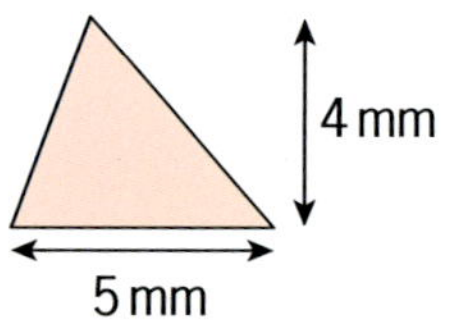

Orientation

What I should know	What I will learn	What this leads to
C18	Draw plans and elevations and nets Calculate surface area	C23

RICH TASK

Here are the front and side views of a 3-D shape made from cubes.

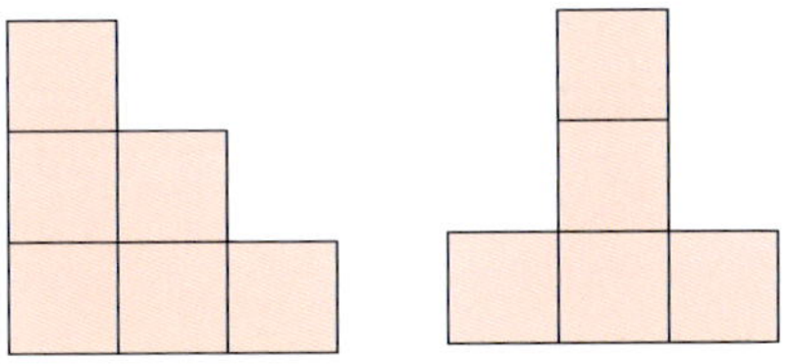

Left Front

Draw the 3-D shape on isometric dotty paper.
Are you sure – are there any other alternatives?
Investigate.

Three dimensions

C20.1

- Study the properties of 3-D objects

You can use **isometric paper** to draw 3-D shapes.
Here is a cuboid on isometric paper.

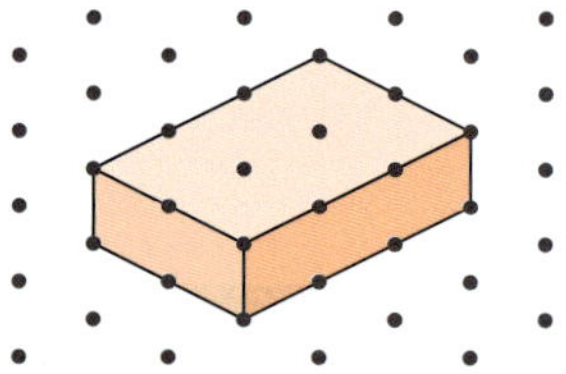

Take care to position the paper correctly.
The lines of dots run **down** the page.

RICH TASK

For the pyramid, **cross-sections** A and B are squares of different sizes.

For the cylinder, cross-sections C and D are circles of the same size.

What shape is cross-section E?

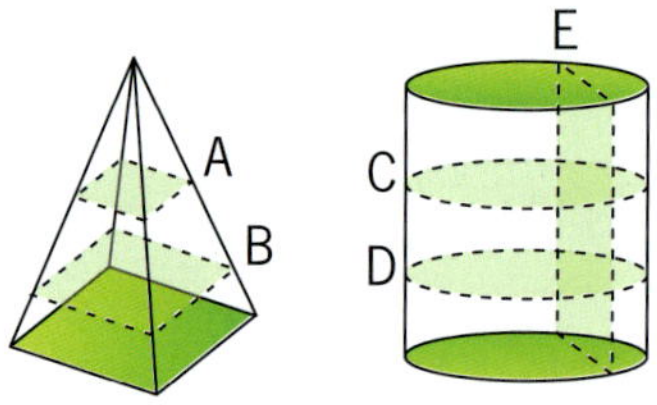

The shape of a cross-section depends on where you imagine the cut through a 3-D object.

EXAMPLE

How many faces, edges, and vertices has this wedge?
Describe the cross-section AA′ shown by the dotted lines.

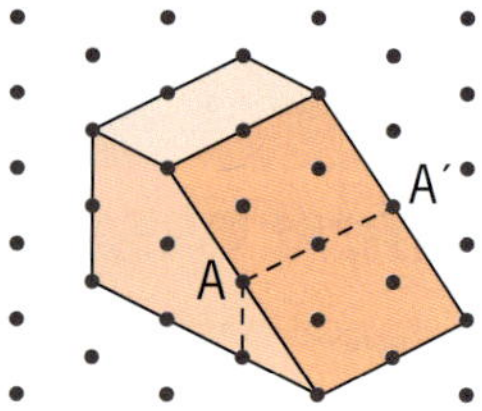

ANSWER

The wedge has 6 faces, 12 edges and 8 vertices.
From the diagram, you can see that the cross-section AA′ is a rectangle.

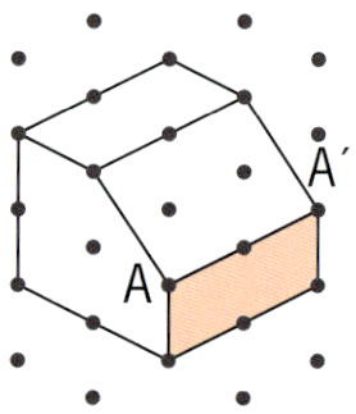

Check these answers by counting them mentally.

Exercise C20.1

LOW

1 a Draw these shapes on isometric paper.
How many faces and edges does each object have?

b Draw some objects of your own, giving the number of their faces and edges.

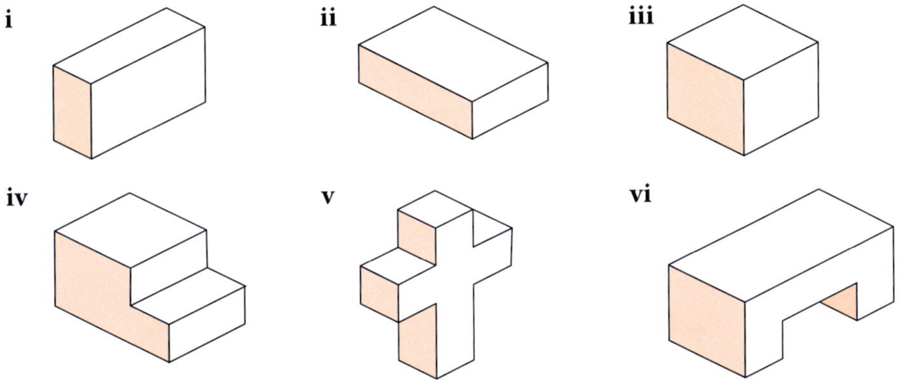

2 The dotted lines indicate cross-sections of these solids.
Draw and name the shape of each lettered cross-section.

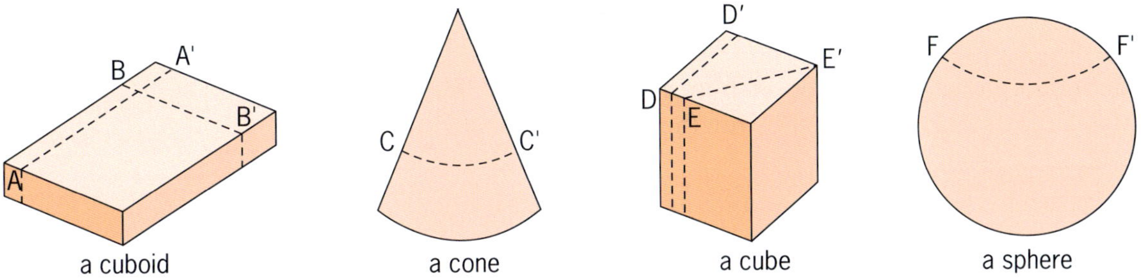

3 These shapes are turned over so that their shaded faces rest on a table.
Use isometric paper to draw the shapes in their new positions.

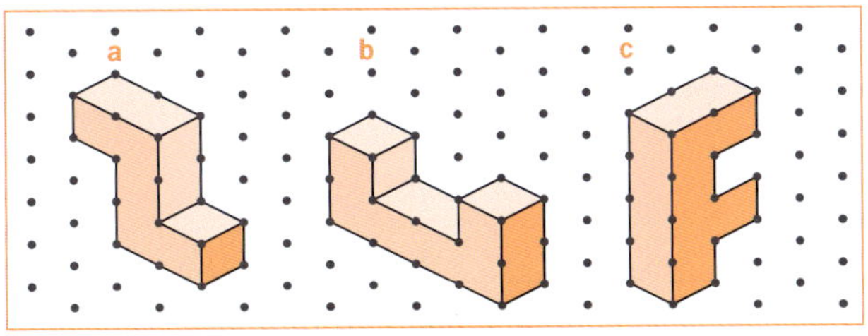

4 ***RICH TASK***

A 3-by-3-by-3 cube is made from small cubes.
All the faces of the large cube are painted.
Explore how many of the small cubes have painted faces.
Repeat with larger cubes of different sizes.

Unit C

Plans and elevations

C20.2

- Understand how plans and elevations can represent 3-D objects

Architects and builders use plans and elevations to represent buildings.

When you look at an object from above, you get a bird's-eye view. This view from above is called the **plan**.

When viewed from the side, what you see is called the **side elevation**.

When viewed from the front, what you see is called the **front elevation**.

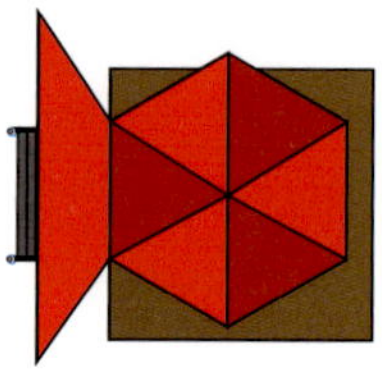

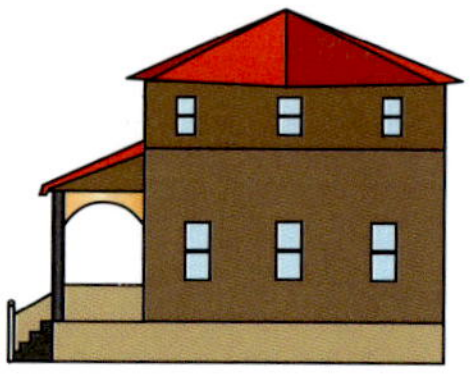

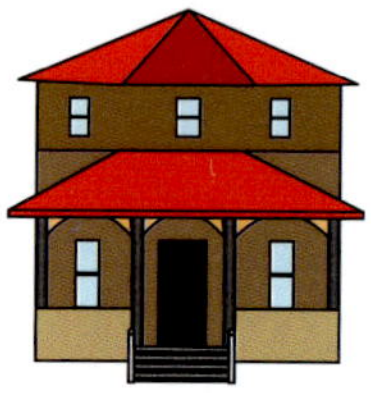

Exercise C20.2

MEDIUM

1 This house and church are viewed from above, front and side.
Match the directions A, B and C with the views 1, 2 and 3.

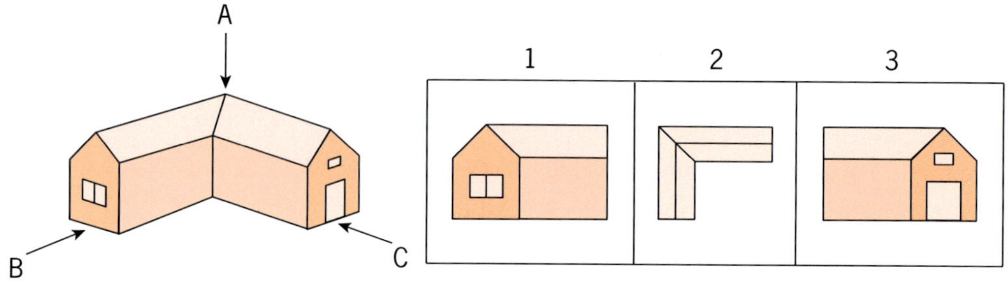

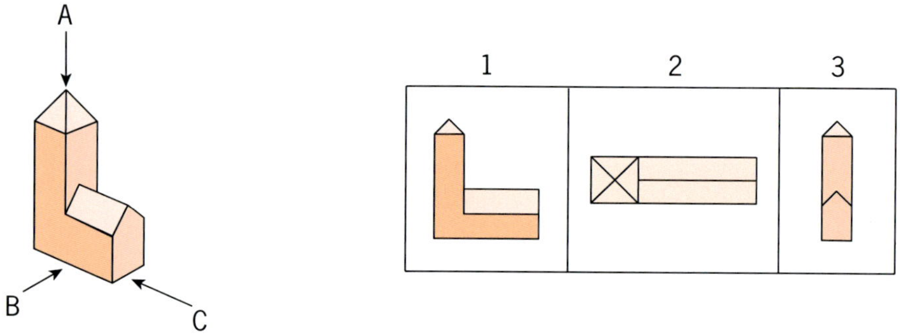

2 For each of these objects, draw a plan view and an elevation in the direction of the arrow.

a

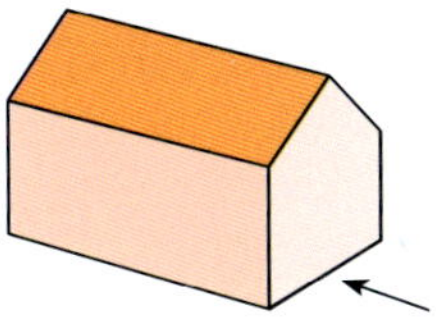

b

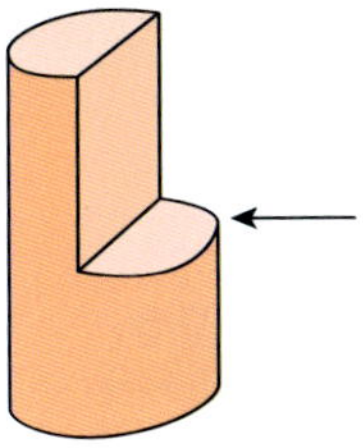

c

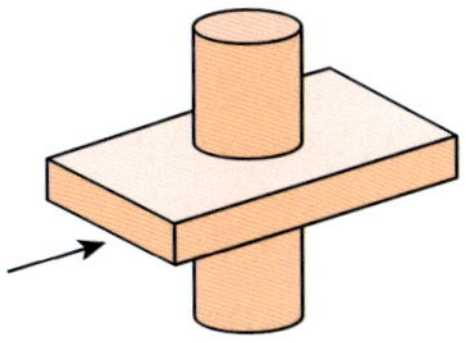

d

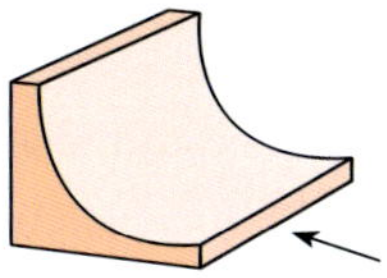

3 Match the solids in the box with the plans and elevations a to d and describe the solids represented by e and f.

cone | **cylinder** | **sphere** | **pyramid**

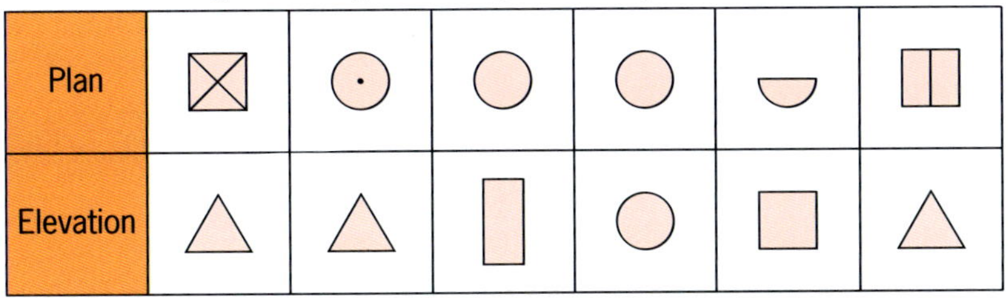

4 Two of these solids have the same side elevation in the direction of the arrow. Which solid is the odd-one-out? Give your reason.

a

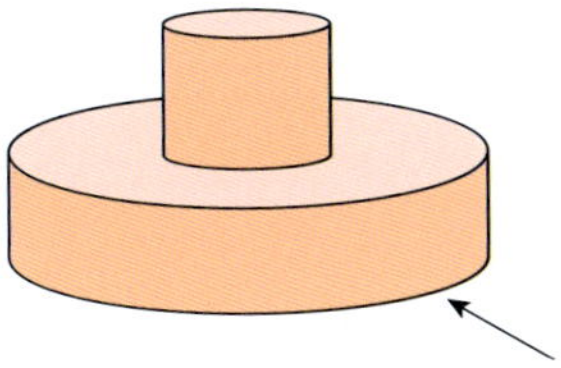

b

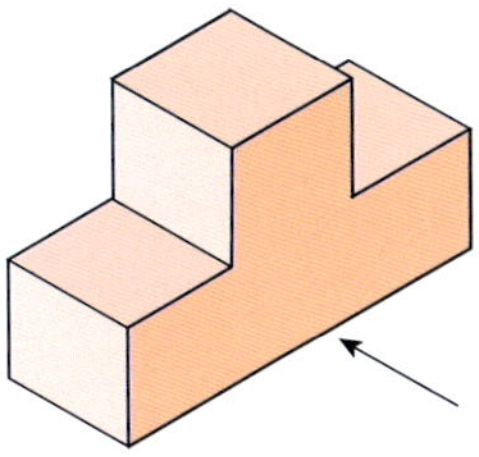

c

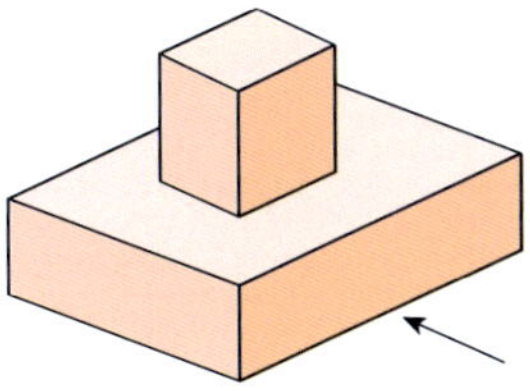

5 In each case, sketch a solid which has the given plan and front elevation.

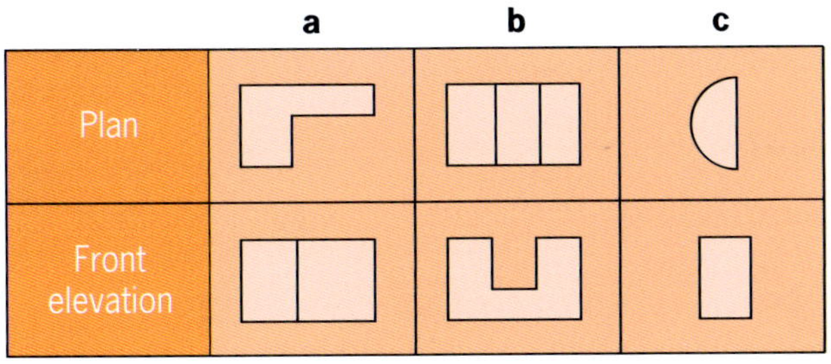

6 Imagine a solid object. Draw its plan and elevations and see if your partner can draw your object from them.

*7 **DISCUSSION**

Can you find a solid object which has plan, side elevation and front elevation all the same?

Unit C

Nets

C20.3

- Understand, use and construct nets of 3-D shapes

RICH TASK

Find as many different nets as you can for a cube.
Share the task with a partner.

→ **A net is a 2-D shape which can be folded to make a hollow 3-D object.**

For example, the box which is used to pack cornflakes can be opened out and flattened. The flat shape that you get is the **net** of the box.

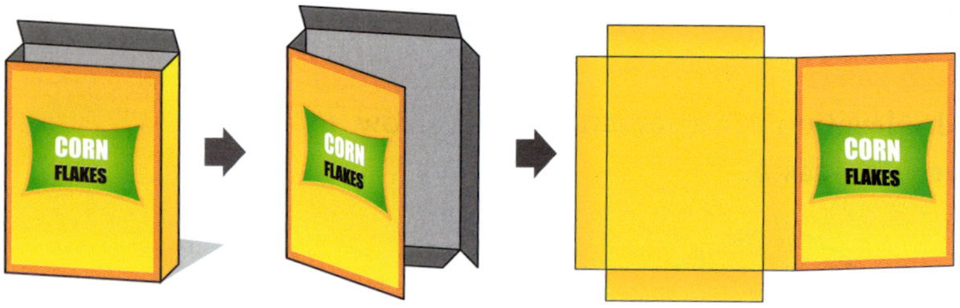

Take the box   Open the seams   Flatten it out to give the net

Exercise C20.3

LOW

1 These three nets are used to make boxes for chocolate bars.
   a Which two of the boxes are the same?
   b What is the name of the shape of this box?
   c What is the name of the shape of the third box?

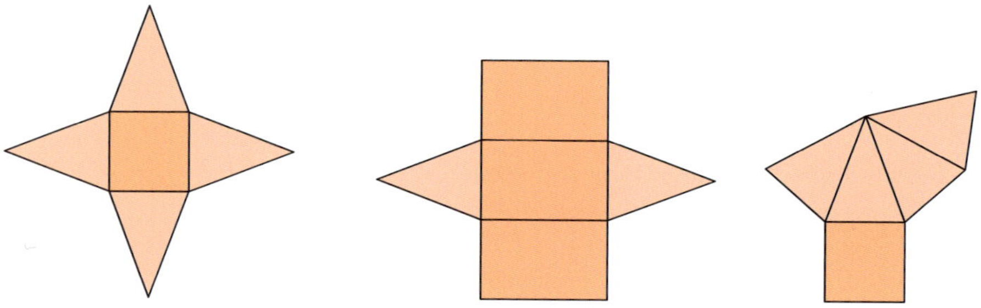

2 Which of these three nets is *not* the net of a cuboid?
Give a reason for your choice.

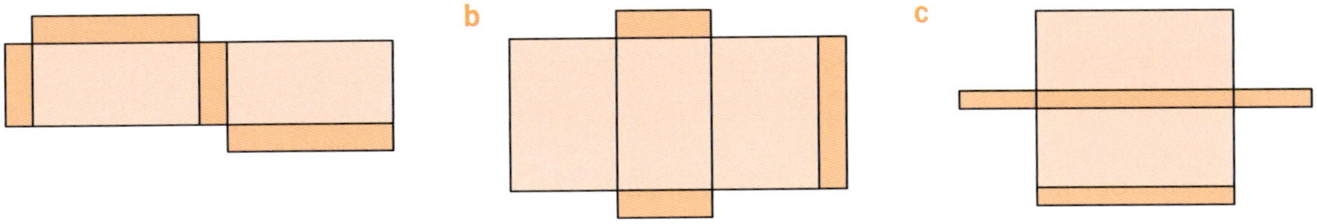

3 These five faces of a net of a 3-D shape have all been separated from each other.
The dimensions are all in cm.

a Arrange them (on square grid paper) to make the net. What is the name of the solid?

b Arrange them another way to make the same solid.

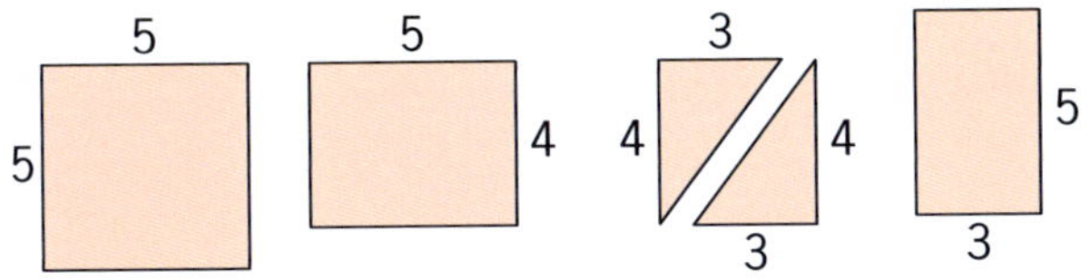

4 Two cubes are being made from these nets.
For each net, say which face will be opposite

a face p   b face q   c face r.

d If the cubes are used as dice and the dots on opposite faces must add up to 7, decide on which faces you could place the dots.

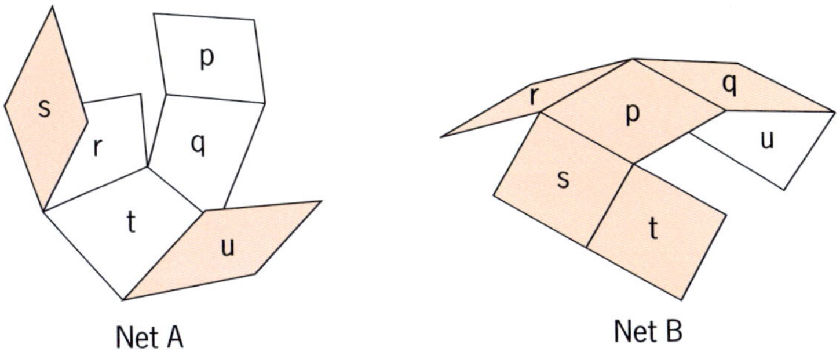

5 Make a net from four identical equilateral triangles.
What 3-D object does it make? Draw it.

6 **RICH TASK**

This cube is cut in half as shown.

a Draw a net for one of the halves.

b Cut a cube into halves in different ways.
For each way, draw a net for one of the halves.

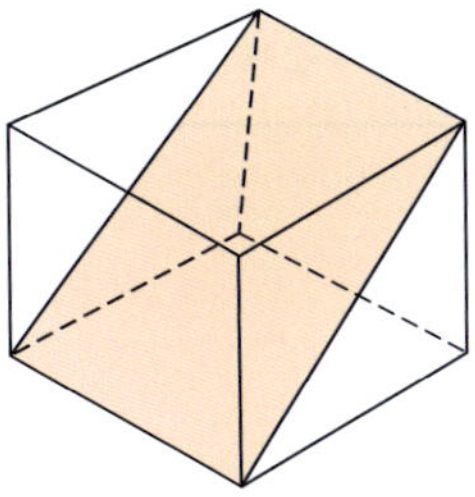

Unit C

Surface area

- Calculate and use the surface area of simple 3-D shapes

The surface area of a 3-D object is the total area of all its faces.

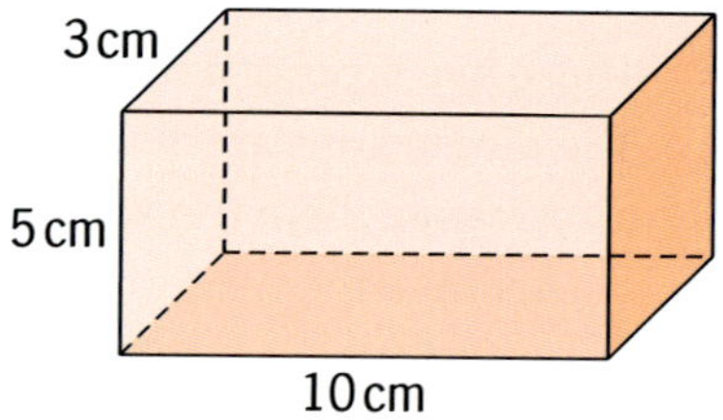

For example, a cuboid has six faces.
The area of

- the top and bottom faces are each $10 \times 3 = 30\,\text{cm}^2$
- the front and back faces are each $10 \times 5 = 50\,\text{cm}^2$
- the two end faces each are each $3 \times 5 = 15\,\text{cm}^2$.

So the surface area of this cuboid is
$30 + 30 + 50 + 50 + 15 + 15 = 190\,\text{cm}^2$

EXAMPLE

This net is used to make a bread bin.
Find the surface area of the bread bin.

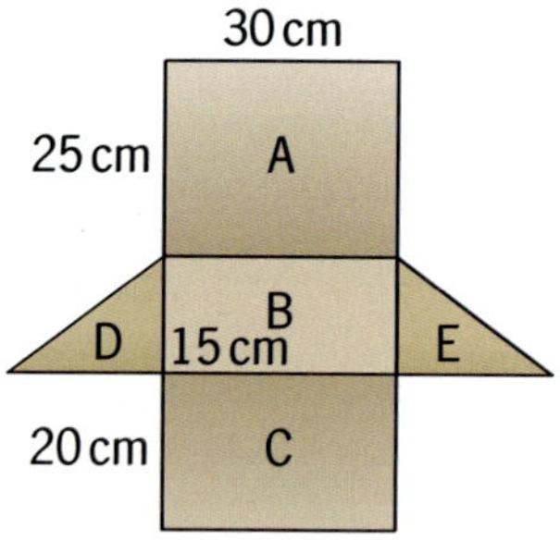

ANSWER

The surface area of the bin is the same as the area of the net.
Here are the areas of the faces.

Face	A	B	C	D	E
Size, cm by cm	30×25	30×15	30×20	$\frac{1}{2} \times 20 \times 15$	$\frac{1}{2} \times 20 \times 15$
Area, cm²	750	450	600	150	150

The surface of the bread bin is
$750 + 450 + 600 + 2 \times 150 = 2100\,\text{cm}^2$.

Exercise C20.4

MEDIUM

1 Find the surface areas of these cuboids.

a
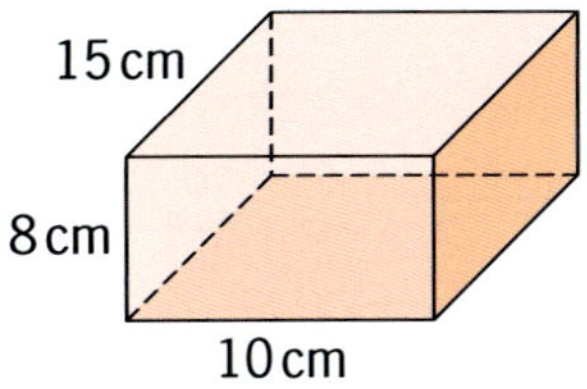

b
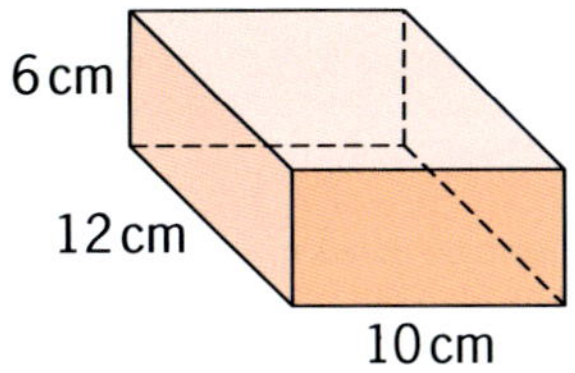

c
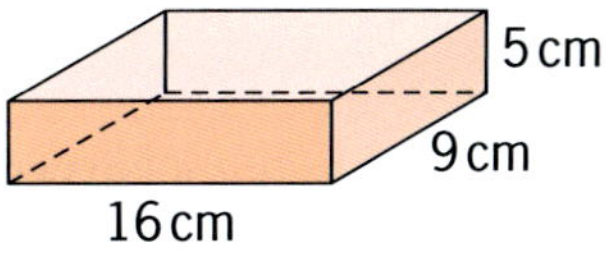

2 Here are the nets of a cuboid and a square-based pyramid.

a
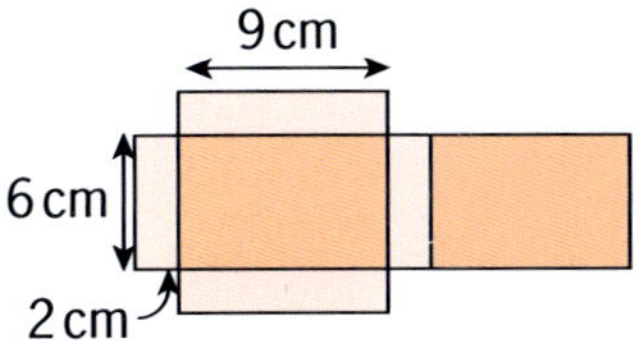

b
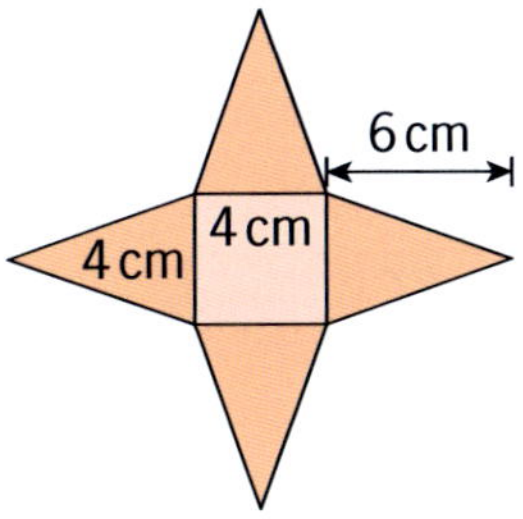

Find the surface area of both objects.

3 A gardener makes a frame for his seedlings from sheets of wood. The frame has four sides and is open at the top and the bottom.

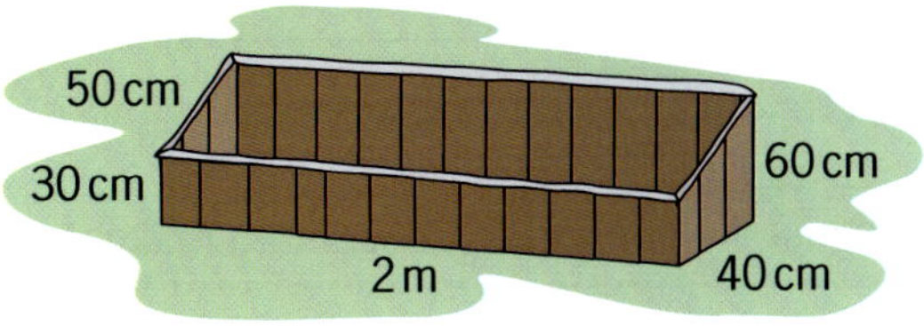

   a Find the area of the wood that he needs.
   b If the wood costs £12.45 per square metre, how much will the wood that he needs cost?
   c He makes a glass top for the frame from glass costing £6.50 per m^2. How much will the glass cost?

4 Draw a net for these two prisms and find their surface areas. The dimensions are given in centimetres.

a
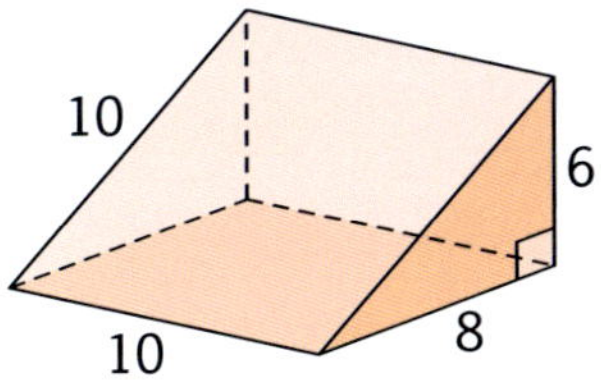

b
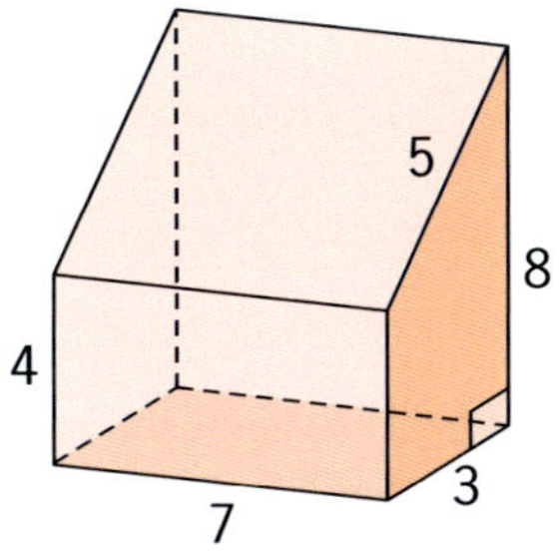

5 An open box (without a lid) is 15 cm long, 10 cm wide and 5 cm high. It is made from a net which is cut in one piece from a larger rectangular sheet of cardboard.
   a Find the area of the net and the dimensions of the smallest possible rectangular sheet.
   b What percentage of the sheet is wasted in making the net?

Unit C

Surface area of cylinders

- Calculate and use the surface area of cylinders

A **hollow cylinder** has a curved surface.
The curved surface can be opened out flat to make a rectangle.
The circumference, C, of the curved surface becomes the length of the rectangle.

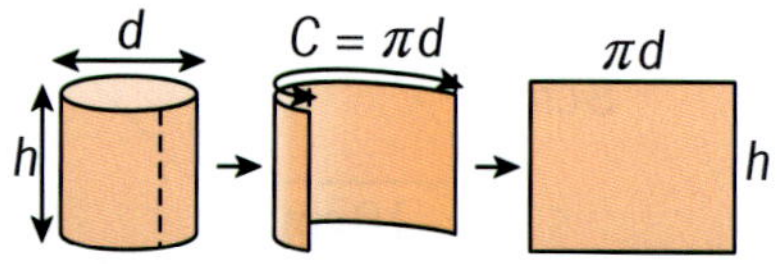

So the area of the curved surface $= \pi d \times h$
$= \pi dh$

Because the diameter is twice the radius, πdh can be written $2\pi rh$

A **solid cylinder** has three surfaces.
It has a curved surface and also two circular ends.
Each circular end has a radius r and an area πr^2.

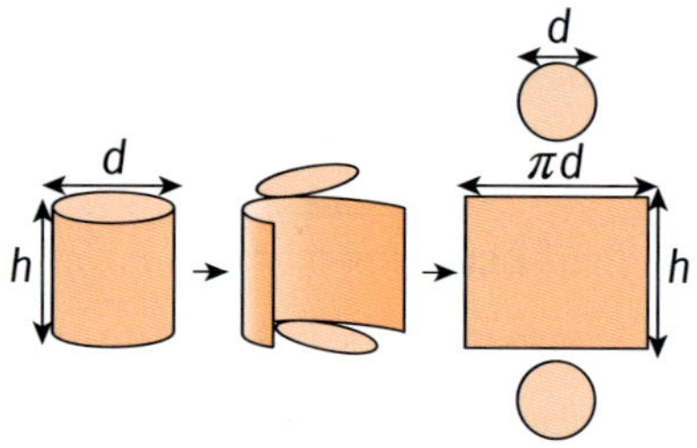

So the total surface area $= \pi dh + \pi r^2 + \pi r^2$
$= \pi dh + 2\pi r^2$

Because $d = 2r$, this formula can be written $A = 2\pi rh + 2\pi r^2$ or $= 2\pi r\,(h + r)$

EXAMPLE

A tin of chopped tomatoes has a radius of 4 cm and a height of 12 cm.
What area of metal is needed to make the tin?

ANSWER

The diameter of the end of the tin $= 2 \times 4 = 8$ cm
The curved surface area $= \pi dh = 3.14 \times 8 \times 12 = 301.44\ \text{cm}^2$
The area of one end $= \pi r^2 = 3.14 \times 4^2 = 50.24\ \text{cm}^2$
The total surface area $= 301.44 + 50.24 + 50.24$
$= 401.92 = 402\ \text{cm}^2$ to the nearest cm^2.

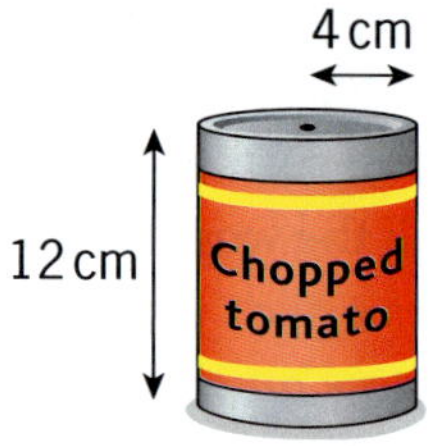

You could measure some tins of groceries and calculate their surface area.

Exercise C20.5

MEDIUM

1 Find the surface area of these tubes and pipes.

a
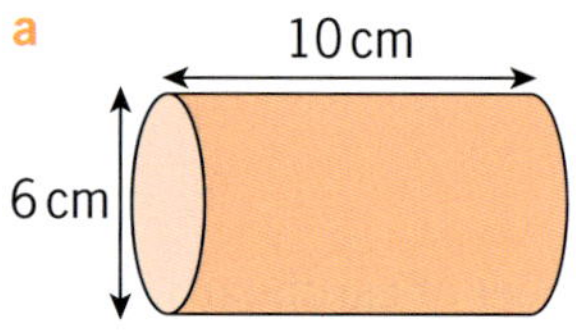

b
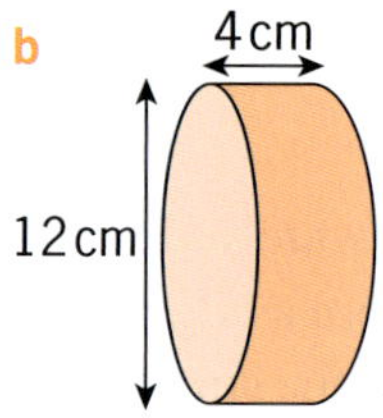

c
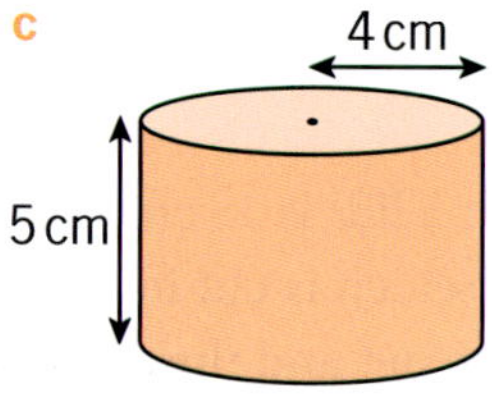

d
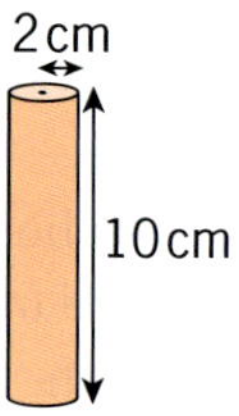

2 A hosepipe is 5 metres long with a diameter of 2 cm.
Calculate its surface area.

3 Find the total surface area of these solid cylinders.

a
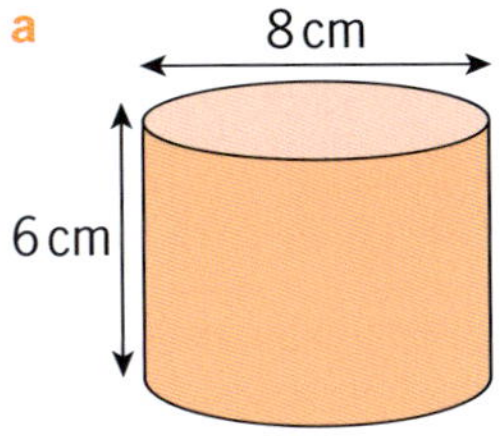

b
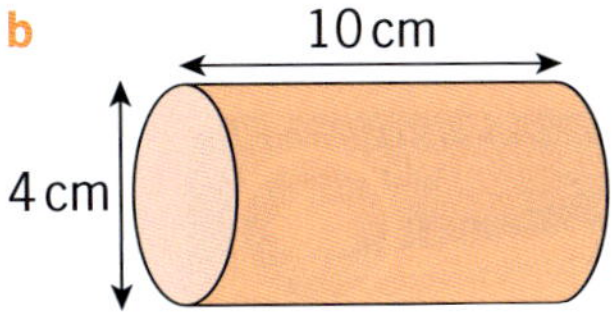

c
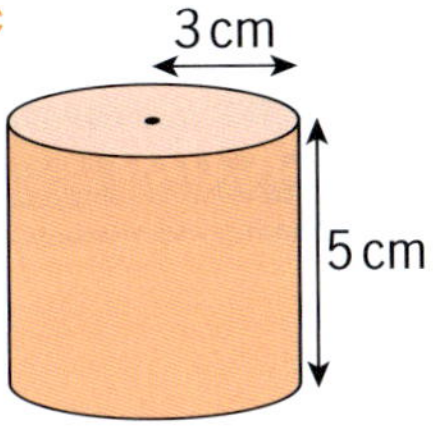

d
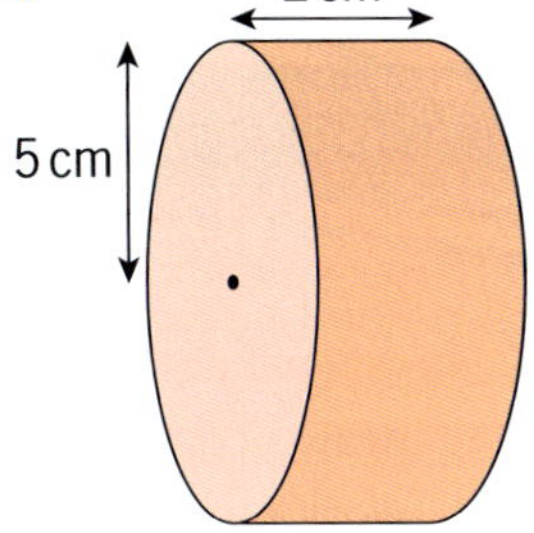

Unit C

4 Find the total surface area of a tin of rice pudding with

a a diameter 10 cm and height 5 cm

b a radius 6 cm and height 10 cm.

5 This diagram shows the flywheel of an engine.
Find its total surface area.

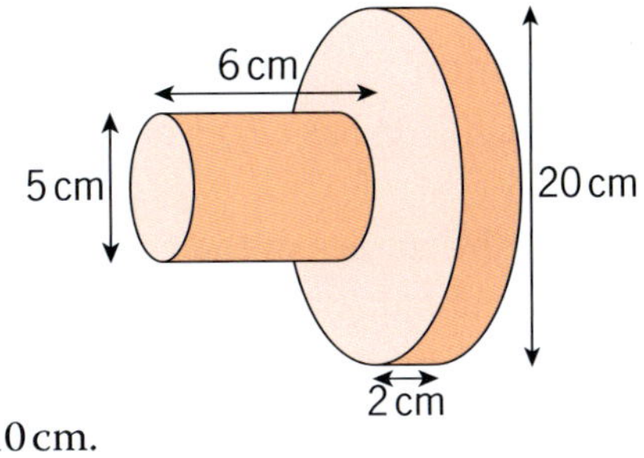

6 A roll of paper has a radius of 30 cm and a height of 10 cm.
As the paper is used, the radius reduces.
When the radius is halved to 15 cm, by what percentage has the curved surface area of the roll been reduced from its original value?

7 Two identical solid cylindrical wooden blocks are sawn in half as shown.
Which of the halves has the smaller surface area?

a
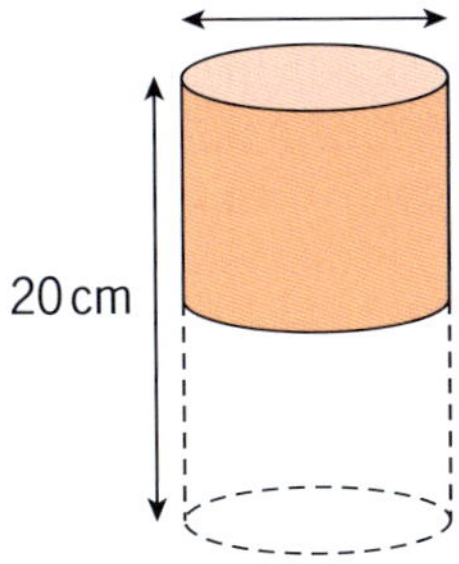

b
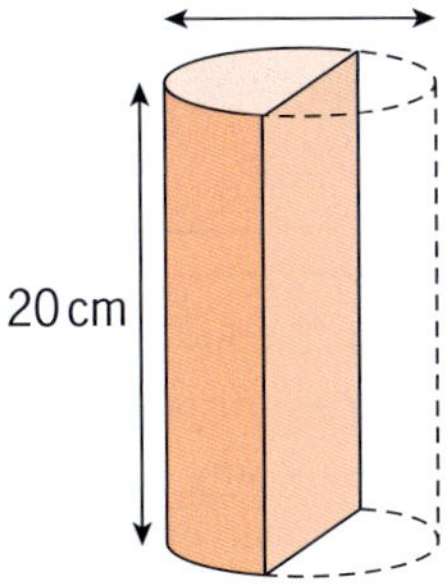

8 **RICH TASK**

The curved surface and two ends of a cylindrical tin (of diameter 10 cm and height 15 cm) are all cut out of the same flat metal sheet.

What size of sheet gives the most economical way of doing this?

Summary and assessment

C20

CHECK OUT

You should now be able to

- draw nets of 3-D objects
- work out and solve problems involving surface area of simple shapes made from rectangles and triangles.

Exam-style question

The cuboid ABCDEFGH has edges of length 6 cm, 3 cm and 2 cm

A fly walks along the surface of the cuboid.

On the net of the cuboid, mark letters to show the positions of the corners and show the shortest route the fly can take to get from A to G.

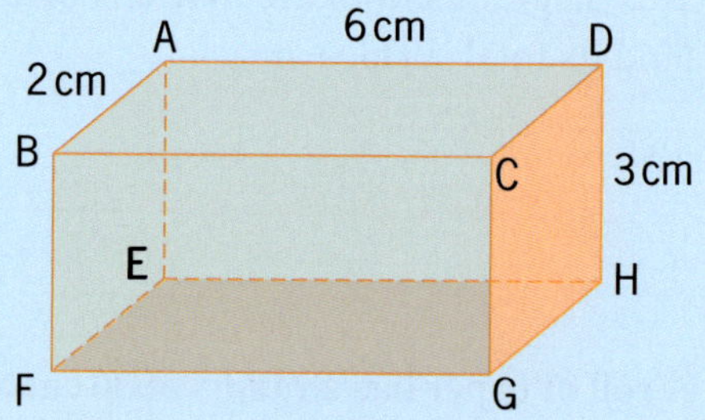

Sian's answer:

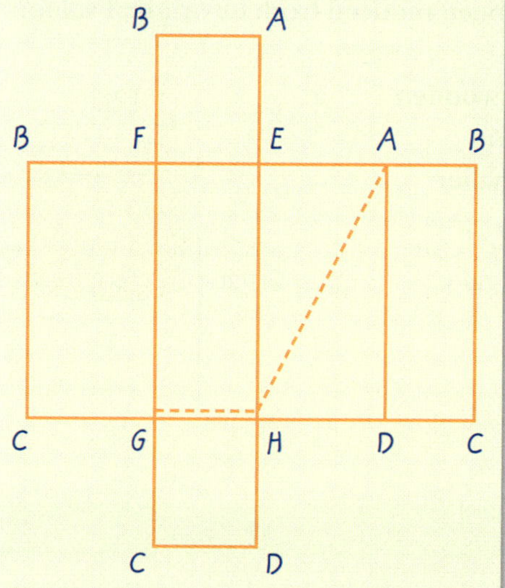

Use the dimensions given to decide where the letters A to H should go.

Look for the shortest distance between any points marked A and G.

Show the position of the letters at all corners of the net.

Draw a dotted line to show where the fly walks.

Exam practice

1 This cuboid has length 4 cm, height 2 cm and width 1 cm.

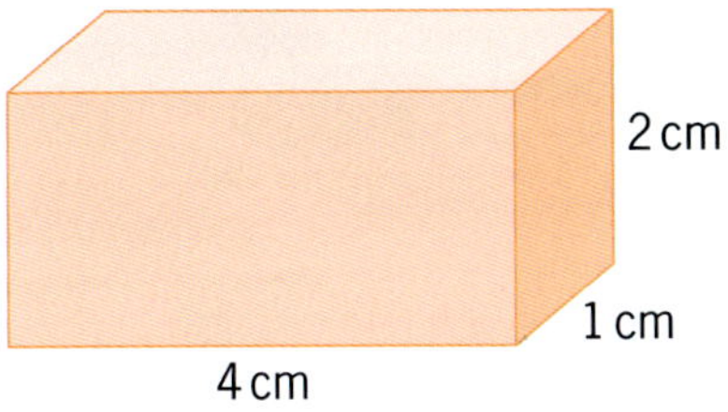

On centimetre squared paper draw a net of the cuboid. (*3 marks*)

2 This is a net for a triangular prism.

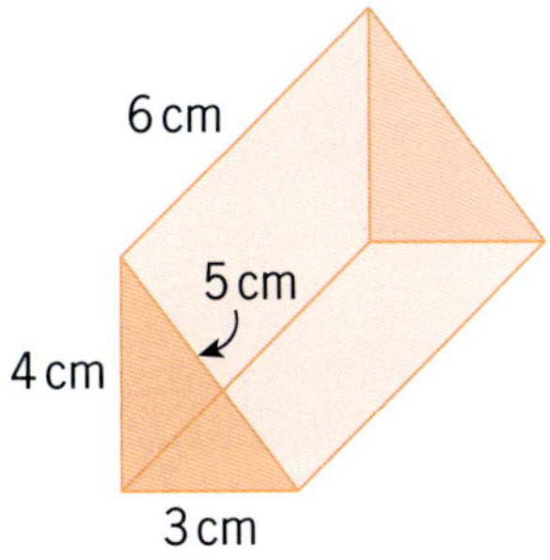

Work out the surface area of the triangular prism. (*4 marks*)

3 Two identical cuboid boxes are joined along one side.
Each box has length 5 cm, width 4 cm and height 2 cm.

These are the three different ways the boxes can be joined.

Decide which box has the smallest surface area and explain your choice.
Work out the surface area of the box with the smallest surface area. (*4 marks*)

21 Graphs

INTRODUCTION

When a skydiver jumps out of an aircraft, she immediately accelerates towards the earth under the force of gravity. She will continue to accelerate until she reaches terminal velocity (speed) at about 55 m/s. When the skydiver pulls her ripcord she will decelerate before reaching a new constant speed.

What's the point?

The skydiver won't be drawing graphs or using a calculator before jumping – but she will need to make a mental calculation based on a formula in order to calculate the height at which to jump, and for how long she can freefall safely.

CHECK IN

1 Work out the value of each expression when $x = 4$.

a $x + 5$ **b** $x - 6$ **c** $3x$ **d** $\frac{x}{2}$

2 Plot these coordinates on a copy of this grid.

a (3, 4)
b (1, 3)
c (−1, 2)
d (4, −5)
e (−3, −4)
f (−3, 2)

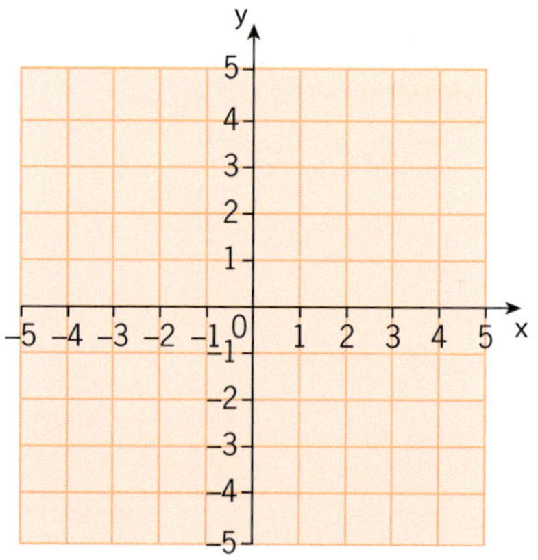

Orientation

What I should know	What I will learn	What this leads to
B14	■ Draw distance-time graphs ■ Interpret graphs ■ Plot graphs of quadratic functions ■ Use coordinates of points in three dimensions	Courses and careers in business and finance

RICH TASK

Here is a graph of a journey.

Write the story of the journey.

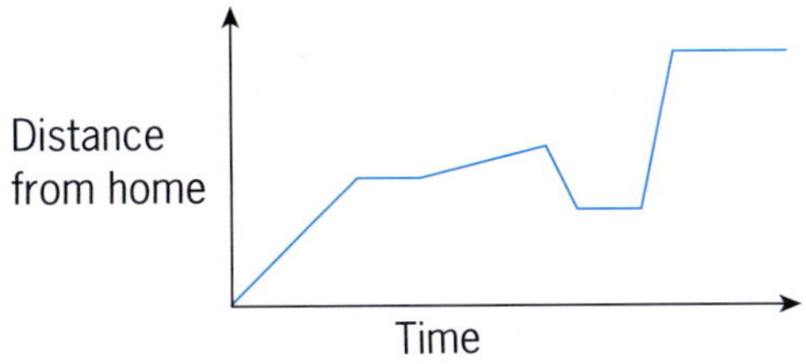

Distance–time graphs

- Interpret distance–time graphs

Journeys can be described using a graph.

ACTIVITY

Jia and Sara Cheng drive the 300 miles from London to Newcastle at a steady speed on the motorway.

The line OA represents their journey.

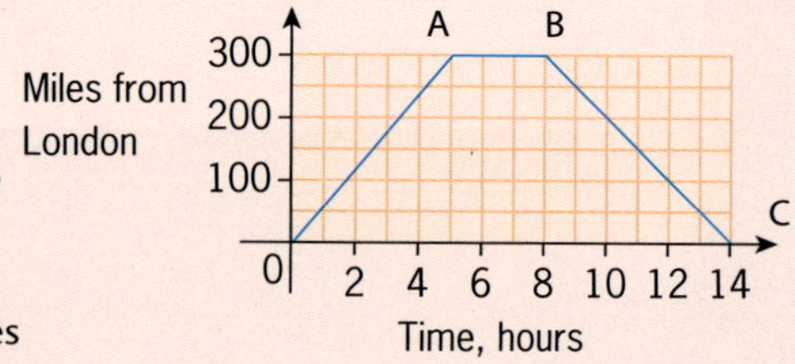

a How does the line tell you that they travelled at a **steady** speed?

b How long did the journey take them and what was their speed?

To find their speed, the graph shows that they travelled 300 miles in 5 hours; that is, $\frac{300}{5}$ = 60 miles in each hour. So their speed was a steady 60 mph. The speed is the gradient of the straight line OA = $\frac{\text{distance travelled}}{\text{time taken}}$

c How do you know that they stayed in Newcastle for 3 hours?

d Which line represents their return journey to London?

e What is their average speed on the return journey?

mph stands for *miles per hour.*

EXAMPLE

A balloon rises vertically from the ground after release.

a Use this graph to find the speed at which it rises

i during the first 5 minutes

ii during the next 10 minutes.

b Find its average speed over the whole 15 minutes.

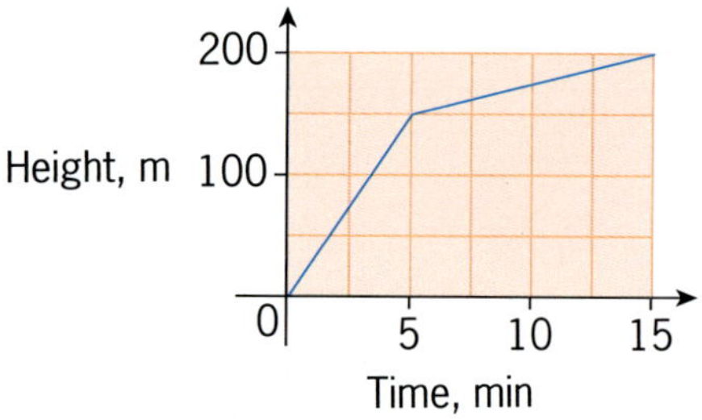

ANSWER

a i In the first 5 minutes, it rises 150 metres.
So it rises a distance of $\frac{150}{5}$ = 30 metres in each minute.
Its speed is 30 metres per minute.

ii In the next 10 minutes, it rises a distance of
200 − 150 = 50 metres.
So its speed is $\frac{50}{10}$ = 5 metres per minute.

b In the whole 15 minutes, it rises 200 metres.
So its overall average speed is $\frac{200}{15}$ = 13.3 metres per minute (to 1 d.p.).

Exercise C21.1

MEDIUM LOW

1 This graph illustrates Liam's walk with his dog.

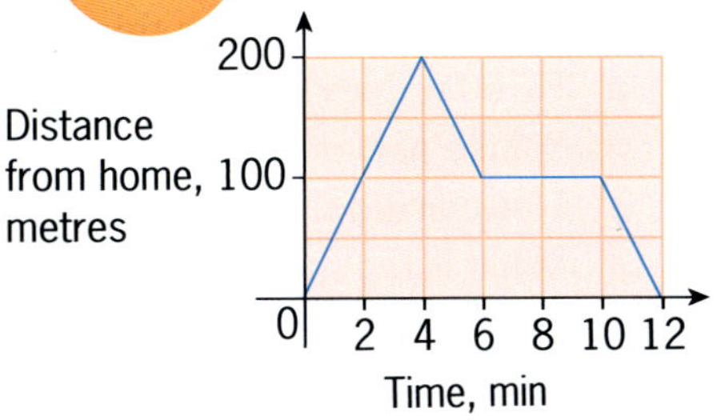

a What is their furthest distance from home?
b How long does it take to get this far?
c For how long does Liam stop on the way back?
d How far from home is he when he stops?
e How long does the whole walk take?

2 Anke is waiting at a bus stop. She first sees her bus 700 metres away. This graph describes its approach.

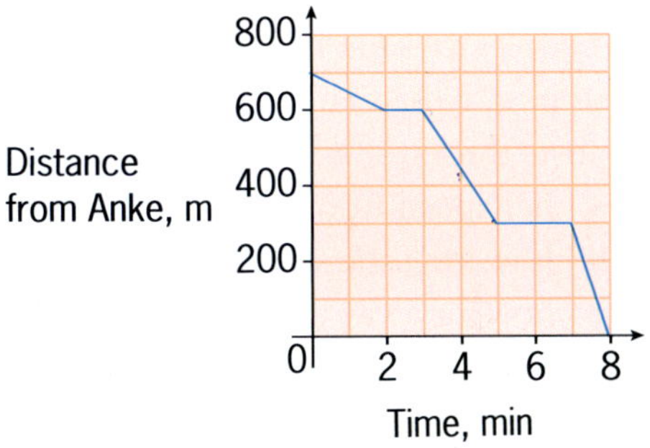

a How many times does it stop before it reaches her?
b How far apart are the first and second stops?
c How long it is at rest at its second stop?
d From first seeing the bus, how long does it take to reach Anke?

3 These graphs show the distance away from you of three cars, X, Y and Z.

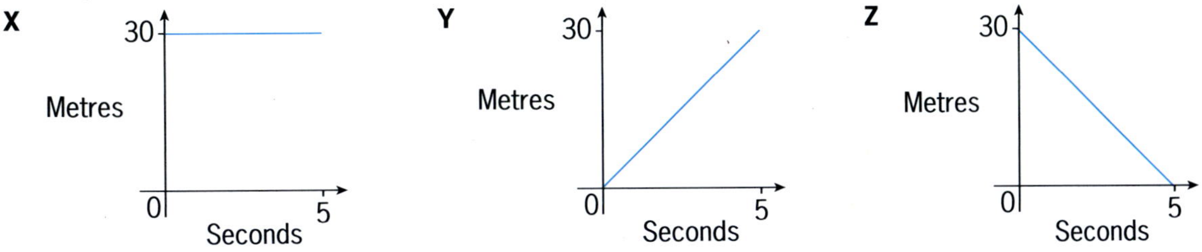

Which of these graphs describes a car which is
a driven away from you at a speed of 6 metres per second
b driven towards you at a speed of 6 metres per second
c at rest 30 metres away from you?

Metres per second can be written as $m\,s^{-1}$.

4 You stand at a point (safely!) where you see many forms of transport, described by these graphs.

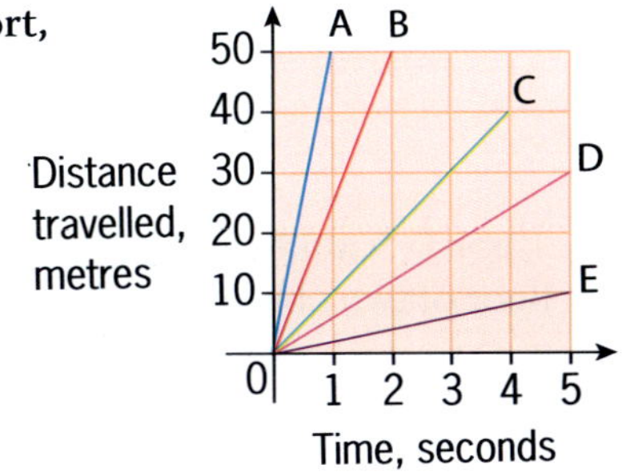

a Match these five graphs A to E to these objects:

an aircraft	a bicycle	a lorry
a pedestrian	a train	

b Calculate their speeds in metres per second.

Continued ➤

5 When Tanya visits her grandma, she walks part of the way and takes a bus part of the way.
Her journey is shown on this graph.

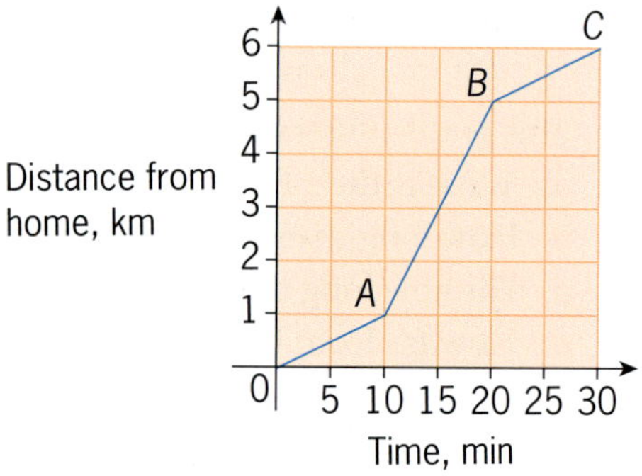

a How far does she walk in the first 10 minutes?
b Which part of the graph represents her bus ride?
c How far did she travel by bus and how long did it take?
d What is the speed of the bus?
e How far away is her grandma's and how long does it take her for the whole journey?
f What is her average speed for the **whole** journey?

6 Xue-Fei uses her bike to post a letter. She leaves home, stops at a road junction on the way to the post office, posts the letter and returns home. Find

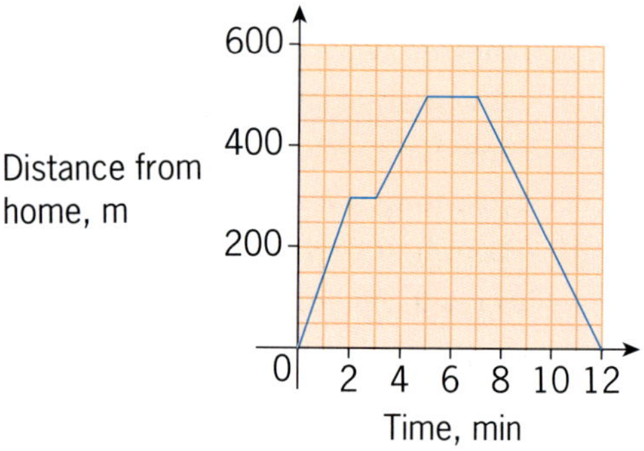

a her speed from home to the road junction
b how long she stopped at the post office
c how far away the post office is from her home
d how long she took to return home
e her speed on the return journey
f her average speed for the entire journey.

7 The line on this graph indicates a steady speed of 30 mph.
Copy the graph and draw two lines on it to indicate a steady speed of
a 15 mph   b 60 mph.

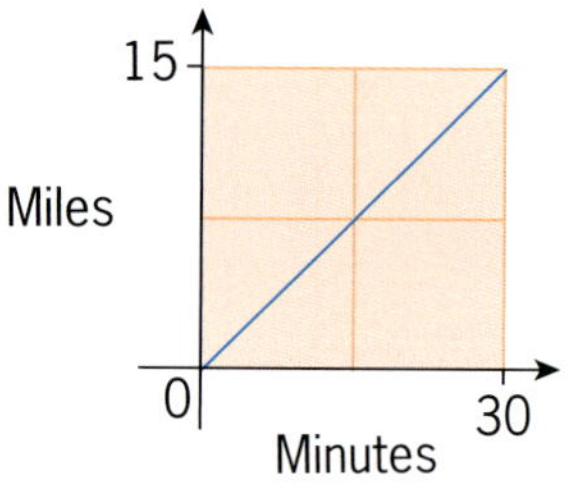

8 David and Stephen cycle on the same road in opposite directions, as shown on this graph.
a Who has a rest on the way? How long is the rest?
b How long is it before they pass each other?
c What is Stephen's speed after he has rested?
d What is Stephen's overall average speed?

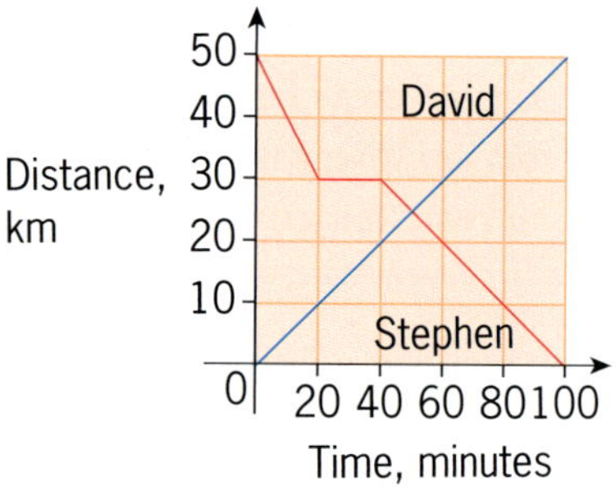

* 9 These two graphs show the heights above the floor of two objects which are dropped onto it. One object is a table-tennis ball and the other is a cricket ball.
a Which object matches which graph?
b Describe what happens to both objects.

X

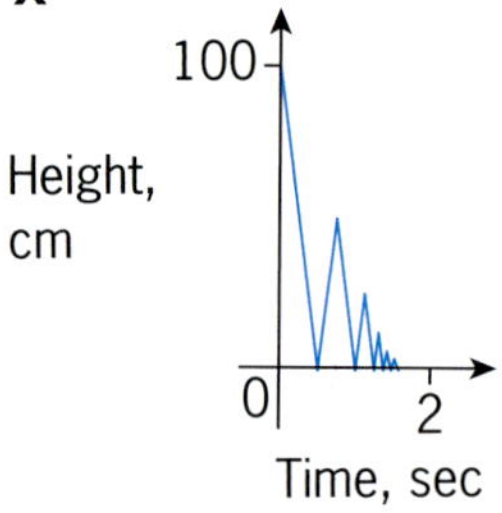

Y

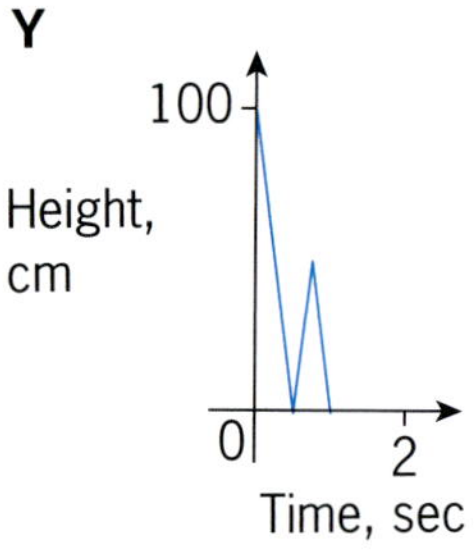

10 An office uses oil for its central heating.
This graphs shows the depth of oil in its oil tank over many weeks.
Describe the changes in the level of oil.
Make some assumptions about what the weather might have been like during this time.

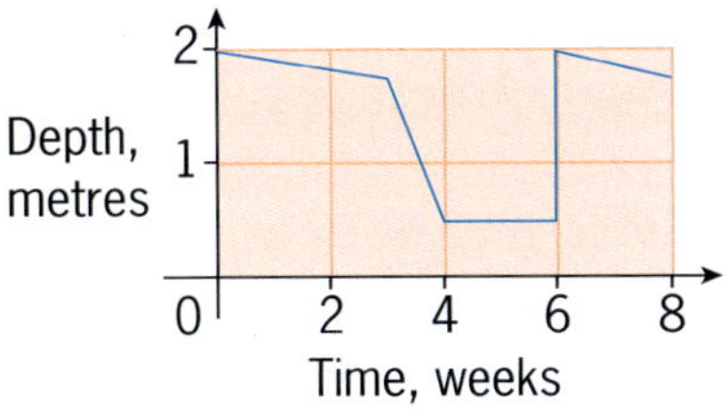

11 This sketch graph shows a jogger's run.
Describe the run, mentioning particularly any changes in direction and speed.

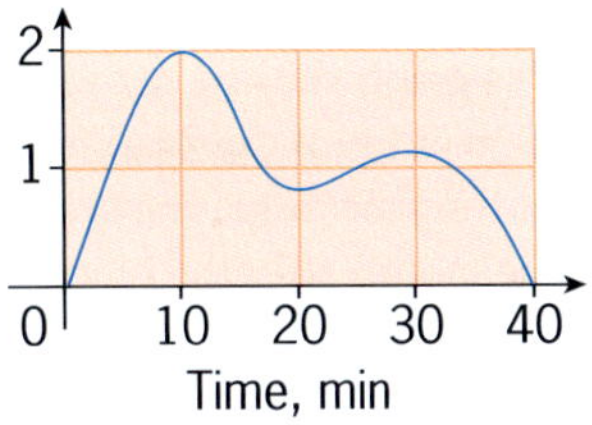

12 An automatic machine runs along a channel so that its distance from one end of the channel is shown by this graph.
Describe the machine's motion during these 10 seconds, giving appropriate detail where possible.

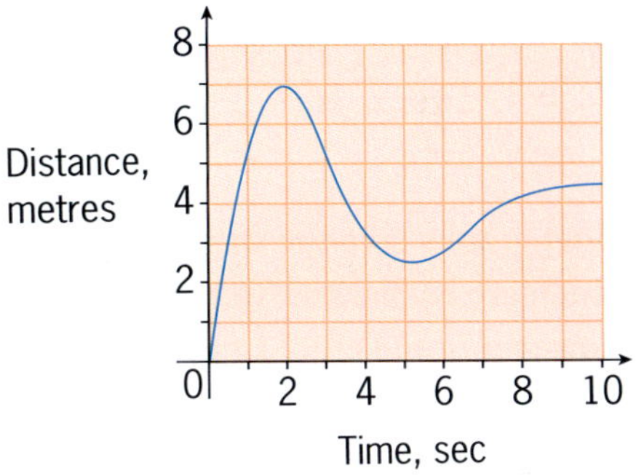

13 This timetable shows details of a coach journey from London to Lincoln.
a Describe the journey in a few sentences.
b Draw a graph of distance against time for the journey, assuming the coach travels at a steady speed between stops.
c Find its speed for the four parts of the journey.
d Find the overall average speed.

London	0 km	*d* 12:00
Scalsby	60 km	*a* 13:00 *d* 13:15
Garthorpe	100 km	*a* 14:15 *d* 14:30
Skeltoft	220 km	*a* 16:30 *d* 17:00
Lincoln	245 km	*a* 17:30

a	arrives
d	departs

14 ***RICH TASK***

Draw a graph for your journey to school in a morning. Show when you are not moving, such as when waiting for a bus.
Give some detail about how far you travel and how long each part of your journey takes.
Can you calculate any speeds for the various parts of your journey?

Unit C

Interpreting graphs

C21.2

- Interpret real-life graphs

*

RICH TASK

This graph shows a 40-year-old's heart rate during 25 minutes of exercise (from 5 minutes to 30 minutes on the time axis). The green line shows the heart rate one month after the orange line was recorded. Discuss the graph with a partner and write what you think is happening.

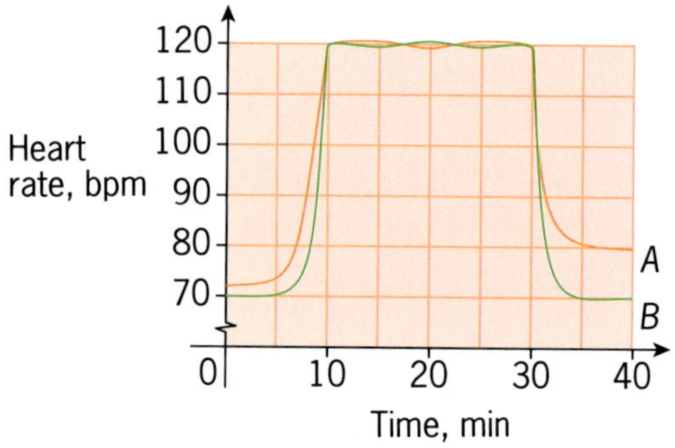

Most graphs take the shape of lines, either straight or curved, which are **continuous**; that is, they have no breaks in them. However, some graphs need to have gaps or jumps because the information they describe is not continuous.

EXAMPLE

Royal Mail delivers large letters by first or second class. The cost of sending a large letter depends on its mass. This graph gives the different costs for second class delivery.

a What is the cost of posting a large letter weighing

i 150 grams **ii** 500 grams?

b You stick stamps worth £1.04 on a large letter weighing 550 grams. Are these enough?

c Within what range can the mass (in grams) of a large letter be, if the valid postage is 76 pence?

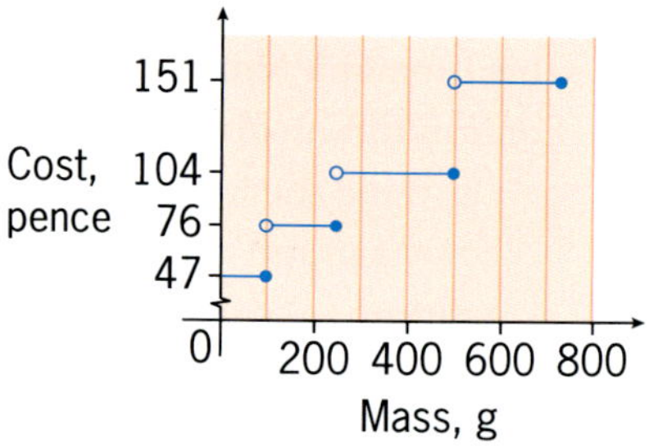

Remember that a solid dot on a graph includes that value but an empty dot does not.

ANSWER

a **i** A 150 gram large letter costs 76p.

ii A 500 gram large letter is the maximum mass which can be sent for £1.04.

b £1.04 is not enough postage for 550 grams. The correct postage is £1.51, so an extra 47p is needed.

c 76p is the valid postage for all masses from (but not including) 100 grams up to and including 250 grams.

Exercise C21.2

MEDIUM LOW

1 FlyBy is an airline which charges £15 to check in your baggage up to a limit of 20 kg.
Any excess baggage above 20 kg is charged at £10 per kg.

a How much are you charged for baggage weighing **i** 17 kg **ii** 20 kg **iii** 25 kg?

b If the total cost to check in your baggage is £45, what is the mass of all your baggage?

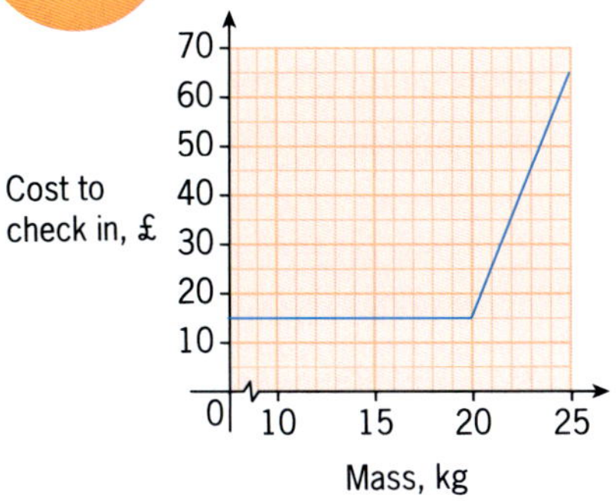

2 Car-hire firm A charges £1 per mile.
Car-hire firm B charges 50p per mile plus a basic charge of £10.

a Match these graphs L and M to the two firms A and B.

b How much do the two firms charge for a journey of 10 miles?

c Which company is cheaper for longer journeys and at what mileage does it become cheaper?

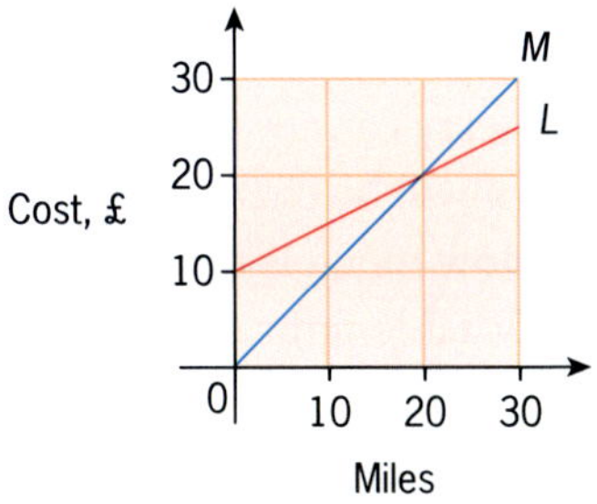

3 Two companies sell electricity. Elex charges 20p per unit.
Trizity charges 15p per unit plus a standing charge of £20.

a Match the graphs P and Q to the two companies.

b If Emma Hooper uses 300 units, how much would the two companies charge her?

c For how many units do the two companies charge the same amount?

d Which company is cheaper for large users of electricity?

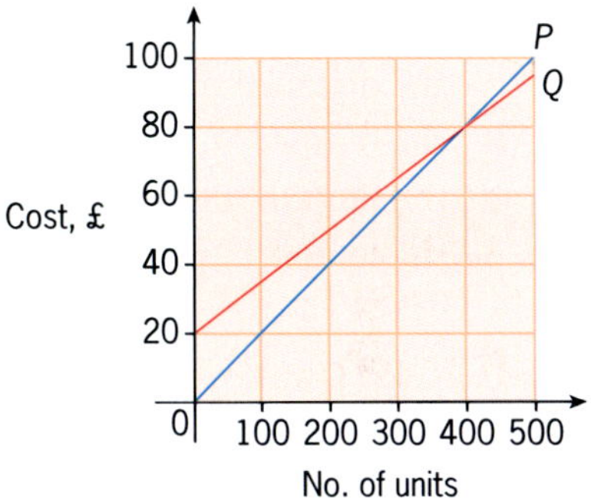

* 4 These graphs show how economical cars are at different speeds when they have petrol or diesel engines.

a Describe what the graphs tell you about

i the most efficient speed to drive a car

ii what happens when speeds increase above 60 mph

iii cars with petrol engines compared to diesel engines.

b Say what effect you think each of the following might have on a car's fuel efficiency.

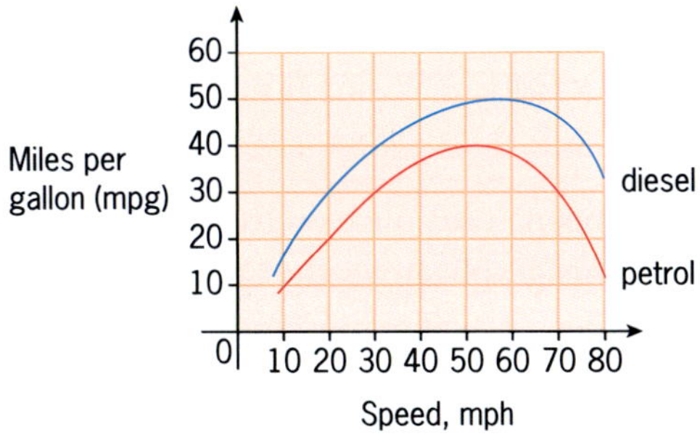

removing a bike rack	accelerating fast
eating while you drive	ensuring tyres are at correct pressure
having a clean air filter	braking hard and often

Continued ➤

Unit C

5 A mobile phone company has three different contracts, A, B and C as shown on the graph. Contract B costs £30 a month for 1000 minutes of 'free' calls and then charges 20p per minute after using all the 'free' minutes.

a How many 'free' calls does contract A give?

b What is the monthly cost of contract A?

c How many calls do you have to make before contract B is the cheaper option?

d Contract C is simply 'pay-as-you-go'. When is C the most economic contract?

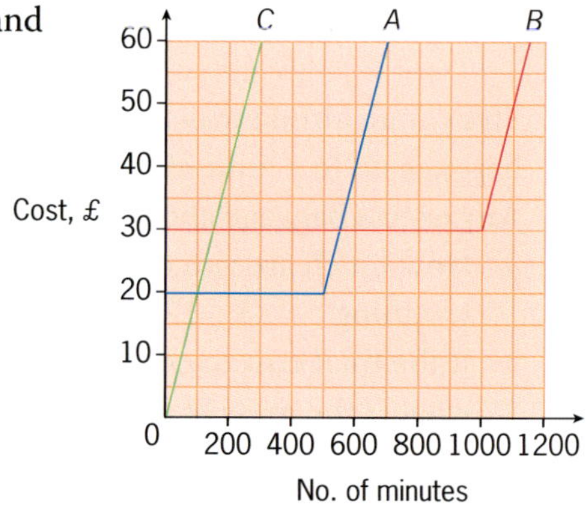

6 A car ferry's charges depend on the length of the vehicle. This graph shows the different rates for different lengths of car. The ferry cannot take vehicles over 8 metres long. A van will be charged 50% more than a car.

a How much is the charge for a car which is 5.42 m long?

b What lengths of car are charged £90?

c A van is 7.28 m long. How much will the charge be?

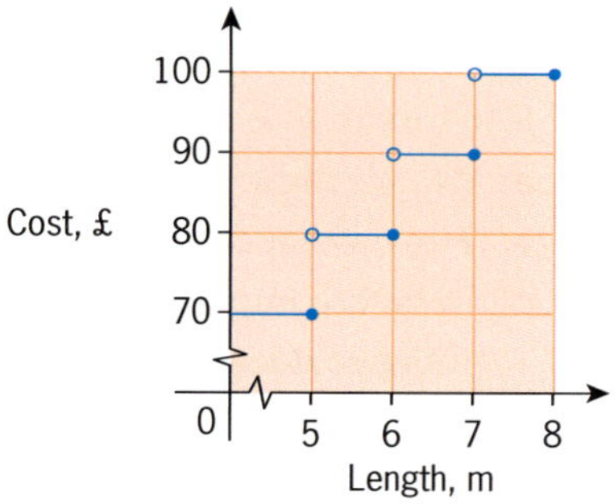

7 These four graphs show changes over time t. Match the graphs P, Q, R and S to changes in these variables, y.

a the distance travelled by a car driven at constant speed

b the temperature of a bowl of soup left uneaten

c the amount of money left in a purse on a shopping trip

d the temperature outdoors on a cloudy day.

P

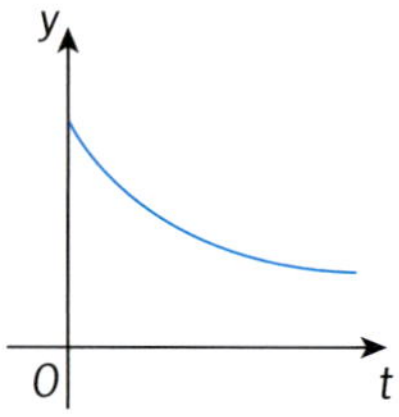

Q

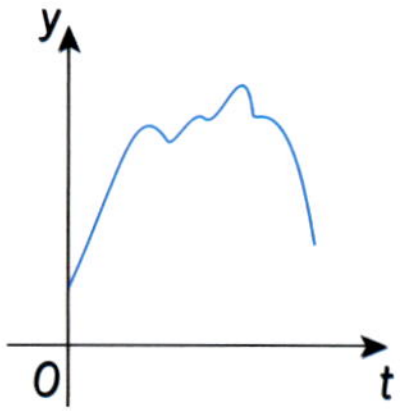

R

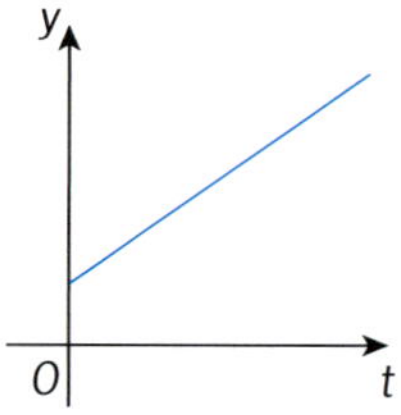

S

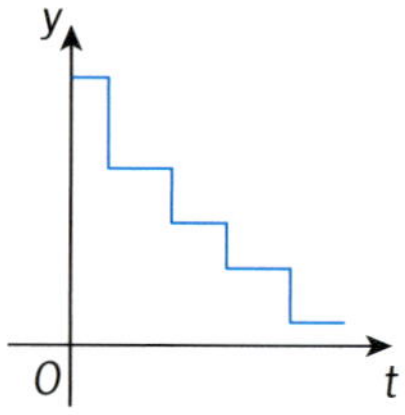

* 8 Sheena uses olive oil from a 1-litre bottle when she cooks. This graph shows the amount of oil in the bottle over a week. Give an explanation for the shape of the graph.

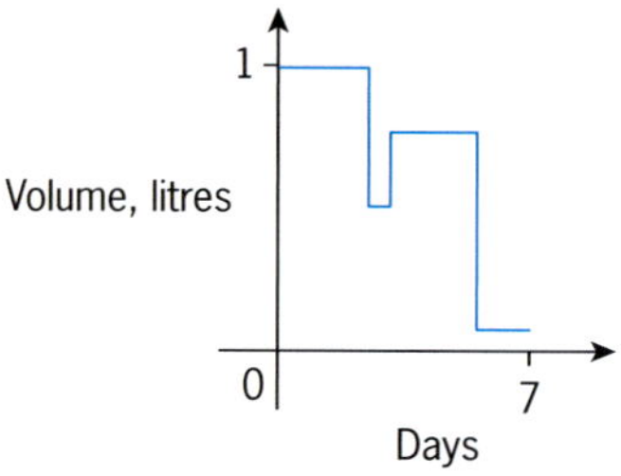

Unit C

9 A cup of coffee can be drunk in different ways.

The graphs shown the amount of coffee, A, in the cup over a period of time, t.

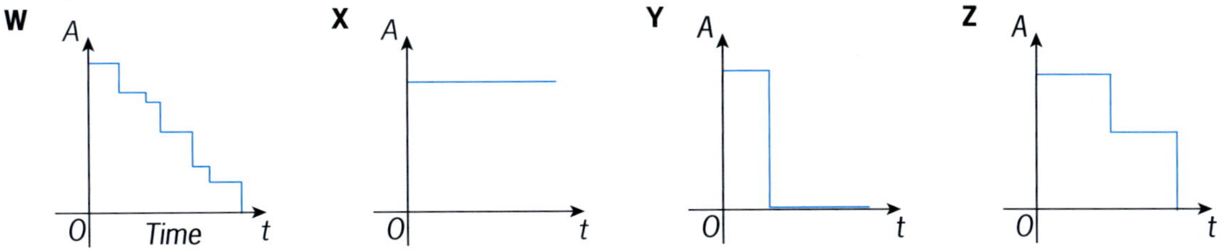

Match the four graphs to these descriptions. The coffee is drunk

a all at one go **b** in two gulps, half a cup each time

c sip by sip **d** not at all.

10 How much a baby weighs after birth changes over time.

This chart shows the mass of a baby over the first 40 days of its life.

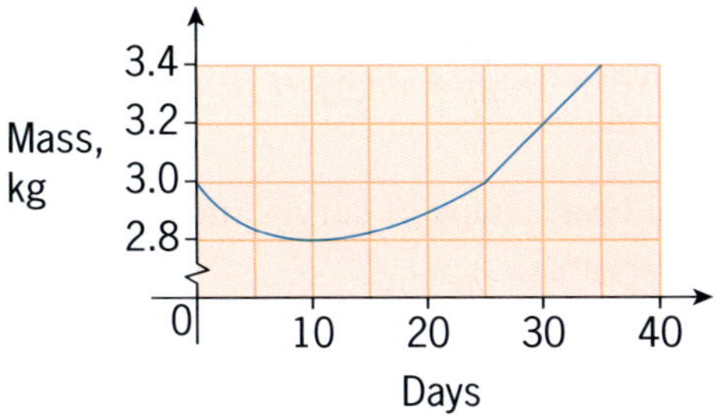

a Describe the change in mass over time.

b What did the baby weigh at birth?

c When does the baby weigh least and start to regain what it has lost?

d How long does it take for the baby to get back to what it weighed at birth?

e How many grams does it then gain in the next 10 days?

***11** Water drips into three beakers of different shapes. The graphs show how the depth, d, of water increases with time, t. Match the beakers to the graphs. Explain why you matched them as you did.

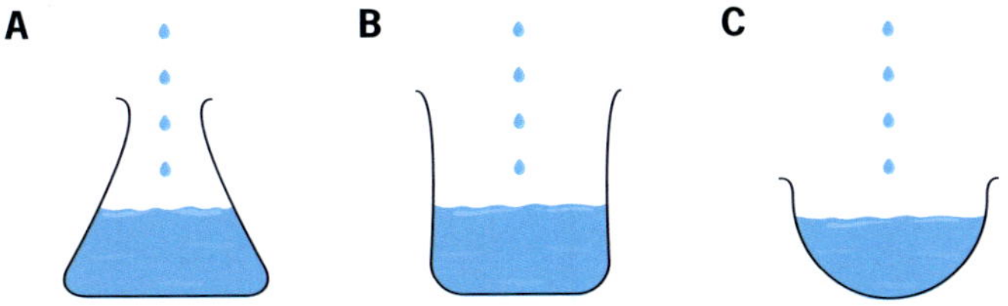

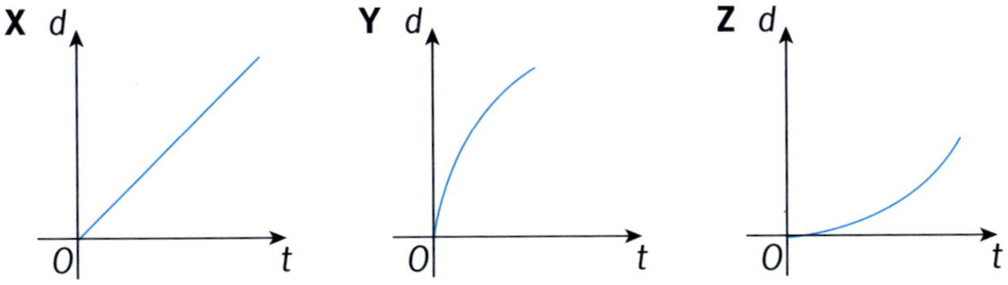

***12** Sketch three graphs, labelling the axes with words, to illustrate:

a the selling price of a car over its lifetime when it was bought new

b the annual cost of repair and maintenance of a car, bought new, until it is scrapped

c the annual mileage of a car over its lifetime if it has several different owners.

Give explanations as to why the graphs have the shapes you have given them.

Parabolic curves

- Generate points and plot graphs of quadratic functions

RICH TASK

Use computer software to draw the graphs of the family of equations $y = x^n$ for different positive integer values of n.
For example, when $n = 1$, $y = x^1$ or simply $y = x$.
When $n = 2$, $y = x^2$. And so on.

Describe the different shapes of the curves.
When n is odd, do the shapes have anything in common?
What happens when n is even?

What happens when $y = -x^n$?

Now look at what happens when $y = x^2 + c$ for different values of c.
What happens when $y = -x^2 + c$?

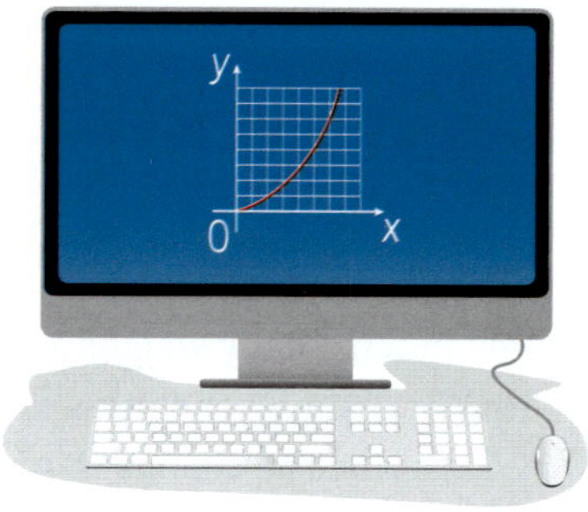

The basic shape of curves such as the graph of $y = x^2$ is called a **parabola.**

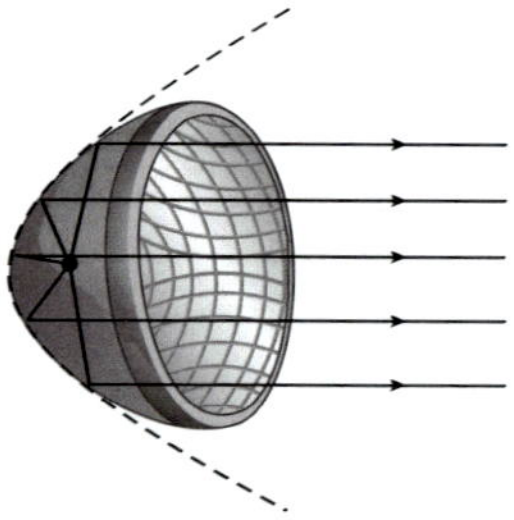

Parabolic reflectors are used to gather light to a point or to create a parallel beam of light as in a car headlight.
Can you think of any other uses?

EXAMPLE

Draw the graph of the function $y = x^2 - 3$ for values of x from -4 to 4. Use the graph to solve these equations

a $x^2 - 3 = 0$ **b** $x^2 - 3 = 8$

ANSWER

Substitute values of x into the equation $y = x^2 - 3$ to find values of y. Set the working out in a table of values.

x	−4	−3	−2	−1	0	1	2	3	4
x^2	16	9	4	1	0	1	4	9	16
−3	−3	−3	−3	−3	−3	−3	−3	−3	−3
y	13	6	1	−2	−3	−2	1	6	13

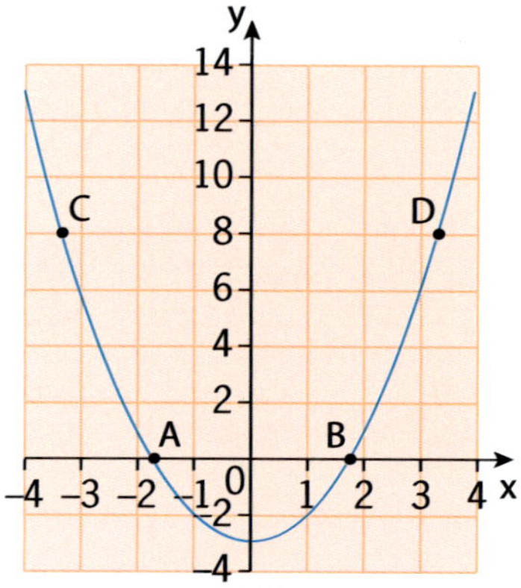

a When $x^2 - 3 = 0$, you find points A and B on the line where $y = 0$. So the two solutions of the equation $x^2 - 3 = 0$ are $x = -1.7$ and $x = 1.7$

b When $x^2 - 3 = 8$, you find points C and D on the line where $y = 8$. So the two solutions of the equation $x^2 - 3 = 8$ are $x = -3.3$ and $x = 3.3$

Tips for drawing curves:

- join points with a smooth curved line
- avoid sketching – provide a single clear line
- avoid missing points
- check for symmetry.

Exercise C21.3

MEDIUM

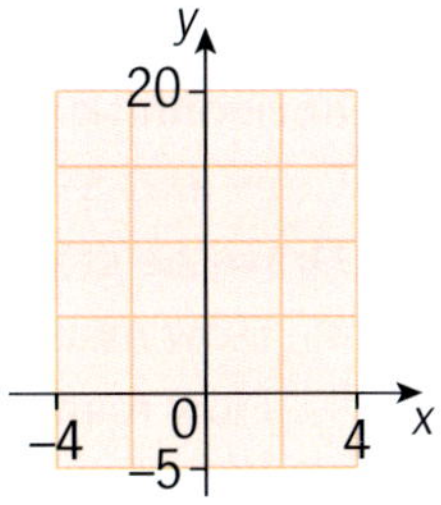

1 Copy and complete each table for values of x from –4 to 4.
Use the tables to plot points and draw the graphs of the three functions on one diagram with axes as shown.
How can the graph of $y = x^2$ be transformed into the other two graphs?

a $y = x^2$

x	–4	–3	–2
y			

b $y = x^2 + 3$

x	–4	–3	–2
x^2			
+ 3			
y			

c $y = x^2 - 4$

x	–4	–3	–2
x^2			
– 4			
y			

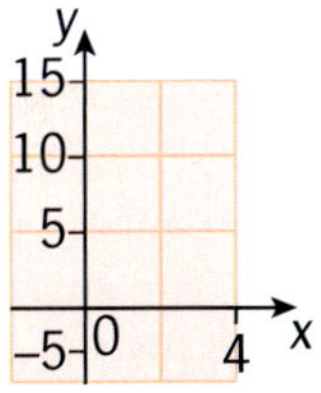

2 These two flow diagrams have the same operations but in a different order. Copy and complete the tables and draw the two graphs on the same axes.

a $y = x^2 - 2$

x → square → –2 → y

x	0	1	2	3	4
x^2					
– 2					
y					

b $y = (x - 2)^2$

x → –2 → square → y

x	0	1	2	3	4
x – 2					
y					

3 Copy and complete these two tables and draw the graphs of $y = x^2 - 5$ and $y = 5 - x^2$ on the same axes.

$y = x^2 - 5$

x	–3	–2	–1	0	1	2	3
x^2							
– 5							
y							

$y = 5 - x^2$

x	–3	–2	–1	0	1	2	3
x^2							
$-x^2$							
+ 5							
y							

a What transformation maps one graph onto the other?

b What are the values of x at the points of intersection of the two graphs?

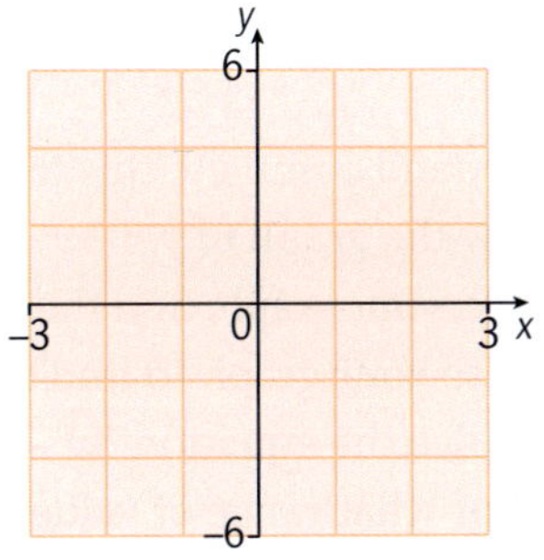

Continued ➤

4 A military aircraft fires a missile horizontally so that its height, h km, above ground is given by $h = 9 - x^2$, where x is the horizontal distance in km that it travels.
Copy and complete this table of values.
Draw the graph of h against x.

x	0	1	2	3
h				

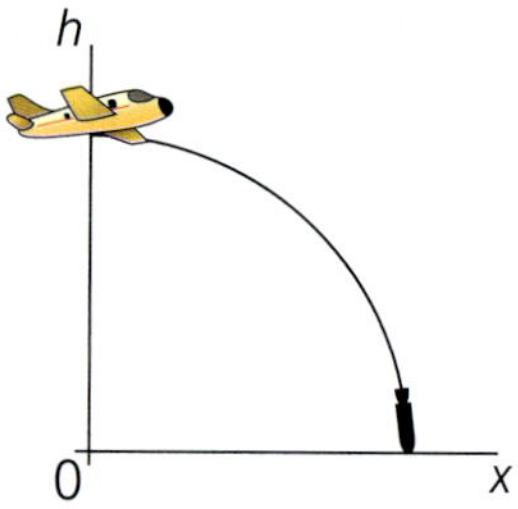

a How many kilometres high is the missile when it is fired?
b How many kilometres does it travel horizontally before it hits the ground?
c When the missile is 4 km above ground, how far has it travelled horizontally?

5 A science experiment lasts 3 hours. A liquid is heated so that its temperature, y °C after a time of t hours, is given by $y = 5 + t^2$.
Copy and complete this table of values.
Draw a graph of y against t.

t	0	1	2	3
y				

a What was the temperature of the liquid
i at the start of the experiment ii at the end of the experiment?
b At what time was the temperature of the liquid 10 °C?

6 In warm weather, algae on the surface of lake grow rapidly. The surface area, A m², covered by algae grows over three days according to the equation $A = 2t^2 + 4$, where t is the time in days.

When $t = 0$, $t^2 = 0$
$2t^2 = 2 \times 0 = 0$
$2t^2 + 4 = 0 + 4 = 4$
So $A = 4$.

Use this table to help you draw the graph of A against t.

t	0	1	2	3
$2t^2$				
+ 4				
A				

a What was the initial area covered by algae?
b What area was covered by algae after $2\frac{1}{2}$ days?
c How long did it take for 10 m² to be covered by algae?
d Is this function a realistic model of the algae's growth over a long period of time? Give your reasons.

7 This flow diagram calculates values of y for the function $y = 3x^2 - 8$.

$x \rightarrow$ square $\rightarrow$ ×3 $\rightarrow$ −8 $\rightarrow y$

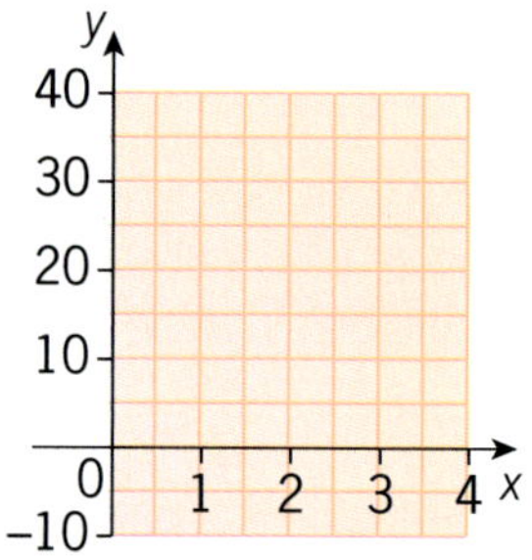

Copy and complete this table of values.
Draw the graph of $y = 3x^2 - 8$ on axes as shown.

x	0	1	2	3	4
$3x^2$					
−8					
y					

a On your graph, find the value of x (as a decimal) where $3x^2 - 8 = 0$.
b Use your graph to find a solution of the equation $3x^2 - 8 = 10$.
c Solve the equation $3x^2 - 8 = 30$.
d For what range of values of x is $3x^2 - 8$ negative on the graph?

8 Copy and complete this table of values for the function $y = 5x^2 - 10$. Draw a graph of y against x on axes as shown here.

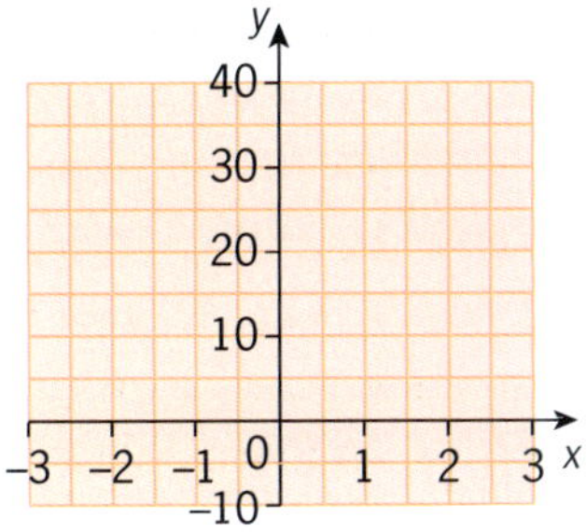

x	−3	−2	−1	0	1	2	3
$5x^2$							
−10							
y							

a Find from the graph the value of x such that $5x^2 - 10 = 0$.

b Use your graph to find the solutions of the equation $5x^2 - 10 = -8$.

c Solve the equation $5x^2 - 10 = 30$.

d For what range of values of x is $5x^2 - 10 < 0$?

9 Draw the graph of $y = \frac{1}{2}x^2 - 3$ by copying and completing this table.

x	−6	−4	−2	0	2	4	6
$\frac{1}{2}x^2$							
− 3							
y							

a Use your graph to write down the solutions of the equation $\frac{1}{2}x^2 - 3 = 0$.

b Draw the straight line $y = 10$. Use this line and the curve to write down solutions of the equation $\frac{1}{2}x^2 - 3 = 10$.

c By substituting your answers to part b into $\frac{1}{2}x^2 - 3$, check the accuracy of your answers.

10 By constructing your own table of values, draw the graph of $y = 30 - 2x^2$ for values of x from −5 to 5.

a Use your graph to solve these equations

i $30 - 2x^2 = 0$ **ii** $30 - 2x^2 = 20$

b For what values of x is $30 - 2x^2 > 20$?

11 ★ **ICT CHALLENGE** ★

Use computer software to draw the graphs of this exercise and so check your answers by finding solutions from the computer displays.

12 ★ **ICT CHALLENGE** ★

Use computer software to construct spreadsheets to generate tables of values for some of the problems in this exercise. Check them with your previous tables.

Unit C

Points in three dimensions

- Understand and use coordinates of points in three dimensions

RICH TASK

Compare a road atlas for planning a car journey, the maps in a school atlas, and an OS map in the Explorer or Landranger series.
How do these maps fix positions?
How do they use squares and grid lines?
Which maps give you heights – of some points, of all points?
What do some maps have on them to give heights of all points?
Discuss which maps are best for which purpose.

The position of a point in **1-D** needs just one coordinate on one axis.
The coordinate of point P is $x = 2$.

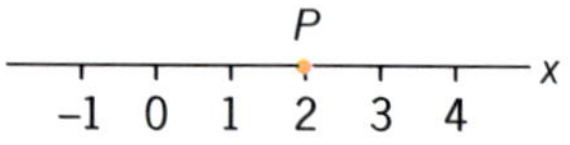

The position of a point in **2-D** needs two coordinates on two axes. The coordinates for point Q are (2, 3). Which is the x-coordinate and which is the y-coordinate?

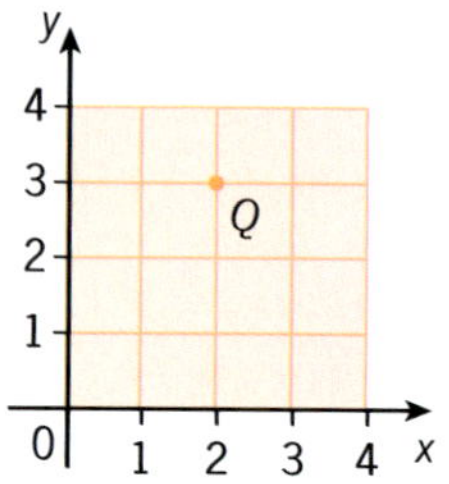

The position of a point in **3-D** needs a third axis which gives the vertical height off the base plane. The coordinates of point R are (2, 3, 4). Which of these is the z-coordinate?

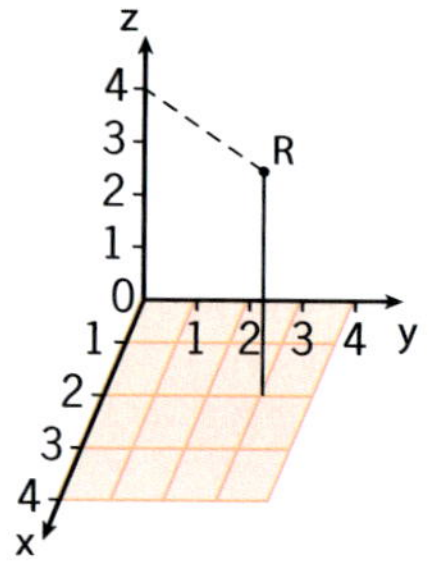

EXAMPLE

Write down the coordinates of

a the top T of the church tower which is 60 metres above ground level

b the end E of the nave roof which is 40 metres above ground level.

ANSWER

T is the point (30, 20, 60)
E is the point (30, 70, 40)

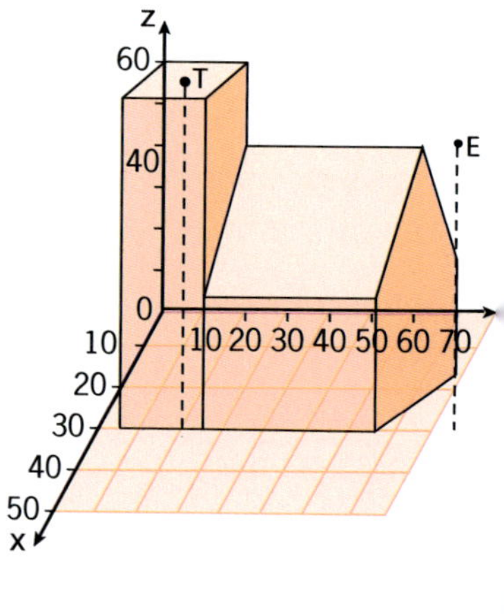

Exercise C21.4

MEDIUM HIGH

1 This diagram shows the tall buildings in a town.
Write the three coordinates for the lettered points A to F indicated by the red dots.
Clock tower A is 30 metres high.
Office block B is 50 metres high.
Crane C is 60 metres high.
Church spire D is 20 metres high.
Factory chimney E is 75 metres high.
Radio mast F is 40 metres high.

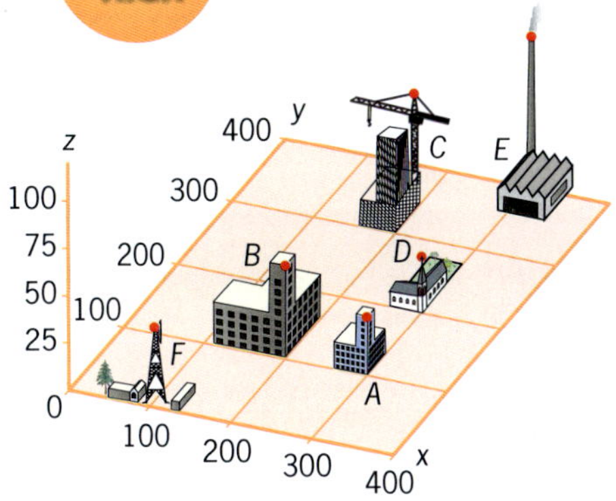

2 This map shows a coastline. The contours on land are at 10-metre intervals.
Write the 3-D coordinates of the lettered points A to L.

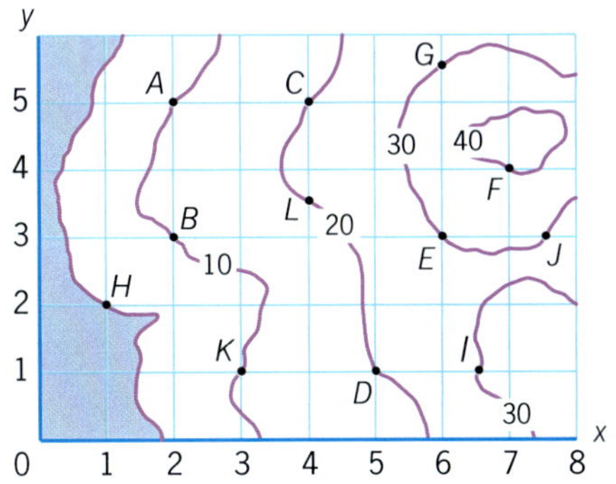

3 The next map also shows a coastline.
Its contours at 50-metre intervals are both above and below sea-level.

a Write the 3-D coordinates of the lettered points A to J.

b Copy and complete these coordinates for the unlettered points.

(40, 50, ...) (80, 20, ...)
(20, ..., 100) (5, ..., 150)
(...., 5, 0) (...., 50, 50)
(75, ..., –50) (50, 5, ...)

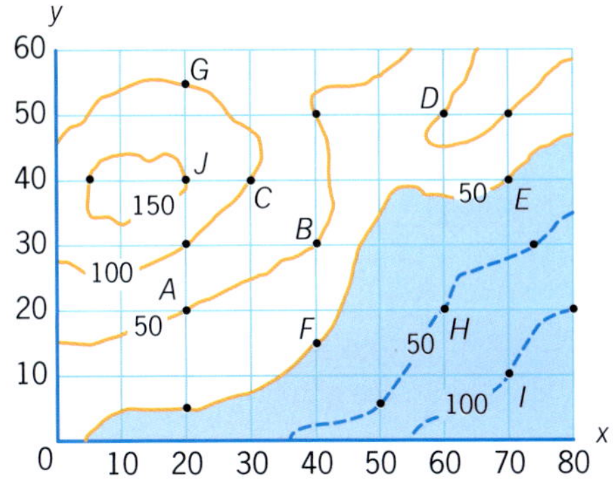

4 **RICH TASK**

Learn how to give a 6-figure grid reference for any place in the UK on the National Grid. You could use the internet or geography reference books.

5 **RICH TASK**

Learn how to play 3-D noughts and crosses.

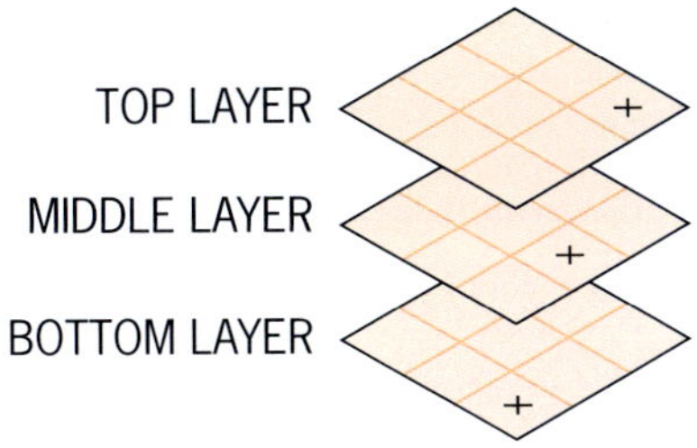

You have to say the three coordinates each time you place a nought or a cross.

Unit C

Summary and assessment

CHECK OUT

You should now be able to

- draw and interpret distance–time graphs and graphs modelling real situations
- find points and draw quadratic curves
- identify 3-D points.

Exam-style question

The graph shows a cyclist's ride.

The cyclist said after the ride that he only stopped once during the ride, reached a top speed of 60 km/h and did not travel more than 16 miles from home.

Comment on the cyclist's claims.

Justify your answers.

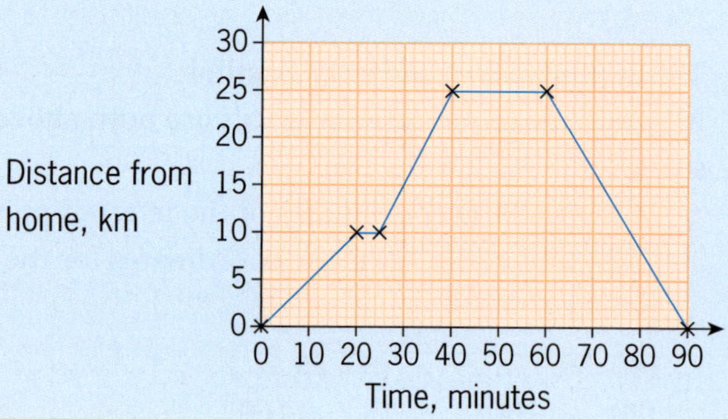

Monika's answer:

Stopping during the ride will mean travelling zero distance. On the graph this will show as horizontal lines. There are 2 horizontal line sections so the cyclist stopped twice, not once.

Speed on a distance–time graph is found from calculating the gradient of line segments

Top speed means gradient of steepest line, this is between 25 and 40 minutes

$$\text{Gradient} = \frac{D}{T} = \frac{25 - 10 \text{ km}}{40 - 25 \text{ minutes}}$$

$$\text{Speed} = \frac{15 \text{ km}}{15 \text{ mins}} = \frac{15 \text{ km}}{\frac{1}{4} \text{ hour}} = 60 \text{ km/h}$$

This claim is correct

The furthest distance from home is 25km

Convert to miles: $25 \times \frac{5}{8} = 15.625$ *miles*

The cyclist did travel less than 16 miles from home

Explain your interpretation of the graph.

Justify your answers with calculations and information that is read directly from the graph.

Write explanations that are complete, clear and concise.

Units are important: ensure you use appropriate units.

Exam practice

1 a Copy and complete the table of values for $y = x^2 - 3x$.

x	−1	0	1	2	3	4
y	4	0	−2	−2		4

b Draw the graph of $y = x^2 - 3x$ for $-1 \le x \le 4$. *(4 marks)*

2 Two taxi firms, Fargo and Speedy have different tariffs for journeys.
The graph shows the charges for journeys up to 12 miles.

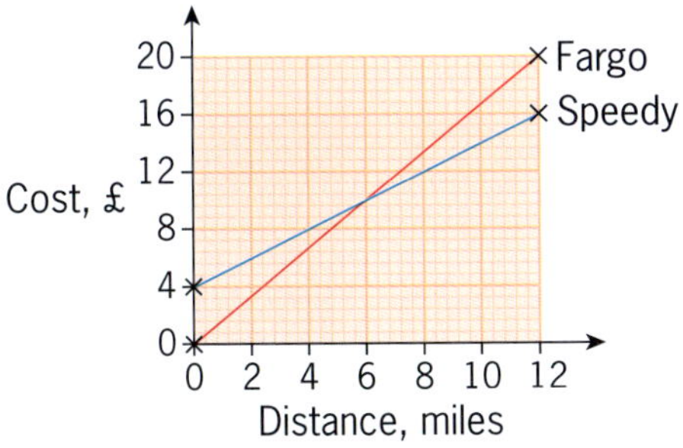

a For a journey of 8 miles, which taxi firm is the cheaper, and by how much?
b What distance would your journey be when both taxi firms charge the same amount? *(3 marks)*

*3 The stopping distance of a car can be estimated using the formula $y = \frac{x^2}{25} + \frac{3x}{25}$
where y is the stopping distance in m and x the speed of the car in km/h.
The table gives stopping distance for speeds up to 50 km/h.

x	0	10	20	30	40	50
y	0	5·2	18·4	39·6	68·8	106

a Draw the graph of $y = \frac{x^2}{25} + \frac{3x}{25}$ for speeds up to 50 km/h.
b Use your graph to estimate the stopping distance for a car travelling at 35 km/h.
c A driver claimed that he was travelling at 44 km/h.
Skid marks of 80 m were measured on the road.
Is the driver's claim correct? Justify your answer. *(5 marks)*

Case study 7: Radio maths

Mathematics can be used to explain how radio transmission works.

Radio transmitters use continuous sine waves to send and receive information such as music or speech.

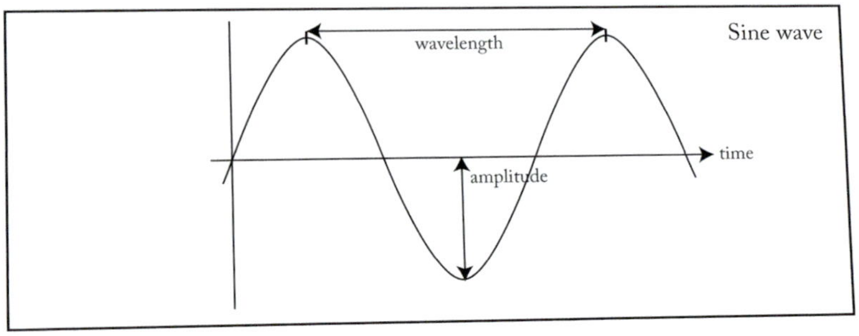

The Frequency is the number of waves transmitted per second, measured in hertz (Hz).

1 kilohertz	= 1 kHz	= 1,000 Hz
1 megahertz	= 1 MHz	= 1,000,000 Hz
1 gigahertz	= 1 GHz	= 1,000,000,000 Hz

Express a. 2,000 Hz in kilohertz
b. 2 GHz in megahertz.

For a sound wave, the larger the amplitude, the louder the sound.
A higher frequency gives a sound with a higher pitch.

Compare the sound produced by these waves. Comment on their pitch and volume, referring to the diagrams.

a. b. c.

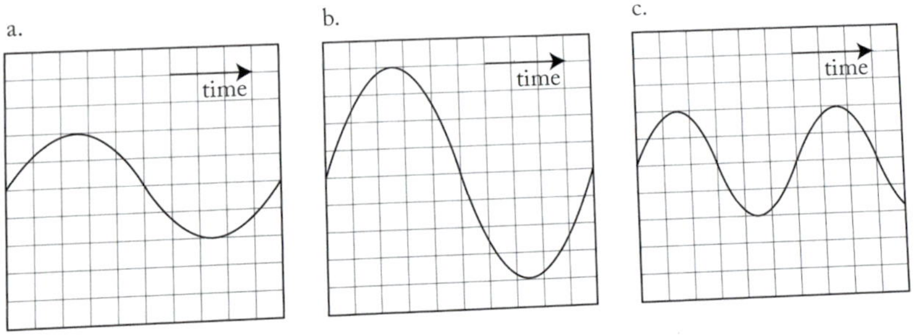

The original sound wave carrying the music or speech is not a sine wave.
The radio transmitter must encode the information on to a sine wave before it can be sent.
Two ways of doing this are by varying the sine waves amplitude (AM radio) or its frequency (FM radio).

AM radio stations transmit on frequencies between 535 kHz and 1700 kHz.

FM radio stations transmit on frequencies between 88 mHz and 108 mHz.

640 on the AM dial stands for 640 kHz.

88.7 FM stands for 88,700,000 Hz.

Wave speed (m/s) = frequency (Hz) × wavelength (m)

Maths FM transmits on the frequency 100.0 FM with a wavelength of 3m.

a. What is the frequency of the radio station in
   i. MHz ii. Hz?
b. Use the formula at the top of the page to calculate the speed of the waves that are being transmitted.
c. Rearrange the formula to give an equation for frequency in terms of wave speed and wavelength.
d. Use your answers from parts b and c to calculate the frequency (in Hz) of a radio station that transmits waves of wavelength 280m.
e. Is the radio station in part d on the AM or FM dial? Justify your answer and write down its AM or FM frequency.

Investigate the frequency and wavelengths used by the radio stations that you and your friends and family listen to.

Mathematics can also be applied to plan and produce radio programmes.

DJ Cool uses this wheel diagram to plan his hour-long show:

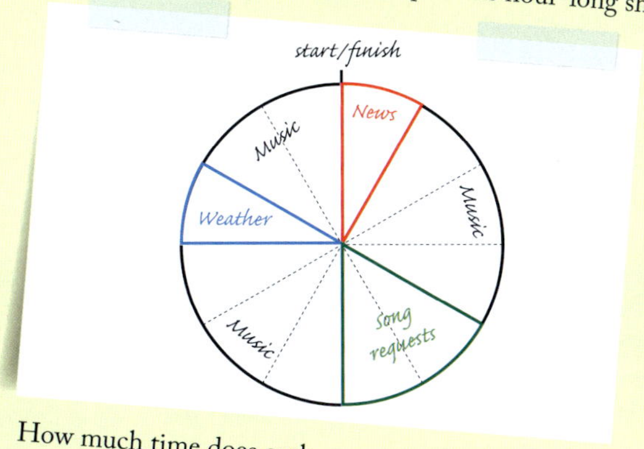

a. How much time does each segment of the diagram represent? Give your answer in minutes.
b. How many minutes of the show are taken up by
   i. news ii. weather iii. music (including requests)?
c. DJ Cool's show starts at 3pm. What time is the
   i. weather forecast ii. news report iii. requests slot?

The manager of the radio station decides that DJ Cool should include a 5-minute travel report at 3:20pm. The weather forecast and news report must not be moved.

d. Draw a wheel diagram to show how DJ Cool's show could look with the travel report included.

Consider some of the radio shows that you and your friends and family listen to. Do they use a format that could be shown on a wheel?

22 Everyday arithmetic and bounds

INTRODUCTION

In many professions it is vital to perform mental checks on answers that have been calculated. These include doctors working out the dose of medicine to give a patient, pilots checking the fuel required for a flight, and civil engineers calculating the amount of material required to construct a building.

What's the point?

If doctors, pilots and engineers don't check that their answers are sensible, things could go badly wrong in a way that will affect other people.

CHECK IN

1 Calculate $5 \times 3 + 4 \times 8$

2 Calculate

a $13 \div 10$ b 0.03×10

3 Round each of these numbers to 2 decimal places.

a 3.5624 b 8.03292 c 0.0551

Orientation

What I should know	What I will learn	What this leads to
A3	■ Solve problems in real-life contexts ■ Use common measures ■ Use compound measures	Applying mathematical skills in everyday life

RICH TASK

The numbers 3 and 6 are multiplied together and then the answer is doubled. This gives 36 (a 2-digit number made from the two original numbers).
Find two 2-digit numbers that when multiplied together and doubled give a 4-digit number made from the two original 2-digit numbers.

Everyday arithmetic

C22.1

- Explore and solve problems in real-life contexts

EXAMPLE

This table shows information concerning inland postal charges in Britain in 2009.

Format	Max weight	Max size	Max thickness
LETTER	100 g	240 mm × 165 mm	5 mm
LARGE LETTER	750 g	353 mm × 250 mm	25 mm
PACKET	over 750 g	over 353 mm × 250 mm	over 25 mm
Format	**Weight up to**	**1st class**	**2nd class**
LETTER	100 g	39p	30p
LARGE LETTER	100 g	61p	47p
	250 g	90p	76p
	500 g	£1.24	£1.04
	750 g	£1.77	£1.51

Format	Weight up to	1st class	2nd class
PACKET	100 g	£1.28	£1.08
	250 g	£1.62	£1.41
	500 g	£2.14	£1.85
	750 g	£2.65	£2.24
	1000 g	£3.25	£2.70 max
	1250 g	£4.45	
	1500 g	£5.15	
	1750 g	£5.85	
	2000 g	£6.55	
	4000 g	£8.55	
	Each additional 2 kg or part there of	£2.80	
Recorded Delivery 1st or 2nd class postage + 75p			

Find the cost of posting these items:

a Iain wishes to post an item 230 mm × 200 mm, of thickness 4 mm that weighs 225 g.

b Nick has an item to post 220 mm × 150 mm, 4 mm thickness weighing 780 g.
Nick reduces the weight to 740 g using less tape.
How much is saved on 1st class?

ANSWER

a One dimension is too large for a letter so look for large letter, under 250 g: £0.90 (1st class) or £0.76 (2nd class).

b The weight is too great to be letter or large letter.
The item size is within maximum packet size: £3.25 (1st class) or £2.70 (2nd class).
This is now large letter 1st class, saving £3.25 – £1.77 = £1.48

Exercise C22.1

LOW

1 a Find the temperature shown by this thermometer in
 i degrees Celsius ii degrees Fahrenheit.
 b Divers use a depth gauge to tell them how deep they are.
 What is the reading on this depth gauge?

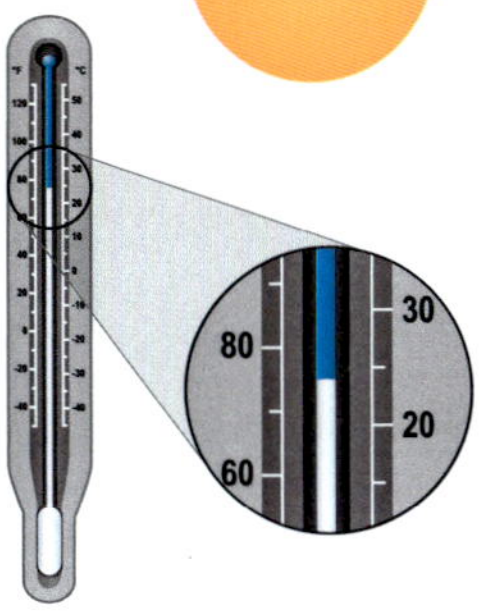

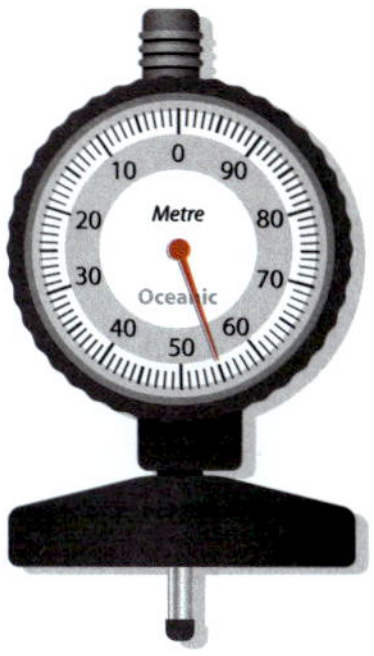

2 A car is travelling along a road which has a speed limit of 70 mph.
Its speed is shown on the speedometer.
 a Is the car being driven within the speed limit?
 b Write down the reading shown on the speedometer.

▲ Question 2

3 The table shows the distances in miles between some towns on the Stage 1 route of the 2007 Tour de France cycle race.

London					
19	Dartford				
48	29	Maidstone			
98	79	50	Tenterden		
111	92	63	13	Ashford	
127	108	79	29	16	Canterbury

 a How far was the cycle route from London to Canterbury in
 i miles ii km?
 b What is the distance covered by the cyclists from Dartford to Ashford?

*4 This section of railway timetable shows the train times from London Victoria to Bromley South on Saturdays after 11 p.m. in 2009.

London Victoria	2303	2318	2312	2321	2339	2351	2342	2353	0035
Herne Hill	---	---	---	2330	---	---	---	0001	0043
Beckenham Junction	---	---	---	2341	---	---	---	0013	0051
Shortlands	---	---	2340	2344	---	---	0010	0016	---
Bromley South	2318	2334	2343	2347	2358	0007	0012	0018	0055

 a How long does the 2321 train from London Victoria take to reach Bromley South?
 b i What is the first and last train I can catch from London Victoria to Shortlands?
 ii How long are the journeys?
 c Why do you think the train leaving London Victoria at 2351 is listed before the train leaving London Victoria at the earlier time of 2342?

Unit C

Using formulae

C22.2

- Explore and solve problems in real-life contexts

RICH TASK

Two different formulae are proposed to calculate the required child's dosage for a medicine.

$$\text{Child dose} = \frac{\text{Age (years)} \times \text{adult dose}}{\text{Age (years)} + 12} \qquad \text{Child dose} = \frac{\text{Age (months)} \times \text{adult dose}}{150}$$

Suppose you prescribed medicine for a 6-year-old, where the adult dose is 30 mg. Will it make any difference which formula you used?
Suppose the child was aged 10 years, what difference will there be?
Try some other ages – can you find the age of the child for which there is no difference?

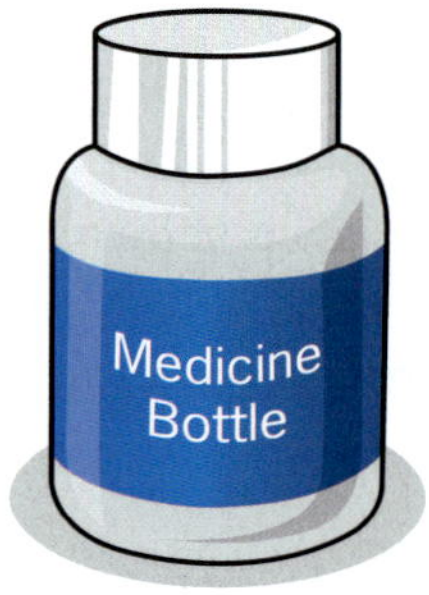

Formulae are used in everyday life to perform calculations.

p. 98

This formula changes temperature in °F to °C.

$$C = \frac{5}{9}(F - 32)$$

You may need this to convert cooking temperatures from old recipe books.

This formula calculates interest on savings or money borrowed.

$$A = P\left(1 + \frac{r}{n}\right)^{nt}$$

You may need this if you want to borrow money to buy a car.

EXAMPLE

Davina is making a circuit for her amplifier.
She puts two resistors in parallel in the circuit.
The total resistance for two resistors in parallel is found using the formula

$$R_{\text{Total}} = \frac{R_1 \times R_2}{R_1 + R_2}$$

R_1 is the value of one resistor
R_2 the value of the other resistor

Calculate the total resistance in Davina's circuit if she uses resistors with values 1.5 ohms and 2.2 ohms.

ANSWER

$$R_{\text{Total}} = \frac{1.5 \times 2.2}{1.5 + 2.2} = \frac{3.3}{3.7} = 0.89 \text{ to 2 decimal places.}$$

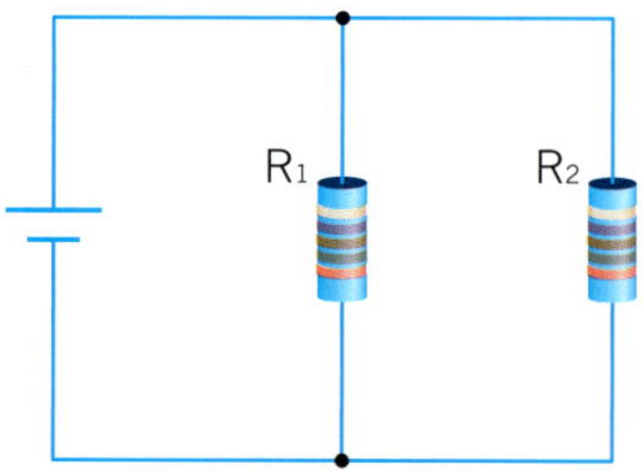

The units of resistance are ohms.

To use formulae:
- → **substitute in the given values**
- → **perform calculations according to the order of arithmetical operations (BIDMAS).**

Exercise C22.2

MEDIUM

1 a A formula for changing degrees Fahrenheit (°F) into degrees Celsius (°C) is

$$C = (F - 32) \times \frac{5}{9}$$

Calculate the temperature in °C for a temperature in °F of

i 40° **ii** 98.4° **iii** 30°

b A formula for changing °C into °F is

$$F = \frac{9}{5} \times C + 32$$

Calculate the temperature in °F for a temperature in °C of

i 22° **ii** −5° **iii** −2°

2 The formula to calculate a person's body mass index (BMI) is

$$\text{BMI} = \frac{m}{h^2}$$

where *m* is the person's mass in kilograms
h is the person's height in metres

1 stone	= 14 pounds
2.2 pounds	= 1 kilogram
1 foot	= 12 inches
1 inch	= 2.54 centimetres

For a healthy person the BMI should be in the range 19–25 inclusive.
John is 5 feet 9 inches tall and his mass is $12\frac{1}{2}$ stones.
To estimate his BMI, John just doubles his mass in stones.

Calculate John's BMI using the formula and show how John's method puts him in the healthy range, when his true BMI is not healthy.

3 The stopping distance of a car can be estimated using the formula

$$D = \frac{S^2 + 3S}{25}$$

S is the speed of the car in km/h
D is the stopping distance in m

After an accident, skid marks are measured on the road.
Estimate the speed of the car correct to the nearest 1 km/h for skid marks of a 60 m b 100 m.

4 A formula to calculate a child's medicine dose based on the child's body surface area (BSA) in m^2 is

$$\text{child dose} = \frac{\text{BSA}}{1.7} \times \text{adult dose}$$

Calculate the child dose for a child with BSA = 0.67 where the adult dose is 40 mg.

Unit C

Everyday arithmetic: common measures

C22.3

- Explore and solve problems that use common measures

Numbers are everywhere: in the media, on the internet, on food packaging and so on. You need to make sense of the information – you won't need all of it!

EXAMPLE

The table gives the guideline daily amounts of calories needed by boys and girls of different ages.

Gender \ Age	11–14	15–18	Adults
Male	2220	2750	2550
Female	1845	2110	1940

You burn 750 calories for each hour of exercise.
You lose one pound of fat for each 3500 calories burnt.

Alex takes in an average 3050 calories each day.
How long does Alex need to exercise each day if

a Alex is a 21-year-old man and doesn't wish to lose any weight

b Alex is a 17-year-old girl and wants to lose $\frac{1}{2}$ pound of fat each day?

Recommended guidelines for weight loss are only 1–2 lb per week.

ANSWER

a A male aged 21 needs 2550 calories.
$3050 - 2550 = 500$
500 extra calories need to be burnt to maintain weight.
$500 \div 750 = \frac{2}{3}$ of an hour $\frac{2}{3}$ of 60 = 40 minutes
Alex needs to do 40 minutes of exercise each day.

b A female aged 17 needs 2110 calories.
$3050 - 2110 = 940$
Alex wants to lose $\frac{1}{2}$ lb of fat daily $\frac{1}{2}$ of 3500 = 1750
Alex needs to burn an extra $1750 + 940 = 2690$ calories each day.
$2690 \div 750 = 3.586...$ $\quad 0.586... \times 60 = 35.2$ minutes
Alex needs to do 3 hours 35 minutes of exercise each day.

A sensible answer here would be around $3\frac{1}{2}$ hours.

When answering problems involving everyday arithmetic

- Read the question twice — **work out what you need to do**
- Perform one step at a time — **you cannot always go directly to an answer**
- Use the numbers in the question — **rounding at the start often leads to errors**
- Avoid rounding intermediary answers — **if you have a calculator**
- Give your final answer to a sensible degree of accuracy
- Use sensible units — **relate back to the problem**

Exercise C22.3

MEDIUM LOW

Unit C

1 Josh painted his bedroom. Copy and complete the three missing values in his paint bill.

Purchase No: 1 Red: 22

Name of Item	Rate Per	Amount
4 tins silk mulsion	@ £18.99	£..........
.... tins gloss	@ £ 8.50	£..........
Total cost		£ 101.46

Where appropriate give final answers to a sensible degree of accuracy.

2 Prices in a shop are given on the price tag in two currencies, pounds and euros. A handbag is priced as £39 and €55; a pair of shoes is priced as £64 and €90. What exchange rate is being used?

3 A pair of jeans is priced as €125 in Paris. The identical jeans are priced as $180 dollars in New York and £110 in London. On a day when the exchange rates are £1 = €1.15 and £1 = $1.75, which place is the cheapest to purchase the jeans?

4 ☆ **PROBLEM** ☆

San Francisco uses Pacific Daylight Time (PDT).
PDT is eight hours behind British Summer Time (BST).
A flight leaves London on Monday at 11:15 BST and arrives in San Francisco at 14:00 PDT the same day.
On the return flight to London, a tail wind means that the flight time is one hour less.
The return flight leaves San Francisco on Tuesday at 15:40 PDT.
What time does the flight arrive in London?

5 ***RICH TASK***

Is it possible to celebrate New Year in two different capital cities on the same night?

If you leave London after the midnight celebration, would you be able to arrive in a different country before midnight to celebrate New Year again?

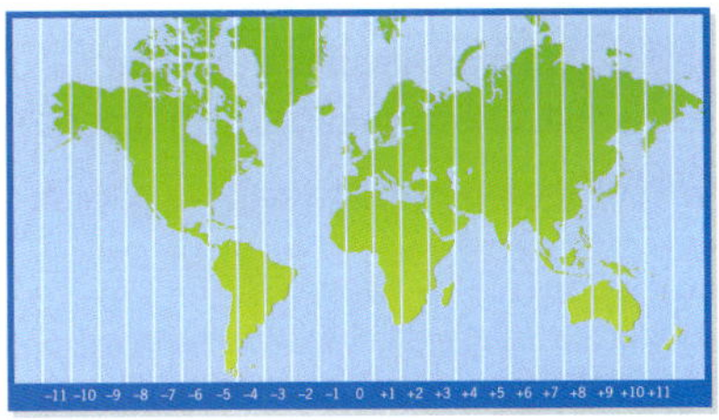

Compound measures

- Explore and solve problems that use compound measures

PROBLEM

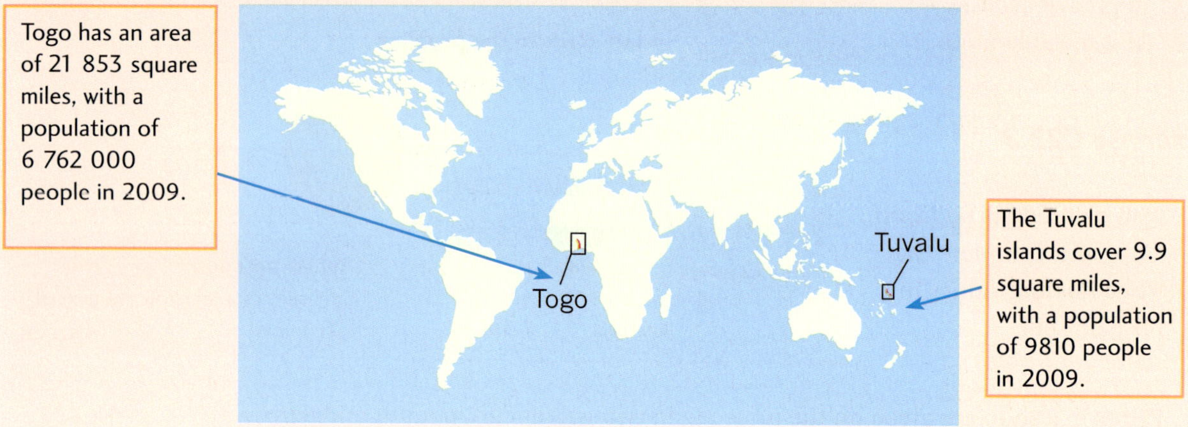

Which country is more crowded?

p. 292

You can use **compound measures**, such as speed, density and rates, to make comparisons.

EXAMPLE

Gareth and Eric drive 150 miles in separate cars.
Gareth begins the journey 30 minutes after Eric.
Eric travels at an average speed of 50 miles per hour.
How fast does Gareth need to go to arrive at the same time as Eric?

ANSWER

$\text{Speed} = \frac{\text{distance}}{\text{time}}$

$50 = \frac{150}{\text{time}} \Rightarrow \text{time taken} = \frac{150}{50} = 3 \text{ hours.}$

$3 \text{ hours} - 30 \text{ minutes} = 2\frac{1}{2} \text{ hours} \quad \frac{150}{2\frac{1}{2}} = 60$

Gareth needs to travel at an average speed of 60 miles per hour.

RICH TASK

Imagine identical sized boxes, one filled with marshmallows and the other filled with sugar. Which box is heavier?

Density compares mass with volume.

→ Density = $\frac{\text{mass}}{\text{volume}}$

EXAMPLE

a Work out the density of a 3 kg lump of metal with volume 1.5 m^3.

b Work out the density of a pane of glass weighing 1.8 kg with a volume 600 cm^3.

c The density of cork is 0.25 g/cm^3. Find the volume of a piece of cork with mass 110 g.

ANSWER

a Density = $\frac{m}{v} = \frac{3}{1.5} = 2$ kg/m^3

b 1.8 kg = 1800 g

Density = $\frac{18000}{6000} = 3$ g/cm^3

c $V = \frac{m}{d} = \frac{110}{0.25}$

$= 440$ cm^3

Exercise C22.4

MEDIUM

1 a Work out the average speed of a cyclist riding

i 48 km in 2 hours 30 minutes ii 30 km in 1 hour 45 minutes.

b Work out the average speed of a car travelling

i 100 miles in 1 hour 15 minutes ii 60 miles in 50 minutes.

2 What is the average speed of a runner who runs

a a marathon (26 miles) in 3 hours 30 minutes

b a half-marathon (13 miles) in 2 hour 40 minutes?

3 The winner of the 2006 Tour De France completed the 3639 km course in 89 hours 39 minutes. What was his average speed?

4 A cheetah runs at a speed of 120 km per hour. At this speed how far does it go in 2 minutes?

5 a Work out the volume of a gold bar with mass 0.6 kg (the density of gold is 19 300 kg/m^3).

b The density of sunflower oil is 0.92 g/cm^3. What is the mass of 2 litres of oil?

6 Armenia had a population of 2 967 004 in 2009, and covers an area of 11 484 square miles. What was the population density of Armenia in 2009?

7 Water drips out of a tap at the rate of 2 ml per second. How long does it take before 5 litres of water is lost?

Unit C

Understanding error in measures

- Recognise that measurements given to the nearest whole unit may be inaccurate by up to $\frac{1}{2}$ in either direction

★ PROBLEM ★

These are the distances, given to the nearest mile, between junctions on a motorway.

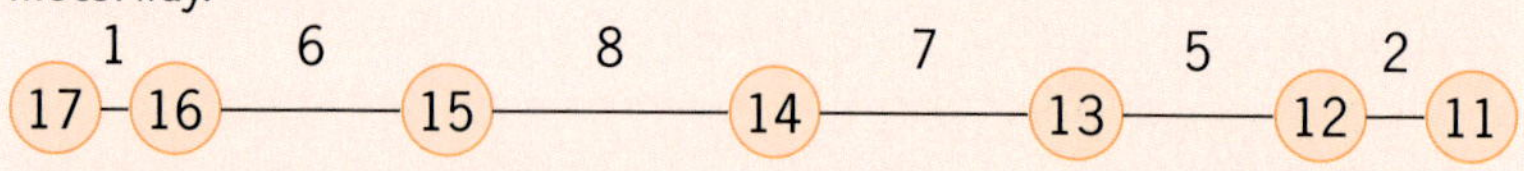

Write down the maximum and minimum possible distances between each motorway junction.

What effect do you think giving these distances to the nearest whole mile may have on planning a journey?

Numbers in real life are often **rounded**. Suppose you wanted to know what the number was before rounding. There will be a **range** of possible values.

A number written as...	could actually be...	
82 to the nearest whole number	smallest 81.5	largest 82.5

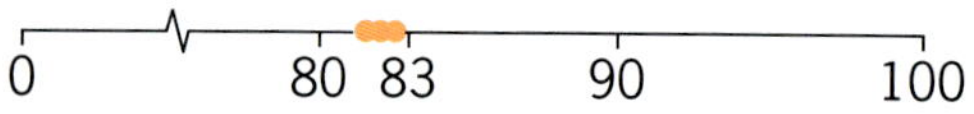

540 to the nearest 10	smallest 535	largest 545

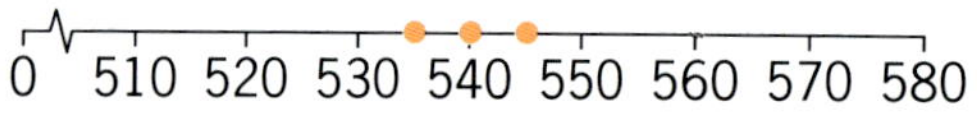

3.7 to 1 decimal place	smallest 3.65	largest 3.75

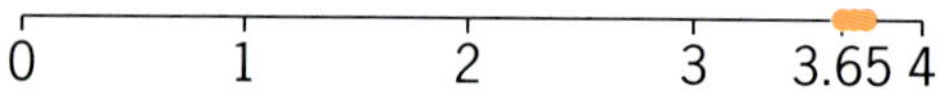

Measurements are only as accurate as the measuring instrument you use. All measurements contain some degree of **error**.

You can measure with a ruler to the nearest mm.

Kitchen scales measure ingredients to the nearest gram.

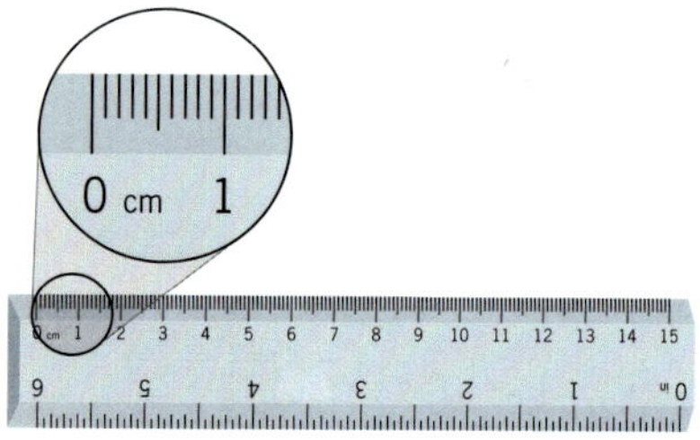

EXAMPLE

a A newborn baby weighs 7 lbs 2 oz to the nearest oz. Write down the range of the baby's possible weights.

b A 50 m swim time is recorded as 25.8 seconds to the nearest 10th of a second. Write down the range of the possible times.

ANSWER

a The newborn could weigh 7 lbs 1.5 oz – 7 lbs 2.5 oz $\pm \frac{1}{2}$ of 1 oz

b The time could be 25.75 seconds to 25.85 seconds $\pm \frac{1}{2}$ 0.1 second

→ **The error in measuring a quantity is $\pm \frac{1}{2}$ of the unit that the measurement is accurate to.**

Unit C

Exercise C22.5

MEDIUM

1 What is the smallest and largest these measures could have been before rounding?

a 68 cm given to the nearest whole number
b 3800 given to the nearest 100
c 730 m given to the nearest 10
d 4950 given to the nearest 10
e 17.6 seconds given to 1 decimal place
f 23.4 seconds given to 1 decimal place

2 Write down the range of possible values of these measurements.

a 28 cm given to the nearest cm
b 48 g given to the nearest gram
c 112 g given to the nearest gram
d 560 cm given to the nearest 10 cm
e 9350 km given to the nearest 10 km
f 9350 km given to the nearest 50 km
g 520 miles given to the nearest 20 miles
h 420 m given to the nearest 10 m
i 43.6 sec given to the nearest 10th of a second
j 5.2 g given to the nearest 10th of a gram
k 7.3 cm given to the nearest 10th of a cm
l 40 m given to the nearest m
m 100 cm given to the nearest cm

*3 Swinley School had a sports day at the end of the school year.

a In the high jump competition, a height jumped was recorded as 154 cm to the nearest cm. Could the height jumped have been higher than 154 cm? Explain your answer.

b The 100 m sprint was won in a time of 16.7 seconds measured to the nearest 10th of a second. Between what two values does this time actually lie?

c In the long jump competition, a distance jumped was recorded as 4.37 m to the nearest cm. Could the distance jumped have been longer than 4.37 m? Explain your answer.

4 162 students sat an exam.
Just over 48% of the students passed the exam.
The percentage was rounded to the nearest whole number.

a What is the largest value that the percentage could have been, to 1 decimal place?

b What is the largest possible number of students that could have passed the exam?

Summary and assessment

C22

CHECK OUT

You should now be able to

- use arithmetic to solve problems involving measures such as time, money, mass, length, area and volume
- use arithmetic to solve problems involving compound measures such as speed and density
- understand error in measurement.

Exam-style question

A company pays its employees these rates for business travel per year.

Vehicle used	First 10 000 business miles	Each mile over 10 000 miles
Cars and vans	40p	25p
Motorcycles	24p	24p
Cycles	20p	20p

Jed and Mike each travel 12 000 business miles in one year; Jed travels by car and Mike travels by motorcycle.
Jed spends £3418 and Mike spends £1542 on their vehicles in that year on insurance, and other fixed costs.
Based on this information, is Jed or Mike financially better off?

Layla's answer:

Jed
10 000 × 40 = 400 000p = £4000
2 000 × 25 = 50 000p = £ 500
Total £4500
Less car expenses 4500 − 3418 = £1082

Mike
12000 × 24 = 288000p = £2880
Less bike expenses 2880 − 1542 = £1338
1338 − 1082 = £256
Mike is better off by £256

With a table use the column and row headings to pick the correct information.

Remember to answer all of the question.

You may need to use information from different parts of the question.

You may need to do more than one calculation. Show all your working clearly.

Exam practice

1 This report was written in 1954.

> Roger Bannister, a 25-year-old British medical student, has become the first man to run a mile in less than four minutes. His time was 3 mins 59.4 seconds.

The time given is to the nearest 10th of a second.
What is the smallest and largest this time could be? (*2 marks*)

2 A family go to the seaside by car. They travel 160 miles in 2 hours 40 minutes, using 18 litres of petrol.

a What was their average speed?

b What was their petrol consumption in miles per gallon? (*6 marks*)

3 A brand of washing powder costs £4.50 for a 2 kg box.
On the box is written

> 'Enough powder for 25 washes'

A new monthly box size is produced to contain enough powder for 30 washes.
The cost of the new box size is 1p per wash cheaper than the 2 kg box.
How much is the new monthly box size and what will it cost? (*5 marks*)

Show your working.

23 Area and volume

INTRODUCTION

Many of the man-made shapes we see in everyday life are cuboids, particularly in packaging.

What's the point?

Cuboids are easy to assemble, and most importantly, they stack up without leaving any gaps. This makes them more economical and practical for packaging and transporting goods.

CHECK IN

1 Calculate the area of each shape.

a

b

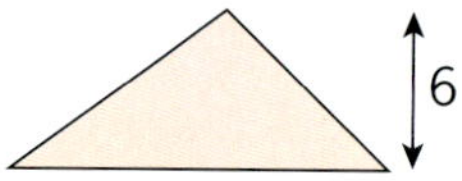

2 Evaluate

a 25×10 b 63×100 c 4.1×10

d 2.5×100 e 3.5×1000 f $40 \div 10$

g $56 \div 10$ h $4000 \div 100$ i $410 \div 100$

j $5200 \div 1000$

Orientation

What I should know	What I will learn	What this leads to
C18	■ Find volumes of 3-D shapes ■ Find scale factors of 3-D shapes	Careers in design and manufacturing

RICH TASK

You will need a circular pinboard.
A triangle has been made on a circular geoboard. Measure its angles.
Investigate the angles in different triangles you can make on the pinboard.
Can you find some rules?

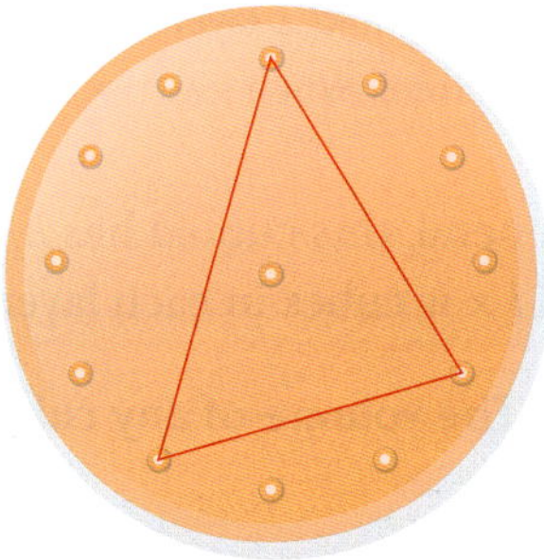

Volume of cuboids

C23.1

- Find and use the volume of cuboids

RICH TASK

Breakfast cereals come in boxes of different sizes and different masses. Here are three:

	Base of box	Height	Mass of cereal
Miniwheats	19 cm by 6 cm	27 cm	500 grams
Oatiebisks	19 cm by 10 cm	25 cm	860 grams
Wheatcakes	8 cm by 8 cm	17 cm	300 grams

Explore which cereal has the most efficient packaging for its mass.

The volume of a liquid that a hollow container will hold is called its **capacity**.

You could find more data to add to this list.

→ **The volume of a solid shape is a measure of the amount of space that it takes up.**

This cube with edges of 1 cm has a volume of 1 cm^3.

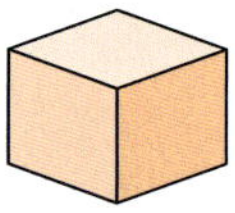

Compare cm^3 for volume with cm^2 for area.

Centimetre cubes can be fitted together to make a solid shape.

This shape uses 8 cubes. Its volume is 8 cm^3.

Centimetre cubes can be packed into boxes, layer by layer.
The first layer needs $3 \times 4 = 12$ cubes.
The box will hold 2 layers, so it needs $2 \times 12 = 24$ cubes in total. When full, the volume of the box is 24 cm^3.

In general, this cuboid has h layers with $l \times w$ cubes in each layer.

→ **The volume of any cuboid = $l \times w \times h$**

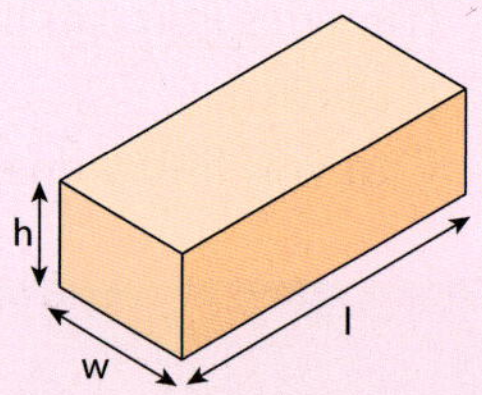

A cube is a cuboid with all its edges the same length, l.

So, the volume of a cube = $l \times l \times l = l^3$.

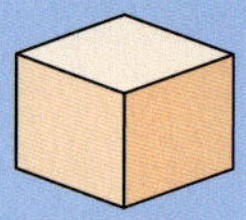

EXAMPLE

This office block has its maintenance equipment in a space on its roof.
Find the total volume of the building.

ANSWER

Volume of office block $= 30 \times 15 \times 10$
$= 4500\ \text{m}^3$.
Volume of roof space $= 10 \times 5 \times 3 = 150\ \text{m}^3$.
Total volume of building $= 4500 + 150$
$= 4650\ \text{m}^3$.

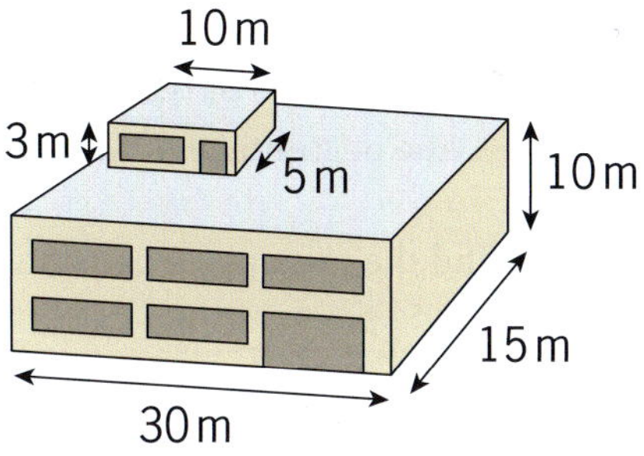

Exercise C23.1

MEDIUM LOW

1 For each of these open boxes, how many centimetre cubes are needed **i** to cover the base **ii** to fill the box?

a

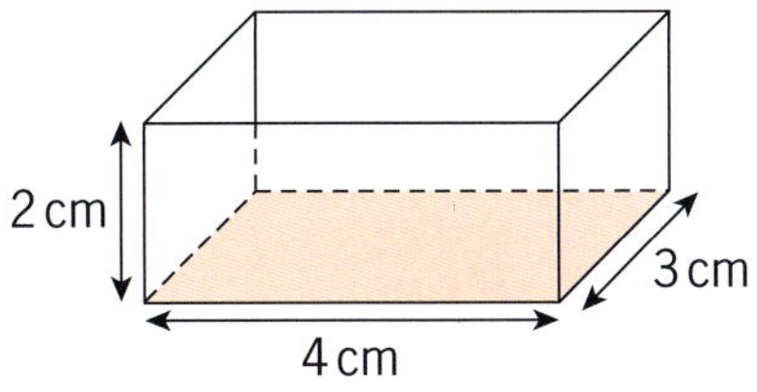

b

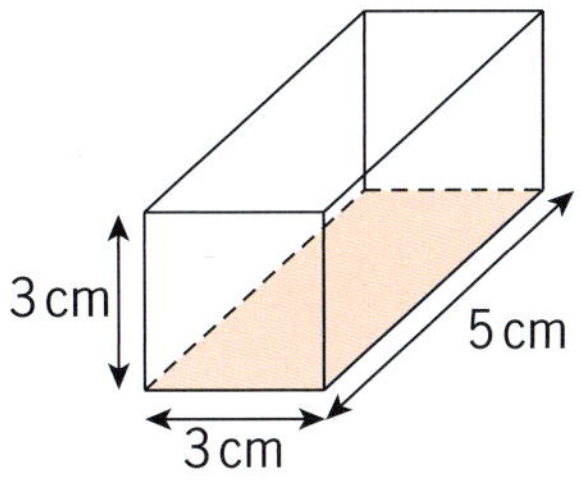

c

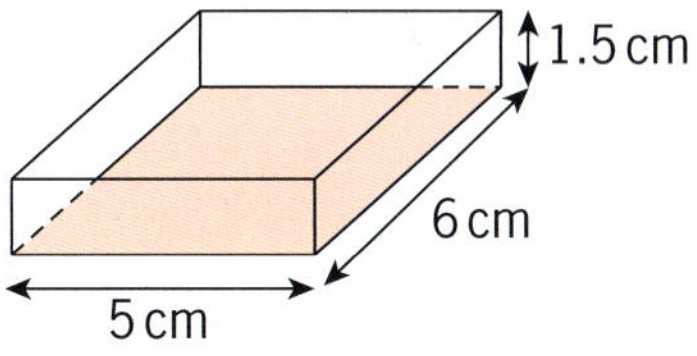

2 Find the volume of these cuboids.

Length	Width	Height
7 cm	4 cm	27 cm
10 cm	12 cm	4 cm
8.5 cm	3 cm	12 cm

3 A full tin of olive oil stands 12 cm tall on a square base of side 9.5 cm.
How much oil does it contain in **a** cm^3 **b** litres?

$1000\ \text{cm}^3 = 1$ litre.

4 An open tank 12.5 metres by 8.2 metres catches rainwater.
If the level rises 3 cm during a storm, how much water has it collected in
a cm^3 **b** litres?

5 A vacuum cleaner is packed in this box.
Find the capacity of the box
a in cm^3 **b** in litres.

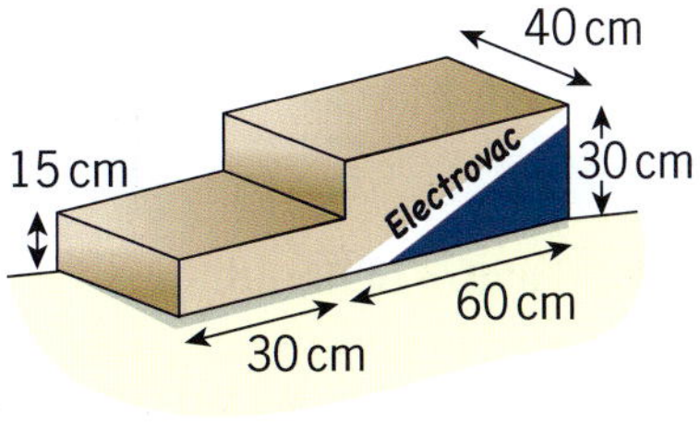

Continued ➤

Unit C

6 A solid piece of wood is hollowed out to create an open-topped container.
Calculate
a the volume of the wood making the container
b the percentage of the solid piece of the wood that has been cut away.

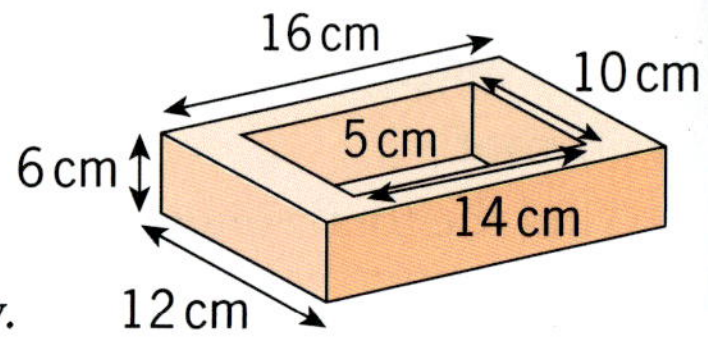

7 Find the missing dimension in each of these cuboids.

a
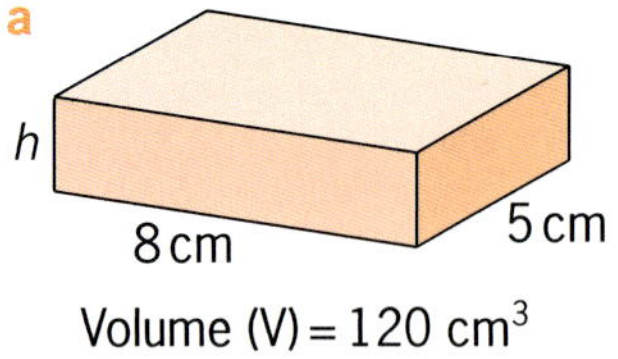

b
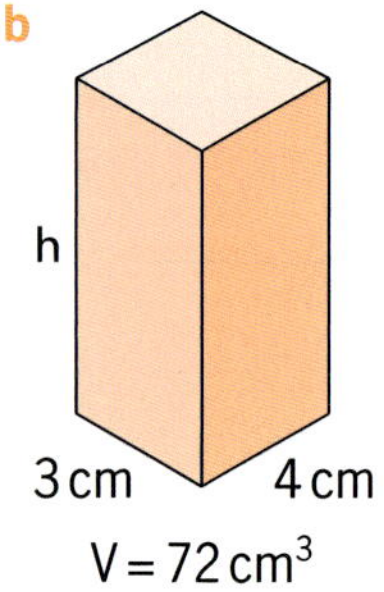

c
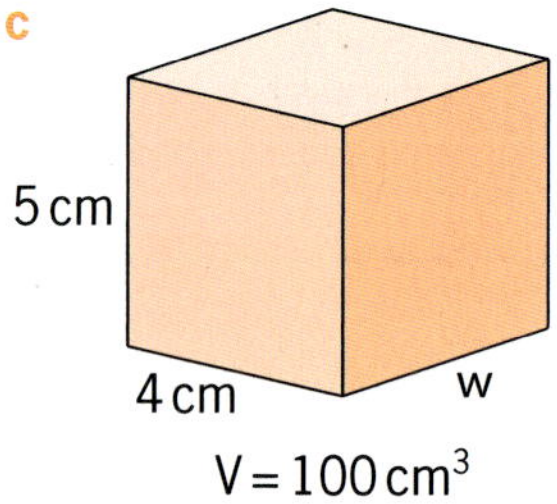

d
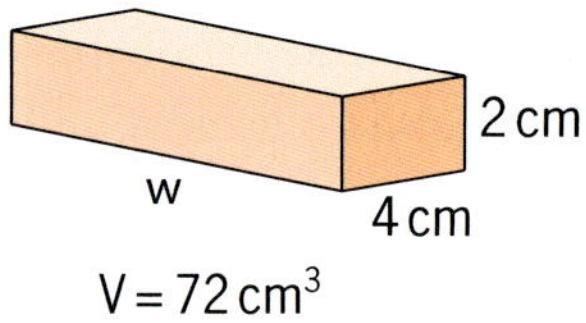

8 A carton holds one litre of orange juice and it is filled with juice to the top of the cuboid.
It has a square base of side 7 cm.
What is the height, h, to the nearest 0.1 cm?

9 Youssou needs a new fridge. It must have a base 50 cm by 40 cm.
The model he likes has a volume of 160 000 cm^3.
Will this model fit under a worktop which is 85 cm above the kitchen floor?

10 **RICH TASK**

Find the dimensions of all the different cuboids which each have
- a volume of 1 litre
- dimensions which are a whole number of centimetres
- a base with both length and width in multiples of 10 cm.

Which of your answers would suit a pack of ready-made custard?

11 **RICH TASK**

A hollow cuboid must hold a litre of liquid.
Suggest different possible dimensions for the cuboid.
Which of your suggestions has the smallest surface area?
What is special about this particular cuboid?

Volume of prisms

C23.2

- Find and use volume of prisms

RICH TASK

Chocolate is packed in triangular prisms of different sizes.
Here are the dimensions (in centimetres) and mass of four sizes.

	Mass, grams	Base, b	Height, h	Length, l
Small	100	2	1.7	10
Standard	200	2.5	2.2	12.5
Large	350	3	2.6	15
Jumbo	800	4	3.4	20

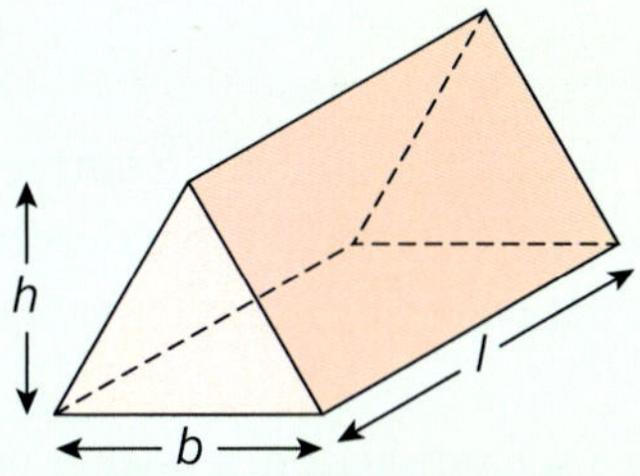

Explore the efficiency of this packaging.

A prism is a 3-D shape which has the same cross-section all along its length.

If its cross-section has an area A, its volume can be built up layer by layer.

If a layer is 1 cm thick, then
the volume of one layer is $A \times 1\text{ cm}^3$,
so the volume of h layers is $A \times h\text{ cm}^3$.

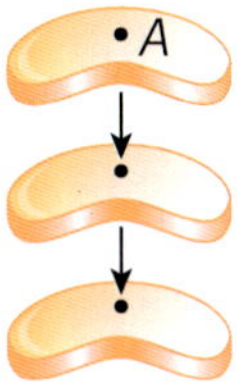

→ **The volume V of a prism with cross-sectional area A and height h is**

$$V = A \times h.$$

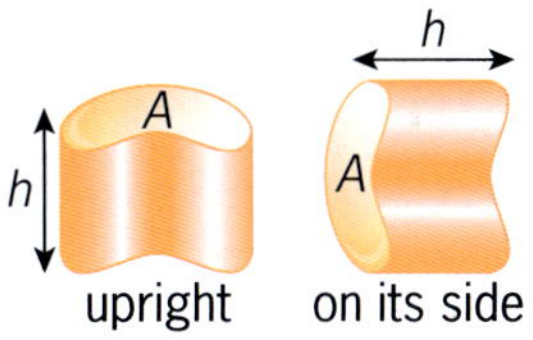

EXAMPLE

This ducting in a warm-air heating system has a length of 85 cm.
Find its volume in cm^3 and in litres.

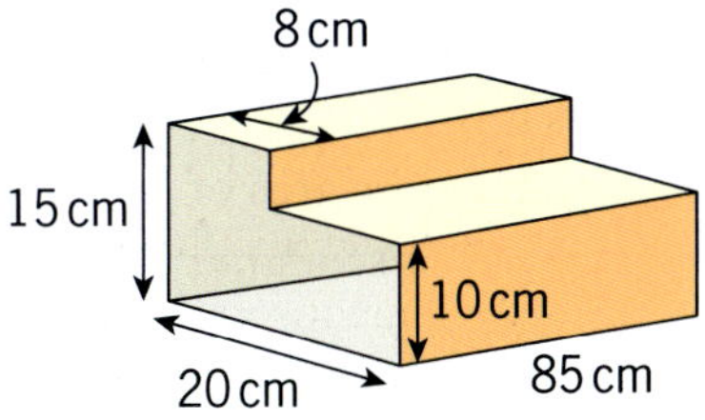

ANSWER

Its cross-section has this shape made from two rectangles.
The area of cross-section is $(20 \times 10) + (5 \times 8) = 200 + 40$
$= 240\text{ cm}^2$.

The volume of the ducting is $A \times l = 240 \times 85$
$= 20\,400\text{ cm}^3$
$= 20.4$ litres.

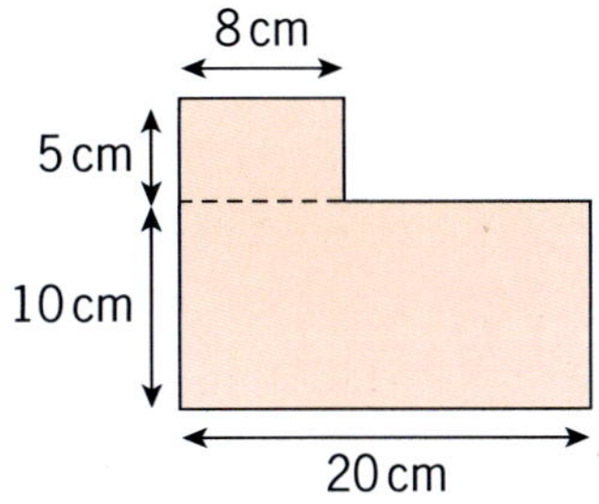

Remember that area is measured in cm^2 and volume in cm^3.

1 litre = 1000 cm^3.

Exercise C23.2

MEDIUM

1 Find the volume of these two prisms.

a
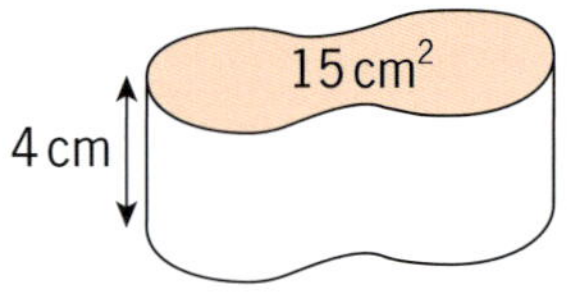

b
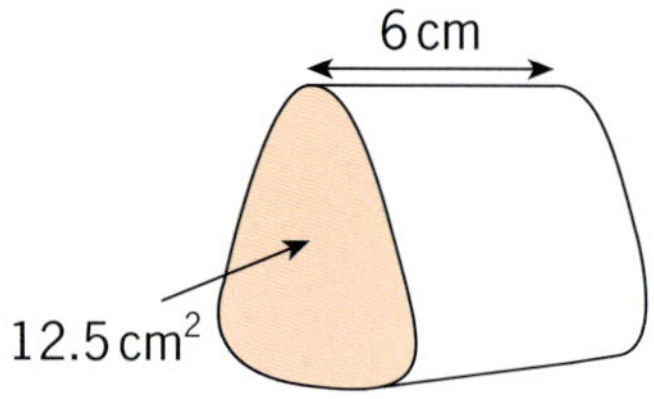

2 A small reservoir has a surface area of 850 m^2.
One night, rain fell to a depth of 0.12 metres.
What volume of rainwater was added to the reservoir overnight?

3 In a science lesson, a beaker of water has a surface area of 250 cm^2.
A lump of iron falls into the beaker and the water level rises by 1.4 cm. What is the volume of the iron?

4 For each prism, calculate
i the darker shaded areas of cross-section
ii the volume of the prism.
All measurements are in centimetres.

a
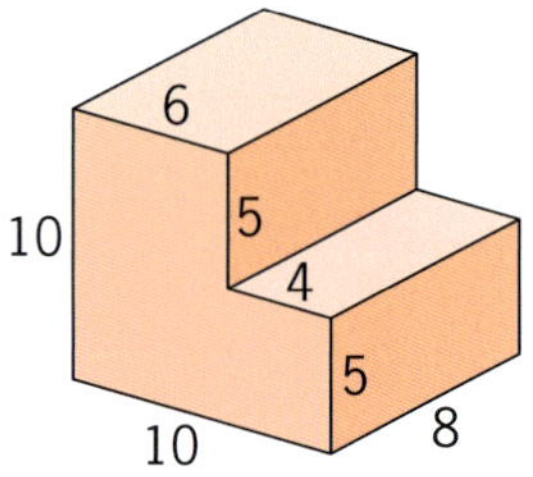

b
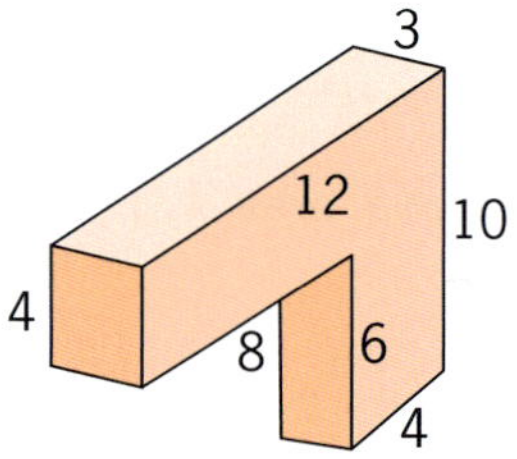

c
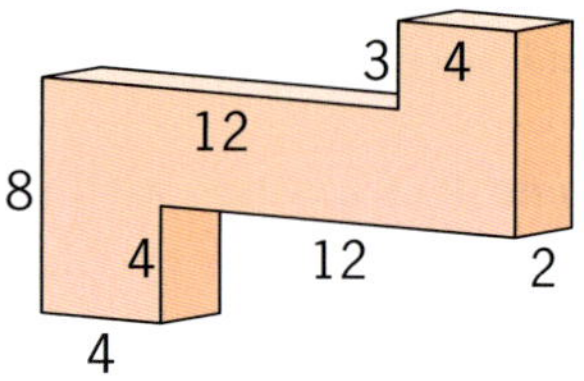

5 This lean-to shed is built alongside a wall.
Find **a** the area of the triangular end
b its volume.

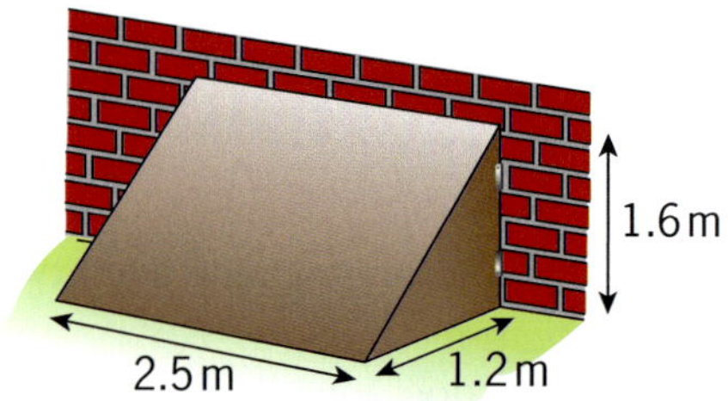

6 A cuboid is 10 cm long, 6 cm wide and 4 cm tall.
How many different ways can you find of cutting it in half with one straight cut?
Give names to the shapes of the halves that you find.

7 Find the area of cross-section of each of these prisms.

a

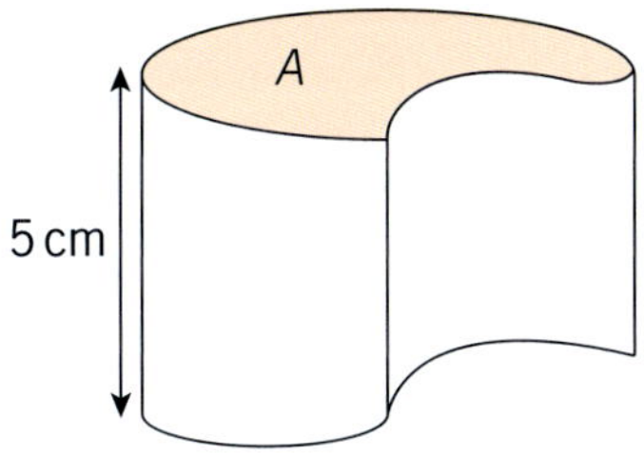

Volume (V) = 50 cm^3

b

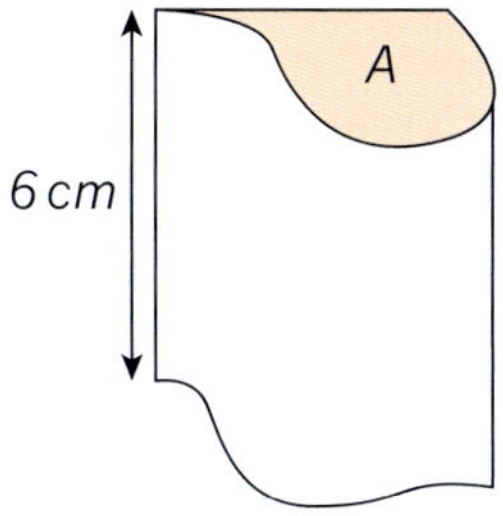

V = 42 cm^3

c

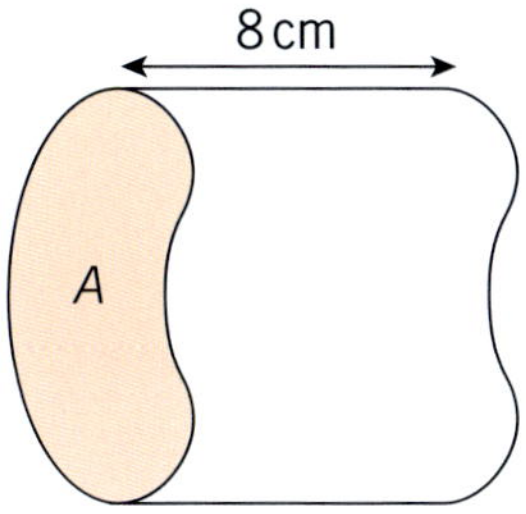

V = 40 cm^3

d

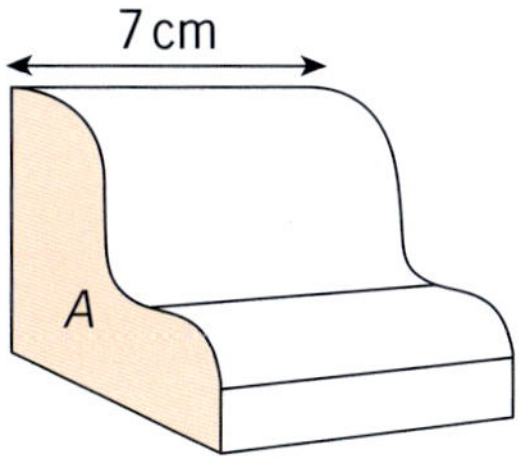

V = 42 cm^3

Unit C

Volume of cylinders

- Calculate and use volume of cylinders in various contexts

RICH TASK

You have two identical beakers.
One contains water 7 cm deep.
The other contains a lump of metal.
The only equipment you have is a ruler and calculator.
How can you find the volume of the metal?

→ **A cylinder is a prism with a circular cross-section.**
The area of the cross-section, $A = \pi r^2$.
The volume of the cylinder, $V = A \times h$.
→ **For a cylinder with height h and radius r, $V = \pi r^2 h$.**

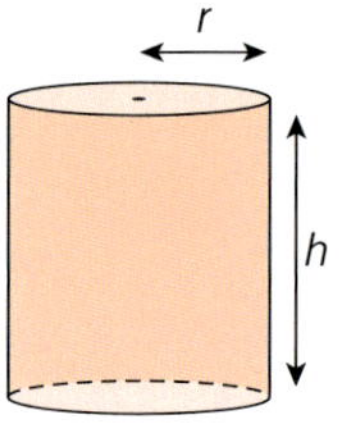

EXAMPLE

Find the capacity of a cylindrical tin of rice pudding which is 10 cm high with a radius of 4 cm.

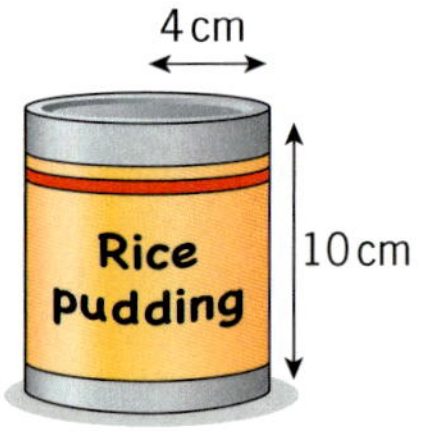

Take π as 3.14 or use the value in your calculator.

ANSWER

The capacity of the tin $= \pi r^2 h$
$= 3.14 \times 4^2 \times 10 = 3.14 \times 16 \times 10$
$= 502\text{ cm}^3$ to the nearest whole cm^3.

This capacity is just over half a litre. It may be written on the label as *500 ml* or *0.5 litres*.

Some cylindrical shapes, like drainage pipes, are hollow in the middle and have thick walls.

EXAMPLE

Find the amount of concrete (in litres) needed to make this drainage pipe.

ANSWER

The total volume $= \pi R^2 h = 3.14 \times 20^2 \times 100$
$= 125\,600\text{ cm}^3$.

The volume of the hole $= \pi r^2 h = 3.14 \times 15^2 \times 100$
$= 70\,650\text{ cm}^3$.

So the volume of the concrete $= 125\,600 - 70\,650$
$= 54\,950\text{ cm}^3$
$= 55$ litres, to the nearest whole litre.

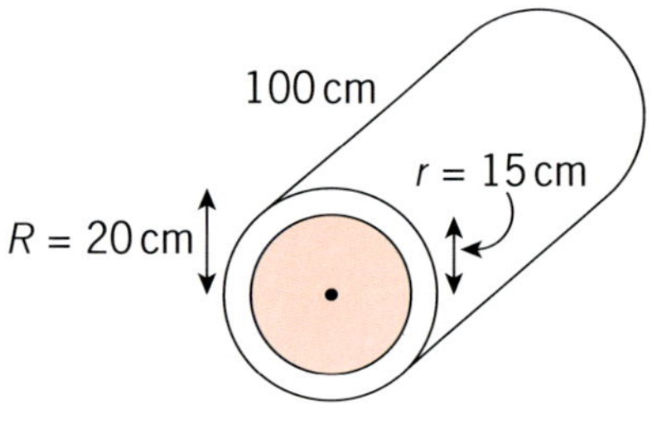

Exercise C23.3

MEDIUM

1 Calculate the volume of these solid cylinders.
Take care with radii and diameters.
Give your answers to the nearest whole cm^3.

a
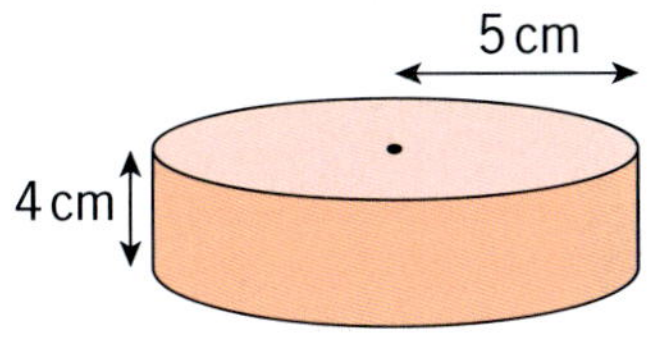

b
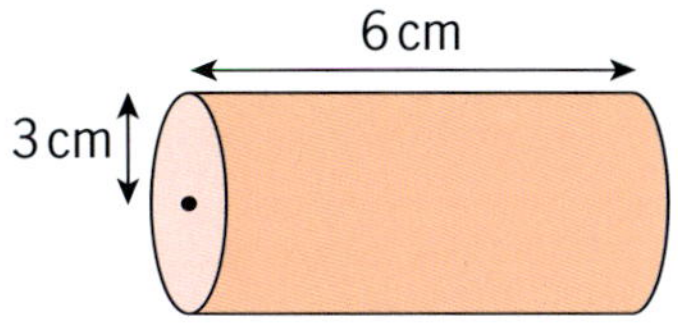

c
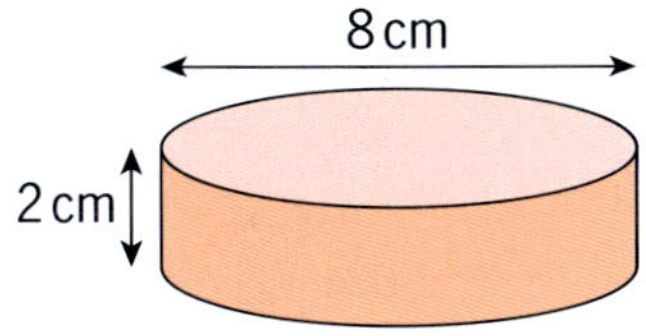

d
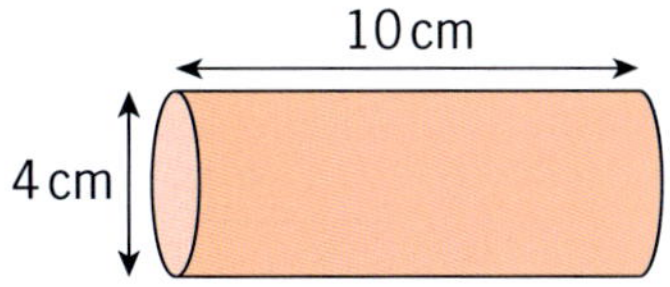

2 Calculate the capacities of these tins of food.
Give your answers in ml to the nearest 10 ml.

	Vegetable soup	Marmalade	Kidney beans	Tomatoes
Radius, cm	3.7	4.0	3.5	5.6
Height, cm	11.0	7.2	7.2	8.5

3 Calculate the volume of concrete needed to make these pipes.

a
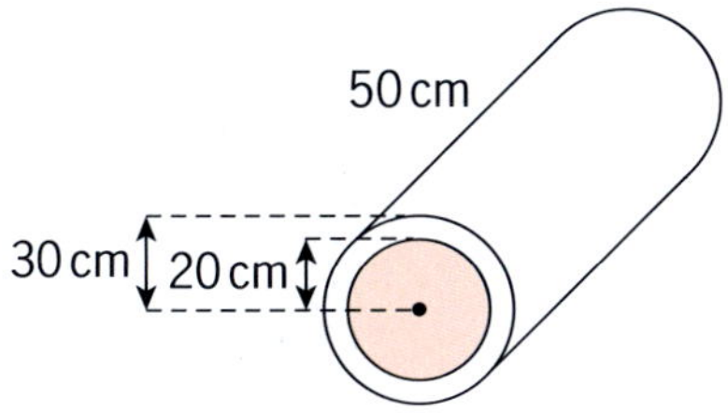

b
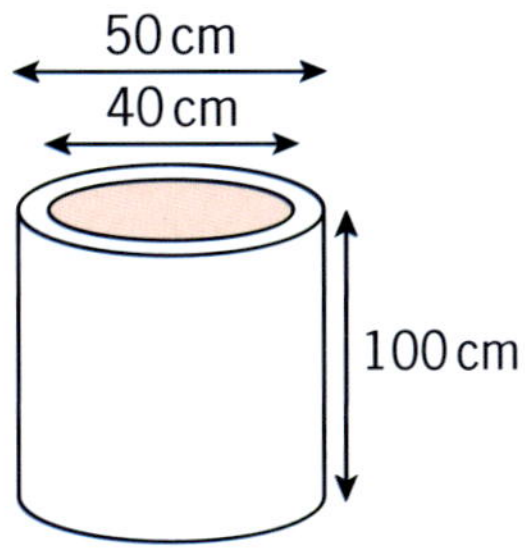

*4 A cylindrical bottle of olive oil says *1 litre* on the label. The bottle's inner diameter is 10.4 cm and its height is 11.8 cm. Should you complain about the labelling?

Continued ➤

5 A car has four cylinders in its engine. Each cylinder has a diameter of 6.0 cm and a length of 17.65 cm.
What do you think will be the total engine capacity quoted in advertisements?

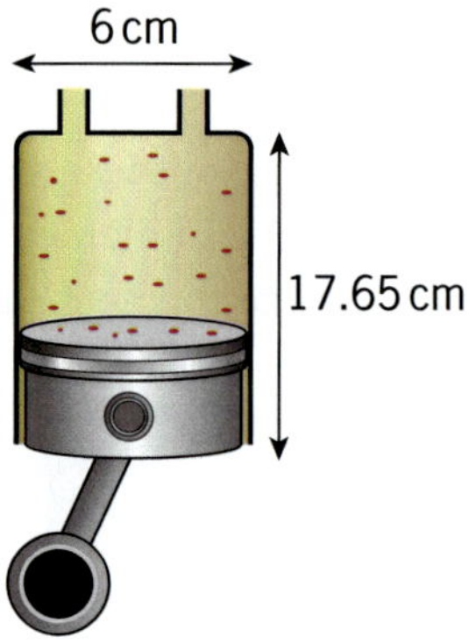

6 Find the heights or lengths of these cylinders (to the nearest cm).

a 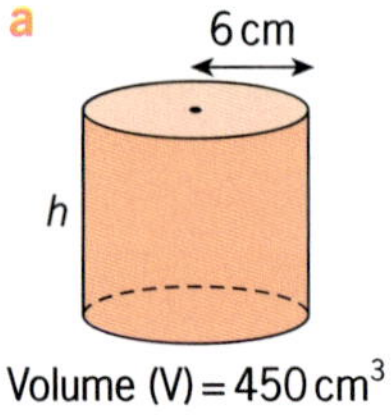

b 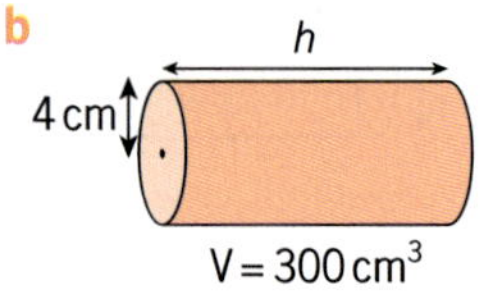

c 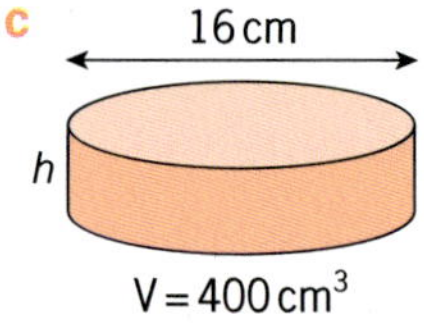

d 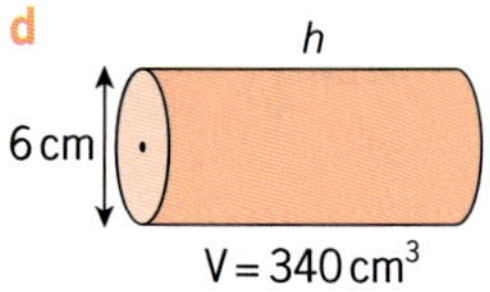

7 Find the radius of each of these cylinders (to the nearest 0.1 cm).

a 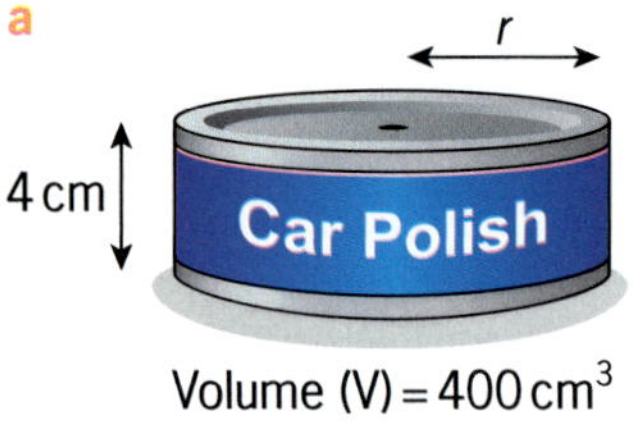

b 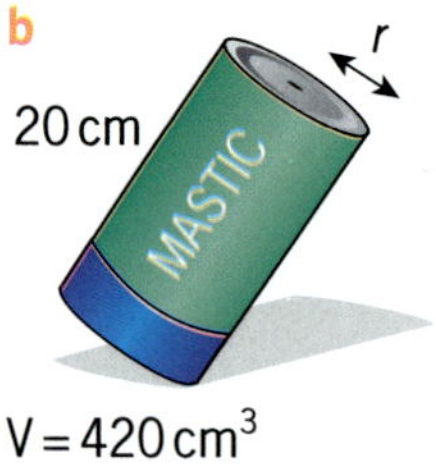

c 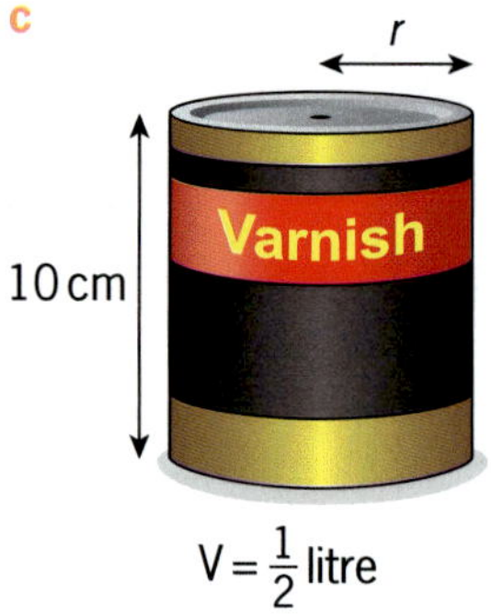

d 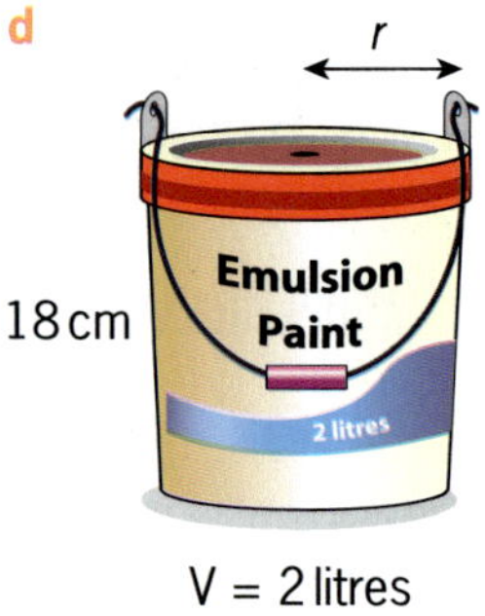

Area and volume scale factors

C23.4

- Find and use area and volume scale factors

RICH TASK

Here is an enlargement in three dimensions.

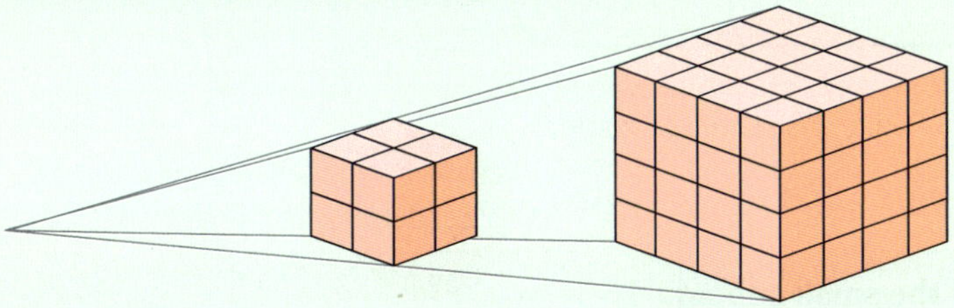

What is the scale factor of enlargement for the length of each edge, the area of each face and the volume of the shape?

1 millimetre (1 mm) enlarged by a factor of 10 becomes 1 centimetre (1 cm).

2 This diagram shows a square with sides of 1 mm enlarged using a length scale factor of 10.

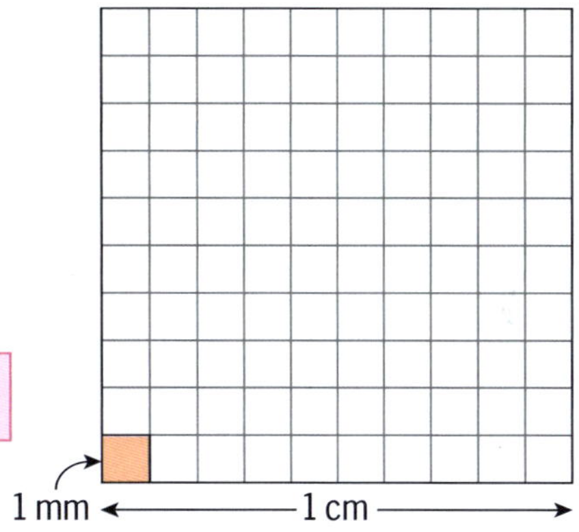

For 10 rows of 10 squares, you need $10^2 = 10 \times 10 = 100$ small squares to make the large square.

→ **$1\ \text{cm}^2 = 10^2\ \text{mm}^2 = 100\ \text{mm}^2$.**

This diagram shows a cube with edges of 1 mm enlarged using a length scale factor of 10.

It needs $10^3 = 10 \times 10 \times 10 = 1000$ small cubes to make the large cube.

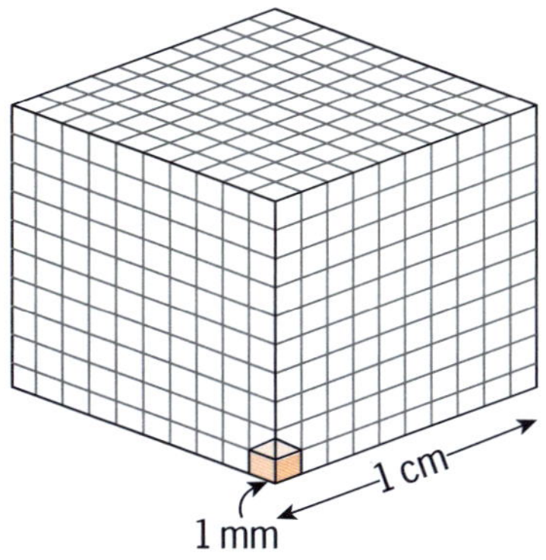

→ **$1\ \text{cm}^3 = 10^3\ \text{mm}^3 = 1000\ \text{mm}^3$.**

When you enlarge a solid shape by a scale factor *m*,
- → **the lengths increase by a factor *m*.**
- → **the surface area increases by a scale factor m^2.**
- → **the volume increases by a scale factor m^3.**

EXAMPLE

Change **a** 6.4 m^2 to cm^2 **b** 350 mm^3 to cm^3.

ANSWER

a $1 \text{ m}^2 = 100^2 \text{ cm}^2 = 10\,000 \text{ cm}^2$

So $6.4 \text{ m}^2 = 6.4 \times 10\,000 \text{ cm}^2$

$= 64\,000 \text{ cm}^2$

b $1 \text{ cm}^3 = 10^3 \text{ mm}^3 = 1000 \text{ mm}^3$

So $350 \text{ mm}^3 = 350 \div 1000 \text{ cm}^3$

$= 0.35 \text{ cm}^3$

There is a similar relationship for metres and centimetres.

1 metre = 100 cm so

- for area, $1 \text{ m}^2 = 100^2 \text{ cm}^2 = 10\,000 \text{ cm}^2$
- for volume, $1 \text{ m}^3 = 100^3 \text{ cm}^3 = 1\,000\,000 \text{ cm}^3$

Unit C

Exercise C23.4

MEDIUM

1 This large cuboid is an enlargement of the small cuboid.

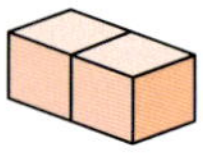

a How many times longer have the edges grown? Write down the length scale factor.

b How many times bigger has the area of the top become? Write down the area scale factor.

c How many times more small cubes are needed for the large cuboid? Write down the volume scale factor.

d Check that the scale factors follow the rule given at the bottom of page 349.

2 Use the conversion 1 cm = 10 mm to change

a i 8 cm to mm	ii 6 cm^2 to mm^2	iii 2 cm^3 to mm^3.
b i 35 mm to cm	ii 140 mm^2 to cm^2	iii 500 mm^3 to cm^3.

3 Use the length conversion 1 km = 1000 metres in this problem. A nature reserve has a lake with a perimeter of 7.4 km and a surface area of 2.1 square kilometres. The lake has an estimated volume of 5 000 000 cubic metres.

Find **a** the perimeter of the lake in metres

b its surface area in square metres

c its estimated volume in cubic kilometres.

4 ★ **CHALLENGE** ★

The new Wembley Stadium has the world's longest single span of 315 metres across its roof. The roof has an area of 44 500 m^2. The stadium encloses a volume of 4 000 000 (4 million) cubic metres.

Find **a** the length of its span in kilometres

b the area of the roof in square kilometres

c the volume of the stadium in cubic kilometres.

5 ★ **CHALLENGE** ★

A hot air balloon with a volume of 1550 m^3 has a maximum diameter of 15 metres and the area of its mouth is 30 m^2.

If you build a smaller version to a scale of 1 : 15, what are the equivalent measurements for your model?

Unit C

Summary and assessment

C23

CHECK OUT

You should now be able to

- solve problems involving volume of
  - cubes and cuboids
  - prisms
  - cylinders
- understand the effect of enlargement on area and volume.

Exam-style question

The diagram shows the cross-section of a swimming pool.

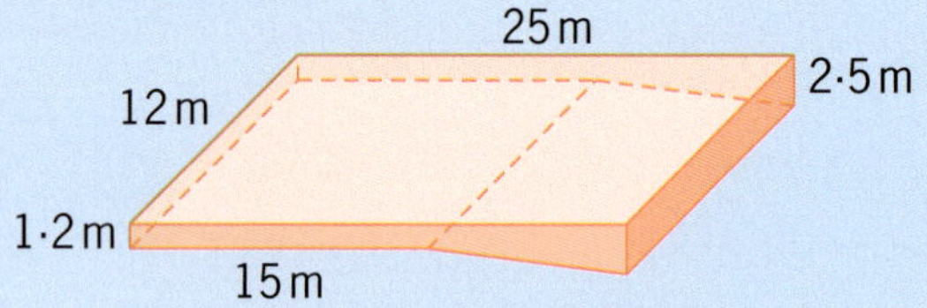

The pool is filled so that the water level is 20 cm from the top of the pool.
Work out the volume of water needed.

Karen's answer:

Depth of water in shallow end is
$1.2 - 0.2 = 1\text{ m}$

Cross section is rectangle plus trapezium

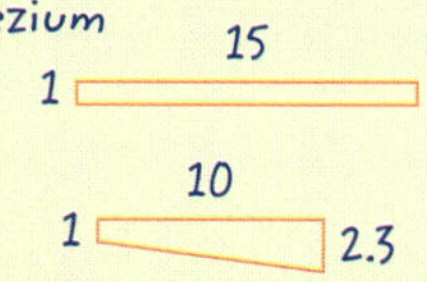

Area rectangle $= 1 \times 15 = 15$

Area trapezium $= \frac{1}{2}(1 + 2.3) \times 10$
$= 16.5$

Total area $= 15 + 16.5 = 31.5$

Volume $= 31.5 \times 12 = 378\text{ m}^3$

Divide the cross-section into shapes you can find the area of.

The width of the pool is the same, so first find the cross-section area.

Always give the units of your final answer.

Work out the lengths of the shapes. Remember to use same units: change 20 cm to 0.2 m.

Exam practice

1 Work out the volume of these prisms.

a

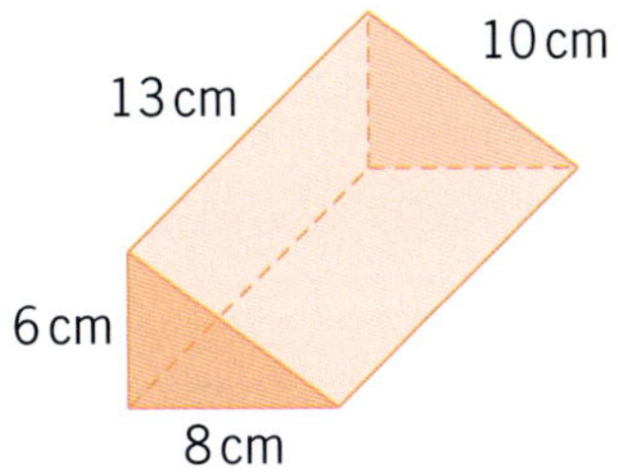

b

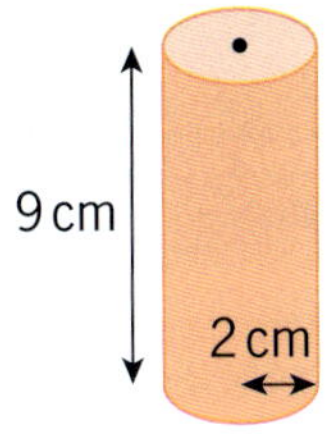

(*6 marks*)

2 This cube has surface area 150 cm^2.
Work out the volume of the cube.

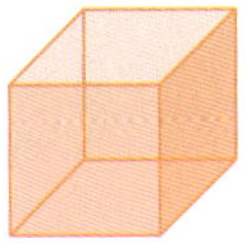

(*4 marks*)

3 Gareth designs and makes a scale-model playhouse for children.
The design is made to scale.
The dimensions of the playhouse will be four times the dimensions of the scale model.

a The scale model uses 100 cm^2 of materials.
What is the total area of materials needed to build the playhouse?

b A chair in the playhouse needs 800 cm^3 of space.
How much space would be needed in the scale model? (*4 marks*)

24 Study of chance

INTRODUCTION

There are many things in life which are uncertain. Will it be sunny tomorrow? Will my football team win the Premier League? Will I be able to afford a house in the future? The mathematics used to deal with uncertainty is called probability.

What's the point?

When the Met Office gives a weather forecast, they use a complex mathematical model to predict the probability of sunshine in a particular region.

CHECK IN

1 Cancel these fractions to their simplest form.

a $\frac{6}{9}$ b $\frac{5}{10}$ c $\frac{15}{20}$ d $\frac{2}{8}$ e $\frac{20}{20}$

2 Calculate

a $\frac{3}{4}+\frac{1}{4}$ b $\frac{7}{10}+\frac{3}{10}$ c $1-\frac{7}{10}$ d $1-\frac{3}{5}$

3 Calculate

a $1 - 0.1$ b $1 - 0.6$ c $1 - 0.15$

4 Convert these fractions into decimals.

a $\frac{1}{100}$ b $\frac{25}{100}$ c $\frac{35}{100}$ d $\frac{5}{100}$ e $\frac{36}{100}$

Orientation

What I should know	What I will learn	What this leads to
B9	■ Calculate experimental and theoretical probability ■ Identify mutually exclusive outcomes	Life choices involving uncertainty

RICH TASK

Two people can play an old game called Rock, Paper, Scissors.

In the game

- rock is a closed fist.
- paper is palm on palm,
- and scissors is the number two horizontally.

The players reveal their 'hand' simultaneously.

A rock beats scissors.

Paper beats rocks, and scissors beats paper.

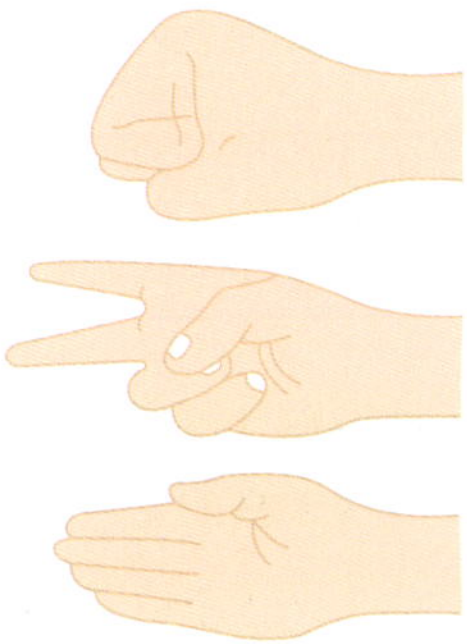

Is the game fair?

Probability and likelihood

- Use the vocabulary of probability
- Understand and use the probability scale

Probability is a measure of how likely a particular **outcome** is to happen in the future.

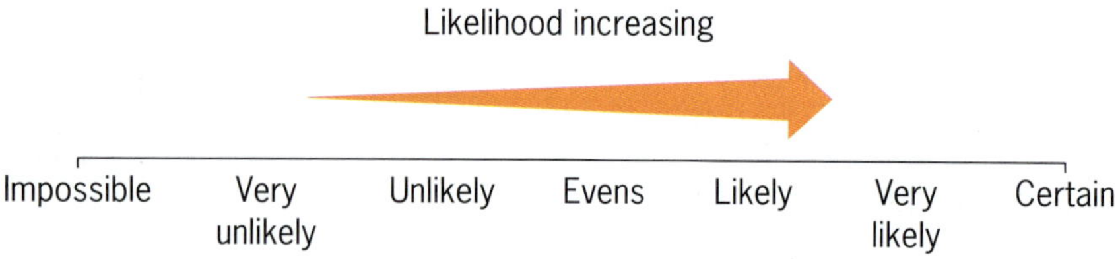

EXAMPLE

Write these outcomes where you think they should appear on a likelihood scale.

a A fair coin landing on tails.
b A score of 9 from an ordinary (six-sided) dice.
c A sunny day occurring in August in the UK.
d The sun rising tomorrow morning.

ANSWER

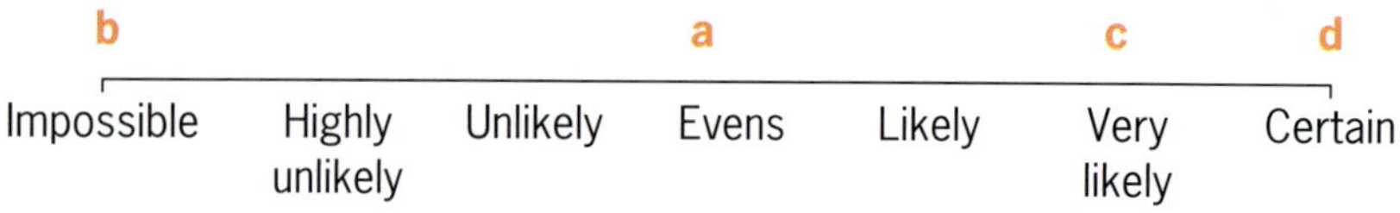

A more exact measure of likelihood is shown on a **probability** scale.

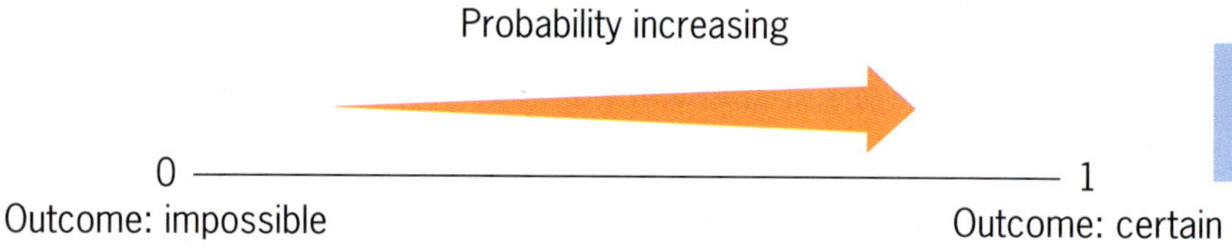

You can write probability as a fraction, decimal or percentage.

- **Outcomes range from impossible (with a probability of 0) to certain (with a probability of 1).**
- **All other outcomes have a probability between 0 and 1.**

Luke predicts that there is an 83% chance that Chelsea will win the football FA cup. He writes

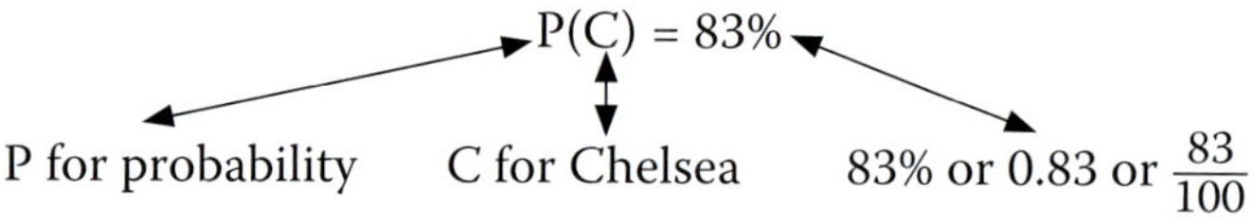

EXAMPLE

Show these probabilities on a probability scale.

a P(R) = 0.7

b The probability of tails on a fair coin = $\frac{1}{2}$

c P(rain tomorrow) = 20%

ANSWER

0	0.1	0.2	0.3	0.4	0.5	0.6	0.7	0.8	0.9	1
		c			b		a			

Exercise C24.1

LOW

Unit C

1 Choose words from the likelihood scale that best fit each of these outcomes.

  a A fair dice when rolled will land on a 4.

  b A fair coin when spun will land on heads.

  c There will be a tornado in a town in England next year.

  d The next Catholic pope elected will be female.

2 Draw a likelihood scale and mark on the scale these events.

  a A day of the week chosen at random will be a Saturday.

  b A fair dice when rolled will show an odd number.

  c A person chosen at random will have
  i blonde hair **ii** red hair.

  d Today, someone somewhere will have a birthday.

3 Write in words

  a $P(X) = 0.6$ b $P(Y) = \frac{1}{4}$ c $P(Z) = 35\%$

4 Draw a probability scale.
Mark on the scale these probabilities:

  a P(choosing a black card from an ordinary pack of playing cards) = 50%

  b P(choosing a spade from a pack of playing cards) = $\frac{1}{4}$

  c P(waking up before the alarm goes off) = 0.9.

  d P(score 2 on fair dice) = $\frac{1}{6}$

DID YOU KNOW...?

Playing cards are thought to be of Chinese origin, and were introduced into Europe in the 14th century. Modern playing cards comprise four 'suits': hearts and diamonds (which are red), and clubs and spades (black).

Equally likely outcomes

C24.2

- Understand measures of probability from theoretical models
- List all outcomes for single events in a systematic way

When you roll a fair dice, there are six possible **outcomes**.
Each outcome is **equally likely** to happen,
so the probability of each outcome = $\frac{1}{6}$.

→ **An event is a collection of outcomes**
With the fair dice, three outcomes (1, 3 and 5) give an odd number.
So the probability of an odd number is $\frac{3}{6}$.

When all the possible outcomes are equally likely you can find the probability of the event happening:

→ **Probability an event happens** $\quad P(\text{event}) = \dfrac{\text{number of favourable outcomes}}{\text{total number of possible outcomes}}$

EXAMPLE

Seven marbles, identical except for their colour, are placed in a bag. Each marble is a rainbow colour: red, orange, yellow, green, blue, indigo or violet. One marble is chosen at random.

Find the probability that the marble is
a blue **b** red **c** not red **d** white **e** a rainbow colour.

ANSWER

a $P(\text{blue}) = \frac{1}{7}$ — only one marble is blue

b $P(\text{red}) = \frac{1}{7}$ — only one marble is red

c $P(\text{not red}) = \frac{6}{7}$ — 6 of the 7 marbles are not red

d $P(\text{white}) = 0$ — none of the marbles is white

e $P(\text{rainbow colour}) = 1$ — all of the marbles are a colour of the rainbow

Notice that

$P(\text{red}) = \frac{1}{7}$ and $P(\text{not red}) = \frac{6}{7}$, and $\frac{1}{7} + \frac{6}{7} = 1$.

→ **Total probability for all possible outcomes = 1**
→ **Probability of an event not happening = 1 − probability of that event happening**

EXAMPLE

One person at a youth group is chosen at random. The probability that the student owns an mp3 player is 0.79. Find the probability the student does not own an mp3 player.

ANSWER

P(not owning an mp3) = 1 – 0.79 = 0.21.

Exercise C24.2

LOW

Unit C

1 A bag contains seven green, eight blue and 10 red counters. Nine of the counters are large, 16 are small. One counter is chosen at random. What is the probability that the counter is

a green b not green c large

d small e red f not blue?

2 There are 30 children in class 10Z: 18 girls and 12 boys. All the children are aged 14 or 15, and all of the children own a mobile phone. A child is chosen at random from class 10Z. Find the probability that the child chosen

a is aged 10 b owns a mobile phone

c is a girl d is a boy.

3 A spinner has either circles or squares or triangles on its sides. The table gives the probabilities of landing on circle and square.

Outcome	Circle	Square	Triangle
Probability	0.25	0.55	

Work out the probability of the spinner landing on triangle.

*4 A bag contains green and yellow marbles. The probability of choosing a green marble is $\frac{3}{4}$.

a Sally thinks there are seven green marbles in the bag. Explain why she must be wrong.

b Tessa takes one marble out of the bag. It is yellow. What is the smallest number of marbles that there could be in the bag?

*5 Harriet has two bags of counters. Harriet takes one counter at random from one of the bags. She wants to choose a green counter. Which bag should Harriet choose from? Explain how you decided on your answer.

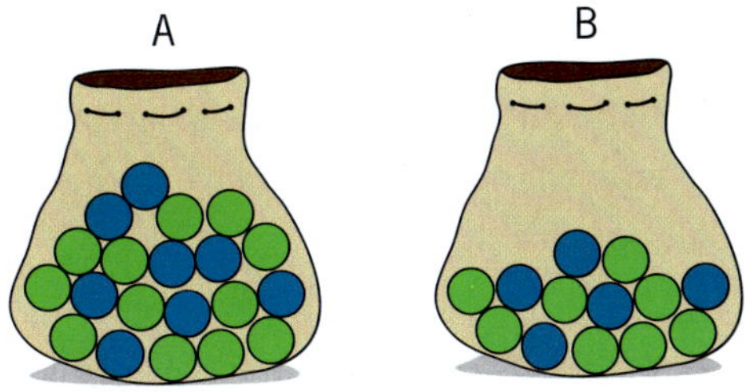

Mutually exclusive outcomes

C24.3

- Identify different mutually exclusive outcomes
- Know that the sum of the probabilities of all mutually exclusive outcomes is 1

→ **Two or more outcomes are mutually exclusive if they cannot happen at the same time.**

→ **When outcomes are mutually exclusive, you can add their probabilities to find the total probability that what happens is one of those outcomes.**

EXAMPLE

A spinner has 12 equal sides:
5 green (G), 4 blue (B), 2 red (R) and 1 white (W).
The spinner is spun.

a What is the probability that the spinner lands on

i green or white **ii** green or blue

iii blue or white **iv** blue or red or white

v not green?

b Why are the answers to parts **iv** and **v** the same?

ANSWER

a **i** $P(G) = \frac{5}{12}$ $P(W) = \frac{1}{12}$

$P(G \text{ or } W) = \frac{5}{12} + \frac{1}{12} = \frac{6}{12}$

ii $P(G) = \frac{5}{12}$ $P(B) = \frac{4}{12}$

$P(G \text{ or } B) = \frac{5}{12} + \frac{4}{12} = \frac{9}{12}$

Add the probabilities.

iii $P(B) = \frac{4}{12}$ $P(W) = \frac{1}{12}$

$P(B \text{ or } W) = \frac{4}{12} + \frac{1}{12} = \frac{5}{12}$

iv $P(B) = \frac{4}{12}$ $P(R) = \frac{2}{12}$ $P(W) = \frac{1}{12}$

$P(B \text{ or } R \text{ or } W) = \frac{4}{12} + \frac{2}{12} + \frac{1}{12} = \frac{7}{12}$

v $P(\text{not } G) = 1 - P(G) = 1 - \frac{5}{12} = \frac{7}{12}$

b The possible outcomes are green, blue, red and white.
So if not green, it has to be blue, red or white.

→ **The sum of the probabilities of mutually exclusive outcomes that cover all possibilities is 1.**

EXAMPLE

When you roll a dice, which two of these four events are mutually exclusive and also cover all possible outcomes?

A – prime numbers B – odd numbers

C – multiples of 2 D – factors of 6

ANSWER

Event B is the numbers 1, 3, 5, and event C is 2, 4, 6.
So events B and C are mutually exclusive and also cover all outcomes.

Exercise C24.3

MEDIUM

1 The table shows information about the type of pet owned by students in class 10A. Each student in class 10A has either one pet or no pets at all.

Pet	Cat	Dog	Hamster	Fish	No pet
Number of students	7	8	2	4	9

a How many students are there in the class?

b One student is chosen at random from the class.
Work out the probability that the student chosen will

i own a cat
ii own a cat or a dog
iii own a dog or a fish
iv own a cat or a hamster
v not own a pet
vi own a pet.

How realistic is it that each student has only one type of pet? What difference, if any, would it make to the answers if their was not true?

2 The table shows the probabilities of getting scores on a biased dice.

Score	1	2	3	4	5	6
Probability	0.2	0.15	0.3	0.1	0.2	0.05

Work out the probability of scoring

a 2 or 3
b 5 or 6
c an even number.

3 The table shows the probabilities of getting colours on a rainbow spinner.

Colour	Red	Orange	Yellow	Green	Blue	Indigo	Violet
Probability	0.1	0.1	0.25	0.13	0.1	0.12	0.2

When the spinner is spun

a which colours are equally likely to happen?

b what colour is the most likely to show?

c what is the probability of

i blue or red
ii indigo or violet
iii orange or green?

*4 ***RICH TASK***

Jonah owns an ordinary six-sided dice numbered 1, 2, 3, 4, 5, 6.

a Describe two events that are mutually exclusive and cover all possibilities.

b Describe three events that together are mutually exclusive and cover all possibilities.

c Describe two events that are not mutually exclusive. Explain why.

Unit C

Experimental and theoretical probability

C24.4

- Compare experimental data to theoretical probabilities

RICH TASK

Patsy makes a five-sided spinner.
Each side is the same size.
They are coloured red, green, blue, white and yellow.
She spins it 100 times and these are the results.

Red	Green	Blue	White	Yellow
27	18	20	16	19

Do you think the spinner is fair or biased?
Give a reason for your answer.

If Patsy's spinner is fair, then the **theoretical probability** of it landing on any colour is $\frac{1}{5}$.

P(green) = $\frac{1}{5}$.

If she thinks her spinner is biased, she can estimate the **experimental probability** of each colour.

P(green) = $\frac{18}{100} = \frac{9}{50}$.

EXAMPLE

A **biased** coin is thrown 200 times. It lands on heads 140 times.

a Estimate the probability of this coin landing on heads on the next throw.

b Estimate the number of heads you expect to get when the coin is thrown 500 times.

ANSWER

a P(head) = $\frac{140}{200} = \frac{7}{10}$.

b Estimate the expected number: $500 \times \frac{7}{10} = 350$.

→ **Expected number = total number × probability.**

EXAMPLE

Tom carries out a survey about the number of people that are left-handed in his village.
He asks 40 people in the village, and five of them are left-handed.

a Estimate the probability that a person in the village is left-handed.

b If 192 people live in his village, estimate how many are left-handed.

ANSWER

a P(left-handed) = $\frac{5}{40} = \frac{1}{8}$.

b Estimated number: $192 \times \frac{1}{8} = 24$.

→ **The closer experimental probability is to theoretical probability, the less likely it is that there is bias.**

Exercise C24.4

MEDIUM

1 The probability that a biased dice will land on a six is 0.35.
The dice is rolled 400 times.
Work out an estimate for the number of times the dice will land on a six.

2 The probability that a biased coin will land on tails is 0.24.
The coin is thrown 500 times.
Work out an estimate for the number of times the coin will land on tails.

3 A biased coin is thrown 80 times. It lands on heads 50 times.
Estimate the probability that this coin will land heads on the next throw.

4 A biased four-sided dice is rolled 120 times.
The table shows the outcomes.

Score	1	2	3	4
Frequency	22	34	44	20

a Why could you say it is twice as likely that the dice will land on a 3 as on 1?
b The dice is rolled once more.
Estimate the probability that it lands on 2.
c The dice is rolled a further 300 times.
How many times would you expect the dice to land on 4?

5 A biased dice is rolled 100 times. The table shows the outcomes.

Score	1	2	3	4	5	6
Frequency	7	20	32	17	13	11

a The dice is rolled once more. Estimate the probability that the dice will land on
i 6 **ii** 3 **iii** 1 or 2
iv 4 or more **v** any number other than 5.
b The dice is going to be rolled a further 400 times.
How many times would you expect the dice to land on
i 2 **ii** 4 or 5?

6 A village has 300 households.
In a survey of 50 households, 29 said they owned two televisions.
How many of the households in the village would you expect to own two televisions?

Experimental probability and relative frequency

- Understand and use estimates of probability from relative frequency

Laura suspects that a coin is biased towards heads.

The first 10 spins of the coin give

H T H H H T H T H H

Laura calculates the **relative frequency** of obtaining a head:

$$\frac{7}{10} = 0.7$$

The next 10 spins of the coin give

T T H T H T H T H T

Laura recalculates the relative frequency from all 20 results:

$$\frac{11}{20} = 0.55$$

There are 11 heads out of 20 total.

This is closer to the expected probability of 0.5, but Laura carries out more trials to find out whether the coin is biased.

Laura spins the coin 40 more times, and finds the number of heads in each group of 10 spins as shown in the table.

Total spins	10	20	30	40	50	60
Total number of heads	7	11	17	22	28	33
Relative frequency	0.7	0.55	0.567	0.55	0.56	0.55

The relative frequency is settling down at around 0.55.
This is the **experimental probability** of obtaining a head with this particular coin.

This may mean that the coin is slightly biased towards heads.

Laura needs to carry out more trials if her estimate is to be reliable.

- **Relative frequency can give an estimate of probability.**
- **The more trials you carry out, the more reliable your estimate will be.**
- **The closer relative frequency is to theoretical probability, the less likely it is that there is bias.**

A **trial** is part of a statistical experiment which is repeated multiple times. In this case, it's a single spin of a coin.

Exercise C24.5

MEDIUM

*1 A spinner has 10 equal sides: five black and five red.
Dave carries out a statistical experiment.
He spins the spinner 280 times.
The spinner lands on black 133 times.
Is the spinner fair? Explain your answer.

*2 Jane carries out a statistical experiment.
She rolls a dice 200 times. The table shows the outcomes.

Outcome	1	2	3	4	5	6
Frequency	32	34	35	31	35	33

Is the dice fair? Explain your answer.

*3 Clara suspects that a coin is biased.
She spins the coin and notes how many heads she gets in each group of 10 spins. Clara spins the coin 100 times in total.
The table shows her results.

Total spins	10	20	30	40	50	60	70	80	90	100
Total number of heads	4	7	11	13	18	22	26	29	32	35
Relative frequency										

a Copy the table and complete the relative frequency row.
b Write down the best estimate of the probability of the coin landing on heads.
c Is the coin biased? Explain your answer.

*4 A four-sided dice is rolled 36 times. These are the outcomes:

3	3	4	2	1	3	2	4	3	1	2	4
2	1	3	4	1	2	3	1	2	4	3	2
3	4	1	2	3	4	1	3	1	4	2	4

a Work out the relative frequency of the dice landing on 1, 2, 3 and 4.
b Is the dice biased? Explain your answer.

*5 Dan suspects that a particular dice is biased towards the number 3.
Dan rolls the dice 30 times and gets these results

4	3	3	3	6	5	1	3	2	5
1	3	4	5	3	3	6	2	5	4
6	3	1	3	6	5	3	2	4	3

a What is the relative frequency of rolling a 3?
b Is the dice biased towards 3? Explain your answer.

Listing outcomes

- List all outcomes from two successive events in a systematic way

EXAMPLE

At a fete there are two silver money trees.
There are five envelopes left on each tree. Each envelope contains either 5p, 10p, 20p, 25p, or 50p.
To choose an envelope from the money tree costs 25p each time.

The amounts are not written on the envelopes. This is to stop people just choosing the 50p envelope each time.

Reuben chooses two envelopes, one from each tree.
He believes that he is likely to receive more money than he paid. Is he correct?

ANSWER

To help him find out if he was wise to play the money tree game,
Reuben lists the **set of all possible outcomes** in a **sample space diagram**.

	5p	10p	20p	25p	50p
5p	10p	15p	25p	30p	55p
10p	15p	20p	30p	35p	60p
10p	15p	20p	30p	35p	60p
50p	55p	60p	70p	75p	100p
50p	55p	60p	70p	75p	100p

Reuben paid a total of 50p.
There are 25 possible outcomes.
Each outcome is equally likely.
Thirteen of these would give a total amount greater than 50p.

The probability that Reuben would receive more than he paid is $\frac{13}{25}$.
This is greater than an even chance, but only just. He may be wise to play.

To solve problems involving probability you need to:
- → **identify the equally likely outcomes**
- → **count how many of these are favourable.**

Exercise C24.6

MEDIUM LOW

1 A coin is spun and a dice is rolled.
  a Use a table to list systematically all possible outcomes.
  b Find the probability of obtaining
    i a head and a 5   ii a tail and an even number.

2 A 10 pence and a 2 pence coin are spun.
  a Draw a table to show all the possible outcomes.
  b Find the probability that the 10p shows tails and the 2p shows heads.

* 3 Deputy school captains are to be chosen from two tutor groups.
In class A, four students were equally likely to be chosen: Helen, Irene, Joe and Kate.
In class B, three students were equally likely to be chosen: Steve, Tom and Ursula.
  a Draw a table to show all possible outcomes of choosing one student from each class.
  b Find the probability that one boy and one girl will be chosen.
  c Is it more likely that two boys will be chosen, or two girls? Explain how you know.

* 4 Two dice are rolled, a red dice and a blue dice. The score is the number showing on each dice multiplied together.
  a Draw a table to show all possible outcomes.
  b Find the probability of getting a score of
    i 18   ii 29   iii greater than 20.
  c Which is more likely, a score that is even or a score that is odd? Explain how you can tell.

* 5 Ed and Gala each make a three-sided spinner using card and a pencil.
Each sector is the same shape and size.
Each spinner is equally likely to land on any one of the sectors.

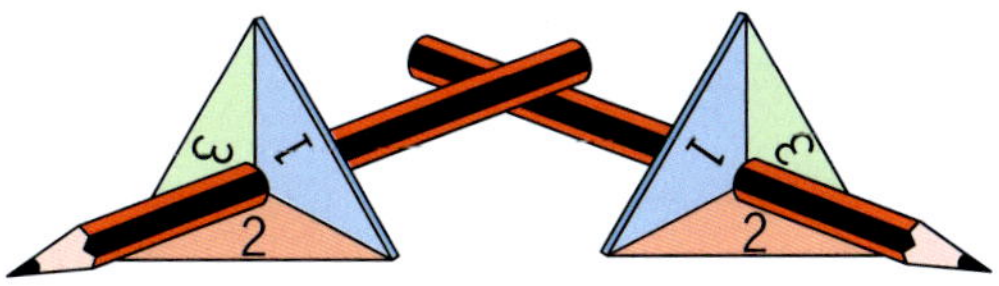

Ed and Gala decide to spin their spinner and add up the scores.
  a Draw a table to show the possible outcomes.
  b Work out the probability of a total score of
    i 2   ii 4.
  c Explain why the probability of a score of 3 and the probability of a score of 5 is the same.

Unit C

Summary and assessment

CHECK OUT

You should now be able to

- find simple probabilities
- compare experimental and theoretical probability
- find relative frequency
- list outcomes systematically.

Exam-style question

a Karl puts two yellow marbles and three green marbles in a bag. He is going to take one marble without looking. What is the probability that the marble will be yellow?

b Karl puts the marble back in the bag and then puts more yellow marbles in the bag. He is going to take one marble without looking. The probability that the marble is yellow is now $\frac{2}{3}$. How many more marbles did Karl put in the bag?

Melissa's answer:

a Total marbles $2 + 3 = 5$
Number of yellow marbles $= 2$
Probability (yellow marble) $= \frac{2}{5}$

b 2 yellow, total 5, $P(Y) = \frac{2}{5}$
3 yellow, total now 6, $P(Y) = \frac{3}{6}$
4 yellow, total now 7, $P(Y) = \frac{4}{7}$
5 yellow, total now 8, $P(Y) = \frac{5}{8}$
6 yellow, total now 9, $P(Y) = \frac{6}{9}$
$\frac{6}{9}$ is the same fraction as $\frac{2}{3}$
2 yellow marbles to begin with, 6 yellow marbles now; Karl added 4 yellow marbles to the bag.

Decide what information you need.

Always show your thinking and your working.

Answers to probability questions are fractions, decimals or percentages.

Finally, answer the question.

Exam practice

1 Jenna spins a five-sided spinner. Copy the spinner for each of these statements and write single-digit numbers in each space to make the statement correct.

a The probability of the spinner landing on a 5 or more is 1.

b It is impossible that the spinner lands on an even number.

c The modal score is 5. *(3 marks)*

2 Gareth buys a large variety pack of cat food pouches for his cats. The box contains 44 pouches in 4 flavours as follows

Chicken and Turkey	11
Tuna	11
Turkey and Ham	11
Duck	11

Gareth chooses one pouch at random.
What is the probability that the pouch contains
a tuna b turkey? *(2 marks)*

3 Two ordinary fair dice are thrown.
One of the dice is green, the other dice is yellow.
Find the probability that the score shown on the green dice is less than the score shown on the yellow dice. *(3 marks)*

Case study 8: Weather

Weather forecasters collect data from all over the world to give information about the current conditions. A supercomputer and knowledge about the atmosphere, the Earth's surface and the oceans are then used to create the forecast.

Write down the temperature shown on each of these thermometers:

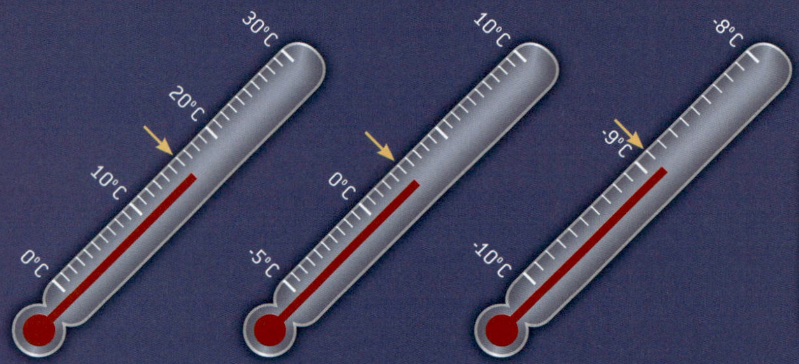

Each day, the Met Office receives and uses around half a million observations.

This chart shows the average mean temperature for January in the UK over the 30 year period from 1971 to 2000:

Write down an approximate value for the average temperature at

a) each of the cities labelled

b) your home town

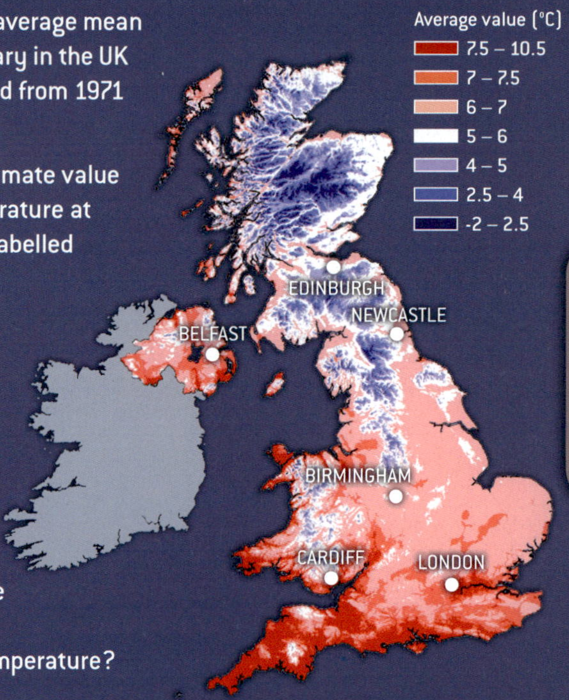

Which area(s) had the

c) highest

d) lowest average temperature?

What is the difference (approximate, in °C) between these temperatures?

How do these average temperatures compare to the temperatures in the UK this January?

Wind direction is measured in tens of degrees relative to true North and is always given from where the wind is blowing. In the UK, wind speed is measured in knots, where 1 knot = 1.15 mph, or in terms of the Beaufort Scale.

Research the Beaufort Scale using the internet.

Easterly wind, 090°

Describe the wind speed (in knots) and direction (in tens of degrees and in words) shown by each of these arrows:

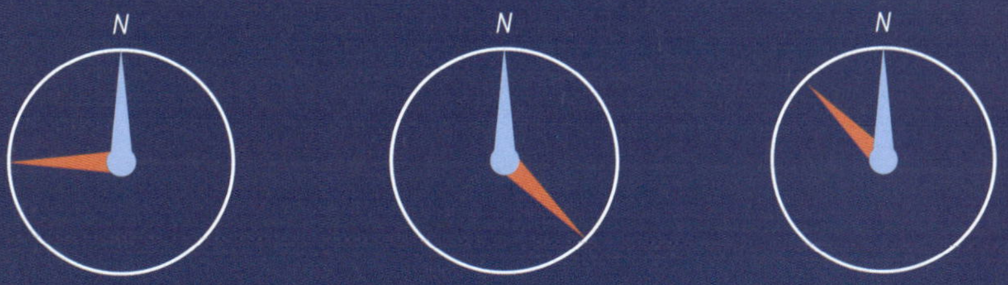

Observed data can be used to make predictions, but there is always some level of uncertainty. This graph shows the range of uncertainty in temperature in Exeter with some indication of the most probable values:

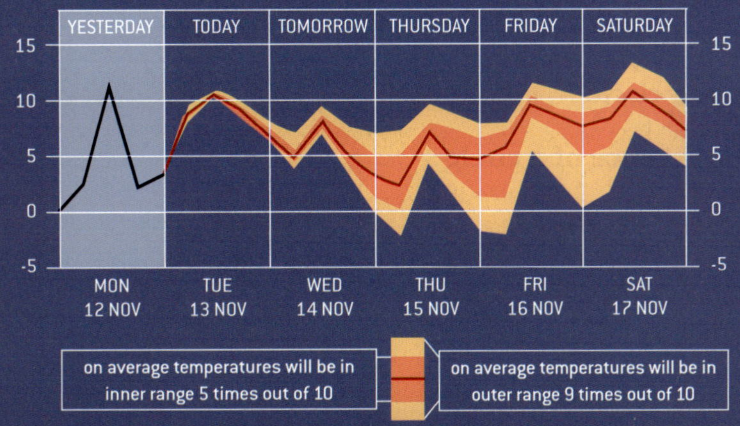

What predictions do you think a weather forecaster would have made about the temperature in Exeter during the week shown?

Justify your response by referring to the graph.

GCSE formulae

In your OCR GCSE examinations you will be given a formula sheet.
Here are the formulae that you are given in your exams.

Area of a trapezium $= \frac{1}{2}(a + b)h$

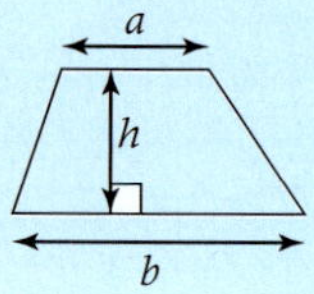

Volume of prism = area of cross section × length

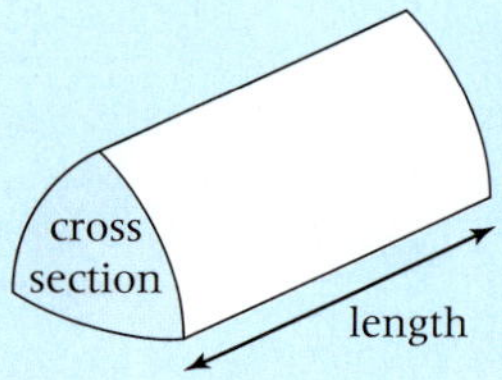

Here are some other formulae that you should learn.

Area of a rectangle = length × width

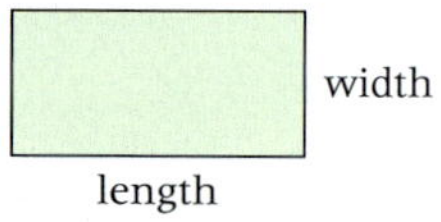

Area of a triangle $= \frac{1}{2} \times$ base × height

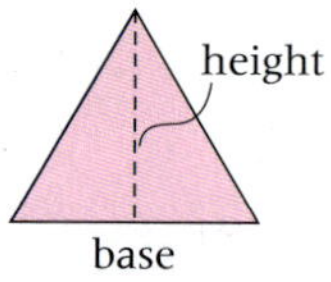

Area of a parallelogram = base × perpendicular height

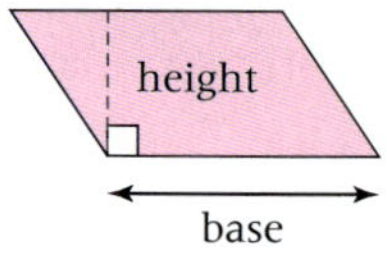

Area of a circle $= \pi r^2$

Circumference of a circle $= \pi d = 2\pi r$

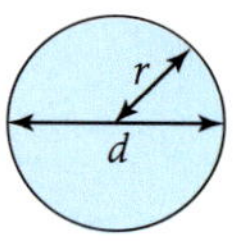

Volume of a cuboid = length × width × height

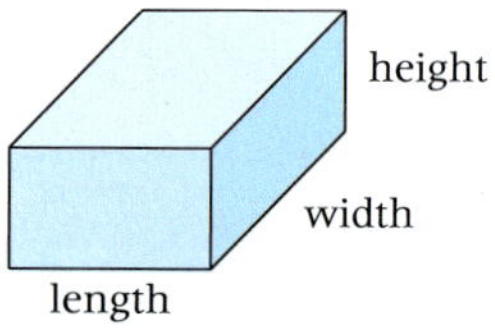

Volume of a cylinder = area of circle × length

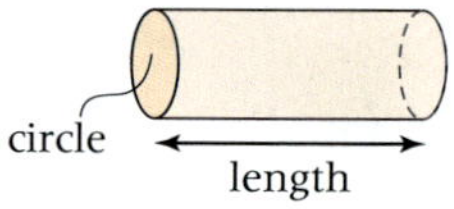

Pythagoras' theorem states,

For any right-angled triangle, $c^2 = a^2 + b^2$
where c is the hypotenuse.

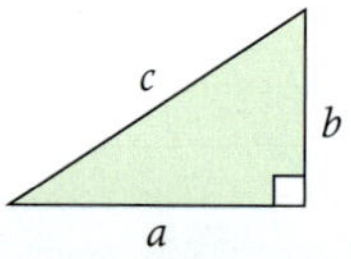

Answers

A1 Integers and decimals

Check in

1 −8, −5, −3, −1, 2, 4

2 a 7 b −10 c −12 d −3

3 1, 2, 3, 4, 6, 8, 12, 16, 24, 48

Ex A1.1

1 a one thousand three hundred and seven
b twenty nine thousand and six
c three hundred thousand
d six hundred and five thousand and thirty

2 a 8043 b 70 000 000
c 200 051 d 2010

3 a 8000 b 40 c 70

Ex A1.2

1 a 1 b −2 c −4 d 1
e −12 f −15 g −9 h −10
i 7 j 15 k −19 l −12
m −8 n −5 o 5 p 8

2 a −12 b −35 c −32 d −27
e 20 f 36 g 20 h 21
i −9 j −3 k −2 l −3
m 3 n 4 o 8 p 9
q 77 r 12 s −5

3 a 16, 23, 7; 23, 22, 1 b 8°C

4 a 36, −24 b $4 \times 3 + 6 \times -2 = 0$
c eg. Answer 4, all correct; Answer 9, 6 correct.

Ex A1.3

1 a 630 m b 220 m c 200 m d 200 m

2 a 7700 g b 5500 g c 3100 g d 2600 g

3 a 40 730, 40 700 b 310, 300
c 590, 600 d 150 900, 150 900

4 302 km 254 km

5 a 19.37 b 19.4 c 0.00 752 d 0.008
e 153.262 f 153.26 g 34.16 h 4.90

6 a 76 000 b 76 300 c 23 400 d 58 000
e 46 000 f 23.40 g 23.4 h 7.184

7 a 165 m to 175 m b 50, 51, 52, 53, 54 kg

Ex A1.4

1 a 61 b 23.83 c 1.5625 d 0.35
e 1.33 f 156.25 g 500 h 7.5

2 a 1.736 307 692 1.7 b −1.693 877 551 −1.69
c −0.191 345 436 −0.19 d 12.68 283 582 12.7
e 3.839 270 764 3.84 f 2.370 021 097 2.4

3 a 0.25 b $0.\dot{1}$ c 0.4 d $0.2\dot{7}$ e 5
f 6.25 g 2 h 4 i 2.5

4 a 1296 b 78125 c 100000
d 262144 e 410.0625 f 0.004096

Exam practice

1 a 2.71828 b 2.72

2 a 1.716961498 b 2

3 Missing numbers are bottom row −1 middle row −6, 2

A2 Summary statistics

Check in

1 a 44, 47, 48, 55, 56, 59, 61, 65
b 0.5, 0.8, 1.5, 1.7, 2.1, 2.5
c 13.1, 13.2, 21.3, 23.1, 31.2, 32.1

2 a 81 b 179 c 105 d 205 e 441
f 108 g 135 h 177 i 219 j 41

Ex A2.1

1 a i 1 ii 3 b i 5 ii 10
d No. 18 cells, only 16 poems

3 a 11 b % of Welsh with AB for example

Ex A2.2

1 a i cont. ii disc b i cont. ii disc
c i disc ii cont d i disc ii cont

2 a secondary b a survey, primary

Ex A2.3

1 a 75 b 63 c 60.1 d 63

2 a 71 b bimodal c 40.4 d 37

3 a £26 b bimodal c £84.20 d £87.00

4 a i £72 000 ii £27 000
iii £46 900 iv £34 000
b lowest price

Ex A2.4

1 a

4	5 8
5	1 1 2 4 9 9
6	1 2 3 6 9
7	0 1 4 5 7 8
8	1 2 9
9	3

7 | 4 means 74%

b i 48% ii 66%

2 a

2	6 8 9
3	0 3 7 8 9
4	0 1 3 3 4 7 9
5	2 5 8 9
6	0 1 3 6 7 9
7	1 3 5 7
8	0 4

5 | 8 means 58 mins

b i 58 mins ii 52 mins

Unit A

3 a

0	3 5 6 7 8 9
1	0 1 2 2 5 6 7 8 9
2	0 1 2 4 6 7 7 9 9
3	1 2 3 4 5 7
4	0 1

1 | 7 means 17 mins

b **i** 38 mins **ii** 20.5 mins

4 a

3	2 6 7 9
4	2 5 7 8 9
5	1 2 4 6 6 6 7 8 9
6	0 1 3 6
7	0

4 | 5 means 45 kg

b **i** 38 kg **ii** 54 kg

5 a

14	6 7 9
15	0 0 2 2 3 4 5 5 7 8 9
16	0 2 3 5 7 8
17	1 2 2

16 | 0 means 160 cm

b **i** 26 cm **ii** 157 cm

6 a

8	8 9
9	1 2 4 8
10	1 3 4 5 6 7 8 8 9
11	0 2 4 6 6 7 7 8 9
12	1 5 6 7 9
13	1 3

11 | 2 means IQ score of 112

b **i** 45 **ii** 110

7 a

0	3 6 8 9 9
1	0 1 2 5 6 7 8 9
2	1 1 2 3 3 5 7
3	0 2 3

2 | 3 means 23 mins

b **i** 30 mins **ii** 18 mins

Ex A2.5

1 Boys' range is 4.0secs, girls' is 4.2secs so girls' is larger. Median for boys is 5.3 and for girls is 6.3, so on average girls have longer reaction times.

2 a

Puzzle P		Puzzle Z
9 8 8 6	0	8 9
9 8 7 7 4 3 2 1 0	1	1 2 4 5 6 7 7 8 9
9 7 4 2	2	1 3 4 5 8 9
7 1	3	2 3

1 | 3 | 2 means puzzle P 31, puzzle Z 32 minute

b **i** P 31, Z 25 **ii** P 2 modes, Z 17
iii P 17.5, Z 19.5 **iv** P 17, Z 18

c not much difference, but range for Z is larger and Z is more difficult because takes more time on average.

3 a

class X		Class Y
9	8	7
9 8 6	9	3 7 9
9 8 6 5 5 4 3	10	2 3 4 6 7 7 9
8 8 7 7 6 5 5 2 0 0	11	3 4 4 4 6 8 8 8 9 9
9 6 4 3 3 2 1	12	1 1 4 6 6 8 9
4 1 0	13	0 1 2

3 | 10 | 2 means 103 for X, 102 for Y

b X and Y have same range, 45, median for Y is 116 and for X is 115 so Y have slightly higher IQ.

Ex A2.6

1 a **i** 6 **ii** 6 **iii** 5.8 **iv** 4
b **i** 5 **ii** 5 **iii** 5 **iv** 4

2 a and **b** Mean greater for boys (16.2, 16.1) Mode greater for boys (18, 15) Median 16 for both; no evidence girls have better recall.

Ex A2.7

1 a $10 < B \le 20$ **b** £24.20 **c** uses all data
d 11, 45; 12, 46; 13, 47; 14, 48; 15, 49; 16, 50

2 a £88.00 **b** $0 < M \le 40$
c Min value ≤ £40; Max value = £250; Yes, Range 200 – 40 = 160

3 Modal value and median same for both groups; Mean = 11.5 km for teachers, 14.1 km for office workers; conclusion supported.

Ex A2.8

1 a eg. Assumes show is liked
b eg. boxes don't cover all answers, assumes people listen, "per day" not specified.

Exam practice

1 a Charlie **b** Charlie leaves, Ellen joins

2 a 4.13 mins **b** $4 \le m < 5$

3

	Mean	Median	Range
Girls	64.7	62	46
Boys	64	66	44

Not a great deal of difference – boys did slightly better

A3 Constructions

Check in

1 a 45° b 135°
2 Check construction
3 a 65.5 mm b 6.55 cm

Ex A3.1

1 168.2 units

2 a 23 cm b 38 cm c 46 cm
d 162 kg e 177 kg f 189 kg
g 640 g h 780 g i 850 g
j 10.4°C k 11.9°C l 12.6°C
m 3.26 l n 3.38 l o 3.42 l
p $7\frac{1}{4}$ oz q $8\frac{3}{4}$ oz r $9\frac{1}{2}$ oz

3 a 07 : 10, 19 : 10 b 09.35, 21.35

4 17 : 37

5 a 1.25 kg b 2.65 m c 63.5°C
d 36.5 t e 39.5 t f 135 volts
g 155 volts h 0.605 l i 0.685 l

Ex A3.2

1 a i 6 m ii 8 m iii 14 m
b i 18 feet ii 24 feet iii 42 feet

2 a i 8 m ii 6 m iii 8 m iv 20 m
b i 20 feet ii 15 feet iii 20 feet iv 50 feet

3 3 km

4 1.189 m by 0.841 m

5 a i 60 cm ii 24 km iii 2.5 kg
iv 27 L v 2 L
b i 3 feet ii 25 miles iii 18 lbs
iv 36 pints v 4 pints

6 196 lbs and 169 lbs, 27 lbs lost

Ex A3.3

1 AB = 11 km YZ = 12.1 km

2 $a = 70°$ $x = 5.0$ cm $y = 5.5$ cm
$b = 81°$ $c = 61°$ $z = 4.5$ cm

3 AB = 6.6 km BC = 7.3 km

4 7.3 m, 62°

5 2 km

Ex A3.4

1 b W 41° 56° 83°
X 119° 32° 29°
Y 37° 53° 90°
Z 44° 68° 68°
c Y is right-angled
Z is isosceles

2 39°

3 1.2 m

4 d triangular based pyramid (tetrahedron)

5 c parallelogram
d 29° 29° they are the same

Ex A3.5

1 b

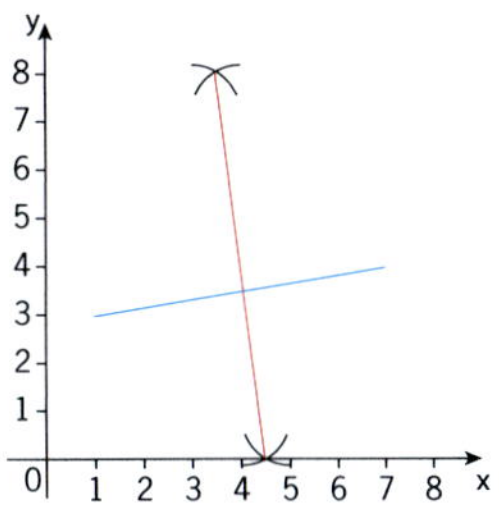

3 b

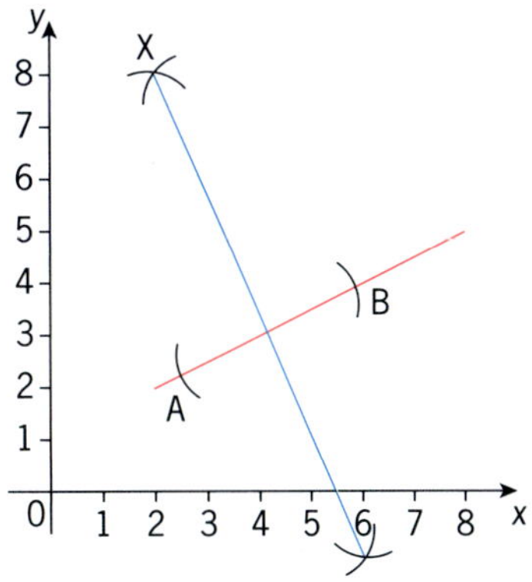

4 a

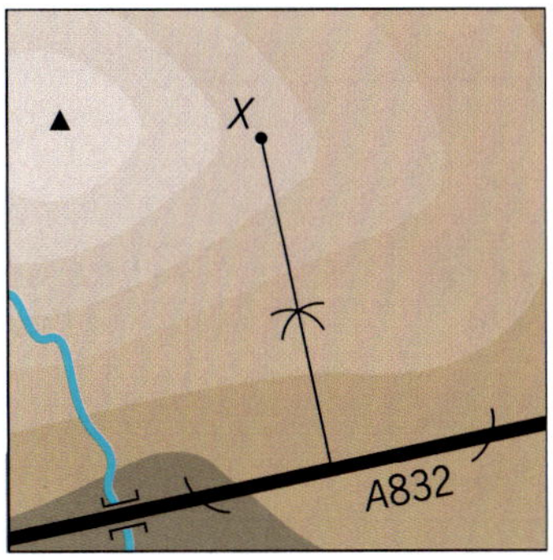

b

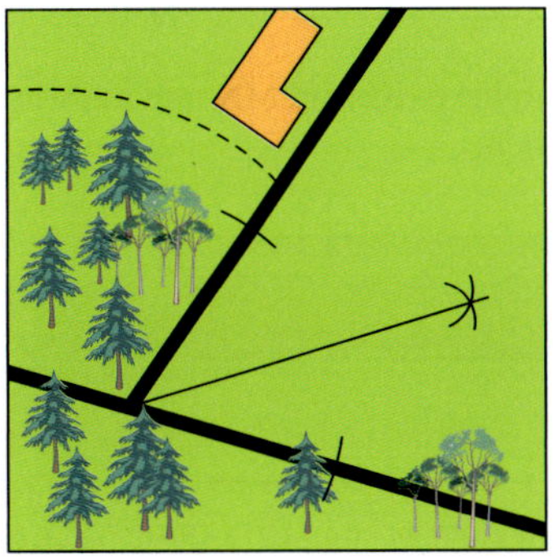

5 a, b

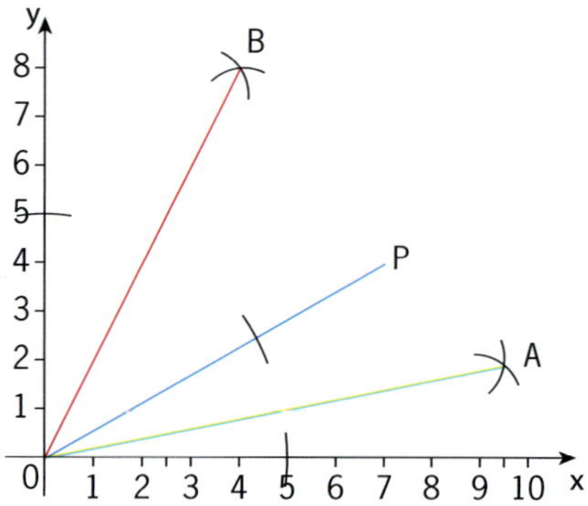

c 45° This is half of the angle between the x and y axes

6 5.7 m

Ex A3.6

1 a

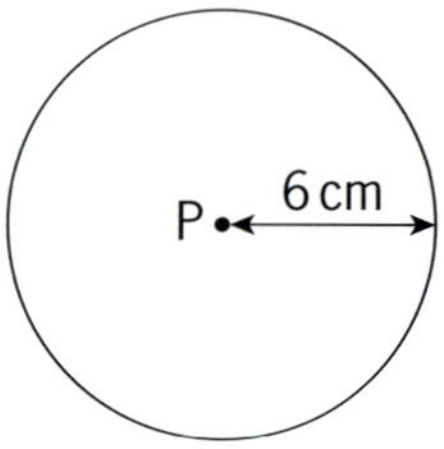

the circle

b

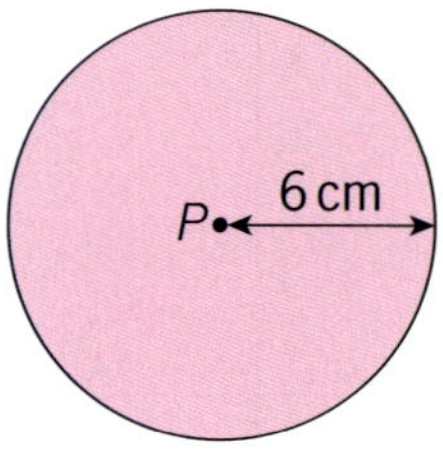

the area inside the circle

c

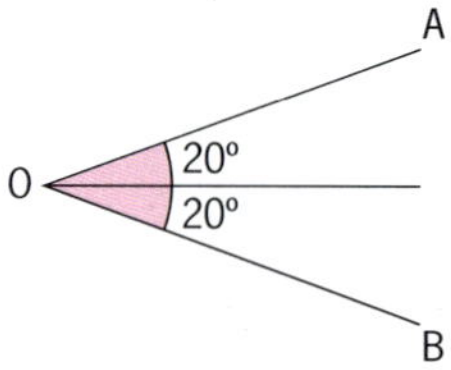

the line bisecting the angle AOB

d

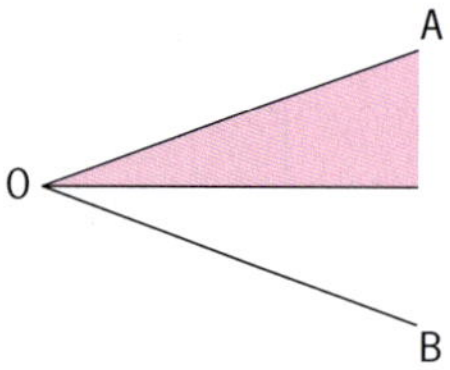

the area between the line OA and the bisector

e

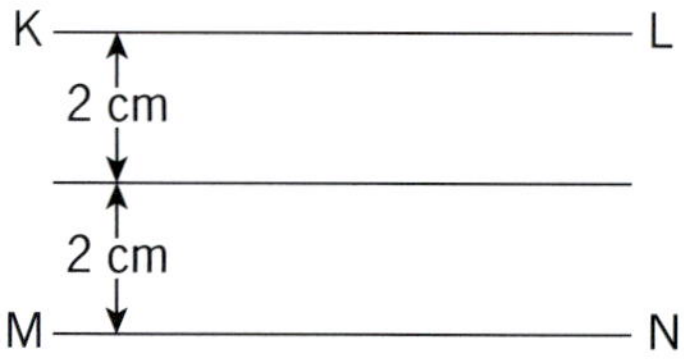

the line parallel to KL and MN and midway between them

f

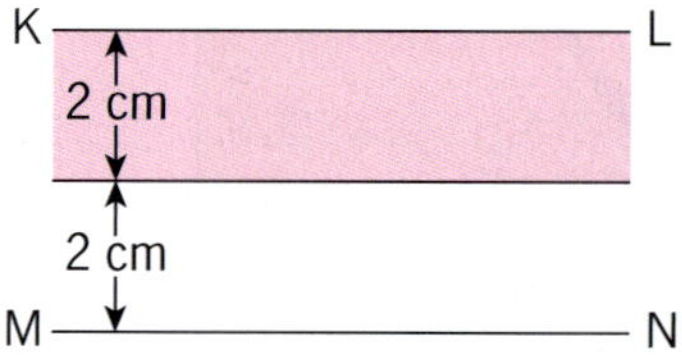

the area between the line KL and the mid-line

g

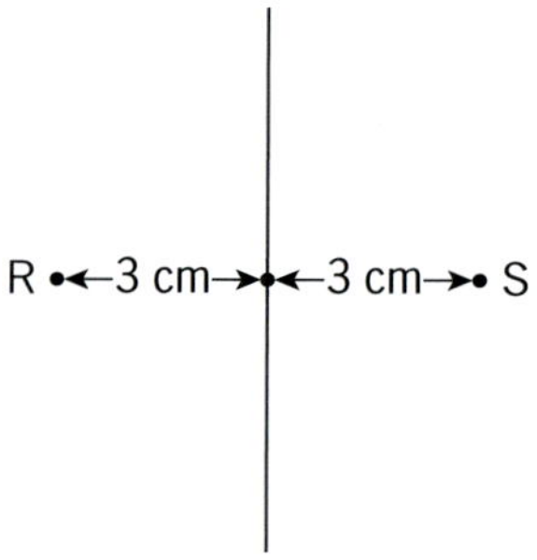

the perpendicular bisector of the line RS

h

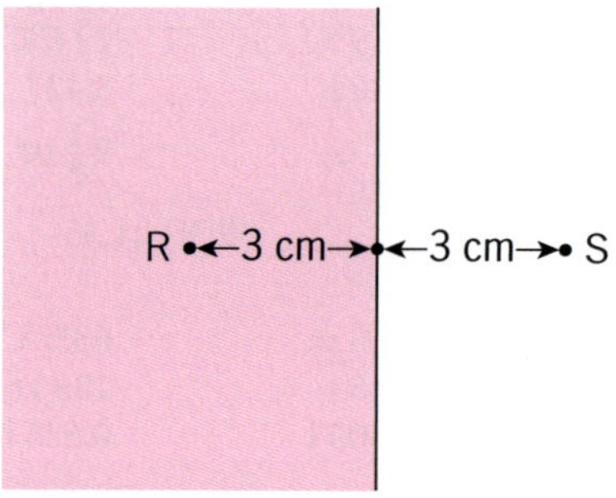

the area on the side of the perpendicular bisector that contains R

2

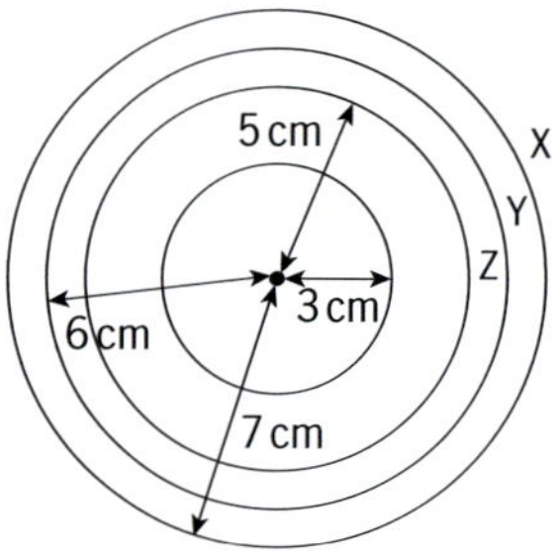

3

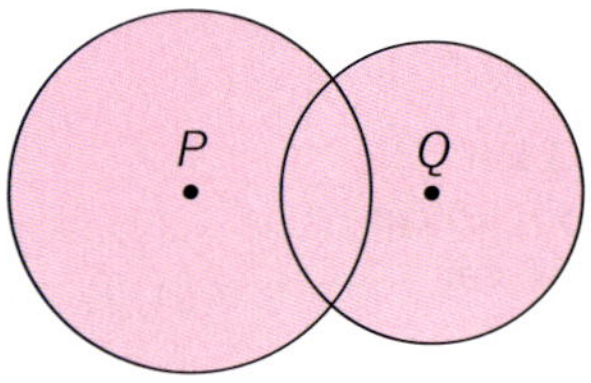

4

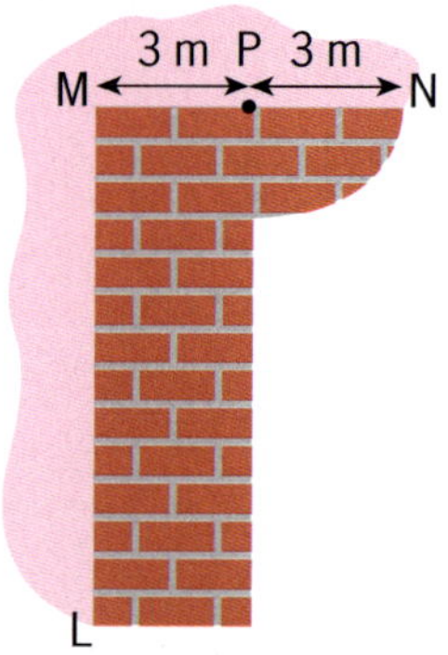

5

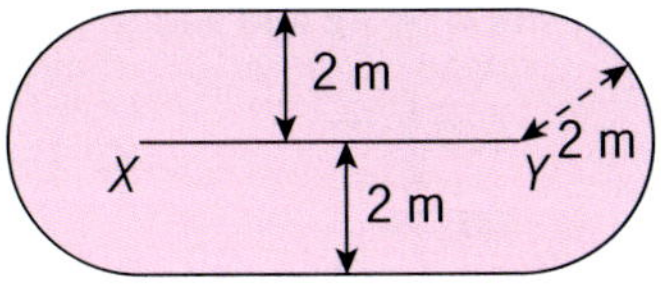

6 a

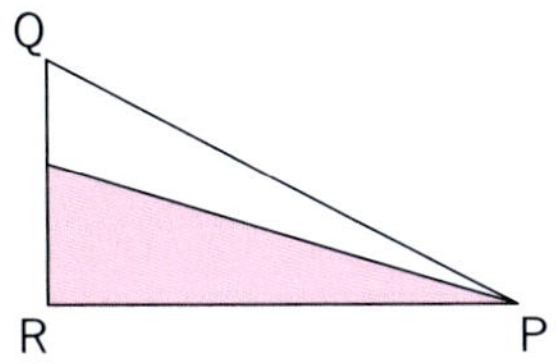

b

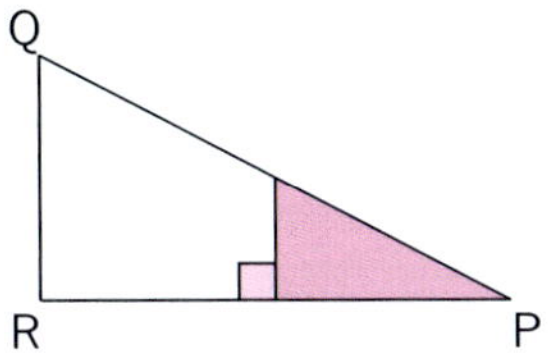

7

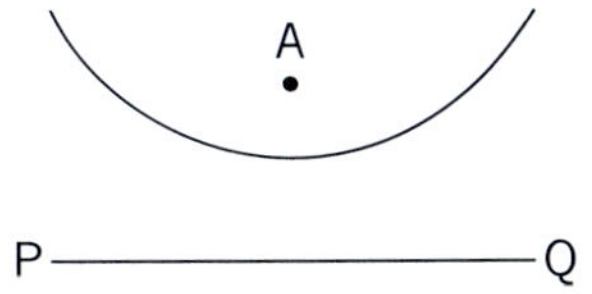

Exam practice

1 3.4 cm, 42°, 88°
2 44 lbs

A4 Factors, multiples and ratio

Check in

1 240 miles
2 15 miles

Ex A4.1

1 a 3, 4 b 5, 25 c 3, 21, 24, 30
d 21, 28 e 4, 25 f 3, 5, 17, 23
2 a 1, 3, 5, 15 b 1, 2, 3, 4, 6, 12 c 1, 2, 11, 22
d 1, 5, 25 e 1, 2, 4, 5, 8, 10, 20, 40
3 55
4 42, 48
5 31, 37
6 a **i** 1, 2, 11, 22 **ii** 1, 2, 3, 6, 11, 22, 33, 66
b 22 is factor of 66
7 a 1, 3 b 1, 2, 3, 6 c 1, 7 d 1, 2, 3, 6
8 a 4, 8, 12 b 2, 4, 6, 8, 10, 12
c 4 is multiple of 2
9 a eg 12, 24 b eg 30, 60 c eg 21, 42 d eg 18, 36
10 a eg 24 b eg 1, 4, 16

Ex A4.2

1 a $2^4 \times 3$ b 3^3 c 2×3^3 d $2^3 \times 3 \times 5$
e $2^3 \times 3^2$ f $2 \times 3 \times 7^2$ g 7×53
2 a 504, 4 b 3960, 3 c 3384, 2
d 196, 7 e 390, 5
3 a 12 b 48
4 The 3 factors are 1, prime number and the number itself

Ex A4.3

1 a 1 : 3 b 2 : 3 c 2 : 3 d 5 : 8 e 3 : 7
2 a 1 : 2.5 b 1 : 1.8 c 1 : 3.8 d 3.3 : 1 e 2.5 : 1
3 a £4.95 b £24.30 c £25.08 d £23.45

Ex A4.4

1 a £8 : £40 b £56 : £8 c £24 : £84
d £56 : £40 e £60 : £120 : £240 f £75 : £150 : £200
2 £45.30
3 £8.40
4 93.75
5 137.5
6 680 g
7 a 6 : 5 b 7 : 6 c 6 and 3 years old

Exam practice

1 a **i** 24 **ii** 25 **iii** 23
b $2^2 \times 7$
2 45
3 a $\frac{3}{5}$ b 5 : 3

A5 Sequences

Check in

1 a 3 b 6 c 5 d 6
2 a 3 b 4 c 5 d 4
3 a 4, 8, 12, 16, 20, 24 b 3, 6, 9, 12, 15, 18
c 5, 10, 15, 20, 25, 30 d 6, 12, 18, 24, 30, 36

Ex A5.1

1 a 80, 75, 70, 65, 60, 55, 50 b 16
2 a 5, 9, 13, 17 b 21, 25, 29 Add 4 each time c 81
3 a

No. of columns	1	2	3	4	5
No. of tiles used	2	5	8	11	14

b 59. Number of tiles = Treble the number of columns and subtract 1
4 a 18, 20, 22. 48 b 20, 23, 26. 62

Ex A5.2

1 a

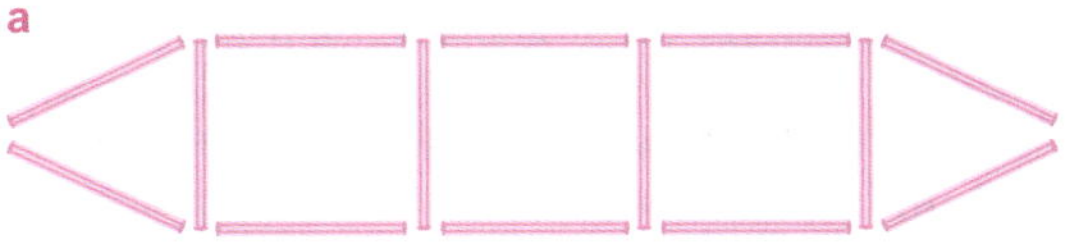

5, 8, 11, 14
b Start with 5 and add 3 each time. 17, 20, 23
2 a 3, 8, 18, 38, 78 b 1, 6, 21, 66, 201
c $\frac{1}{4}$, $1\frac{1}{2}$, 4, 9, 19 d 20, 17, 14, 11, 8
e 600, 280, 120, 40, 0 f 4, 4, 4, 4, 4
3 a 2, 44, 128, 296 b 50, 140, 320, 680
c −30, −20, 0, 40

4 Term → +4 → ×2 → −1 → Next term

Term → ×2 → +4 → −1 → Next term

Term → +4 → −1 → ×2 → Next term

Term → −1 → ×2 → +4 → Next term

5 a Start with 8 and add 3 each time. 20, 23, 26
b Start with 30 and subtract 3 each time. 18, 15, 12
c Start with 1, then double and add 1 each time. 31, 63, 127
d Start with 3, then double and subtract 1 each time. 33, 65, 129
e Start with 1, then treble and add 1 each time. 121, 364, 1093
f Start with 10, then subtract 3 each time. −2, −5, −8
g Start with 200, then halve each time. 12.5, 6.25, 3.125
h Start with $\frac{1}{2}$, then double and add 2 each time. 38, 78, 158
i Start with −7 and add 3 each time. 5, 8, 11

6 5, 11, 23, 47, 95

Ex A5.3

1 a 4
b

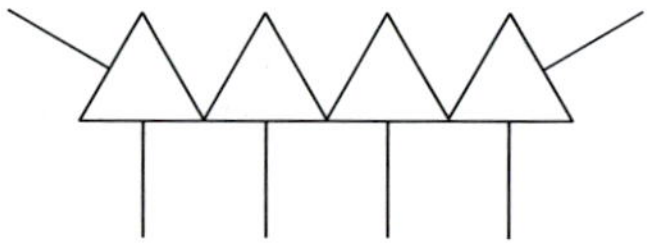

c Start with 6 and add 4 each time
d

Position	1	2	3	4	5
4 × table	4	8	12	16	20
Term	6	10	14	18	22

Multiply position by 4 and add 2
e 202

2 a ii Multiply the position by 4 and subtract 1.
iii 23, 27, 399
b ii Multiply the position by 3 and add 1.
iii 19, 22, 301
c ii Multiply the position by 2 and subtract 1.
iii 11, 13, 199
d ii Multiply the position by 4 and add 1.
iii 25, 29, 401
e ii Multiply the position by 5 and subtract 2.
iii 28, 33, 498
f ii Multiply the position by 6 and add 1.
iii 37, 43, 601
g ii Multiply the position by $\frac{1}{2}$ and add 3.
iii 6, $6\frac{1}{2}$, 53
h ii Multiply the position by −2 and add 8.
iii −4, −6, −192

3 a 19, 37, 301 b 5, $9\frac{1}{2}$, 77

4 a 00139 b 00243

5 Multiply the position by 3 and add 12, 102 mm

6 Because it turns 6 revolutions every minute. Multiply the position by 6 and add 325, 685

Ex A5.4

1 a 4
b

Position	1	2	3	4	5	.....	n
4 × table	4	8	12	16	20		$4n$
Term	3	7	11	15	19		$4n-1$

c $4n - 1$ d 23, 199

2 a $2n + 1$, 201 b $3n + 1$, 301
c $4n$, 400 d $4n + 1$, 401

3 a Multiply the position by 2 and add 5.
$T(n) = 2n + 5$, 17, 85
b Multiply the position by 3 and subtract 1.
$T(n) = 3n - 1$, 17, 119
c Multiply the position by 1 and add 4.
$T(n) = n + 4$, 10, 44
d Multiply the position by 5 and subtract 2.
$T(n) = 5n - 2$, 28, 198
e Multiply the position by 6 and add 1.
$T(n) = 6n + 1$, 37, 241
f Multiply the position by 5 and add 1.
$T(n) = 5n + 1$, 31, 201
g Multiply the position by 4 and subtract 5.
$T(n) = 4n - 5$, 19, 155
h Multiply the position by 7 and add 3.
$T(n) = 7n + 3$, 45, 283
i Multiply the position by 3 and subtract 7.
$T(n) = 3n - 7$, 11, 113
j Multiply the position by −2 and add 10.
$T(n) = -2n + 10$, −2, −70

4 a 21 31 501 b 32 62 1002
c 15 23 399 d 29 45 797

5 4 6 8 10 12. 19

Ex A5.5

1 a i 10 12 14 16 18
ii 6 4 2 0 −2
iii The first sequence is increasing, the second sequence is decreasing.
b i 11 13 15 17 19
ii 7 5 3 1 −1
iii The first sequence is increasing, the second sequence is decreasing.

c $T(n) = 2n + 5$
d 0 2 4 6 8
It starts with zero and increases by 2 metres each time.

2 a 1 4 9 16 25 36 b $T(n) = n^2$
c i 81 ii 144 iii 400 d 8

3 a 1 3 6 10
b $T(1) = 1$
$T(2) = 1 + 2 = 3$
$T(3) = 1 + 2 + 3 = 6$
$T(4) = 1 + 2 + 3 + 4 = 10$
$T(5) = 1 + 2 + 3 + 4 + 5 = 15$
c $T(1) = \frac{1}{2} \times 1 \times (1 + 1) = 1$
$T(2) = \frac{1}{2} \times 2 \times (2 + 1) = 3$
$T(3) = \frac{1}{2} \times 3 \times (3 + 1) = 6$
$T(4) = \frac{1}{2} \times 4 \times (4 + 1) = 10$
$T(5) = \frac{1}{2} \times 5 \times (5 + 1) = 15$
d $T(6) = \frac{1}{2} \times 6 \times (6 + 1) = 21$
$T(6) = 1 + 2 + 3 + 4 + 5 + 6 = 21$

4 a 1 b 3 c 6
5 people = 10 handshakes, 6 people = 15 handshakes.
It is the same sequence of triangle numbers.
4950 handshakes for 100 people.

5 Using the second method, Sue would get
10p + 20p + 40p + 80p + 160p + 320p + 640p = £12.70 per week.
Sue should opt for this method.

Exam practice

1 a 16, 19 b double previous term c 2, 5, 10
2 a i 21 ii 61 b 4n – 3
3 10 more minutes

A6 Representing and interpreting data

Check In

1 a 4, 5.5, 6.5, 7, 8
b 2.3, 2.4, 3.2, 3.4, 4.2, 4.3
c 7.5, 7.9, 8.3, 8.6, 9.1
2 a 6, 6, 7, 8, 8, 9, 9, 9, 9, 9
b 100, 100, 101, 102, 102, 104, 104, 104, 104, 104

Ex A6.1

1 Bar chart
2 a Bar chart
b Higher proportion of girls recalled the Rabbit first recall similar otherwise.
3 eg. most popular; total number sold; preferences by gender
4 a i car ii microwave

Ex A6.2

1 Histogram
2 a Histograms b $15 \le t < 20$ $20 \le t < 25$
c $15 \le t < 20$ $15 \le t < 20$
d eg. boys quicker – modal class
3 eg range is same, modal class on patio is $12 < m \le 16$, but greenhouse is $4 < m \le 8$ so weights larger on patio.

Ex A6.3

1 Pie chart
2 a pie chart b i England ii 54%
3 a 6.5 million tonnes b 21.6°

Ex A6.4

1 a baths, showers, hand-washing
b 4100 megalitres
2 a leisure, recreation, holidays b rail
c no connection between the sets of data.
3 pie charts

Ex A6.5

1 a polygon b i $20 < t \le 30$ $30 < t \le 40$
ii 50 mins, 50 mins
c office workers take longer
2 a polygon b i $40 < m \le 60$, $40 < m \le 60$,
ii 100 miles, 80 miles
c eg smaller range in Jan because no short journeys
3 a polygons
b eg modal class D $0 < t \le 5$, P $5 < t \le 10$; range same for both, median in class $10 < t \le 15$ for both; Powerblast better buy because of shape of distribution and modal class.
4 eg 2002–2003 may have range 67, and median in class 50–60, so could be $52\frac{1}{2}$
5 eg higher proportion of 40 – 60 year olds, the highest spending group, on weekdays, so more money spent per person on weekday.

Exam practice

1 a i 9 ii –6 b i 7 ii –3
c i –10 ii 10
2 a 15 b £0–£10
c Because we do not know the exact amount spent
3 a polygon
b the modal class for both months is 5–10 mins. The range is larger for August than July.

A7 Formulae and equations

Check in

1 16 cm
2 a x b m c $3n$ d $2p$

Ex A7.1

1 a formula b equation c expression

2 a 14 b 2 c 14
3 a formula b equation c expression
4 a formula b expression c equation
5 a i 16 ii 5 b i 2 ii 5
c i $5x + 3$ ii $5x - 1$

Ex A7.2

1 $4x + 5y$
2 a $6x + 8y$ b $8x + 8x^2$ c $4x + 3y$
d $x^2 + 8$ e $8m + 2n$ f $6p + 2q + 7$
g $6a - 2b + 9$ h $9x^2 + 8x + 7$
3 a $6x + 10$ b $10x + 20$ c $18y - 12$
d $8z - 12$ e $15 + 20a$ f $10 - 6c$
g $2a + 10b$ h $8x - 12y$ i $18a - 6b + 3c$
j $12 + 6x^2$ k $15y^2 + 10y$ l $4 + 8x + 12x^2$
4 a $23x + 12$ b $18x + 10$ c $9p + 4$
d $18q + 8$ e $13m + 3$ f $19n$
5 a $2(3x + 4)$ b $3(3x + 2)$ c $4(2x + 3)$
d $5(x + 2)$ e $3(x + 3)$ f $6(x - 3)$
g $4(3y + 2)$ h $5(3a - 2)$ i $2(2z - 1)$
j $2(4c + 3d)$ k $5(2x + 3y)$ l $3(3s - 2t)$
6 a $5(2x + 3)$ b $4(2y + 3)$ c $16(z + 1)$
d $4(2p + q)$ e $3(3a + 2b)$ f $2(4s + t)$
g $8(1 + m)$ h $6(3c - 1)$ i $7(2 + z)$
7 3 ways using whole numbers. If fractions are used there are an infinite number of ways.
8 Length could be $(12x + 24)$, $(6x + 12)$, $(4x + 8)$, $(3x + 6)$, $(2x + 4)$, or $(x + 2)$ with corresponding widths 1, 2, 3, 4, 6, and 12
9 $4x + 3y$.
The different lengths are:

$3x + 3y$	$2x + 3y$	$x + 3y$	$3y$	
$4x + 2y$	$3x + 2y$	$2x + 2y$	$x + 2y$	$2y$
$4x + y$	$3x + y$	$2x + y$	$x + y$	y
$4x$	$3x$	$2x$	x	

Ex A7.3

1 a £22 b £31
2 a £636 b £1237.75
3 a X 41.5 mpg
Y 38.8 mpg
Z 39.9 mpg
b Z likely to be more accurate as it uses a longer distance and therefore is likely to be based on more varied driving conditions. Yes, the journeys are consecutive so you could use the total mileage and the total fuel used to get a more accurate result.
4 a 42 b 30 c 19.6 d 9.4
5 a 18 b 11 c 8.4 d −2
6 a 48 b 10 c 37.5
7 25.5

Ex A7.4

1 a 2 b 1.5 c 4
d 5.5 e 4 f 10
2 a 2 b 6 c 5 d 4
e 10 f 7 g 4 h 4
3 a 5 b 4 c 32 d 4.56
4 $x = 20, q = 56, a = 5\frac{2}{3}$

Ex A7.5

1 a $n = x + 1$
b 5 Yes, if you like strong tea
2 a Area = $20 + 2x$ cm² b Area = $8 + 6x$ cm²
c Area = $10 + 7x$ cm² d Area = $24 + 5x$ cm²
3 a 45910 b $45670 + 50x$
4 a 8 hours b $5 + x$ hours
5 100 × integer + my age

Ex A7.6

1 a 5 b 13 c 5 d 20
e 1 f 5 g 6 h 8
i −3 j 15 k 3 l 4
2 a 4 b $\frac{1}{2}$
3 a 2 b 5 c 1 d 3
e 9 f $1\frac{1}{2}$ g $1\frac{1}{4}$ h 0
i 4 j $\frac{1}{2}$ k $\frac{1}{2}$ l $16\frac{1}{2}$
4 7
5 a 9 b 7
6 12 years old
7 $x = 13$
$y = 12$

Ex A7.7

1 a 2 b 1 c 4 d 3 e 1
f 3 g 2 h 4 i 4 j 6
k $\frac{1}{2}$ l $1\frac{1}{4}$
2 a 2 b 1 c 0 d 3 e 2
f 3 g 1 h 0 i 10 j 4
3 a 5 b $3\frac{1}{2}$ c $1\frac{1}{2}$ d $2\frac{1}{3}$ e 3
f 6 g 9 h 3 i 2 j $1\frac{1}{4}$
k $3\frac{1}{2}$ l $1\frac{1}{4}$ m $2\frac{1}{2}$ n $2\frac{2}{3}$ o $6\frac{1}{2}$
p 2 q 1 r 5
4 $3(x - 4) = 24$
$x = 12$
5 $30x + 40 = 220$
£6 an hour
6 $2(x - 12) - 10 = 16$
$x = 25$
£13
7 B $2x + 6$
C $x + 8$
$2x + 6 + x + 8 = x + 100$
$x = 43$

Ex A7.8

1 a 20 b 15 c 16 d 30
e 7 f 3 g 5 h 5
i 4 j 6 k 4 l 2

2 a 3 b 4 c 6 d $1\frac{1}{2}$
e $\frac{1}{2}$ f $1\frac{1}{4}$ g 3 h $3\frac{1}{3}$
i 3 j 4 k $3\frac{1}{2}$ l $2\frac{1}{4}$
m −2 n −2 o −1 p −3
q $-1\frac{1}{2}$ r −7 s $-1\frac{1}{4}$ t $-1\frac{1}{3}$

3 a 2 b 1 c 3 d 5 e 1
f 2 g 4 h 2 i −2 j −3

4 a 3 b −1 c 5 d −2
e $1\frac{1}{2}$ f −3 g $\frac{1}{2}$ h $1\frac{1}{4}$
i $\frac{1}{4}$ j 0 k $1\frac{1}{2}$ l $-1\frac{1}{4}$

5 £40

6 16 years old

Ex A7.9

1 a (2, 3) b (3, −2) c (−3, 2)
d (−4, −3) e (−3, −5) f Post Office
g Rail bridge over road
h Crossroads **i** School

2 a Yes b Yes
c **i** (2, 4) **ii** (5, 7) d S(3, 2.5) T(6.5, 3)

3 a (10, 10) b (4, 9)

4 (30, 25) (30, 25) It is a parallelogram

5 a (14, 14) b (16, 9) A quarter of the total length

Exam practice

1 a $3a + 4b$ b **i** $5(x - 3)$ **ii** $x(x + 4)$
c $4\frac{3}{5}$

2 $b = \frac{4}{5}(e - 32)$

3 a No
since $3 + 2 \neq 3$
b $2n - 3$

A8 Bearings and Pythagoras

Check in

1 a 180° b 180° c 360°

2 a 50° b 70°

Ex A8.1

1 a **i** 035° **ii** 130° **iii** 207°
iv 225° **v** 035° **vi** 0°
b **i** 067° **ii** 247° c 15 km
d **i** 57 km **ii** 69 km **iii** 33 km **iv** 73 km

2 a 180° b 045° c 225°

3 a 33 m b 14 m

4 a 24 km b 51 km

5 a 49 km b 70 km

6 a Drogheda to Douglas: bearing 066°, distance 140 km
Douglas to Heysham: bearing 097°, distance 110 km
b 260°

7 95 m

8 150°, 240° 330°, just over 1 hour

9 a 500 km b 2500 km c 3200 km

Ex A8.2

1 a $p = 5$ cm b $q = 13$ cm c $c = 10$ cm
d $r = 6.4$ cm e $s = 7.8$ cm f $t = 7.3$ cm

2 6.8 m

3 5.57 m

4 6.5 miles

5 2.28 km

6 Jim's error is in his first line as $x^2 = 3^2 + 5^2$
Jan's error is in her last line as $x = \sqrt{34} = 5.83$ cm

Ex A8.3

1 a $w = 9.2$ cm b $x = 8.7$ cm
c $y = 9.4$ cm d $z = 9$ cm

2 a $p = 12$ cm b $q = 6$ cm
c $r = 4.9$ cm d $s = 7$ cm

3 3.04 m

4 1.4 km

5 21 m

Exam practice

1 9.49 cm

2 a 110 km, 100° b 170 km, 71° c 95 km, 246°

3 31

B9 Fractions, decimals and percentages

Check in

1 0.1, $\frac{1}{4}$, $\frac{1}{2}$ = 0.5, 23% = $\frac{23}{100}$

2 a 0.375 b $0.8\dot{3}$

Ex B9.1

1 a $\frac{2}{3}$ b $\frac{1}{3}$ c $\frac{3}{8}$ d $\frac{3}{5}$ e $\frac{2}{3}$ f $\frac{5}{7}$

2 a $\frac{1}{4}$ b $\frac{1}{3}$ c $\frac{3}{5}$

3 a $\frac{3}{8}$ b $\frac{2}{3}$ c $\frac{2}{5}$

4 $\frac{9}{30} = \frac{3}{10}$

5 $\frac{30}{40}$, $\frac{6}{8}$, $\frac{9}{12}$

6 a $\frac{1}{9}$ $\frac{1}{4}$ $\frac{1}{3}$ $\frac{1}{2}$ b $\frac{2}{7}$ $\frac{4}{7}$ $\frac{5}{7}$
c $\frac{3}{8}$ $\frac{1}{2}$ $\frac{5}{6}$ d $\frac{2}{3}$ $\frac{3}{4}$ $\frac{5}{6}$

7 $\frac{21}{40}$

Unit A

Ex B9.2

1 a 25% b 60%

2 0.5 50%, $\frac{1}{4}$ 25%, $\frac{3}{4}$ 0.75, $\frac{1}{10}$ 0.1, 0.2 20%, $\frac{7}{10}$ 70%, $\frac{23}{100}$ 0.23, 0.6 60%

3 a $\frac{1}{4}$ 40% 0.7 b 40% 0.45 $\frac{4}{5}$
c 0.09 15% $\frac{2}{5}$ d 7% $\frac{11}{20}$ 0.6

4 0.625 62.5%

5 $\frac{2}{5}$

6 15%

7 a 45% b 80%

8 32%

9 60%

Ex B9.3

1 a 0.205 0.3 0.43 0.61 b 0.205 0.701 0.84 0.9
c 0.405 0.41 0.415 0.45 d 0.809 0.89 0.908 0.98
e 0.2 0.205 0.25 0.5 0.52

2 a 9.3 b 10.44 c 9.49 d 9.25
e 4.2 f 22.89 g 141.65 h 15.14

3 a 4.3 b 11.2 c 27.7 d 3.2
e 1.5 f 3.4 g 2.45 h 4.74

4 a 0.06 b 0.48 c 1.15 d 0.48
e 0.6 f 0.72 g 0.126 h 0.345

5 a 3 b 52 c 200 d 72.8
e 32 f 272.5 g 100 h 49

6 a 0.2, 0.25 b 2 ÷ 0.2 = 10

7 c 19.38

Ex B9.4

1 a 53 b 21 c 8 d 22.5
e 4.2 f 50 g 5 h 3.2

2 a 18 b 800 c 10.8 d 25.5
e 8.4 f 50 g 20 h 3

3 a 67.2 b 45 c 66 d 10.4
e 1.26 f 98 g 60 h 28.8

4 25% of 60

5 50% of 30

6 25

7 a 400 b 350

Ex B9.5

1 a £22.50 b £21 c £10 d £16

2 a $\frac{4}{5}$ b $\frac{1}{2}$

3 a $\frac{1}{2}$ b $\frac{11}{16}$ c $\frac{27}{70}$ d $\frac{59}{60}$

4 a $\frac{1}{4}$ b $\frac{16}{27}$ c $\frac{1}{10}$ d $\frac{5}{12}$

Ex B9.6

1 a $\frac{1}{6}$ b $\frac{3}{35}$ c $\frac{24}{35}$ d $\frac{1}{6}$

2 a $\frac{1}{10}$ b $\frac{1}{4}$ c $\frac{3}{20}$ d $\frac{3}{20}$
e $\frac{2}{5}$ f $\frac{8}{9}$ g $\frac{15}{16}$ h $\frac{3}{4}$

3 a i $\frac{1}{10}, \frac{3}{8}$ ii $\frac{1}{2}, \frac{2}{5}, \frac{1}{10}$
b i $\frac{1}{5}, \frac{2}{7}$ ii $\frac{1}{2}, \frac{1}{5}, \frac{3}{10}$

Ex B9.7

1 a $1\frac{3}{5}$ b $7\frac{1}{2}$ c $7\frac{1}{2}$ d $\frac{3}{4}$
e $9\frac{1}{6}$ f $2\frac{2}{3}$ g $12\frac{1}{2}$ h $\frac{6}{7}$

2 6 cans

3 a 5 b 6 c 13

4 a $3\frac{31}{40}$ b $11\frac{3}{4}$ c $8\frac{1}{12}$ d $5\frac{2}{5}$
e $2\frac{5}{26}$ f $1\frac{13}{20}$ g $-\frac{5}{12}$ h $-\frac{11}{12}$

Ex B9.8

1 a $0.\dot{1}$ $0.\dot{2}$ $0.\dot{3}$ $0.\dot{4}$ $0.\dot{5}$ $0.\dot{6}$ $0.\dot{7}$ $0.\dot{8}$
b All digits the same as numerator

2 3.14, π

3 a i $0.1\dot{6}$ ii $0.8\dot{3}$ b i 0.167 ii 0.833

4 a i 4 ii 3.996 iii 3.6 b i 0.333 ii 0.3

Exam practice

1 a 13 b £80

2 $\frac{7}{30}$

3 a i $\frac{1}{8}$ ii $\frac{5}{8}$
b i 0.125 ii 0.625

B10 Circles, angles and lines

Check in

1 a Diameter b Circumference c Radius

2 360°

Ex B10.1

1 a semicircle b quadrant

2 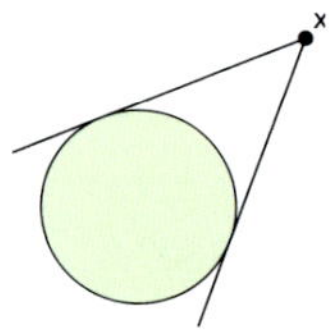

AX = BX. Symmetry about the line joining X to the centre of the circle.

3 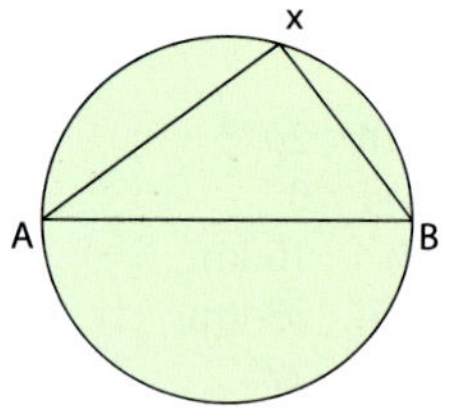

90°, yes

4 Cone, yes

5 a i 2 ii 4 iii 7 iv 11

6 They are all prehistoric constructions based on circles and connected with the movement of the sun. Stonehenge is another famous example.

Ex B10.2

1 a acute 20° b right-angle 90°
c acute 60° d obtuse 140°
e Straight line 180° f reflex 220°

2 p right-angle 90° q acute 70°
r right-angle 90° s obtuse 110°
t acute 20° u reflex 210°
v acute 20° w obtuse 110°

3 a obtuse 180° b right-angle 90°
c reflex 270° d obtuse 135°

4 a $x = 70°$ b $y = 160°$ c $z = 130°$
d $u = 180°$ e $v = 80°$ f $w = 100°$

5 120°

6 4 angles 72° 144° 216° 288°

Ex B10.3

1 a 50° b 100° c 55°
d 110° e 70° f 70°
g 30° h 150° i 150°

2 p 45° q 25° r 125° s 55°
t 60° u 40° v 70° w 60°
x 120° y 80°

3 p 50° q 130° r 90°
s 130° t 200° u 220°

Ex B10.4

2 a $a = 50°$ $b = c = 130°$ b $d = 120°$ $e = f = 60°$
c $h = 70°$ $g = i = 110°$ d $l = 65°$ $j = k = 115°$

3 $a = 120°$ $b = c = 60°$ $d = e = f = 110°$
$g = h = 70°$ $i = 120°$ $j = 50°$ $k = 130°$

4 $a = 120°$ $b = 140°$ $c = 60°$ $d = 130°$
$e = 50°$ $f = 125°$ $g = 55°$
$h = j = 120°$ $i = 60°$ $k = l = 70°$
$m = n = 50°$ $p = q = 65°$ $r = s = 115°$

5 $p = 135°$
$q = 45°$

6 $x = 145°$
$y = z = 35°$

7 $p = q = t = 125°$
$r = s = u = 55°$

8 e.g.

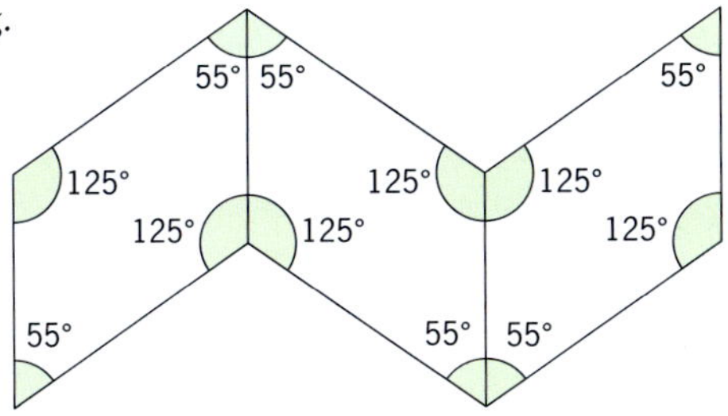

9 $y = 180° - x$

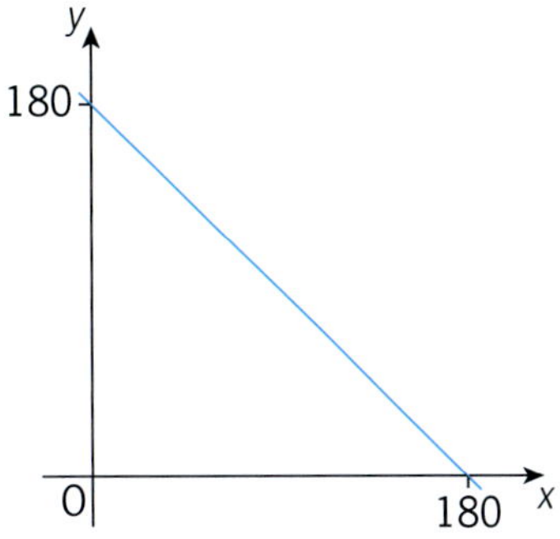

Closed when
$x = 0$ or 180
$y = 180$ or 0

Exam practice

1 No, because $66° + 71° + 53° \neq 180°$

2 $x = 24°$
$y = 48°$

3 102°

B11 Straight lines and inequalities

Check in

1 a 6 b 11 c 9 d 6

2 a $3x + 3$ b $2x - 2$
c $8x + 12$ d $12x - 6$

3 a 4 b 6

4 a $4(x + 2)$ b $3(y + 6)$ c $3(y - 3)$

Ex B11.1

1

No. of seconds, x	5	10	15	20	25	30
No. of miles away, y	1	2	3	4	5	6

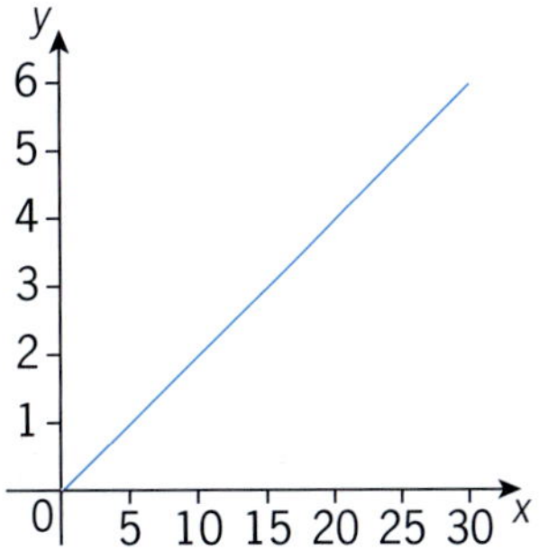

2

No.of hours, x	1	2	3	4	5
Cost, £C	6	8	10	12	14

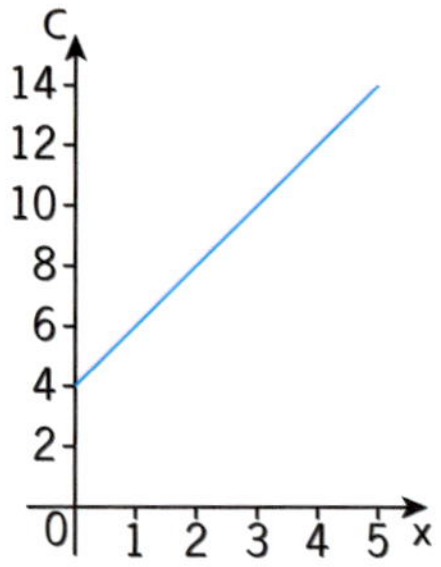

a 3.5 hrs

b 8.5 hrs

3

x, miles	50	100	200	300
C, £	21	22	24	26

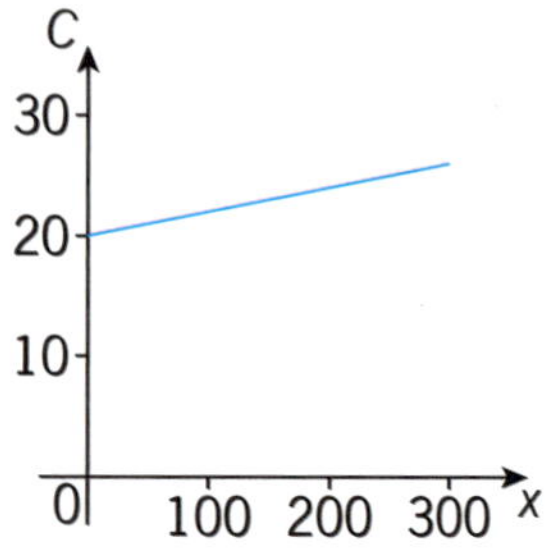

£23

4 For example:

a, age in years	0	4	8	12	16
h, no. of hours sleep required	17	15	13	11	9

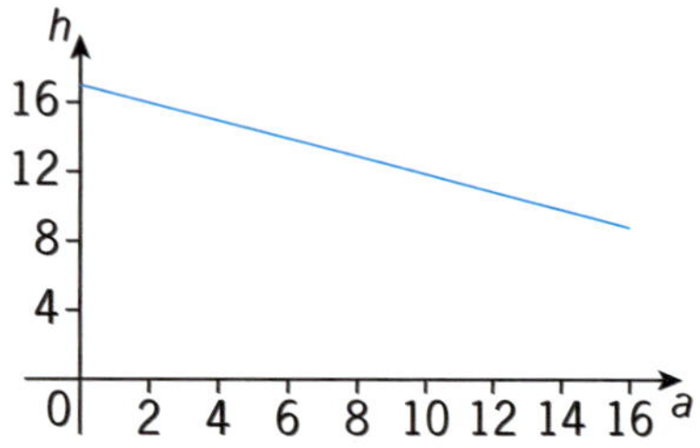

5 For example:

h, height in cms	160	170	180	190	200
m, mass in kg	54	63	72	81	90

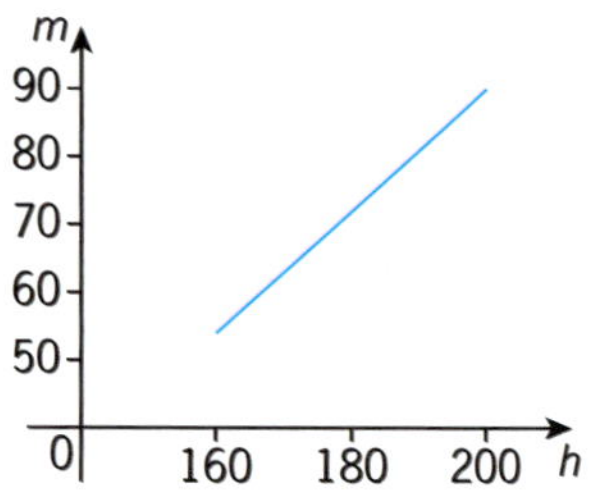

Ex B11.2

1 a

Length of sides, x	5	6	7	8	9	10
Length of edging, y	20	24	28	32	36	40

b

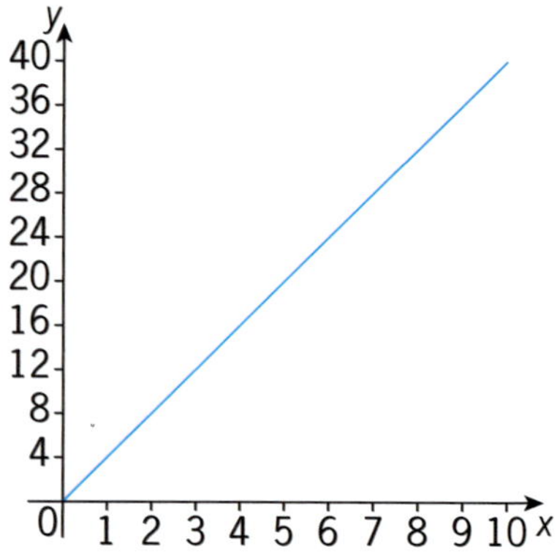

$y = 4x$

c 100 m

2 a

Amount used, x litres	0	10	20	30	40
Amount left in tank, y litres	40	30	20	10	0

b

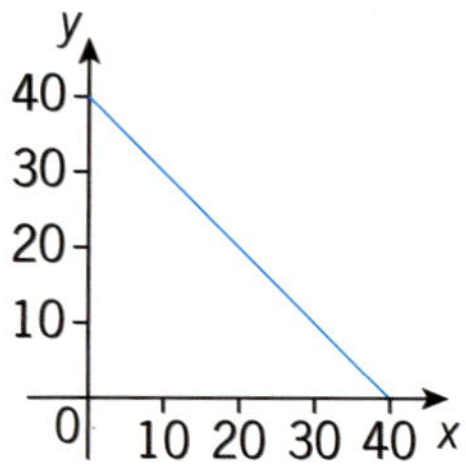

$y = 40 - x$

c 5, 420 miles

3 a

Length of journey, x km	1	2	3	4	5
Cost, £y	5	8	11	14	17

b

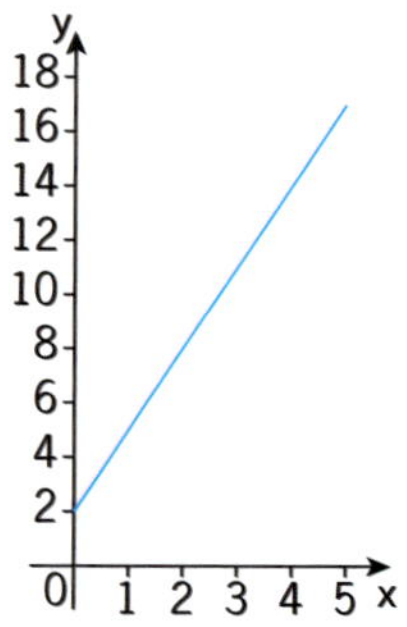

$y = 3x + 2$

c 23, No

4 For example:

x, in degrees	10	20	30	40	50	60	70	80
y, in degrees	160	140	120	100	80	60	40	20

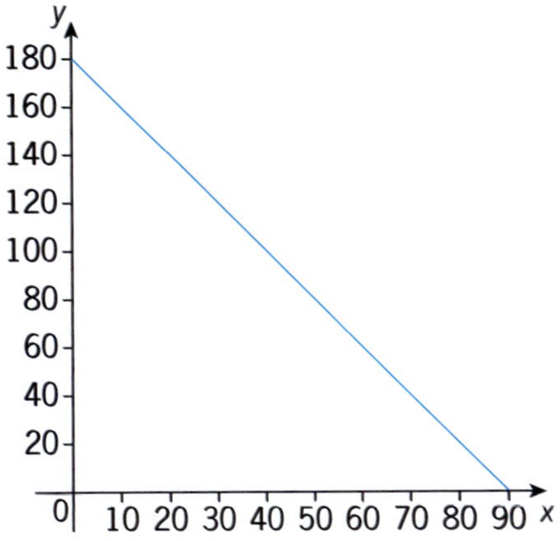

$y = 180 - 2x$

5 a 360° b 180° c 6°

For example:

x, minutes	10	20	30	40	50	60
y, degrees	60	120	180	240	300	360

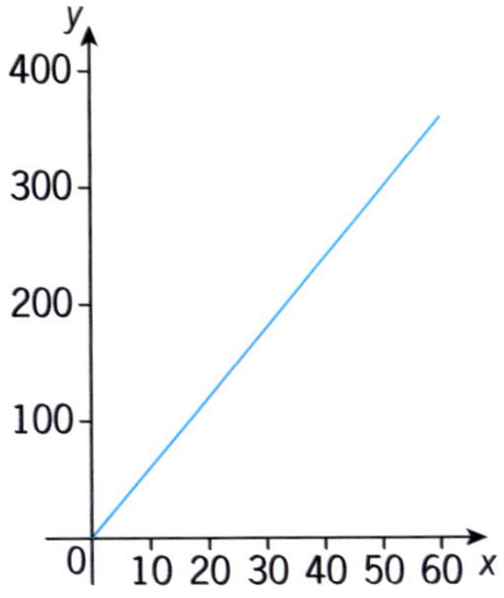

$y = 6x$

Ex B11.3

1 a 20°C b 50°C c 40°C
d 8 mins e 5°C

2 a 750 L b 250 L c 8 weeks
d 16 weeks e After 12 weeks

3 a i $70 ii £25
b No, because the exchange rates are different. She will get less than £40 back.

4 50 miles

a

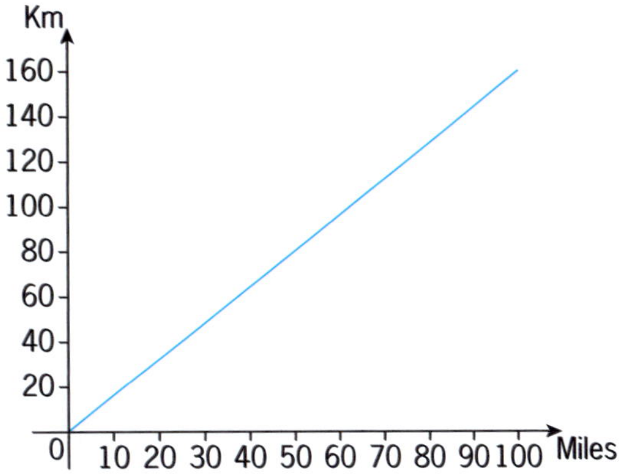

b 48 km, 128 km

Ex B11.4

Rich task

$x > 4$

1 a equation b identity
c inequality d inequality

2 a $x < 7$ b $x > 7$ c $x \geq 7$ d $x \leq 7$

3 a $x > 2$ b $x \geq 0$ c $x \geq 1$
d $x < 1$ e $x < -2$ f $x \geq -1$
g $x < 0$ h $1 \leq x \leq 2$ i $-1 \leq x \leq 2$

4 a

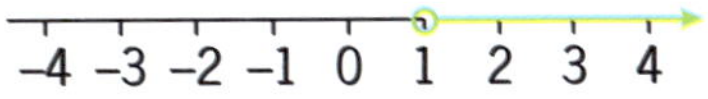

b

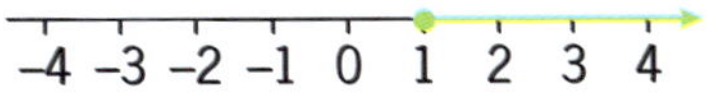

c

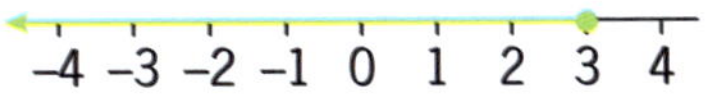

d

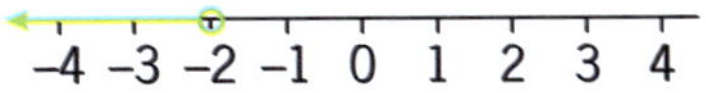

e

f

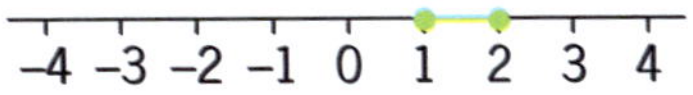

g

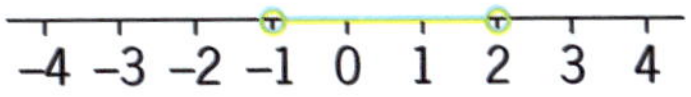

h

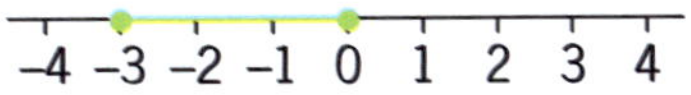

5 a $x > 3$

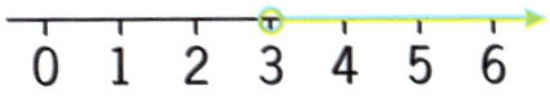

b $x < 2$

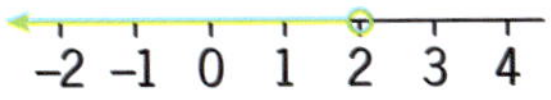

c $x \geq 4$

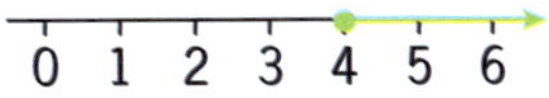

d $x \leq 3$

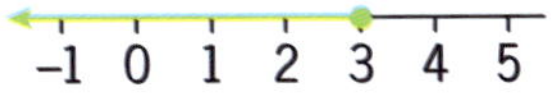

e $x > 5$

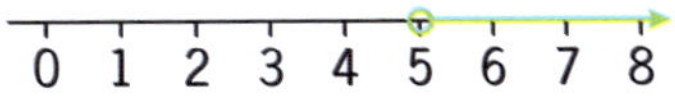

f $x < 5$

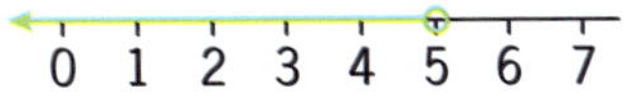

Unit B

g $x > 5$

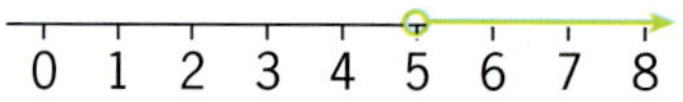

h $x < -2$

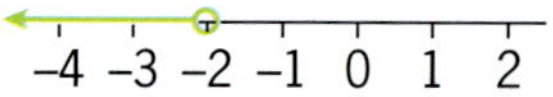

i $x < 3$

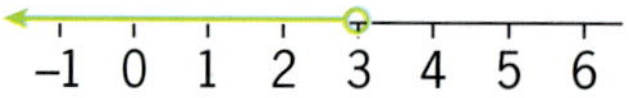

j $x \geq 4$

k $x < 5$

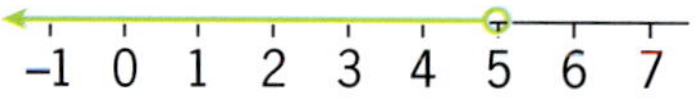

l $x \leq 1$

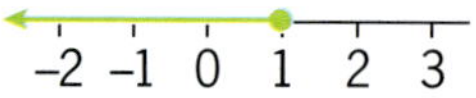

m $x < 2$

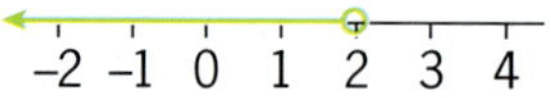

n $x \geq 3$

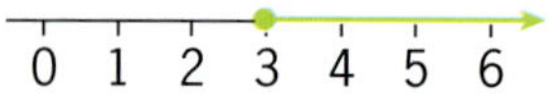

o $x \leq -1$

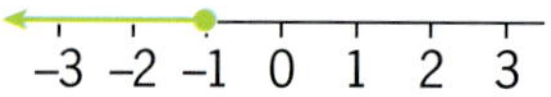

p $x > -2$

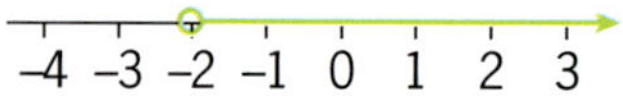

6 a $3 < x < 5$

b $3 \leq x \leq 5$

c $5 < x < 7$

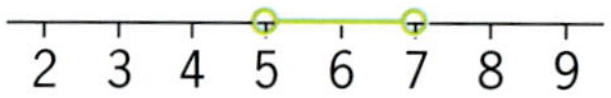

7 a 6, 7, 8 b 5, 6, 7, 8, 9 c –1, 0, 1

8 a 7, 9 b 1, 3, 5 c –3, –1, 1

Exam practice

1 a $x \geq 7.5$

b i

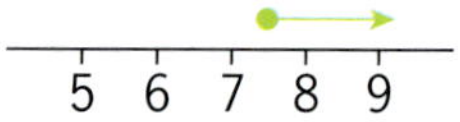

ii 8, 9

2 a graph horizontal for call times between 0 and 1 minute

b $C = 10 + 30\,m$ (only valid when $m \geq 1$)

3 a at least 16 times

B12 Transformations

Check in

1 a 180° b 270° c 90° d 180°

2 a Clockwise b Anti-clockwise

Ex B12.1

1 a

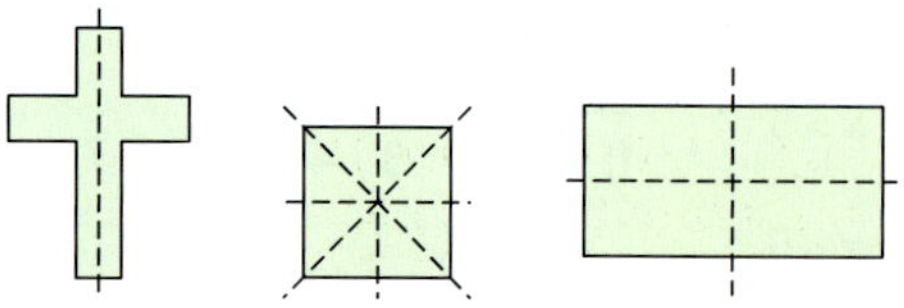

b

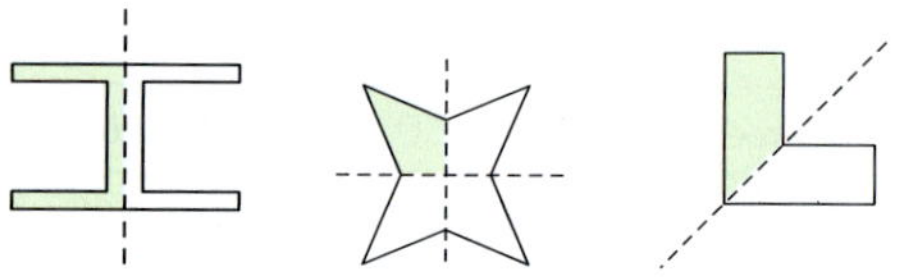

2 a 4 b 2 c 1 d 3

3 1 line: A C D E M T U V W Y K B

0 line: L Q F G J N P R S Z

2 lines: H I X

2 or infinite lines: O

dependent on the shape of font

4

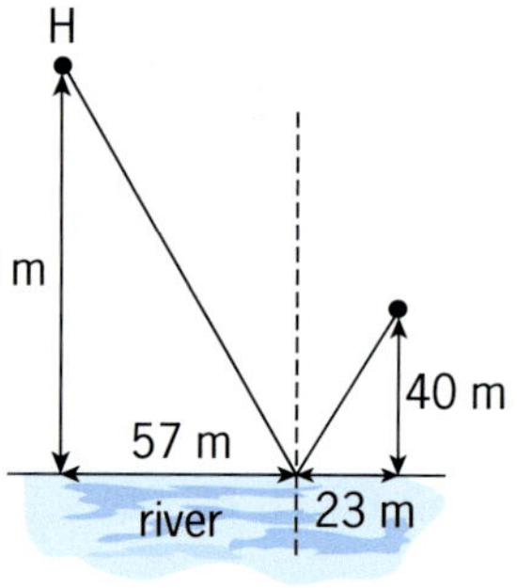

Ex B12.2

1

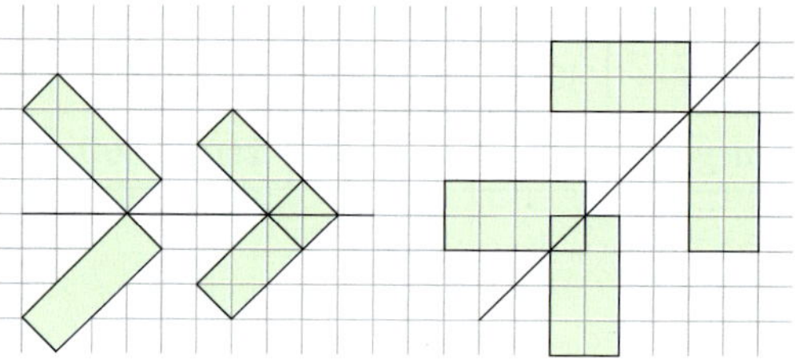

2 a

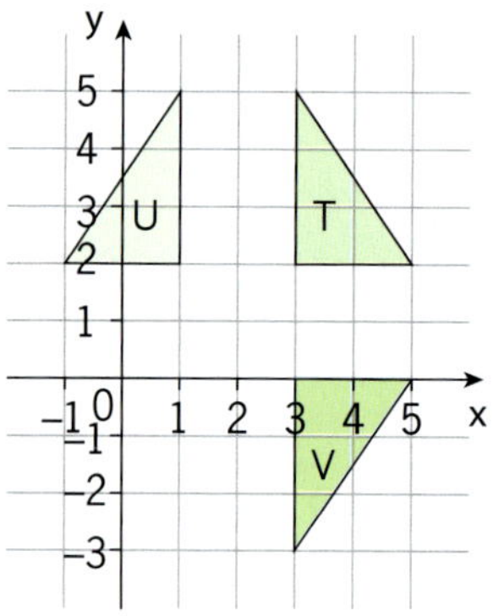

3

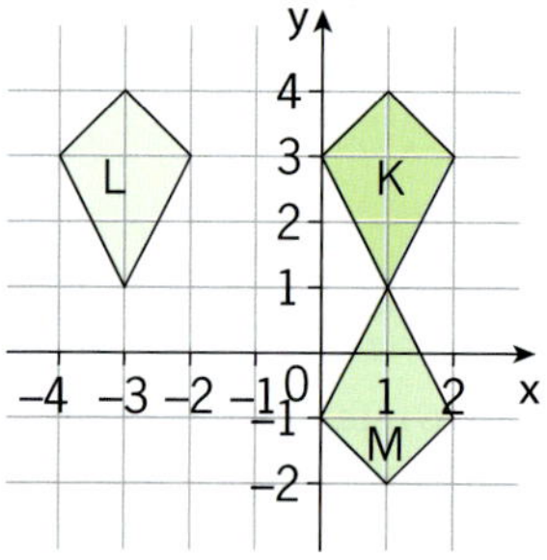

4

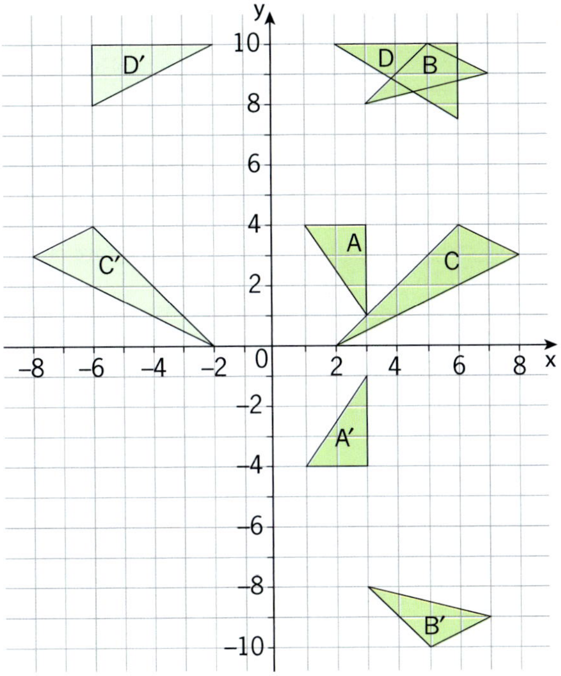

Ex B12.3

1 a 180° b 90° anticlockwise

2

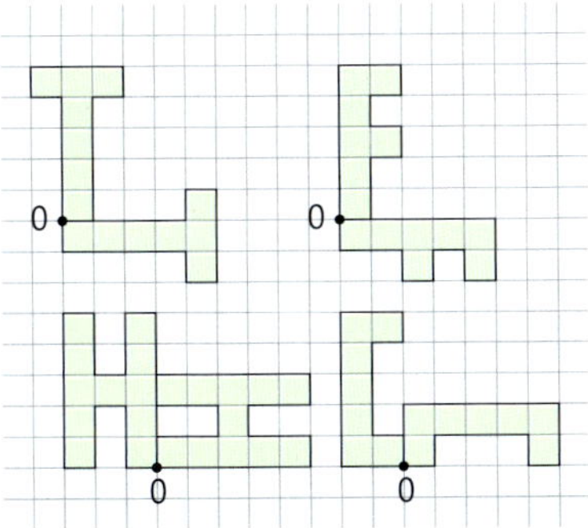

3

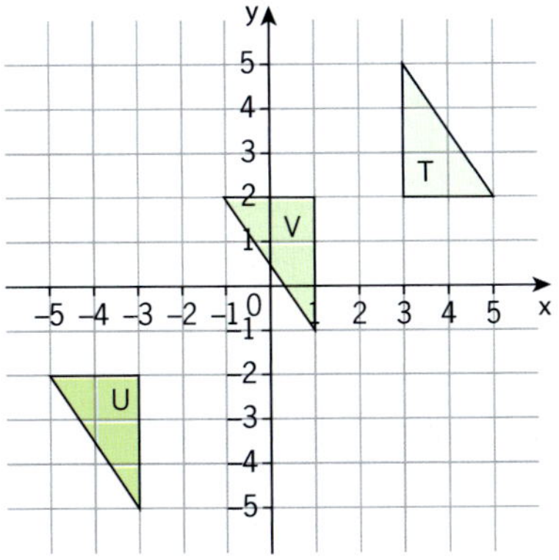

4

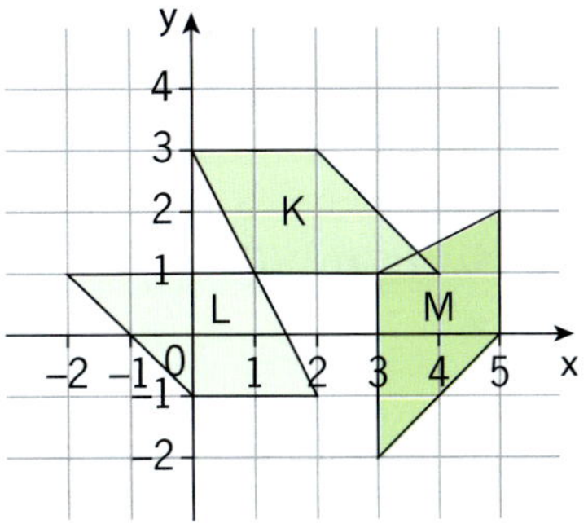

Ex B12.4

1 a (4, 5), (7, 3), (8, 2), (5, 9), (6, −2), (6, −1)

b $\begin{pmatrix}4\\2\end{pmatrix}$, $\begin{pmatrix}2\\-2\end{pmatrix}$, $\begin{pmatrix}4\\0\end{pmatrix}$, $\begin{pmatrix}-7\\-4\end{pmatrix}$, $\begin{pmatrix}-2\\2\end{pmatrix}$, $\begin{pmatrix}-4\\0\end{pmatrix}$

2 a $\begin{pmatrix}3\\2\end{pmatrix}$ b $\begin{pmatrix}5\\1\end{pmatrix}$

c $\begin{pmatrix}2\\-1\end{pmatrix}$ d $\begin{pmatrix}-3\\-2\end{pmatrix}$

3 a

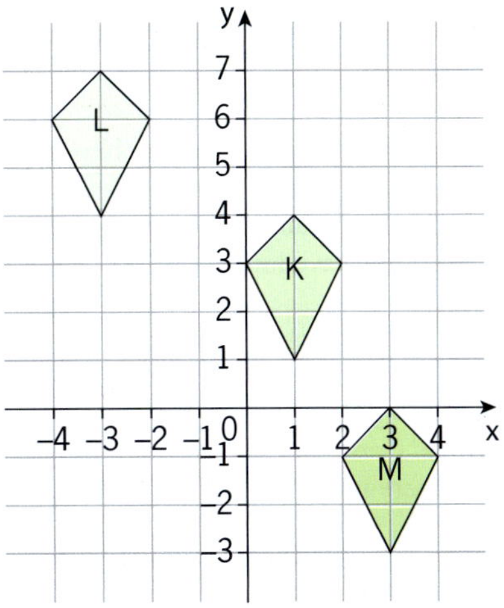

4 $\begin{pmatrix}2\\1\end{pmatrix}$, $\begin{pmatrix}2\\-1\end{pmatrix}$, $\begin{pmatrix}-2\\1\end{pmatrix}$, $\begin{pmatrix}-2\\-1\end{pmatrix}$, $\begin{pmatrix}1\\2\end{pmatrix}$, $\begin{pmatrix}1\\-2\end{pmatrix}$, $\begin{pmatrix}-1\\2\end{pmatrix}$, $\begin{pmatrix}-1\\-2\end{pmatrix}$

Exam practice

1 a **i** no line symmetry
ii rotational symmetry of order 1

b **i** no line symmetry
ii rotational symmetry of order 5

2 a (0, 1) (3, 2) (2, 3) R
b (1, 0) (2, 3) (3, 2) S

3 a S is (1, −2) (1, 0) (2, 1) (4, 1)
b T is (0, 0) (0, 2) (1, 3) (3, 3)

Unit B

B13 Bivariate data and time series

Check in

1

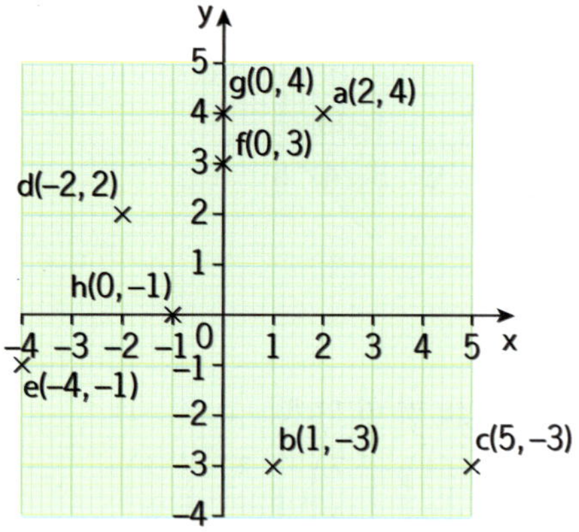

2 a 10 b 25 c 20 d 12.5 e 27.5

Ex B13.1

1 graph
2 graph
3 more alcohol associated with longer reaction times
More time on games associated with shorter reaction times. Longer reaction times at 3.0 am for all subjects

Ex B13.2

1 Positive; people from higher altitudes have larger lung capacity.
2 Positive; good at both subjects
Positive; tall people have long arms
3 b positive
c all except one subject take longer with left hand than right.
4 b no correlation. Number of deaths not related to magnitude. Density of population and type of buildings very important.

Ex B13.3

1 a £108 b £97
2 a 4.9 litres
b i beyond data range
ii relationship only established for men
3 a graph
b If results for black bear and horse are ignored, correlation very weak so estimates are inaccurate

Ex B13.4

1 Highest in June and December, which are holiday months.
2 2005 higher than 2004, but same monthly pattern. Cost per unit may have increased. Highest in Jan-March then Oct-Dec, the coldest months.
3 Highest in warmer months of August, July and June. Also peak in December, a holiday period.
4 Earns most in May-August, maybe school holidays. Earnings increased each year, maybe better rates as older.
5 Highest in April-June, Lowest in Oct-Dec. 2005 less than 2004, maybe economising.

Exam practice

1 No consistent pattern. Average temperature ranged from 13.6°C to 16.7°C
2 More marriages in summer months.
3 Any valid response – data can be compared through scatter graph, mean, median, mode, range or IQR, or box-and-whisker plot.

B14 More straight-line graphs

Check in

1 1, 3, 5, 7, 9

Ex B14.1

1

x	0	1	2	3
y	1	3	5	7

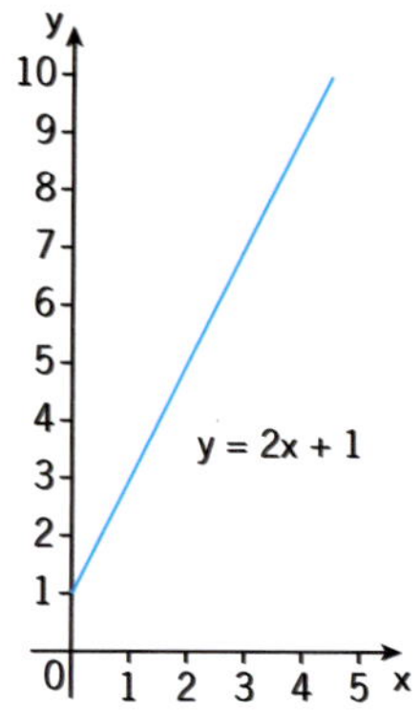

2 a

x	0	1	2	3	4
y	3	4	5	6	7

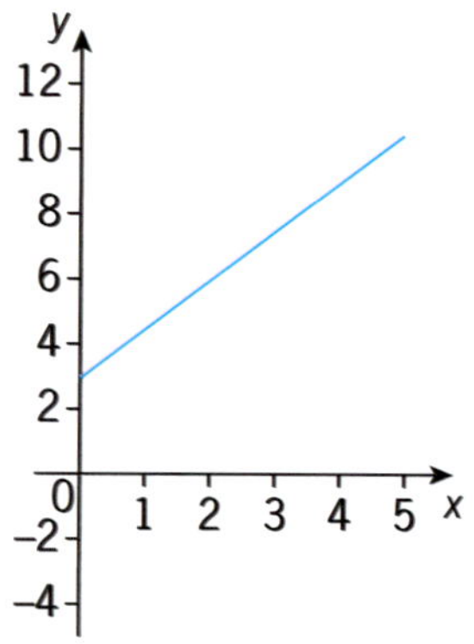

b

x	0	1	2	3	4
y	−3	−2	−1	0	1

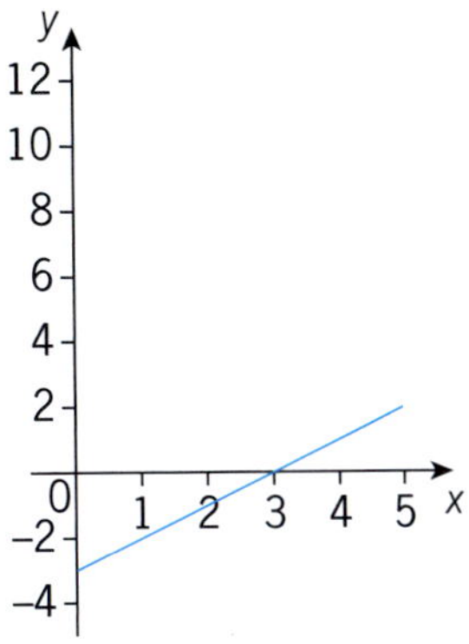

c

x	0	1	2	3	4
y	1	3	5	7	9

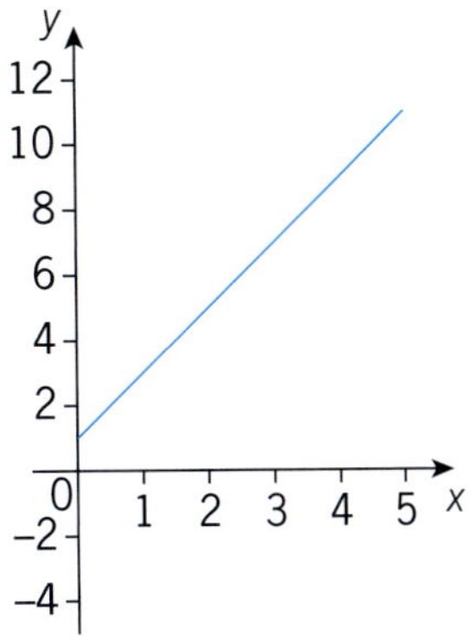

d

x	0	1	2	3	4
y	−2	0	2	4	6

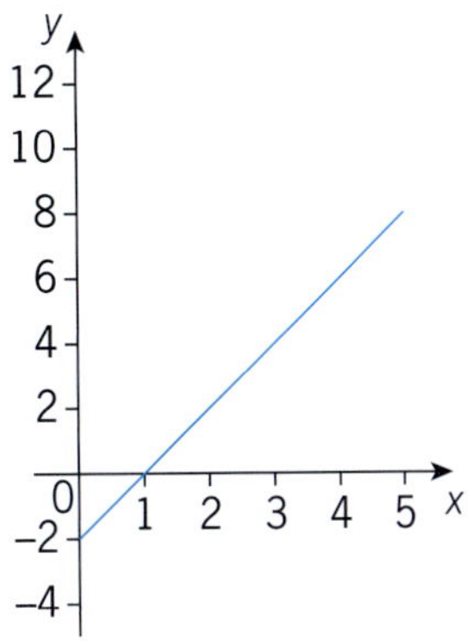

e

x	0	1	2	3	4
y	12	11	10	9	8

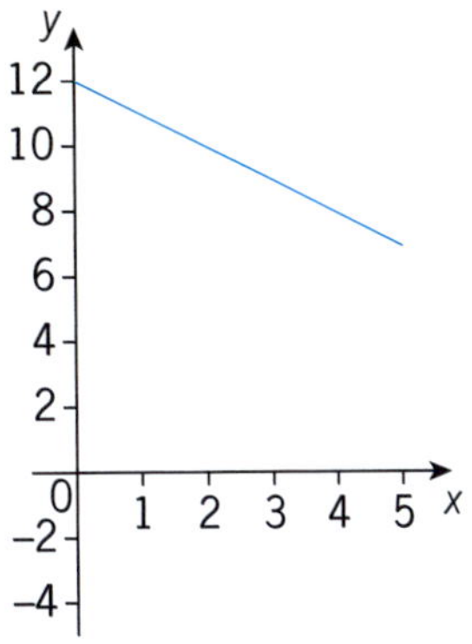

f

x	0	1	2	3	4
y	3	2	1	0	−1

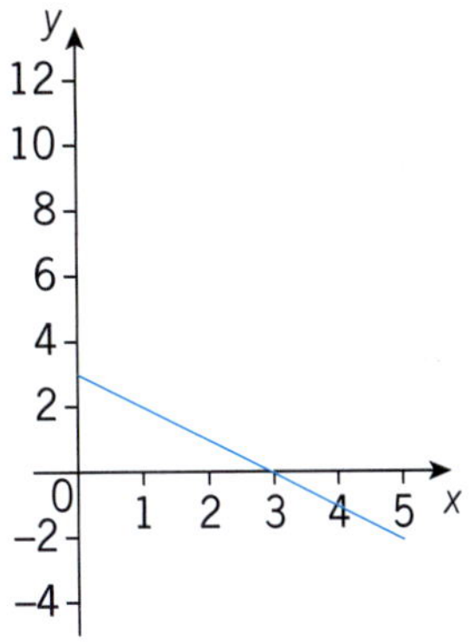

3 **a** Yes **b** Yes **c** No

d Yes **e** Yes **f** No

4 **a**

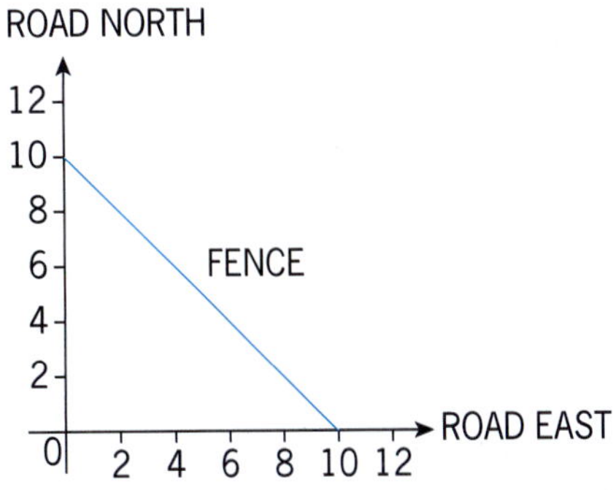

b 50

5 a

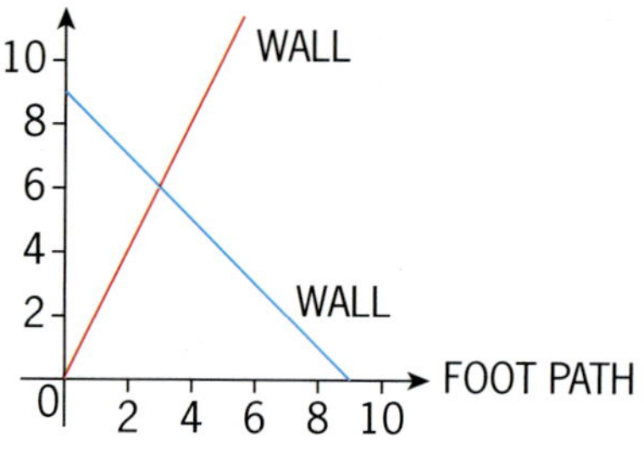

b 27

6

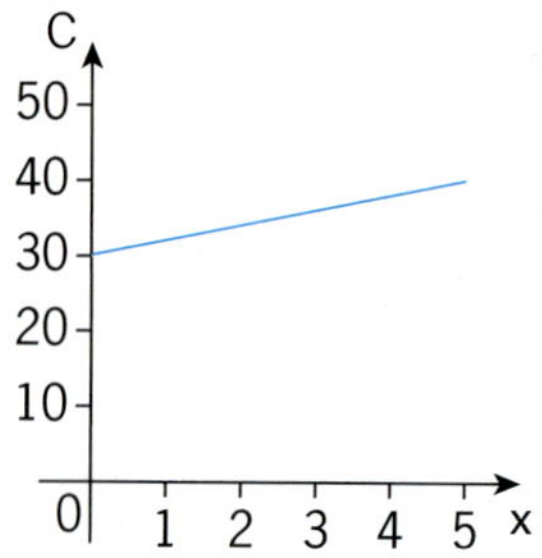

3.5 hours

7 (20, 52)

Ex B14.2

1	P	3	Q	2	R	1	S	−2
	T	0.5	U	−1	V	0		
2	a	2	b	3	c	0.5	d	0
	e	−1	f	−2				
3	a	5	b	4	c	1	d	$\frac{1}{2}$
	e	−6	f	−3	g	−8	h	−8
	i	−4	j	−1	k	$-\frac{1}{2}$	l	3

4 $y = 2x + 1$ and $y = 2x - 1$
$y = 4 - x$ and $y = -x + 5$
$y = \frac{1}{2}x + 3$ and $y = 1 + \frac{1}{2}x$
$y = -2x + 7$ and $y = 6 - 2x$

Rich Task

The graphs of $y = mx + 3$ all pass through (0, 3)
The graphs of $y = 2x + c$ are all parallel lines

Ex B14.3

1 a (4, 0) (0, 6)

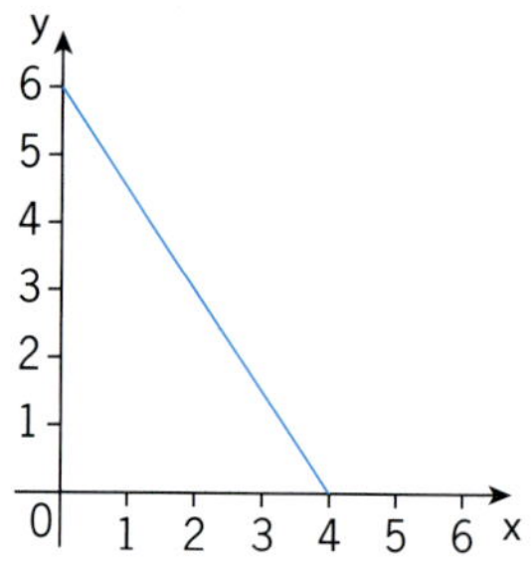

b (5, 0) (0, 2)

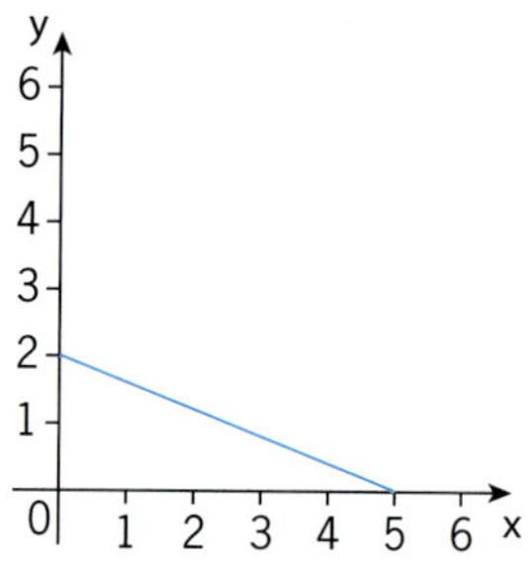

c (5, 0) (0, 3)

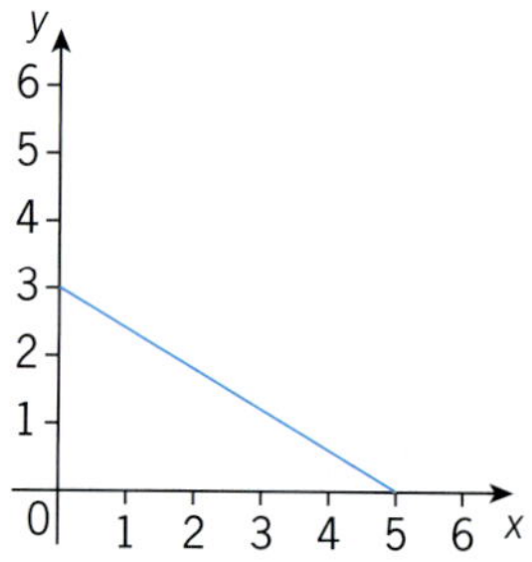

d (3, 0) (0, 4)

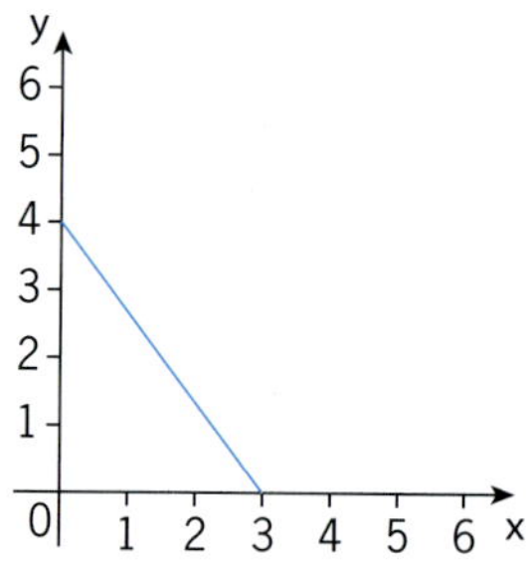

e (4.5, 0) (0, 3)

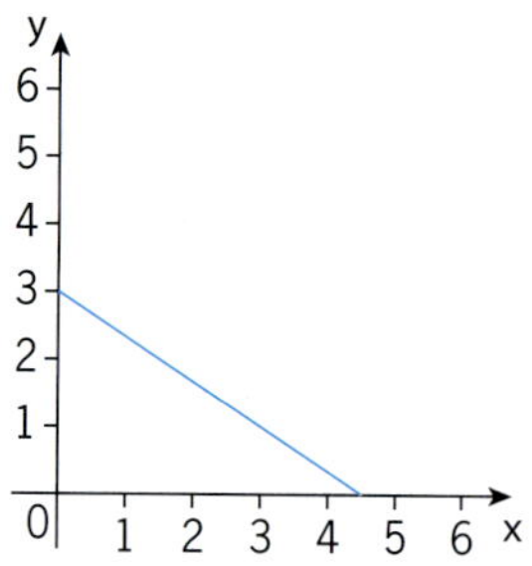

f (6, 0) (0, 2)

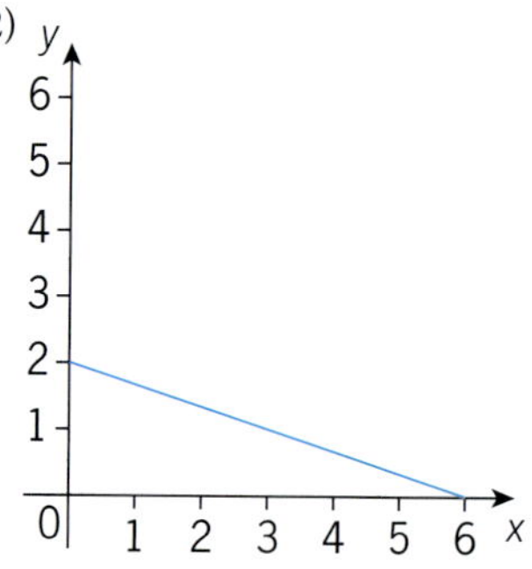

2 **a** (4, 0) (0, −2)

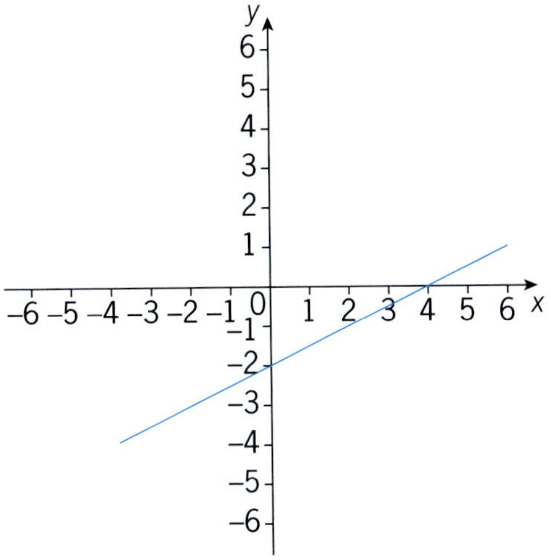

b (5, 0) (0, −6)

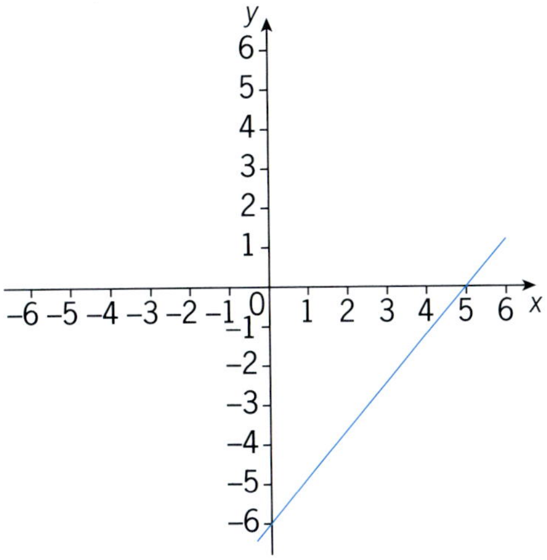

c (−4, 0) (0, −3)

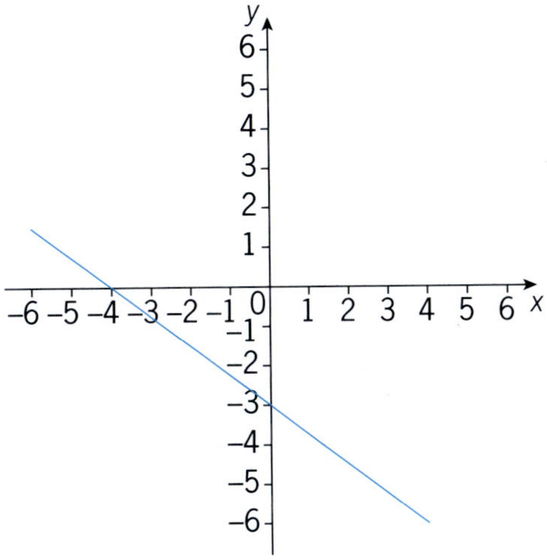

3 **a** gradient = −2

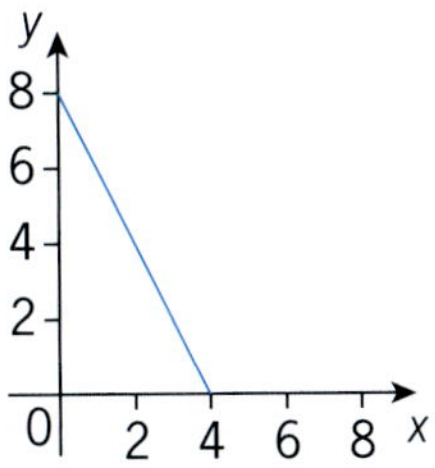

b gradient = −3

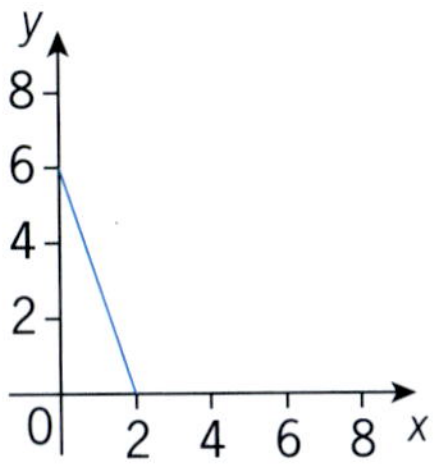

c gradient = $-\frac{1}{2}$

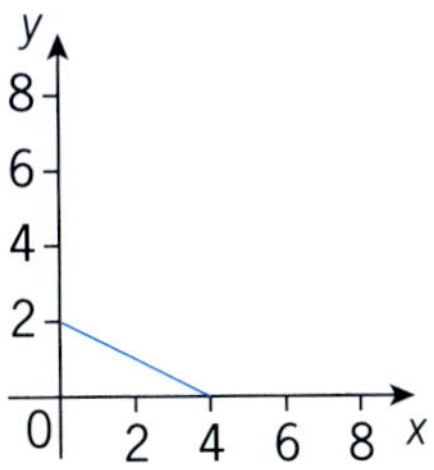

4 **a** gradient = −2

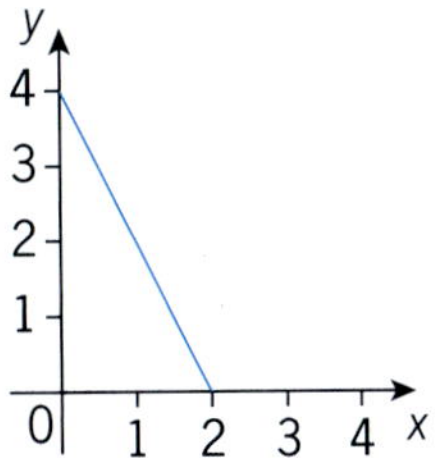

b gradient = −3

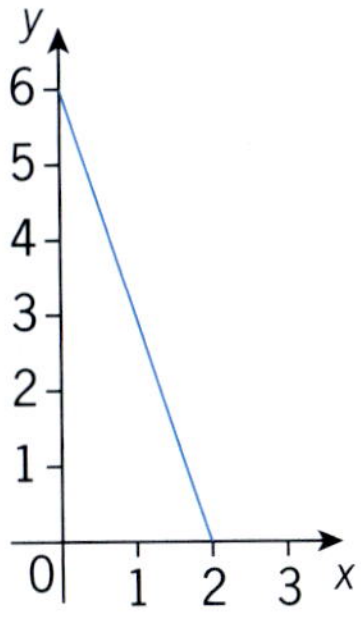

c gradient = 3

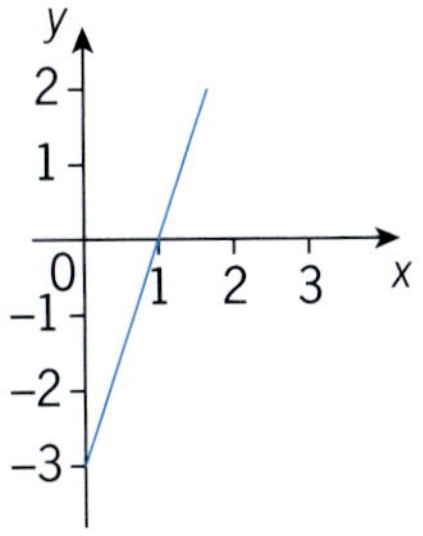

d gradient = −3

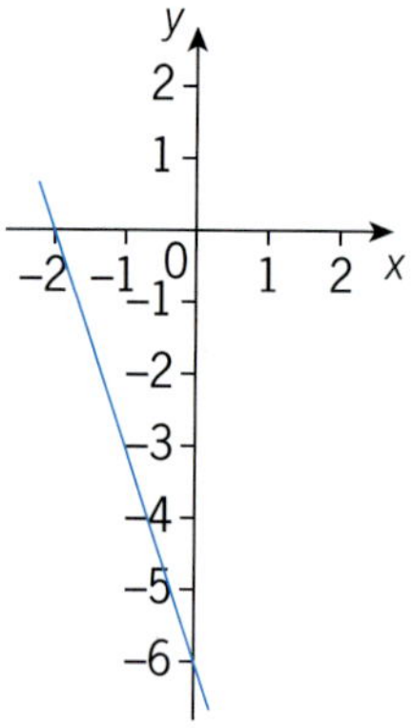

e gradient = $-\frac{1}{2}$

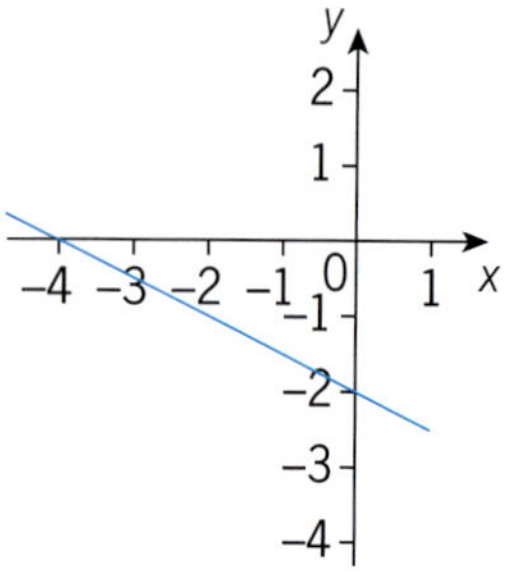

f gradient = 2

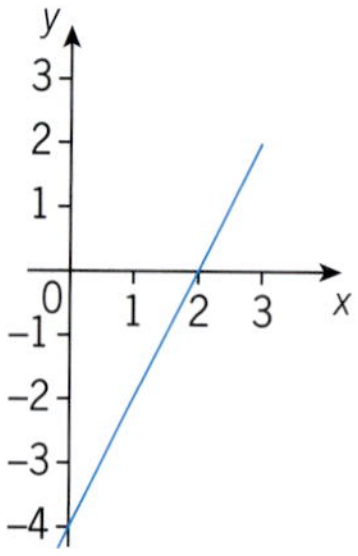

5 Graph will be a straight line through (3, 0) and (0, 6)
Possible pairs are: (1, 4) and (2, 2)

6

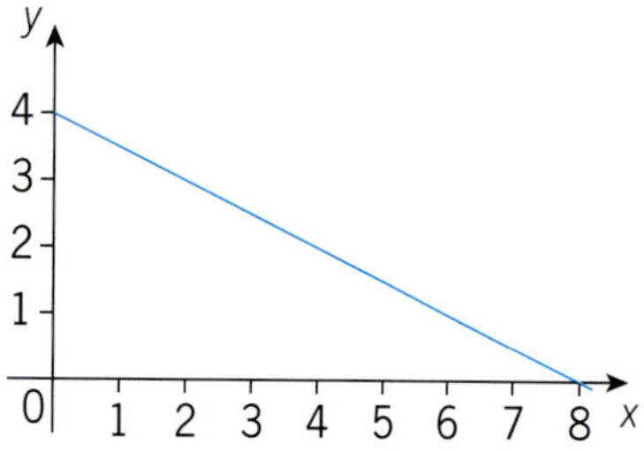

Possible combinations are (0, 4), (2, 3), (4, 2), (6, 1) and (8, 0)
4 Christmas cards and 2 birthday cards

Ex B14.4

1 a $x = 2$ $y = 5$

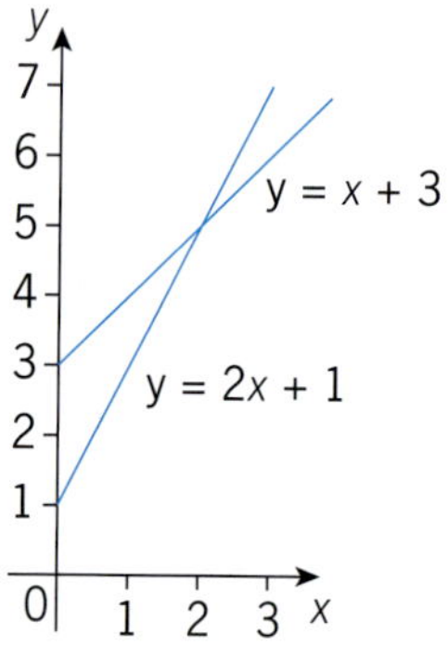

b $x = 2$ $y = 7$

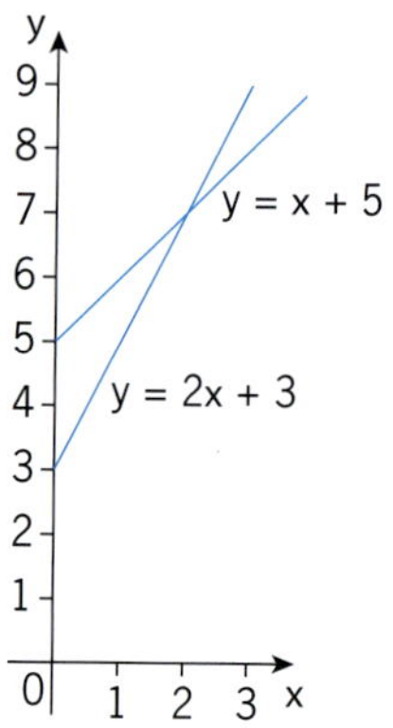

c $x = 3$ $y = 4$

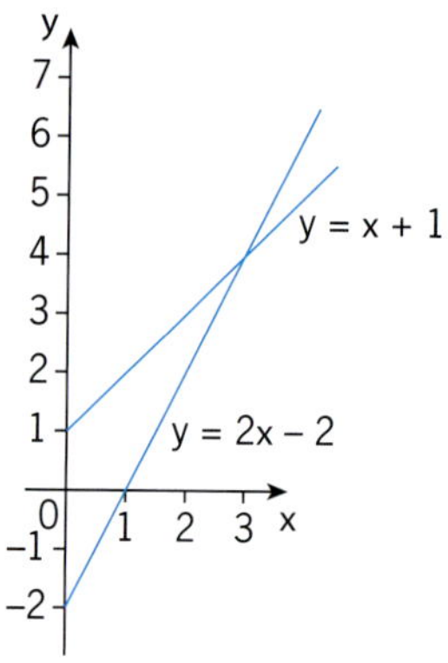

d $x = 3$ $y = 7$

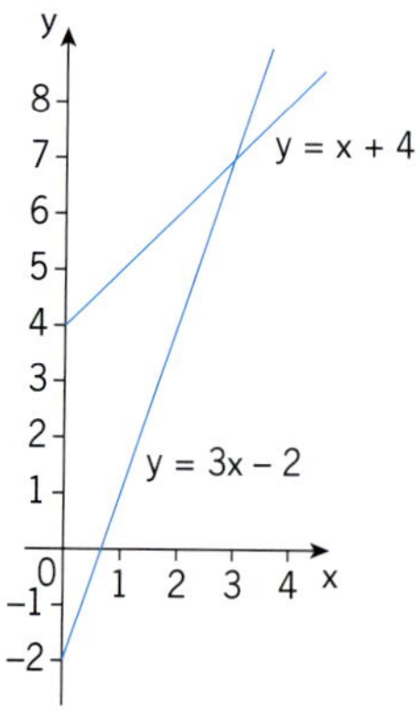

e $x = 2$ $y = 4$

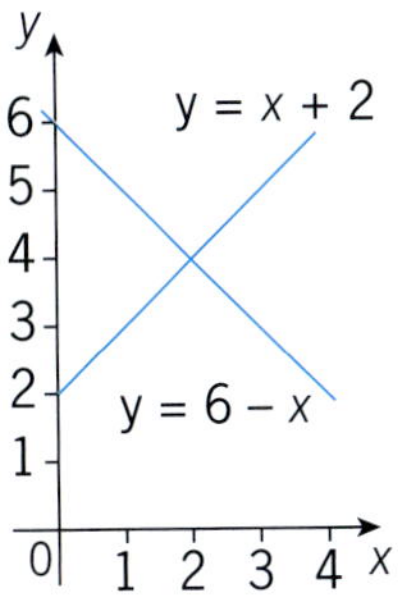

f $x = 4$ $y = 3$

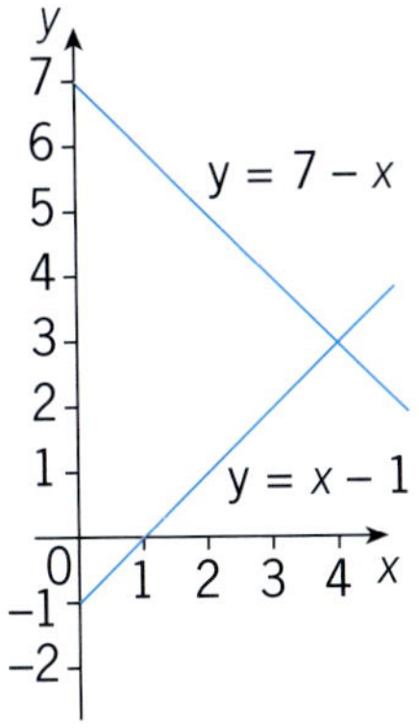

2 a $x = 2$ $y = 3$

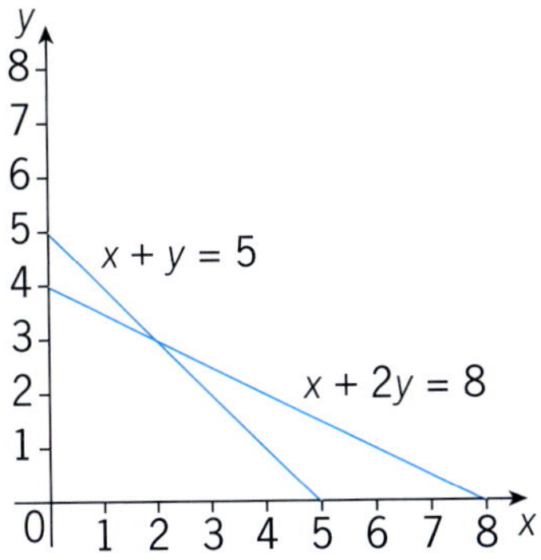

b $x = 3$ $y = 2$

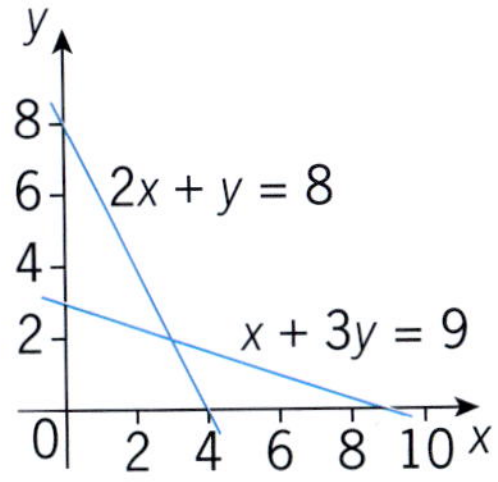

c $x = 3$ $y = 4$

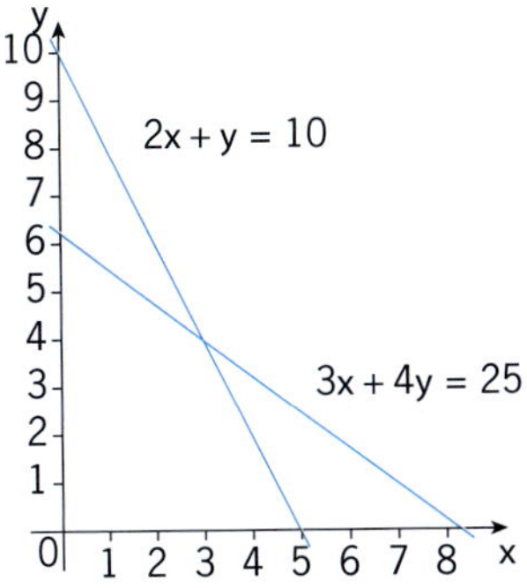

d $x = 3$ $y = 1.5$

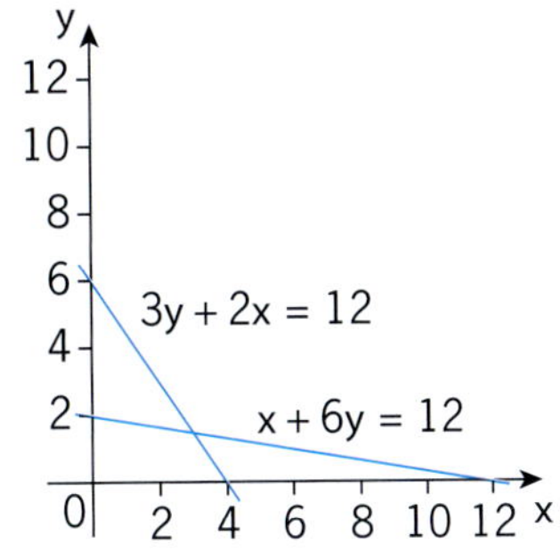

e $x = 6$ $y = 2$

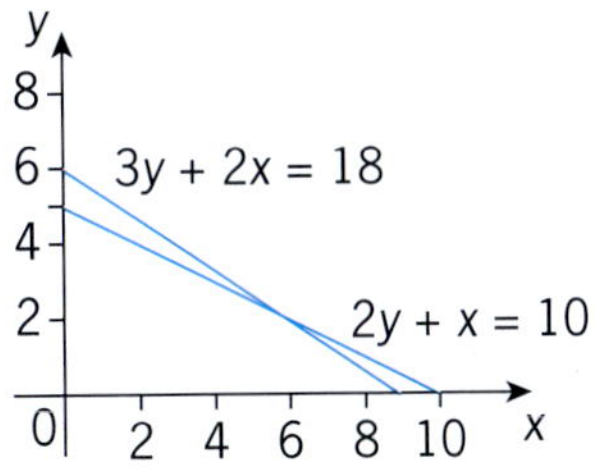

f $x = 2$ $y = 2$

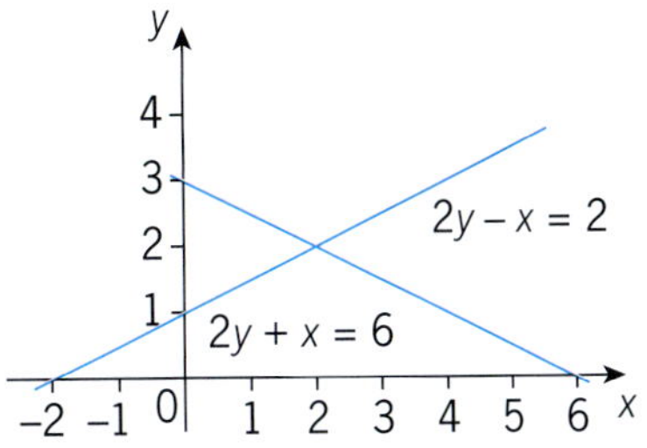

3

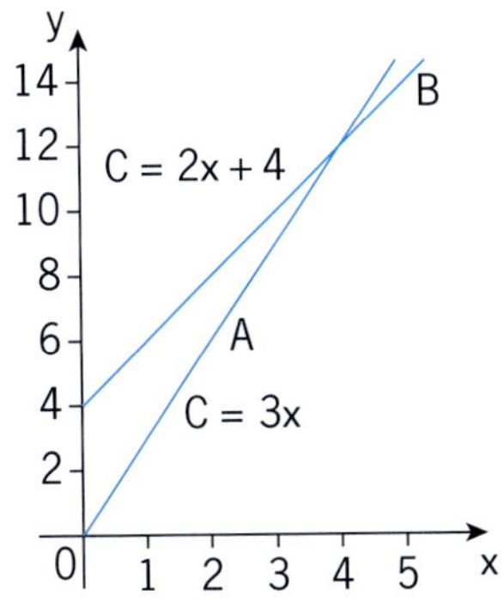

Firm A is cheaper for up to 4 spanners. For more than 4 spanners firm B is cheaper

Exam practice

1 a (2, 5)

b

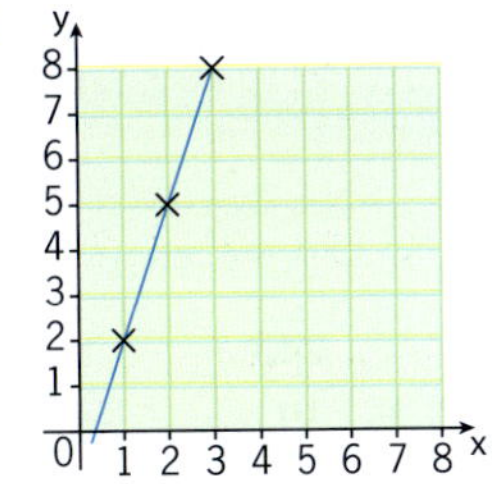

2 a $\frac{1}{2}$

b

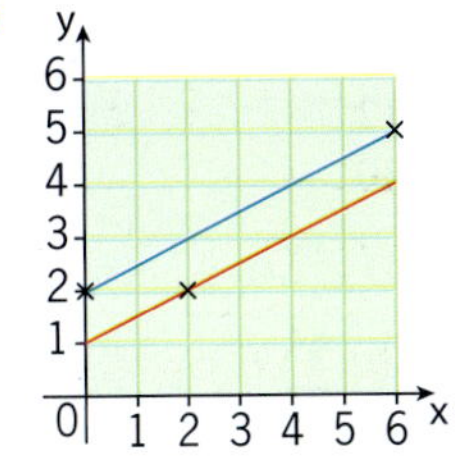

3 $x = 1\frac{1}{3}$ $y = 3\frac{2}{3}$

B15 Estimation and indices

Check in

1 a 49 b 8 c 1000 d 0.32
2 5
3 yes

Ex B15.1

1 a 292 b 601 c 1739 d 273
e 158 f 235 g 20250 h 96
i 2496 j 3.3 k 86.17 l 36.25
2 a i 13.2 ii 0.87
b i 21.4 ii 19.902
c i 7600 ii 0.045
d i 22.5 ii 7.2
3 a £2.76 b £7.77
4 a 33 mins b 35 mins
5 a 49, 51 b 77, 27 or 84, 34
c 98 × 84 = 8232

Ex B15.2

1 a 343 b 400 c 225 d 216 e 125
2 a 7 b 100 c 5 d 4 e 9
3 a 10, −10 b 11, −11 c 3, −3 d 14, −14
4 a i 13 ii 10 b 1.3
5 a i 12 ii 1.2 b i 15 ii 1.5
6 a 10^3 b 10^6 c 10^6
7 a 27 b −198 c 15 d 56 e 7
8

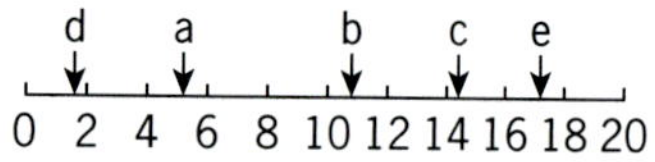

9 from 1 + 8 2 + 7 3 + 6 4 + 5 1 + 15 2 + 14 3+13
4 + 12 5 + 11 6 + 10 7 + 9 7 + 18 8 + 17 9 + 16
10 + 15 11 + 14 12 + 13 1 + 3

Ex B15.3

1 a 600 b 8000 c 16 000 d 100
e 100 f 13 g 8 h 50
2 a 5 b 11
3 a 9 b 10 c 25 d 64
e 7
4 a 4 b 100 c 2 d 10
e 7 f 0.5 g 12 h $\frac{1}{14}$
i 2 j 15 000 k 1.3 l 100
m 0 n 440
5 7
6 12
7 40

Ex B15.4

1 a 64 b 81 c 243 d 625
e 72 f 256 g 16
2 i 8 ii 5
3 a i 5^3 ii 5^3 iii 2^6, 4^3
b i $4^6 = 16^3$ ii all squares iii 4^6, 16^3
4 a $2^4 \times 3$ b $2 \times 3 \times 7$ c $2^3 \times 3 \times 5^2$
d $2^8 \times 5$ e $3 \times 5 \times 7^2$
5 a 7^9 b 3^8 c 8^9
d 15^8 e 4^{12} f 3^6
g 8 h 9^5 i 6^4
6 2, 4

Exam practice

1 9
2 odd (odd × odd = odd)
3 She has enough

B16 Similarity and enlargement

Check in

1

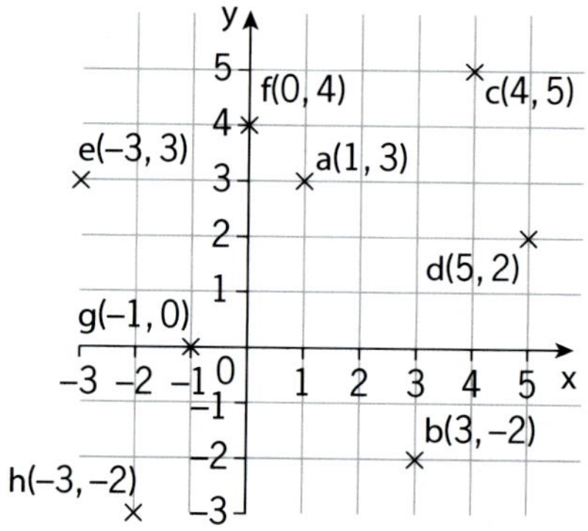

2 a 1 : 3 b 1 : 6 c 2 : 3 d 5 : 2

Ex B16.1

1 a congruent
b not congruent
c congruent
2 a e.g

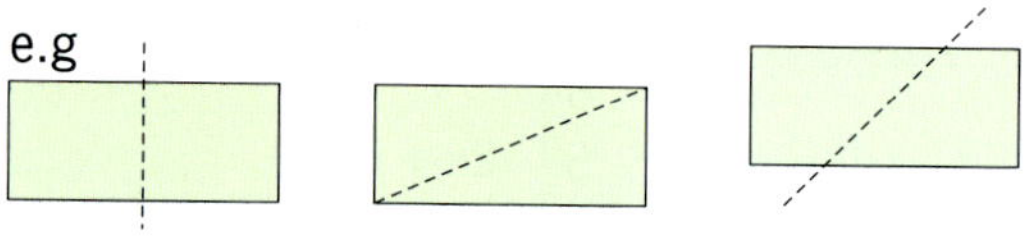

b The vertical line is; the other lines are not
3 a

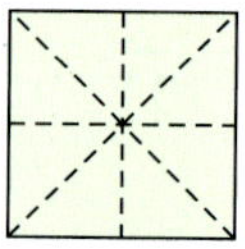

b

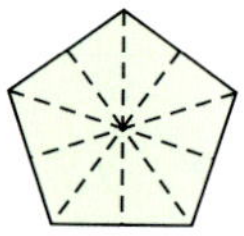

c

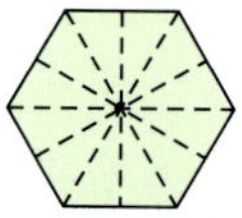

d

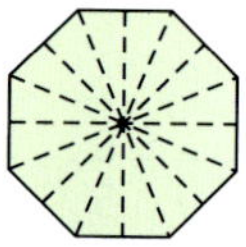

4 a similar b not similar c similar
5 a congruent b congruent c similar
d neither e similar f congruent
g neither
6 4.40 m long, 1.80 m high

Ex B16.2

1 a, b, d

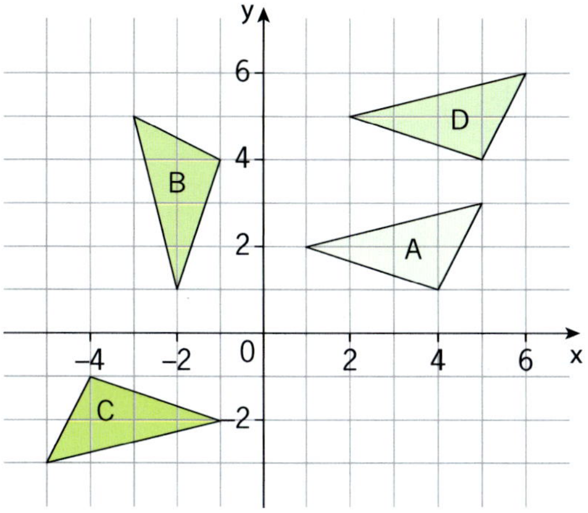

c Rotation of 180° about (0, 0) e Translation $\begin{pmatrix}1\\3\end{pmatrix}$

2 a, b, d

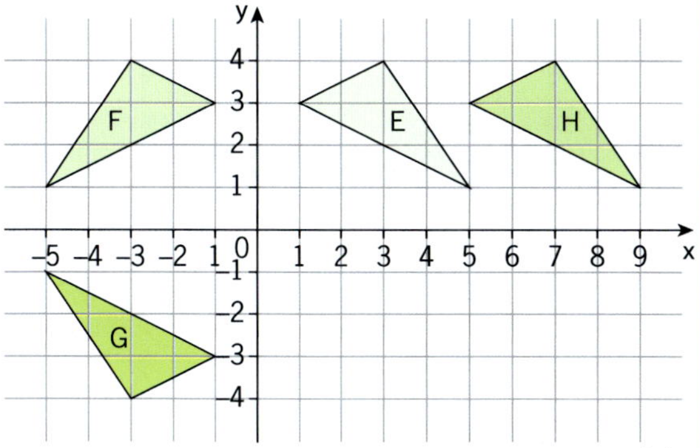

c Rotation of 180° about (0, 0) e Translation $\begin{pmatrix}4\\0\end{pmatrix}$

3 Translation by the vector $\begin{pmatrix}4\\-2\end{pmatrix}$

4 a

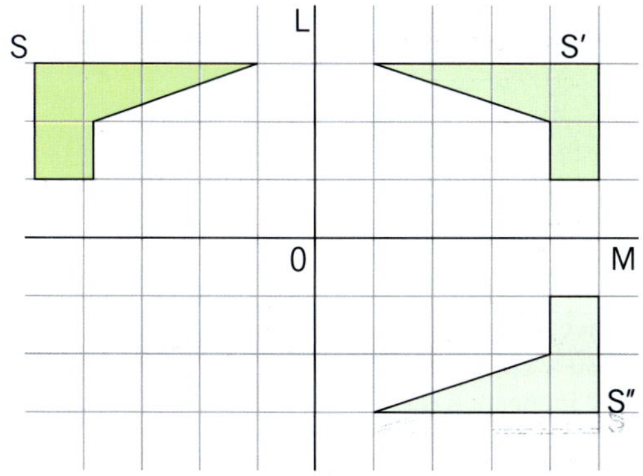

Rotation of 180° about point O

b

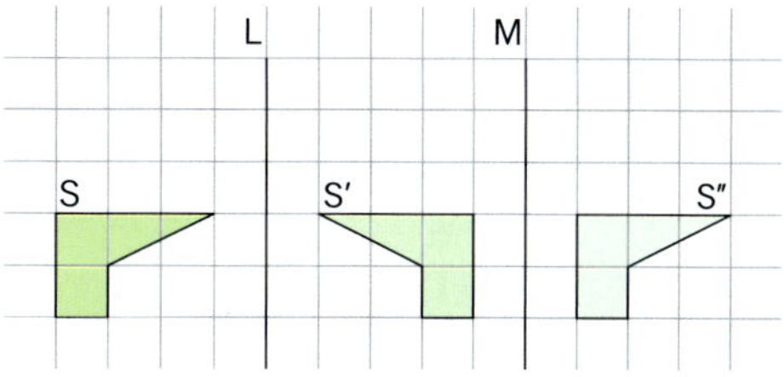

Translation of $\begin{pmatrix}10\\0\end{pmatrix}$

5

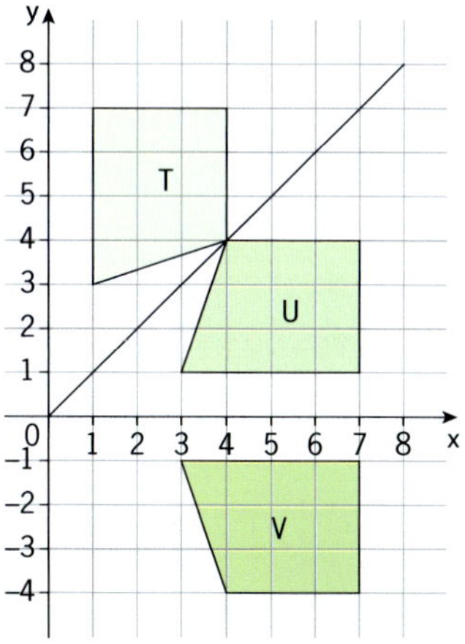

Rotation of 90° clockwise about the origin

6

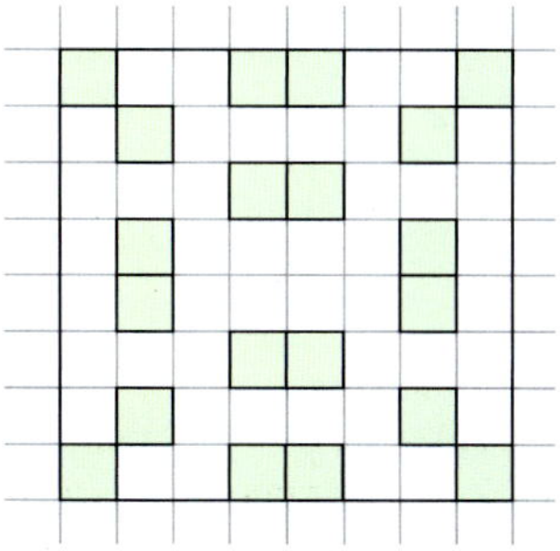

Ex B16.3

1

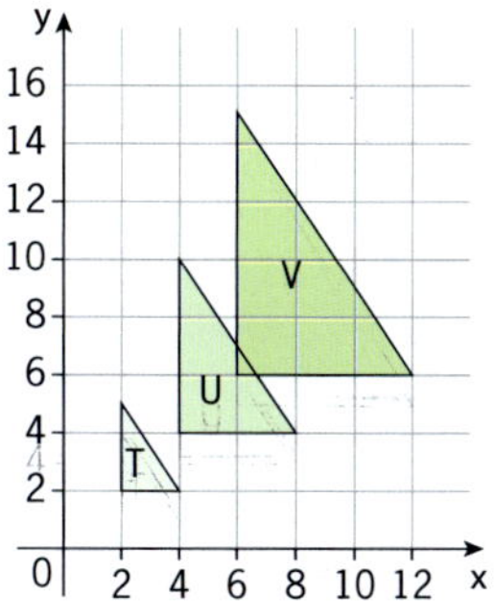

2

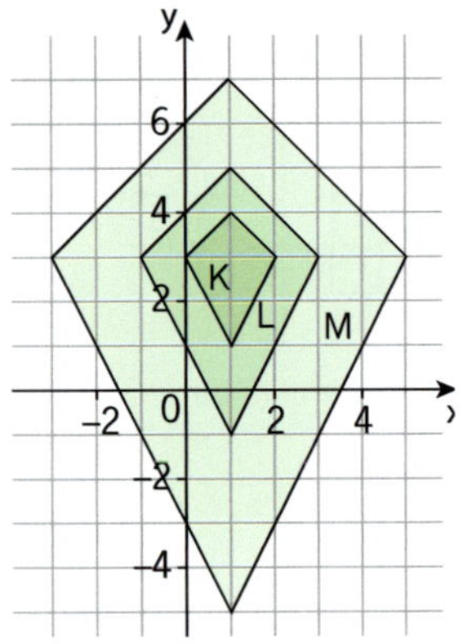

3

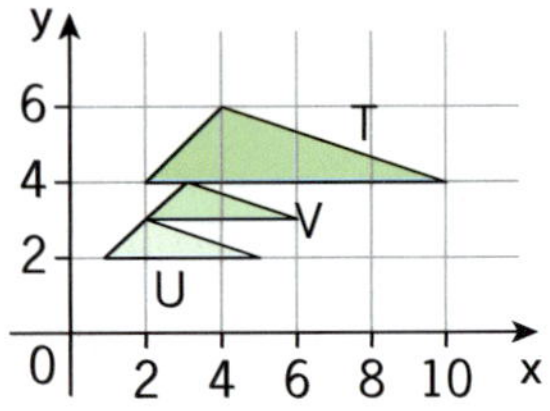

4

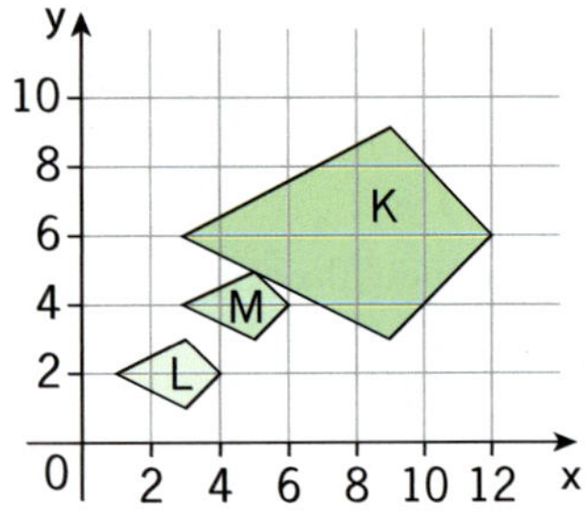

5 Vertices of images are

a (4, 6) (8, 6) (4, 12)

b (18, 0) (24, 6) (15, 9)

c (3, 16) (11, 16) (11, 20) (3, 22)

d (18, 17) (22, 17) (20, 15)

6 a

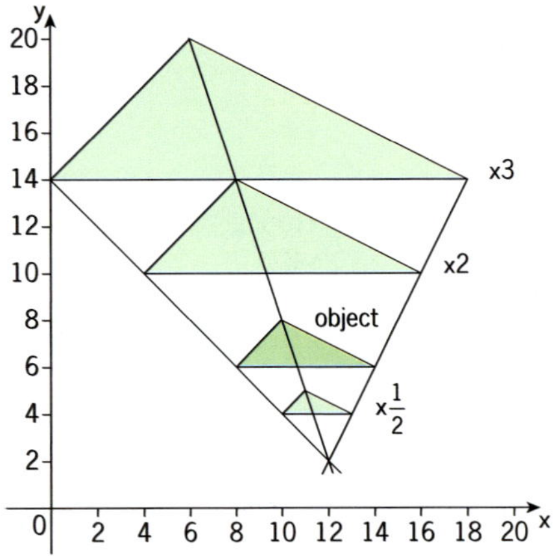

b

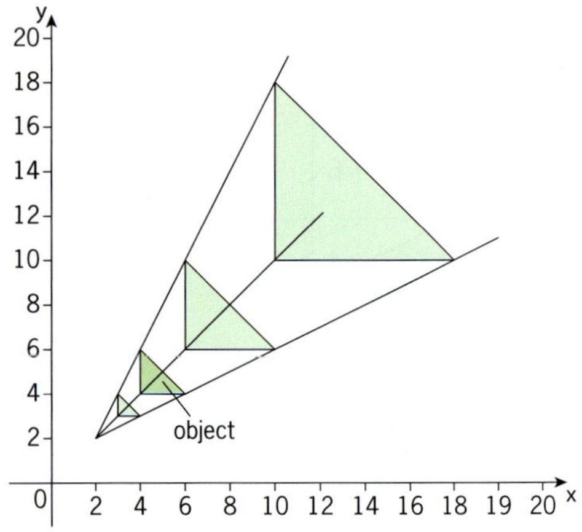

c

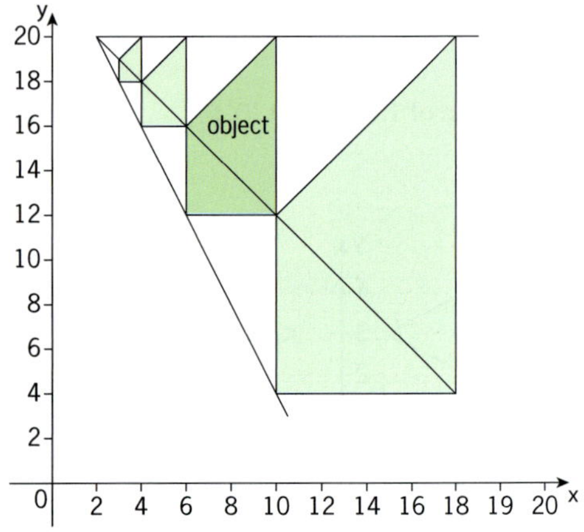

Ex B16.4

1 a (10, 1)

b R scale factor 2; S scale factor 3; Q scale factor $\frac{1}{2}$

2 3

3 a i (0, 3) ii 2

b i (12, 1) ii 3

c i (5, 7) ii 4

d i (1, 24) ii 5

4 a $\frac{1}{2}$ b $\frac{1}{3}$ c $\frac{1}{4}$

5 a (1, 7), $\frac{1}{2}$ b (4, 8), $\frac{1}{4}$

6 200

7 A4 to A3 has a scale factor 2

An enlargement (scale factor 2) is used for each successive paper size

8 180 × 130 mm

9 12 inches × 8 inches and 18 inches × 12 inches are enlargements.

10 inches × 8 inches and 7 inches × 5 inches will need to be cropped.

Exam practice

1

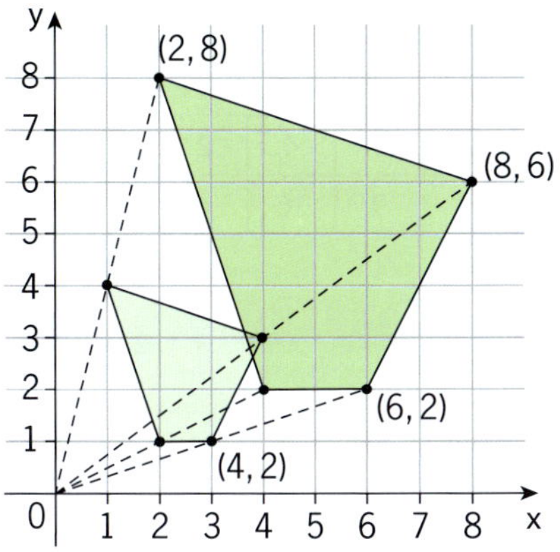

2 a, b

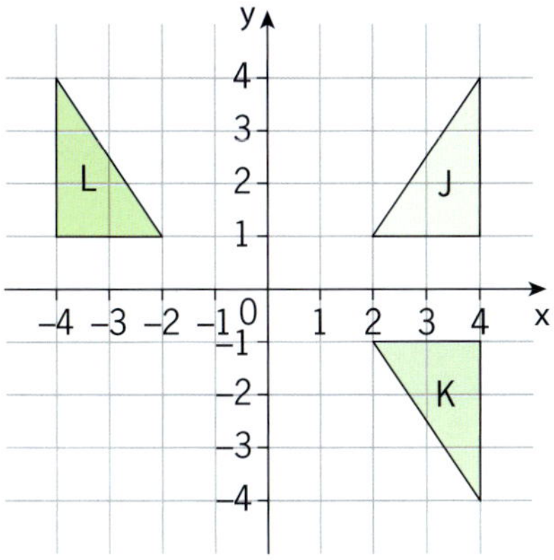

c reflection in y-axis

3 $\begin{pmatrix} 3 \\ -2 \end{pmatrix}$ $\begin{pmatrix} 2 \\ 3 \end{pmatrix}$ $\begin{pmatrix} -2 \\ 2 \end{pmatrix}$

C17 Percentages and proportional change

Check in

1 a 10 b 4
2 3
3 0.7 0.75 0.8 0.875

Ex C17.1

1 a £7 b £18 c £112
2 £58.24
3 a £1905 b £6705 c £502 d £29 130
4 a £74.25 b £6000 c £11 200
d £10 500e£120 f £3.30

Ex C17.2

1 a £374.40 b £48.95
2 a 35p b £7.80
3 £82.68
4 £36 016.80
5 £124.67

Ex C17.3

1 a 13.7% b 14.3%
2 73.3%
3 60%
4 5% profit

Ex C17.4

1 a £4800 b £19 800
2 a £36 b £136
3 a £1008 b 1008
4 £412.50
5 a £200 b £192 option c by £8

Ex C17.5

1 a £40 b £43.20 c £583.20
2 a i £4000 ii £36 000
b i £7600 ii £32 400 c £29 160
3 a 5.4 m b 4.86 m
4 a 1.68 million b 2.36 million (3 sig. figs.)
5 a 25 760 b 21 803
6 a £20 320.50 b £21 732.77 c £26 586.28
7 a £44 184 b £48 898.61 c £56 930.33

Ex C17.6

1 a 76p b 88p
2 a i 3 hours ii 5 hours
iii 7 hours 30 minutes
b i 10.5 days ii 16.8 days
3 a 5 litres for £1.29 b 200 ml for £2.89
4 26p per biscuit
5 14

Exam practice

1 a 160ml b 17.5g
2 1.375%
3 6 eggs

C18 Area and perimeter

Check-in

1 a 600 b 71000 c 48
d 2630 e 4500 f 6
g 75 h 6.5 i 3.2
2 a 9.6 b 22.5 c 10.5
3 a 7.8 b 15.5 c 9.0

Ex C18.1

1 a 9.6 cm b 9.3 cm
2 a 15.3 cm b 17.5 cm

3 a 19.2 metres b £13.00
4 a 5 cm^2 b 7 cm^2
5 3 cm^2 507 km^2
6 a 20 cm^2 b 6 cm^2 c 27 cm^2
d 14 cm^2 e 14 cm^2 f 8.55 cm^2
7 a 4560 cm^2 b 2280 cm^2
8 Colorado, Wyoming
Multiplying length by width which is area is not equal to the actual area, because the shapes of the states are not true rectangles. The shapes are on the earth's sphere and are defined by the earth's meridian and so the shapes taper slightly towards their northern sides.

Ex C18.2

1 a 16 cm b 25 cm c 19 cm d 63 cm
2 a 89.2 mm b 14.1 metres
c 75.4 cm d 21.4 cm
3 a 125.6 cm b 125.6 metres
4 a 40000 km (4 sig. figs.)
b i 584 million miles ii 1.6 million miles
iii 67000 miles
5 a 113 m^2 b 78.5 cm^2 c 707 cm^2 d 5.31 cm^2
e 1260 cm^2 f 7850 cm^2 g 1810 cm^2 h 24.6 cm^2
6 452 m^2
7 16.6 m^2
8 a 615 m^2 b £52300
9 a 1.3 cm, 0.8 cm b 2.01 cm^2 c 3.30 cm^2
10 141 cm^2

Ex C18.3

1 a 56 cm^2 b 55 cm^2 c 48 cm^2 d 38 cm^2
e 80 cm^2 f 36.5 cm^2 g 49 cm^2 h 36 cm^2
2 a 66 m^2 b £534.60
3 36 m^2

Ex C18.4

1 a 20 cm^2 b 30 cm^2 c 21 cm^2 d 15 cm^2
2 a 20 cm^2 b 15.6 cm^2
3 2400 cm^2
4 A square of side $2a$
Trapeziums with parallel sides
i a and $7a$, ii $2a$ and $6a$.

Ex C18.5

1 a w = 5 cm b h = 5 cm c h = 7 cm
d h = 6 cm e r = 3.6 cm f r = 3.0 cm
2 10 metres
3 2.5 metres
4 1.4 cm
5 a i 15.9 cm ii 31.8 cm iii 207 mm
b i 2.5 cm ii 5.2 m iii 5.6 mm
6 333 cm

Ex C18.6

1 a (2, 0) 2 4 b (1, 1) 3 9
2

	a	b	c	d
Length SF	2	3	4	2
Area SF	4	9	16	4

3 135 cm^2
4 a 9 b 3 c 21 cm
5 Yes, 2.5, 6.25

Exam practice

1 a 44 cm^2 36 cm
b 72 cm^2 40 cm
2 a i C ii none
3 65.94 cm^2

C19 Algebraic manipulation

Check-in

1 a 4 b 5.061 c 26.368
2 a $x = 7$ b $x = 10$

Ex C19.1

1 a $2(x + y)$ b $2x + 3y$
2 a $5x + y$ b $7p + 10q + 2$ c $m - 1$
d $3(x - y + z)$ e $5t$ f $4 + 3p - 3q$
g 0 h $5t$
3 a $a = 10$ $b = 10$ b $a = 5$ $b = 3$
c $a = 7$ $b = -5$ d $a = 2$ $b = 6$
e $a = 3$ $b = -1$ f $a = 3$ $b = -2$
g $a = -5$ $b = 8$ h $a = 0$ $b = 2$
4 a $24x + 10$ b $14x + 9$ c $8p + 6$
d $11q + 17$ e $11r$ f $24n$
g $2x^2 + 6x$ h $2z^2 + 4z$ i $x^2 + 8x + 12$
j $y^2 - y + 6$
5 a $2(2x + 3)$ b $3(2y + 3)$ c $4(3z + 2)$
d $3(x + 5)$ e $5(x + 4)$ f $3(6z - 5)$
g $4(2 + 3y)$ h $5(3 - 2y)$ i $2(2 - t)$
j $x(x + 3)$ k $y(y - 5)$ l $z(4 - z)$
m $p(8 - p)$ n $3b(b + 2)$ o $4c(2c - 1)$
6 $4(6x + 3y)$
Any three from $24x + 12y$, $2(12x + 6y)$, $3(8x + 4y)$, $6(4x + 2y)$, $12(2x + y)$

Ex C19.2

1 a 8 b 9 c 7 d 8
e 6 f 8 g 7 h 9

2

x	x^3	Too high or two low?
2	8	too low
3	27	far too high
2.1	9.26	too low
2.2	10.64	too high
2.15	9.93	too low

Therefore the solution lies between 2.15 and 2.2.
The solution is 2.2 to 1 decimal place

3 8.3

4 a 4.3 b 6.3 c 5.4
d 20.5 e 26.4 f 0.6

Ex C19.3

1 a 8 b 8.1

2 $x(x + 4) = 70$
6.6

3 $30x - x^2 = 90$
3.4

4 $n^3 - n^2 = n$
1.6 (note that $n = 0$ and $n = -0.6$ are also solutions)

5 a 10 b 8.3

6 a 5 b 2.9 c 4.8
d 5.9 e 4.5 f −9.7

7 $n + 4$
35, 37, 39

Exam practice

1 if question is $x^3 - 3 = 17$, answer is 2.7
if question is $x^3 - 3x = 17$, answer is 3.0
if question is $x^3 - 3x^2 = 17$, answer is 4.0

2 7.7

C20 3-D shapes

Check-in

1 a a circle radius 6 cm

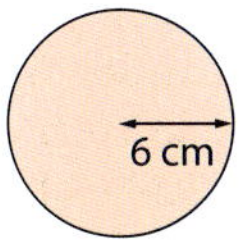

b a circle radius 4.5 cm

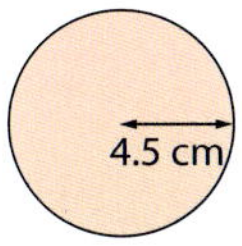

2 a 28 cm^2 b 10 mm^2

Ex C20.1

1 a i

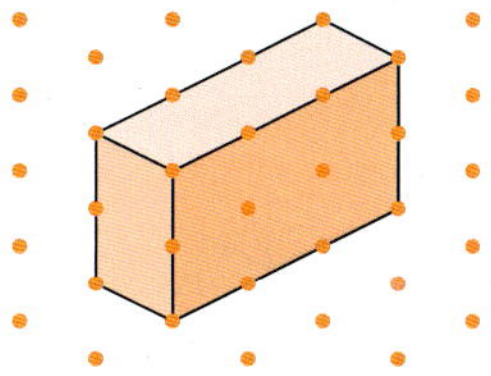

Six faces, twelve edges

ii

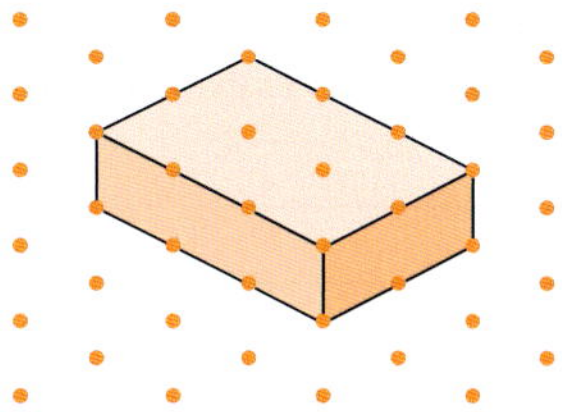

Six faces, twelve edges

iii

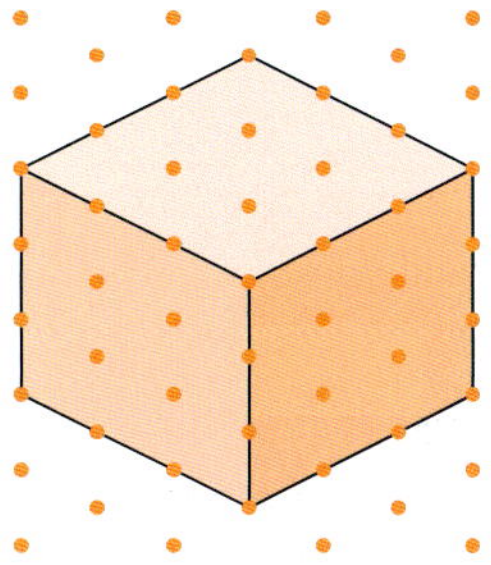

Six faces, 12 edges

iv

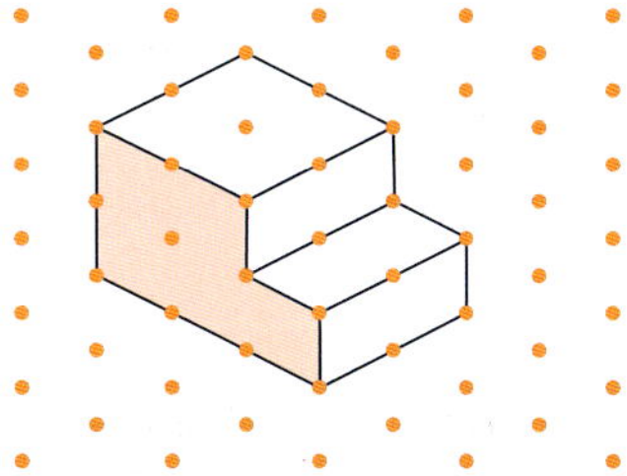

Eight faces, 18 edges

v

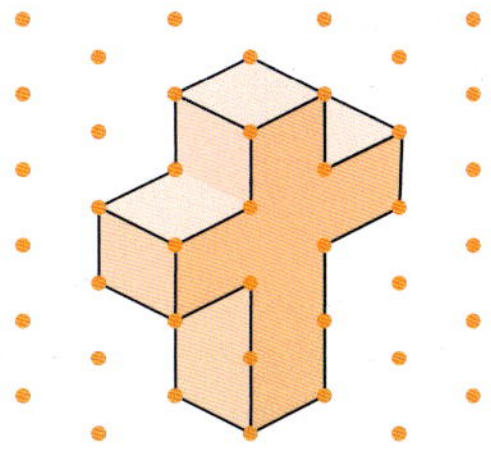

14 faces, 36 edges

Unit C

vi

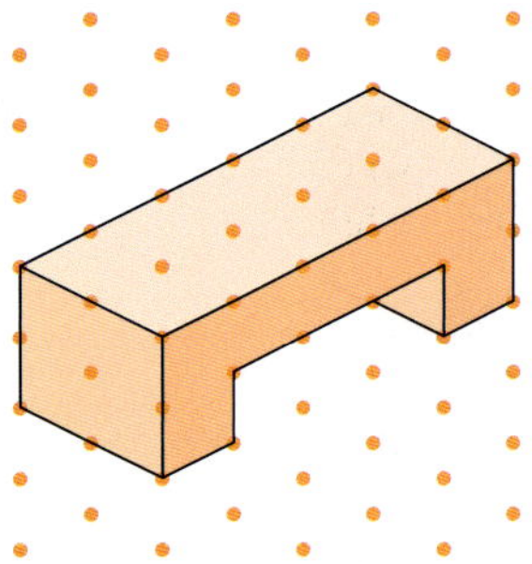

10 faces, 24 edges

2 a AA′

rectangle

BB′

rectangle

CC′ circle

DD′ square

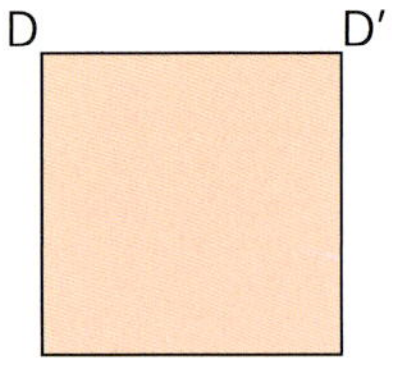

EE′ rectangle

FF′ circle

3 a

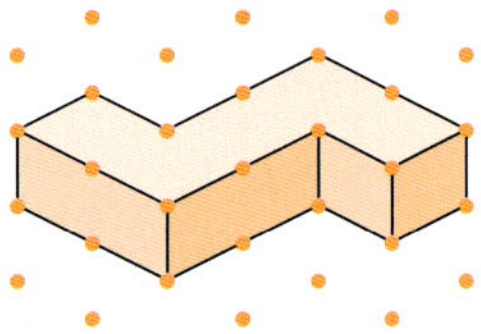

b

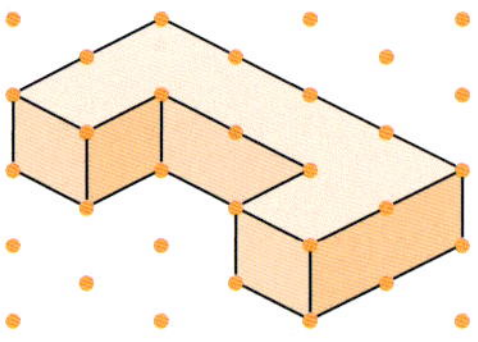

c

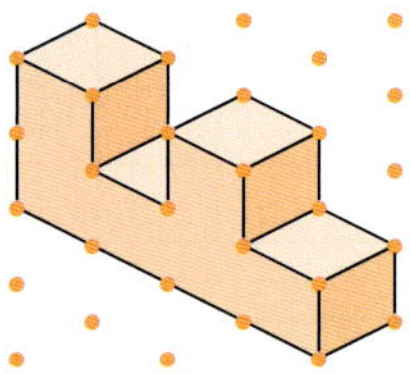

4

3 × 3 cube	26 small cubes have painted faces	
4 × 4 cube	56	-do-
5 × 5 cube	98	-do-

Ex C20.2

1 A view 2; B view 1; C view 3 for both buildings

2 a

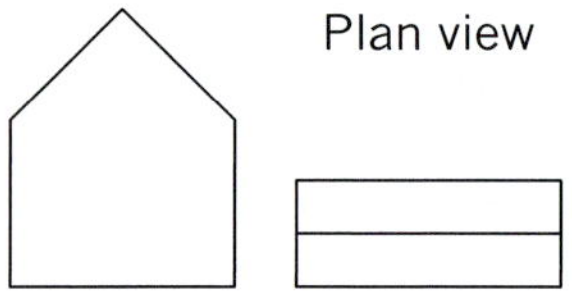

b

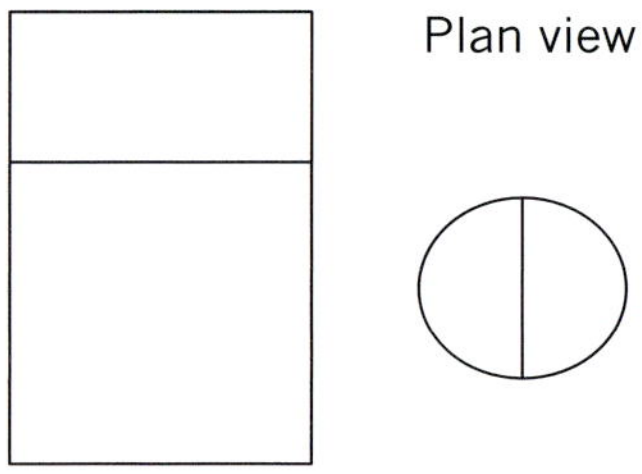

c

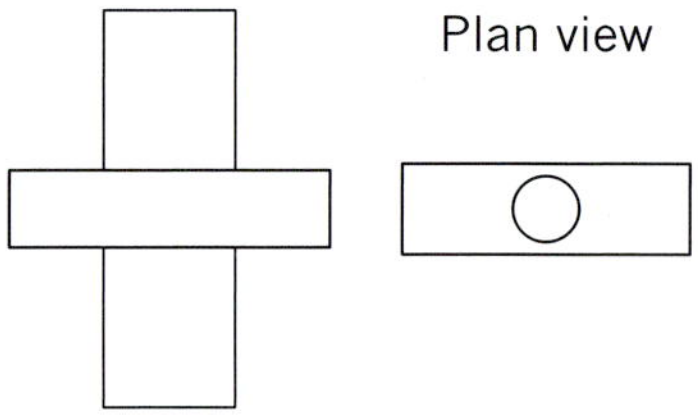

d

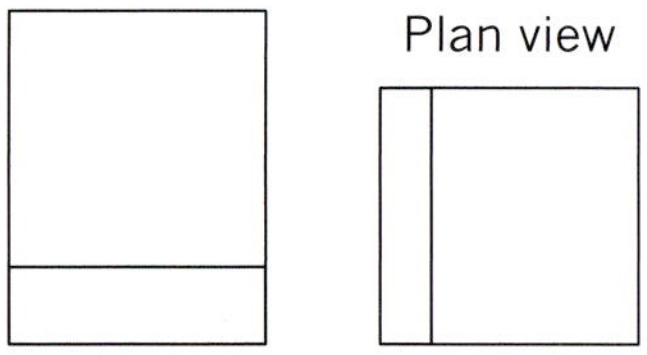

3 a pyramid
b cone
c cylinder
d sphere
e a half cylinder cut along its axis
f a triangular prism

4 B;

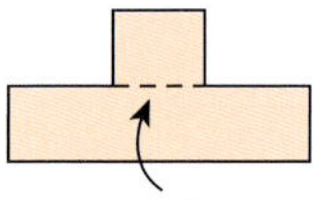

A and C have a line here
B does not.

5 a

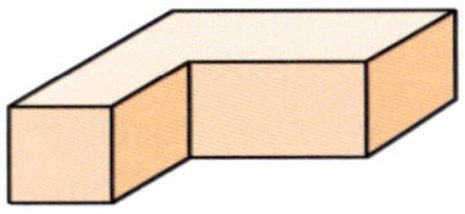

b

c

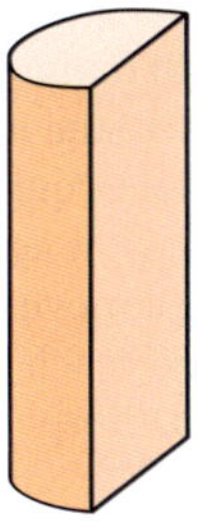

7 cube; sphere

Ex C20.3

1 a 1st and 3rd b pyramid
  c prism with a triangular cross-section
2 b The faces of a cuboid must be in pairs
3 For example,

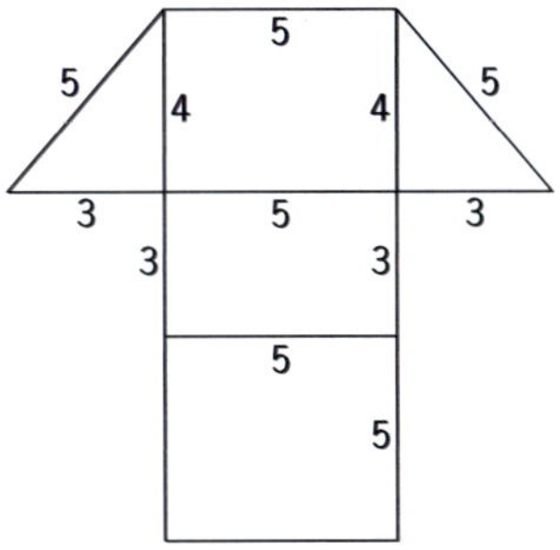

prism with a triangular cross-section
4 Cube A: a t b s c u
  Cube B: a u b s c t
5 For example,

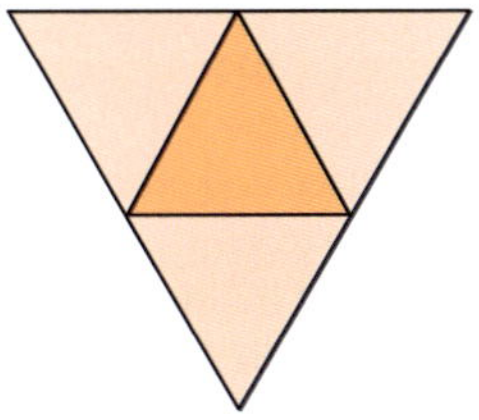

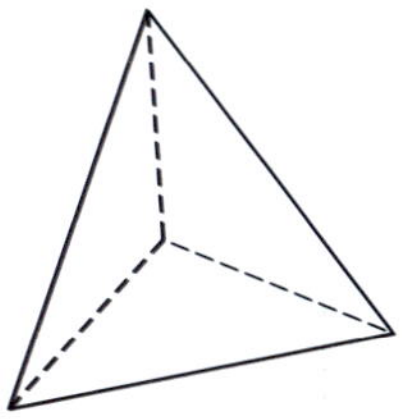

Tetrahedron (pyramid with a triangular base)

6 a For example,

Ex C20.4

1 a 700 cm³ b 504 cm³ c 538 cm³
2 a 168 cm² b 64 cm³
3 a 2.96 m² b £36.85 c £6.50
4 For example,

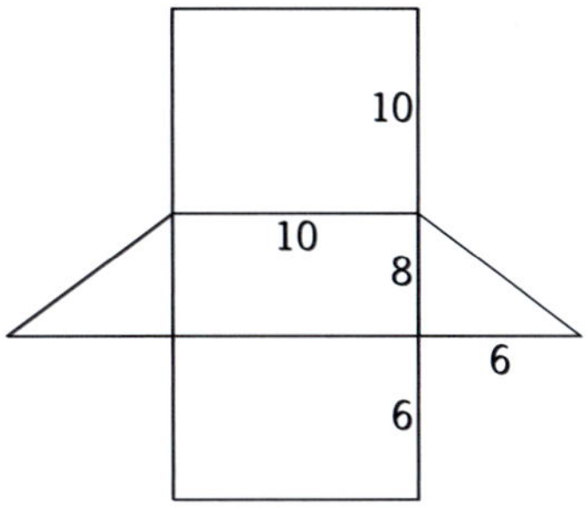

surface area 288 cm²

For example,

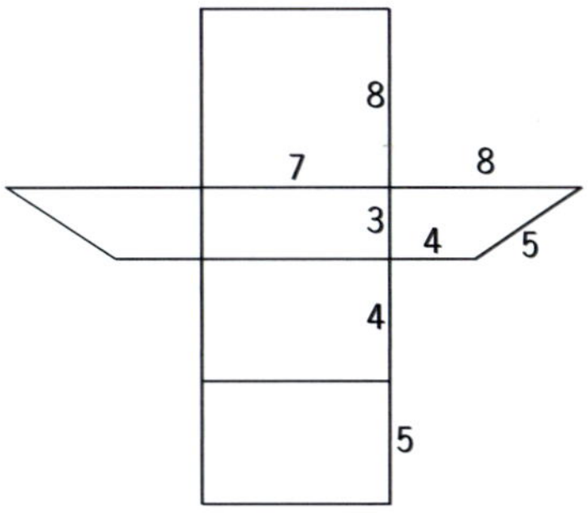

surface area 176 cm²
5 a 400 cm²; 25 cm × 20 cm b 20%

Ex C20.5

1 a 188 cm² b 151 cm² c 126 cm² d 126 cm²
2 3140 cm² or 0.314 m²
3 a 251 cm² b 151 cm² c 151 cm² d 220 cm²
4 a 314 cm² b 603 cm²
5 848 cm²
6 50%
7 a has the smaller surface area of 603 cm².
  b has surface area of 730 cm²
8 A sheet 52 cm × 15 cm (area 780 cm²)

Exam practice

1 for example
2 84 cm²
3 Join faces which are 5 cm × 4 cm 56 cm²

C21 Graphs

Check-in

1 a 9 b −2 c 12 d 2

Unit C

2

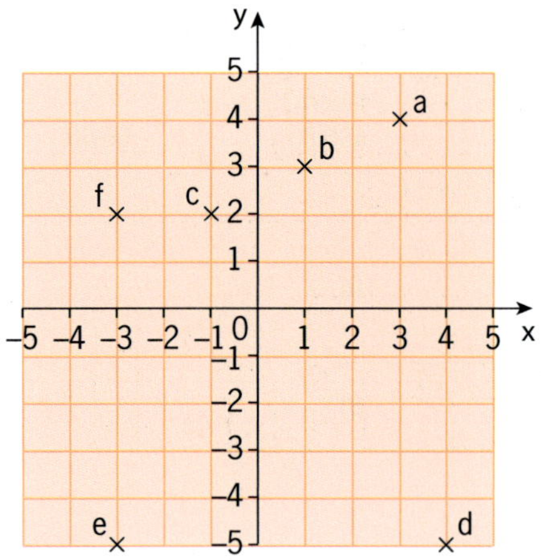

Ex C21.1

1 **a** 200 m **b** 4 mins **c** 4 mins
d 100 m **e** 12 mins

2 **a** Twice **b** 300 m **c** 2 mins **d** 8 mins

3 **a** Y **b** Z **c** X

4 **a** A Aircraft B Train C Lorry
D Bicycle E Pedestrian
b A 50 m/s B 25 m/s C 10 m/s
D 6 m/s E 2 m/s

5 **a** 1 km **b** A to B **c** 4 km, 10 mins
d 24 km/h **e** 6 km, 30 mins **f** 12 km/h

6 **a** 150 m/min or 2.5 m/s **b** 2 mins **c** 500 m
d 5 min **e** 100 m/min or 1.67 m/s
f 83 m/min or 1.4 m/s

7

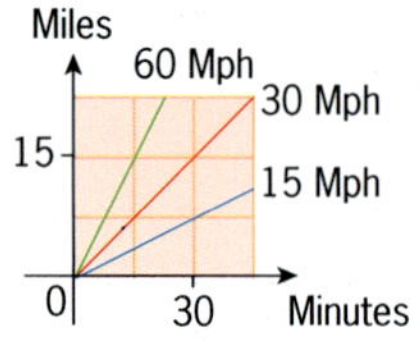

8 **a** Stephen, 20 mins **b** 50 mins
c 30 km/h **d** 30 km/h

9 **a** X Table tennis ball
Y Cricket ball
b The table tennis ball bounces 5 times, bouncing lower each time. The cricket ball bounces twice only.

10 The oil level decreases steadily for 3 weeks, then rapidly for 1 week. Then no oil is used for 2 weeks. Then, after 6 weeks, the oil is refilled to the original level. Then the level reduces steadily again.
The weather was probably a little cold until week 3, then very cold for week 3 to 4, warm for weeks 4 to 6 and then a little cold for weeks 6 to 8.

11 The jogger starts running at about 12 km/h. After 10 mins he turns and runs back at about the same speed. He slows and turns again after 20 mins. He then runs forward more slowly, stopping and turning after 30 mins. He then runs back to the starting point, arriving after 40 mins.

12 The machine moves quickly away for 2 secs, then turns and moves quickly back for 3 secs, then turns and moves away again more slowly. After 10 secs it has stopped moving.

13 **a** The coach leaves London, calling at Scalsby, Garthorpe, Skeltoft and Lincoln, It stops for 15 mins at Scalsby and Garthorpe, and it stops for 30 mins at Skeltoft. The whole journey takes $5\frac{1}{2}$ hours.

b

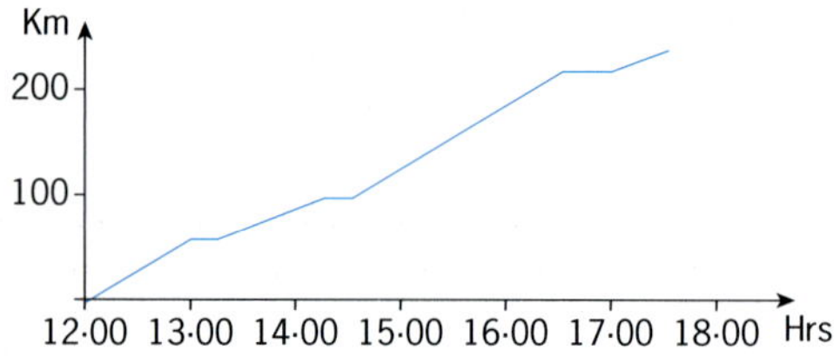

c 60 km/h, 40 km/h, 60 km/h, 50 km/h
d 44.5 km/h

Ex C21.2

1 **a** **i** £15 **ii** £15 **iii** £65 **b** 23 kg

2 **a** L Firm B **b** Firm A charges £10
M Firm A Firm B charges £15
c Firm B is cheaper for longer journeys, 20 miles

3 **a** P Elex **b** Elex charges £60
Q Trizity Trizity charges £65
c 400 units **d** Trizity

4 **a** **i** The most economical speed is about 55 mph for petrol engines and 60mph for diesel engines.
ii The miles travelled per gallon decreases.
iii Diesel engines do more miles to the gallon than petrol engines at all speeds.

b

removing a bike rack	Increases fuel efficiency
accelerating fast	Decreases fuel efficiency
eating while you drive	No effect on fuel efficiency
tyres inflated at correct pressure	Increases fuel efficiency
having a clean air filter	Increases fuel efficiency
breaking hard and often	Decreases fuel efficiency

5 **a** 500 mins **b** £20 **c** 550 mins
d For less than 100 minutes per month.

6 **a** £80
b Over 6 m and less than or equal to 7 m **c** £150

7 **a** the distance travelled by a car driven at constant speed Graph R
b the temperature of a bowl of soup left uneaten Graph P
c the amount of money left in a purse on a shopping trip. Graph S
d the temperature outdoors on a cloudy day Graph Q

8 eg. on day 3 sheena uses half the oil for frying. When the remaining oil is cold she puts it back in the bottle. She uses more oil on day 6 and almost empties the bottle.

9 **a** all at one go — Graph Y
b in two gulps, half a cup each time — Graph Z
c sip by sip — Graph W
d not at all. — Graph X

10 **a** The baby's weight reduces initially and after 10 days starts to increase again. By day 25 the baby has regained the weight it lost and continues to increase more rapidly.
b 3.0 kg **c** Day 10
d By day 25 **e** 400 g

11 A Z B X C Y
The different shapes of the beakers give different graphs. Beaker A is widest at the bottom and so initially the height increases slowly. Beaker B has a constant width, so the height increases uniformly with time. Beaker C is narrower at the bottom so initially the height increases more quickly.

Ex C21.3

1 $y = x^2$

x	−4	−3	−2	−1	0	1	2	3	4
y	16	9	4	1	0	1	4	9	16

$y = x^2 + 3$

x	−4	−3	−2	−1	0	1	2	3	4
x^2	16	9	4	1	0	1	4	9	16
+3	+3	+3	+3	+3	+3	+3	+3	+3	+3
y	19	12	7	4	3	4	7	12	19

$y = x^2 - 4$

x	−4	−3	−2	−1	0	1	2	3	4
x	16	9	4	1	0	1	4	9	16
−4	−4	−4	−4	−4	−4	−4	−4	−4	−4
y	12	5	0	−3	−4	−3	0	5	12

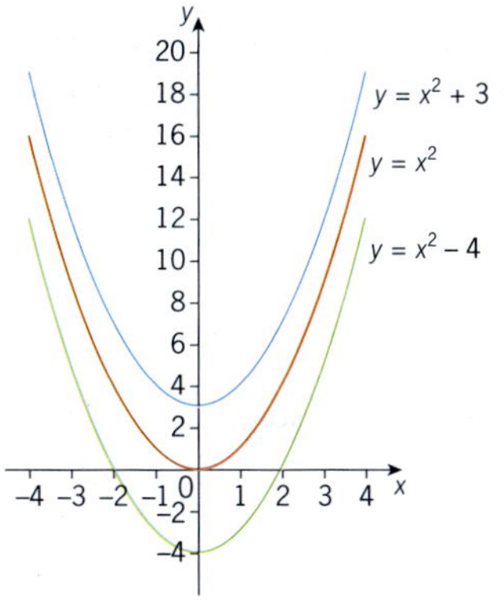

The curves can be transformed by translations: +3 in the y direction, and −4 in the y direction

2 $y = x^2 - 2$

x	0	1	2	3	4
x^2	0	1	4	9	16
−2	−2	−2	−2	−2	−2
y	−2	−1	2	7	14

$y = (x - 2)^2$

x	0	1	2	3	4
x − 2	−2	−1	0	1	2
y	4	1	0	1	4

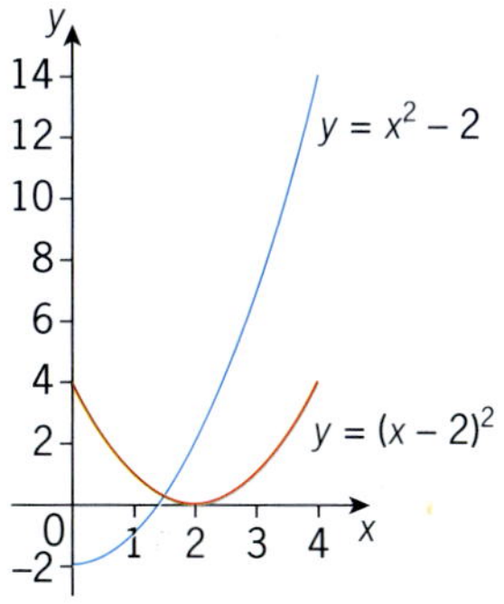

3 $y = x^2 - 5$

x	−3	−2	−1	0	1	2	3
x^2	9	4	1	0	1	4	9
−5	−5	−5	−5	−5	−5	−5	−5
y	4	−1	−4	−5	−4	−1	4

$y = 5 - x^2$

x	−3	−2	−1	0	1	2	3
5	5	5	5	5	5	5	5
$-x^2$	−9	−4	−1	0	−1	−4	−9
y	−4	1	4	5	4	1	−4

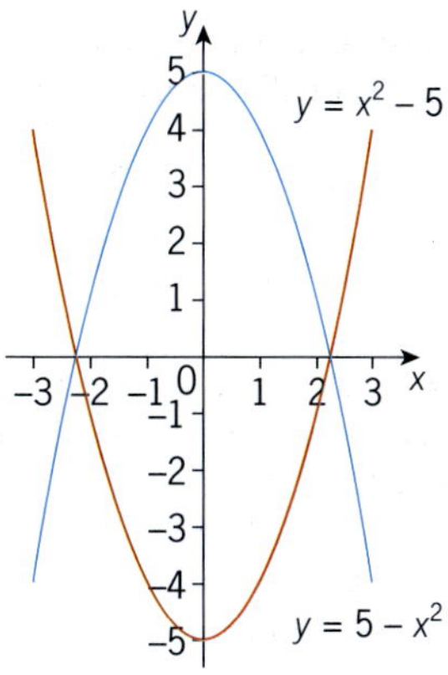

a Reflection in the x-axis **b** (2.2, 0) and (−2.2, 0)

Unit C

4 $h = 6 - x^2$

x	0	1	2	3
h	6	5	2	–3

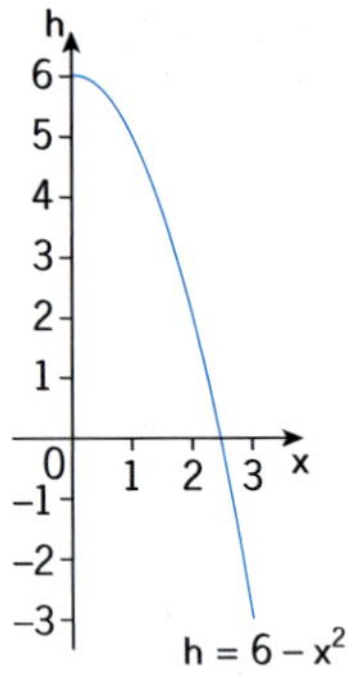

a 6 km b 2.4 km c 1.4 km

5 $y = 5 + t^2$

t	0	1	2	3
y	5	6	9	14

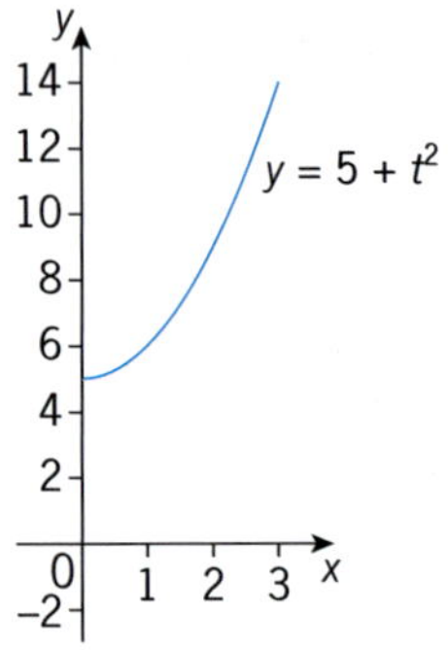

a i 5°C ii 14°C b 2.2 hours

6 $A = 2t^2 + 4$

t	0	1	2	3
$2t^2$	0	2	8	18
A	4	6	12	22

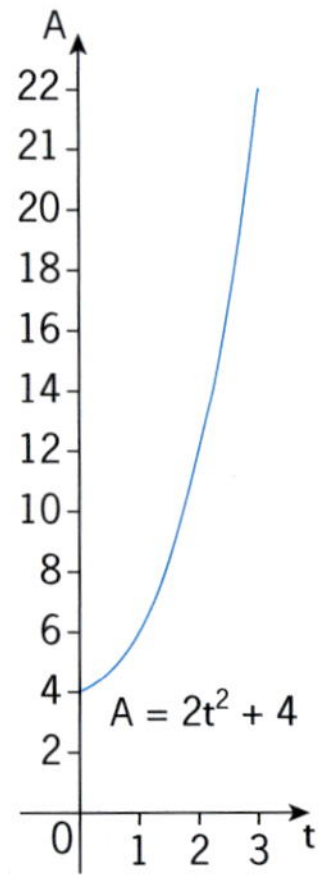

a 4 m² b 16.5 m² c 1.7 days

d No. This function shows the algae increasing rapidly, without a limit, but the lake must have a finite area.

7 $y = 3x^2 - 8$

x	0	1	2	3	4
$3x^2$	0	3	12	27	48
–8	–8	–8	–8	–8	–8
y	–8	–5	4	19	40

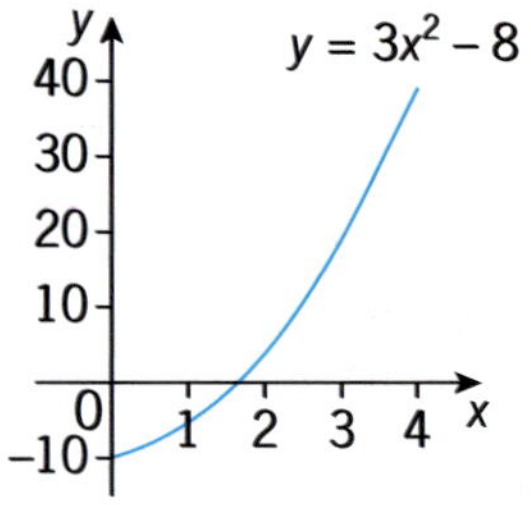

a 1.6 b 2.4 c 3.6 d $x \le 1.6$

8 $y = 5x^2 - 10$

x	–3	–2	–1	0	1	2	3
$5x^2$	45	20	5	0	5	20	45
–10	–10	–10	–10	–10	–10	–10	–10
y	35	10	–5	–10	–5	10	35

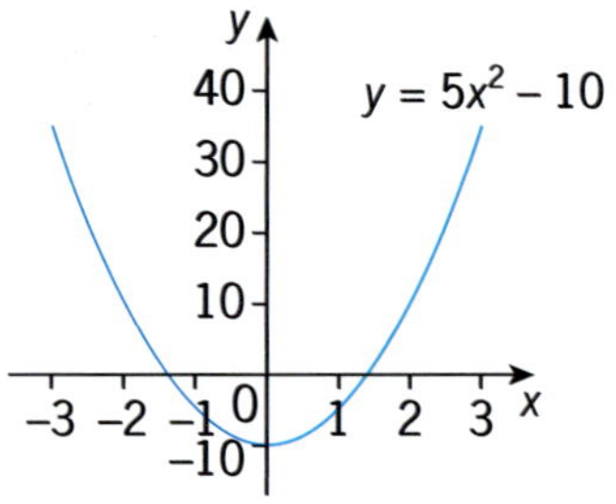

a 1.4, –1.4 b 0.6, –0.6 c 2.8, –2.8

d $-1.4 \le x \le 1.4$

9 $y = \frac{1}{2}x^2 - 3$

x	–6	–4	–2	0	2	4	6
$\frac{1}{2}x^2$	18	8	2	0	2	8	18
–3	–3	–3	–3	–3	–3	–3	–3
y	15	5	–1	–3	–1	5	15

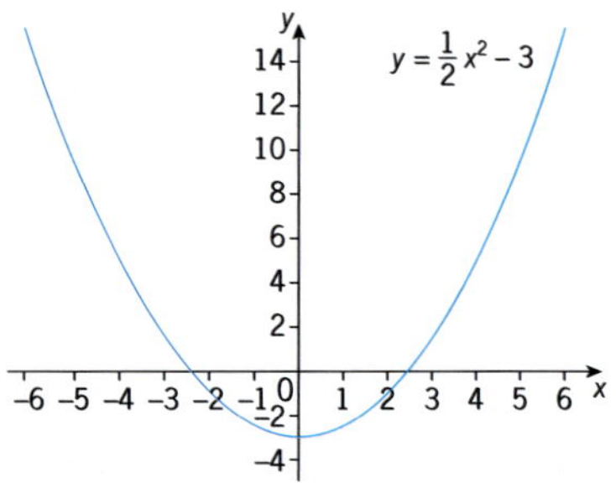

a 2.4, –2.4 b 5.1, –5.1

10

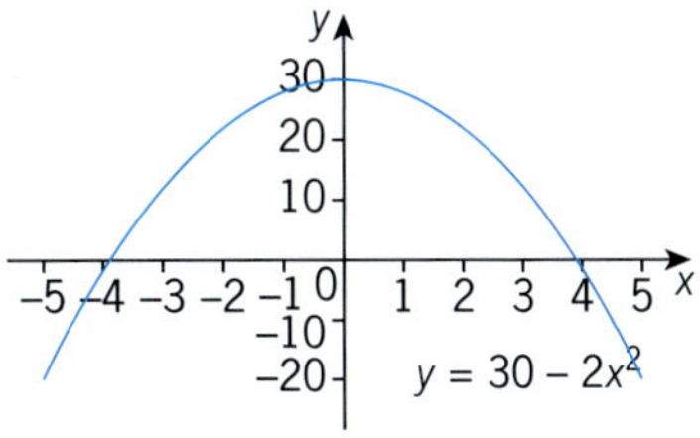

a i 3.9, –3.9 ii 2.2, –2.2 b $-2.2 \le x \le 2.2$

Ex C21.4

1 A (300, 100, 30) B (200, 100, 50)
C (200, 300, 60) D (300, 200, 20)
E (300, 400, 75) F (100, 0, 40)

2 A (2, 5, 10) B (2, 3, 10)
C (4, 5, 20) D (5, 1, 20)
E (6, 3, 30) F (7, 4, 40)
G (6, 5.5, 30) H (1, 2, 0)
I (6.5, 1, 30) J (7.5, 3, 30)
K (3, 1, 10) L (4, 3.5, 20)

3 a A (20, 20, 50) B (40, 30, 50) C (30, 40, 100)
D (60, 50, 50) E (70, 40, 0) F (40, 15, 0)
G (20, 55, 100) H (60, 20, –50)
I (70, 10, –100) J (20, 40, 150)

b (40, 50, 50) (80, 20, –100) (20, 30, 100) (5, 40, 150) (20, 5, 0) (40, 50, 50) (75, 30, –50) (50, 5, –50)

Exam practice

1 a 0 b

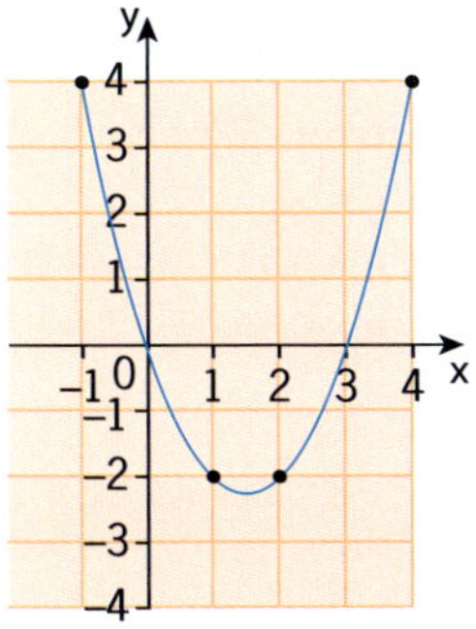

2 a Speedy by $\frac{1}{2}$ 1.60 b 6 miles

3 a

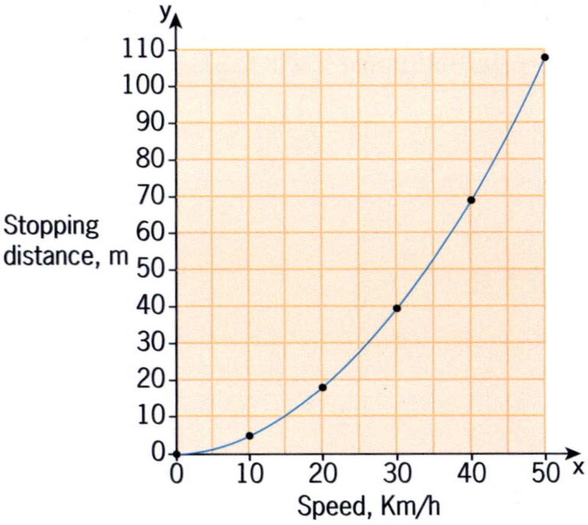

b 53 m,

c Driver's claim is not correct.
Stopping distance of 80 m implies speed of approx 44 km/h from graph, but skid marks are only from application of brakes which does not allow for reaction time. So skid marks of 80 m imply stopping distance greater than 80 m and therefore speed greater than 44 km/h.

C22 Everyday arithmetic and bounds

Check in

1 27

2 a 1.3 b 0.3

3 a 3.56 b 8.04 c 0.06

Ex C22.1

1 a i 24°C ii 74°F b 55 m

2 a yes b 55 mph

3 a i 127 miles ii 203.2 km b 92 miles

4 a 26 minutes
b i 2312, 2353 ii 28 minutes 23 minutes
c arrives earlier

Ex C22.2

1 a i 4.4°C ii 36.9°C iii –1.1°C
b i 71.6°F ii 23°F iii 28.4°F

2 John's estimate = 25, true value = 25.9

3 a 37 km/h b 49 km/h

4 15.8 mg

Ex C22.3

1 £75.96, 3, £25.50

2 £1 = €1.41

3 Paris = £108.70, New York = £102.86, New York

4 9.25 BST next day

Ex C22.4

1 a i 19.2 km/h ii 17.1 km/h
b i 80 mph ii 72 mph

2 a 7.43 mph b 4.88 mph

3 40.6 km/h

4 4 km

5 a 3.1×10^{-5} m³ or 31 cm³ b 1.84 kg or 1840 g

6 258 per square mile

7 2500 sec = 41 minutes 40 seconds

Ex C22.5

1 a 67.5 68.5 b 3750 3850 c 725 735
d 4945 4955 e 17.55 17.65 f 23.35 23.45

2 a 27.5 – 28.5 b 47.5 – 48.5 c 111.5 – 112.5
d 555 – 565 e 9345 – 9355 f 9325 – 9375
g 510 – 530 h 415 – 425 i 43.55 – 43.65
j 5.15 – 5.25 k 7.25 – 7.35 l 39.5 – 40.5
m 99.5 – 100.5

3 a Yes, could be 154.5 cm b 16.65 – 16.75 seconds
c Yes, could be 4.375 m

4 a 48.5% b 78

Exam practice

1 3min 59.35sec 3min 59.45sec

2 a 60mph
b 40miles per gallon (using 4.5 litres = l gallon)

3 2.4 kg +5.10

C23 Trigonometry

Check in

1 a 40 cm^2 b 30 cm^2

2 a 250 b 6300 c 41 d 250
e 3500 f 4 g 5.6 h 40
i 4.1 j 5.2

Ex C23.1

1 a i 12 ii 24
b i 15 ii 45
c i 30
ii 60 (would overfill the box, but cannot fill exactly)
45 (if cubes can be cut in half)

2 a 756 cm^3 b 480 cm^3 c 306 cm^3

3 a 1083 cm^3 b 1.083 litres

4 a 3 075 000 cm^3 b 3075 litres

5 a 90 000 cm^3 b 90 litres

6 a 452 cm^3 b 61%

7 a 3 cm b 6 cm c 5 cm d 9cm

8 20.4 cm

9 yes, it is 80 cm tall

10

	Length, cm	Width, cm	Height, cm
a	10	10	10
b	10	20	5
c	10	50	2
d	10	100	1
e	20	50	1

The 10 cm × 10 cm × 10 cm pack is suitable.

11 The cube with dimensions 10 cm × 10 cm × 10 cm has the smallest surface area

Ex C23.2

1 a 60 cm^3 b 75 cm^3

2 102 m^3

3 350 cm^3

4 A a 80 cm^2 b 640 cm^3
B a 72 cm^2 b 216 cm^3
C a 92 cm^2 b 184 cm^3

5 a 0.96 m^2 b 2.4 m^3

6 An infinite number of ways.
The halves are cuboids or prisms with either triangular or trapezium cross-sections

7 a 10 cm^2 b 7 cm^2 c 5 cm^2 d 6 cm^2

Ex C23.3

1 a 314 cm^3 b 170 cm^3 c 100 cm^3 d 126 cm^3

2 Vegetable soup 470 ml
Marmalade 360 ml
Kidney beans 280 ml
Tomatoes 840 ml

3 a 78 500 cm^3 b 70 650 cm^3

4 no, there is just over 1 litre in the bottle

5 2 litre engine

6 a 4 cm b 6 cm c 2 cm d 12 cm

7 a 5.6 cm b 2.6 cm c 4.0 cm d 5.9 cm

Ex C23.4

1 a 3 b 9 c 27

2 a i 80 mm ii 600 mm^2 iii 2000 mm^3
b i 3.5 cm ii 1.4 cm^2 iii 0.5 cm^3

3 a 7400 metres b 2 100 000 m^2 c 0.005 km^3

4 a 0.315 km b 0.0445 km^2 c 0.004 km^3

5 Volume = 0.459 m^3, diameter = 1 m, area = 0.133 m^2

Exam practice

1 a 312 cm^3 b 113 cm^3

2 125 cm^3

3 a 1600 cm^2 b 12.5 cm^3

C24 Graphs 2

Check in

1 a $\frac{2}{3}$ b $\frac{1}{2}$ c $\frac{3}{4}$ d $\frac{1}{4}$ e 1

2 a 1 b 1 c $\frac{3}{10}$ d $\frac{2}{5}$

3 a 0.9 b 0.4 c 0.85

4 a 0.01 b 0.25 c 0.35
d 0.05 e 0.36

Ex C24.1

1 a Unlikely b Evens c Unlikely
d Impossible

2 a unlikely b evens
c i unlikely ii highly unlikely d certain

3 a probability that X will happen is 0.6
b probability that Y will happen is $\frac{1}{4}$
c probability that Z will happen is 35% scale

4
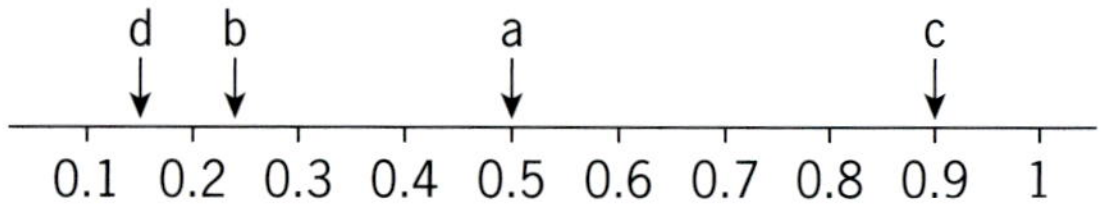

Ex C24.2

1 a $\frac{7}{25}$ b $\frac{18}{25}$ c $\frac{9}{25}$
d $\frac{16}{25}$ e $\frac{10}{25}$ f $\frac{17}{25}$

2 a 0 b 1 c $\frac{18}{30}$ d $\frac{12}{30}$

3 0.2

4 a 3 is not a factor of 7 b 4

5 The second bag, because the fraction of green counters is higher

Ex C24.3

1 a 30

b i $\frac{7}{30}$ ii $\frac{15}{30}$ iii $\frac{12}{30}$

iv $\frac{9}{30}$ v $\frac{9}{30}$ vi $\frac{21}{30}$

2 a 0.45 b 0.25 c 0.3

3 a red and blue and orange b yellow

c i 0.2 ii 0.32 iii 0.23

4 a eg odd and even b eg 1 or 2, 3 or 4, 5 or 6

c eg even and prime, 2 is both

Ex C24.4

1 140

2 120

3 0.625

4 a $22 \times 2 = 44$ b $\frac{34}{120}$ c 50

5 a i 0.11 ii 0.32 iii 0.27

iv 0.41 v 0.87

b i 80 ii 120

6 174

Ex C24.5

1 fair; would expect 140

2 fair; all frequencies between 31 and 35

3 a 0.4 0.35 0.37 0.33 0.36 0.37 0.37 0.36 0.36 0.35

b 0.35 c biased; fair is 0.5

4 a $\frac{8}{36}$ $\frac{9}{36}$ $\frac{10}{36}$ $\frac{9}{36}$

b fair; all approximately equal

5 a $\frac{11}{30}$

b biased; other relative frequencies are $\frac{5}{30}$ or less

Ex C24.6

1 a table b i $\frac{1}{12}$ ii $\frac{3}{12}$

2 a table b $\frac{1}{4}$

3 a table b $\frac{7}{12}$

c 2 girls; P(2 boys) = $\frac{2}{12}$, P(2 girls) = $\frac{3}{12}$

4 a table

b i $\frac{2}{36}$ ii 0 iii $\frac{6}{36}$

c even; P(even) = $\frac{27}{36}$ P(odd) = $\frac{9}{30}$

5 a table b i $\frac{1}{9}$ ii $\frac{3}{9}$

c both have 2 out of 9 outcomes

Exam practice

1 a eg. 5, 6, 7, 8, 9

b eg. 1, 3, 5, 7, 9

c eg. 1, 2, 3, 5, 5

2 a $\frac{11}{44}$ b $\frac{22}{44}$

3 $\frac{15}{36}$

Index